Agricultural Crop Production and Soil Quality

Agricultural Crop Production and Soil Quality

Editor: John Boyd

www.callistoreference.com

Callisto Reference,
118-35 Queens Blvd., Suite 400,
Forest Hills, NY 11375, USA

Visit us on the World Wide Web at:
www.callistoreference.com

ISBN: 978-1-63239-948-9 (Hardback)

Cataloging-in-Publication Data

Agricultural crop production and soil quality / edited by John Boyd.
p. cm.
Includes bibliographical references and index.
ISBN 978-1-63239-948-9
1. Agriculture. 2. Crops. 3. Soils--Quality. 4. Agronomy. I. Boyd, John.
SB98 .A37 2018
630--dc23

Table of Contents

Preface

Agricultural crop production studies optimal conditions required for the growth of crops. Soil quality is a primary aspect of this field. Agricultural soil science studies soil fertility and soil composition with a view to improve agricultural output. Related concerns include soil classification, irrigation, plant nutrition, soil sustainability, water content, etc. Some of the diverse topics covered in this book address the varied branches that fall under this category. The topics included in this book on agricultural crop production and soil quality are of utmost significance and are bound to provide incredible insights to readers.

This book unites the global concepts and researches in an organized manner for a comprehensive understanding of the subject. It is a ripe text for all researchers, students, scientists or anyone else who is interested in acquiring a better knowledge of this dynamic field.

I extend my sincere thanks to the contributors for such eloquent research chapters. Finally, I thank my family for being a source of support and help.

Editor

1

The Effects of Manure and Nitrogen Fertilizer Applications on Soil Organic Carbon and Nitrogen in a High-Input Cropping System

Tao Ren[1,2], Jingguo Wang[1], Qing Chen[1]*, Fusuo Zhang[1], Shuchang Lu[3]

1 College of Resources and Environmental Science, China Agricultural University, Beijing, China, **2** College of Resources and Environment, Huazhong Agricultural University, Wuhan, China, **3** Department of Agronomy, Tianjin Agricultural University, Tianjin, China

Abstract

With the goal of improving N fertilizer management to maximize soil organic carbon (SOC) storage and minimize N losses in high-intensity cropping system, a 6-years greenhouse vegetable experiment was conducted from 2004 to 2010 in Shouguang, northern China. Treatment tested the effects of organic manure and N fertilizer on SOC, total N (TN) pool and annual apparent N losses. The results demonstrated that SOC and TN concentrations in the 0-10cm soil layer decreased significantly without organic manure and mineral N applications, primarily because of the decomposition of stable C. Increasing C inputs through wheat straw and chicken manure incorporation couldn't increase SOC pools over the 4 year duration of the experiment. In contrast to the organic manure treatment, the SOC and TN pools were not increased with the combination of organic manure and N fertilizer. However, the soil labile carbon fractions increased significantly when both chicken manure and N fertilizer were applied together. Additionally, lower optimized N fertilizer inputs did not decrease SOC and TN accumulation compared with conventional N applications. Despite the annual apparent N losses for the optimized N treatment were significantly lower than that for the conventional N treatment, the unchanged SOC over the past 6 years might limit N storage in the soil and more surplus N were lost to the environment. Consequently, optimized N fertilizer inputs according to root-zone N management did not influence the accumulation of SOC and TN in soil; but beneficial in reducing apparent N losses. N fertilizer management in a greenhouse cropping system should not only identify how to reduce N fertilizer input but should also be more attentive to improving soil fertility with better management of organic manure.

Editor: Xiujun Wang, University of Maryland, United States of America

Funding: The authors are grateful to the National Natural Science Foundation of China (No. 31071858), Innovative Research Team of Beijing Fruit Vegetable Industry, the innovative group grant of NSFC (No. 30821003) and Basic Application and Cutting-edge Technology Research Projects of Tianjin City (09JCYBJC08600). The funders had no role in study design, data collection and analysis, decision to publish, or preparation of the manuscript.

Competing Interests: The authors have declared that no competing interests exist.

* E-mail: qchen@cau.edu.cn

Introduction

Soil organic matter plays a key role in soil biological and chemical processes, and changes in soil organic matter strongly influence soil N turnover because of the importance of available C for microbial immobilization [1-3]. Soils with higher organic matter contents may immobilize more N and reduce N loss to the environment. Otherwise, the depletion of available C will cause more rapid N turnover and losses [4]. In addition, changes in N availability can also alter soil C turnover [5]. There is no doubt that higher crop production in response to mineral N fertilizer application results in greater root exudates and more crop residues, thereby enhancing SOC sequestration in agricultural soils [6]. In addition increasing N fertilizer application can stabilize organic matter [7] and retard the mineralization of older soil organic matter [8]. N fertilization plays a positive role in enhancing the SOC [7], [9-12]. However, the addition of N fertilizer has also been reported to have a negative or no effect on SOC accumulation [13-17]. Changes in the decomposability of fresh plant litter and soil organic matter fractions, the stability of soil aggregates, and/or shifts in the microbial community can be used to explain the decreases in SOC attributed to N fertilizer addition [13], [15]. Therefore, achieving a better understanding of the interaction between N fertilizer and SOC in agricultural soils is essential for maximizing SOC storage and minimizing potential N losses.

Intensive vegetable production systems in northern China differ from other ecosystems in which excessive nutrients and water are applied, which far exceed the resources needed for vegetable growth [18-19]. As shown in previous studies [20-21], these practices have resulted in serious N losses to the environment. Therefore, more work was done to understand how to reduce N fertilizer input with optimal N and irrigation strategies, along with catch crops in the intensive greenhouse vegetable cropping system [22-24]. Nevertheless, a recent survey of the largest greenhouse vegetable production region in northern China showed that the soil C/N ratio in greenhouse soils was lower than that of the adjacent open field soils because of the high accumulation rate of soil N as a result of an excessive N input [25]. The low soil C/N ratio implied that the C levels were insufficient for the cropping system, which would limit N immobilization by soil microorganisms [2], [26-27] and may lead to high N losses [28]. This finding

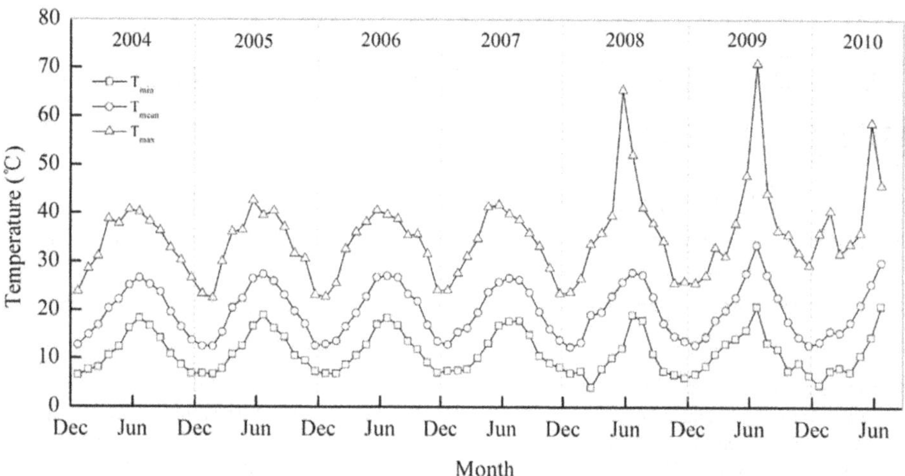

Figure 1. The monthly mean temperature, minimum and maximum temperature inside greenhouse from 2004WS to 2009AW in the year-round greenhouse tomato planting system in Shouguang, northern China.

shows that not only optimal N fertilizer management is needed to be studied, but also the soil organic matter content must be improved to enhance soil N retention capacity, which will reduce N losses to the environment and improve N use efficiency.

In a conventional greenhouse vegetable planting system, most of the plant residues are removed at harvest out of fear of infecting the next crop with fungi and other pathogens. Large amounts of organic manure with a low C/N ratio, such as poultry manure and pig manure, are the main soil carbon supplements in the greenhouse vegetable cropping system in northern China [29]. Excessive N fertilizer application is also a significant characteristic of this cropping system. However, it is unclear how the continuous application of poultry manure and excessive mineral N influences the soil organic matter and total N. Whether lower optimized N fertilizer inputs will alter the accumulation of soil organic matter and total N in the greenhouse field? A greenhouse tomato experiment into which different N management strategies were introduced was conducted from 2004 to 2010 in Shouguang county, the largest greenhouse vegetable production region in northern China. In contrast to common farming practice, optimal N management could reduce mineral N fertilizer input by 72% without decreasing the fruit yield [24]. In this experiment, different types of organic manure, including conventional dry chicken manure and wheat straw, were applied. This environment provided an opportunity to gain a better understanding of the influence of organic manure together with mineral N inputs on the SOC and total N pools, and determine if reducing mineral N fertilizer input will alter SOC and TN accumulation. Moreover, it provided an opportunity to analyse the influence of changes in the SOC pool on N losses in the greenhouse vegetable cropping system. Evaluating this system will be of great assistance in improving management practices for maintaining soil fertility and productivity while minimizing potential N losses from the high-input greenhouse vegetable cropping system.

Materials and Methods

Ethics statement

No specific permits were required for these field studies. No specific permissions were required for these locations/activities because they were not carried out on privately owned or protected areas. The field studies did not involve endangered or protected species.

Site description and crop management

The experiment was established in February 2004 in a traditional unheated commercial solar greenhouse (84×8.5 m) in Luojia (36°55′N, 118°45′E), Shouguang, northern China. The greenhouse was constructed from a vertical clay wall and covered with polyethylene film throughout the year. Thus the air temperature inside greenhouse is higher than in outside. The monthly mean temperature, minimum and maximum temperature inside greenhouse from 2004 to 2010 was showed in Figure 1. Groundwater was used for irrigation with an average of 573 mm over 11 applications for every growing season. During construction in 1999, surface soil in the greenhouse was used to build the clay wall. The silt loam that remained in the field was approximately 60 cm lower than the open field. Chicken manure was applied at 30 t DW $ha^{-1}season^{-1}$ for the first 2 years to improve soil fertility. After 2001, chicken manure application was reduced to 15 t DW ha^{-1} $season^{-1}$ until the experiment was established in 2004. According to FAO classification, the soil at the outset of the trial had 637 g sand kg^{-1}, 323 g silt kg^{-1} and 40 g clay kg^{-1} from 0 to 0.1m soil depth, 620 g sand kg^{-1}, 335 g silt kg^{-1} and 45 g clay kg^{-1} from 0.1 to 0.3 soil depth, 662 g sand kg^{-1}, 301 g silt kg^{-1} and 37 g clay kg^{-1} from 0.3 to 0.6 soil depth, respectively.

Tomato (*Lycopersicon esculentum* Mill.) has been the sole crop since the construction of the greenhouse in 1999. There were two cropping seasons per year, namely a winter-spring (WS) and an autumn-winter (AW) tomato crop. For the WS season, 4-week-old tomato seedlings were transplanted by hand into double rows in the middle of February, harvesting was completed in the middle of June. After a 2-month fallow period, the second (AW) crop was transplanted in early August and the final harvest was taken the following January. The tomato vines were removed from the greenhouse after each final harvest to reduce the infecting of disease carry over into the next crop.

Treatments

From February 2004 the treatments included (1) **CK**, control treatment, where neither organic manure nor mineral fertilizer N was applied. The N from irrigation water was the important source of N input in the CK treatment, ranging from 25 to 177 kg

N ha^{-1} $season^{-1}$ with an average of 102 kg N ha^{-1} $season^{-1}$. The large variation in N input from irrigation water across different growing season was due to different amount of irrigation water and different concentration of NO_3^-–N in irrigation water. (2) **MN**, organic manure treatment, only organic manure was broadcast as a basal fertilizer with no mineral N fertilizer applied. Organic manure was bought from different poultry farms every growing season and C and N content of chicken manure were different across different growing season. From 2004WS to 2006WS only dry chicken manure was used; and the application rates of dry chicken manure were 8, 11, 8, 11 and 5 t ha^{-1} $season^{-1}$, with an average of 271 kg N ha^{-1} $season^{-1}$. During the autumn-winter season in 2006 (2006AW) and onward, additional chopped wheat straw (1-5 cm long) was added to the soil together with dry chicken manure. From 2006AW to 2009AW the application rates of dry chicken manure were 8, 8, 8, 8, 8, 10 and 8 t ha^{-1} $season^{-1}$; and the rates of wheat straw were 2, 2.5, 4, 2, 4, 4 and 4 t ha^{-1} $season^{-1}$, with an average of 22 kg N ha^{-1} $season^{-1}$. The treatment was also labeled as "MN+S" instead of "MN". (3) **CN**, conventional N treatment, organic manure was applied as in the MN treatment (except in 2006AW). N fertilizer was applied as a side-dressing at a rate of 120 kg N ha^{-1} on 4-6 occasions based on local farmers normal management practice which depended on the weather conditions, tomato cultivar and growth stage. The average mineral N fertilizer input was 635 kg N ha^{-1} $season^{-1}$. From 2006AW, the CN treatment plot was split into CN and CN+S sub-treatments with plot sizes of 21.8 m^2 and 32.8 m^2, respectively. For the CN sub plot only dry chicken manure was applied; and for the CN+S treatment both wheat straw and dry chicken manure were incorporated. The chicken manure and wheat straw were applied at the same rates and timings as in the MN treatment. The mineral N fertilizer application was the same as for the original CN plot. (4) **RN**, reduced N treatment, chicken manure was applied at the same rate as in the MN and CN treatment, and mineral N fertilizer was applied as a side-dressing based on an N target value and soil mineral N content in the root zone (0-30 cm soil layer) for different growth stages from 2004WS to 2007WS. The equation used was as follows:

$$\textbf{Recommended fertilizer N} = \textbf{N target value} - \textbf{NO}_3^- - \textbf{N in the top 0.3m of the soil profile before the recommendation} - \textbf{NO}_3^- - \textbf{N from irrigation water} \quad (1)$$

N target value is calculated from crop N uptake, the necessary soil N_{min} residue and soil net mineralization [30], which reflects the synchronization of crop N requirement and soil N supply. Here the initial N target values were 300 kg N ha^{-1} for the side-dressing at each stage of fruit cluster development in 2004. From 2005 the target values from transplanting to the third cluster growth stage were changed to 250 and 200 kg N ha^{-1} for the fourth cluster to the end of harvest in the WS seasons. In the AW season the N target value from transplanting to the fourth cluster growth stage was changed to 200 kg N ha^{-1}, with 250 kg N ha^{-1} for the fifth and sixth cluster growth stages [24]. Considering crop N uptake in different growth period and soil N supply, N from irrigation water, only 2-4 side-dressing events were applied according to the differences between N target values and soil N_{min} content in the root zone. From 2007AW the N recommendation was simplified based on the experiences of preceding years. Three or four side-dressing events with an interval of 7–10 days at a rate of 50 kg N ha^{-1} were required in April and October. The more detail introduction of optimized N management was reported by Ren et al [24]. Compared with the CN treatment, 71.3% of mineral N fertilizer was reduced without influencing fruit yield, with an average of 182 kg N ha^{-1} $season^{-1}$. From 2006AW, the RN treatment plot was split into RN and RN+S sub-treatments with the same straw amendments and plot sizes as in the CN treatment. During 2006AW and 2007WS growing season, the mineral N fertilizer application rates of RN and RN+S treatment were determined based on the N target values and soil N_{min} content in the root zone before side-dressing. There were litter differences on mineral N fertilizer application rates between RN and RN+S treatment. Since 2007AW, the mineral N fertilizer application was the same for RN and RN+S treatment.

All treatments were set up in a randomized block design with three replicates. Plots were separated for each other by plastic film at a depth of 30 cm. The exogenous C and N inputs are shown in Table 1. The average C input from chicken manure was 2779 kg C ha^{-1} $season^{-1}$, with a range of 1114 to 5445 kg C ha^{-1} $season^{-1}$. Beginning in 2006AW, an average of 1119 kg wheat straw C ha^{-1} $season^{-1}$ was supplied, and the total C input was as high as 3483 kg C ha^{-1} $season^{-1}$. The N sources included mineral N fertilizer, organic manure and irrigation water. In the CK treatment, the N from irrigation water was the only source of N input, with an average of 102 kg N ha^{-1} $season^{-1}$. The average exogenous N inputs for the MN, RN and CN treatments were 354, 527 and 984 kg N ha^{-1} $season^{-1}$, respectively.

Urea was the main mineral N fertilizer. All plots received P_2O_5 as calcium monophosphate (12% P_2O_5) and K_2O as potassium sulfate (50% K_2O) during each growing season. The average input during the past 12 growing seasons were 350 kg P_2O_5 ha^{-1} and 563 kg K_2O ha^{-1} per season.

Soil sampling and analysis

Soil organic carbon and total N concentrations. Three soil cores (3.5 cm in diameter) were taken from each plot to a depth of 0.6 m and subdivided into 0-0.1, 0.1-0.3 and 0.3-0.6 m increments on April 18, 2004, and January 20, 2010. Fresh soil cores were taken to the lab immediately, mixed thoroughly to provide a composite sample from each plot, sieved through a 2 mm mesh and then air-dried and stored in plastic bottles. After removing any carbonates with 1 M HCl, the SOC and total N concentrations were determined using a C and N analyzer (vario MACRO CN, Elementar, Germany).

At the same time, soil bulk density in different soil layers was measured by the cutting ring method. In 2004, three samples were collected from the whole greenhouse and the average soil bulk densities were 1420 kg m^{-3}, 1450 kg m^{-3} and 1480 kg m^{-3} for the 0-10 cm, 10-30 cm and 30-60 cm soil layers, respectively. The bulk density of each plot monitored in 2010 is shown in Table 2 and no significant differences were observed among the different treatments.

Soil organic carbon fraction. Changes in the quantity and quality of the SOM pool are generally difficult to detect in the short term following agricultural management. Labile and recalcitrant SOM separated by different methods provide a more sensitive indicator for evaluating the effect of different management strategies on SOM dynamics [31-33]. Here, soil organic carbon fractionation procedures were carried out as described by Blair et al [31]. Air-dried soil samples containing approximately 15 mg of C were oxidized with 25 mL of 333 mmol L^{-1} $KMnO_4$ for 1 h at 25°C on a shaker at 180 rpm. The samples were then

Table 1. Exogenous C and N inputs in the greenhouse tomato production system in Shouguang, northern China (kg ha^{-1} $season^{-1}$).

Growing season	C[1]						N[1]					
	CK	MN(MN+S)[3]	RN	RN+S	CN	CN+S	CK	MN(MN+S)	RN	RN+S	CN	CN+S
2004WS[2]	0	2860	2860	-	2860	-	56	316	644	-	1186	-
2004AW[2]	0	5445	5445	-	5445	-	118	478	638	-	1198	-
2005WS	0	3476	3476	-	3476	-	54	370	497	-	1000	-
2005AW	0	3902	3902	-	3902	-	177	435	636	-	1155	-
2006WS	0	1114	1114	-	1114	-	165	327	465	-	927	-
2006AW	0	4095	3423	4095	5135	5807	133	456	651	671	1200	1212
2007WS	0	2258	1698	2258	1698	2258	144	308	509	486	838	848
2007AW	0	4806	3406	4806	3406	4806	82	417	492	517	872	897
2008WS	0	2110	1606	2110	1606	2110	25	180	321	330	771	780
2008AW	0	3052	1606	3052	1606	3052	39	202	382	402	782	802
2009WS	0	4205	2592	4205	2592	4205	135	441	605	641	1105	1141
2009AW	0	3857	2214	3857	2214	3857	99	322	489	523	769	802
Average	0	3432	2779	3483	2921	3728	102	354	527	510	984	926

[1]C input from chicken manure and wheat straw; N input from mineral fertilizer, organic manure and irrigation water;
[2]WS: winter-spring growing season, AW: autumn-winter growing season;
[3]Since 2006AW chopped wheat straw and dry chicken manure was broadcast as a basal fertilizer and the treatment was labeled as "MN+S" instead of "MN";

Table 2. Soil bulk density in the soil profile in Jan. 2010 in a year-round greenhouse tomato planting system in Shouguang, northern China (kg m^{-3}).

Treatment	Soil layer (cm)		
	0-10	10-30	30-60
CK	1333±41	1545±70	1589±63
MN(MN+S)[1]	1296±118	1548±79	1586±56
RN	1301±82	1502±30	1460±42
RN+S	1285±35	1508±126	1546±87
CN	1227±213	1584±129	1512±127
CN+S	1283±43	1493±114	1542±132
Average	1288	1530	1539

[1]Since 2006AW chopped wheat straw and dry chicken manure was broadcast as a basal fertilizer and the treatment was labeled as "MN+S" instead of "MN";

centrifuged, diluted and spectrophotometrically measured at 565 nm. The oxidized carbon was considered labile C, and the remainder represented the non-labile C.

Apparent N losses. The nutrient balance is often used to estimate potential environmental risks [34-35]. The apparent N losses were calculated according to Equation (2), as described by Ren et al [24], as follows:

$$N_{loss} = N_{min\ initial} + N_{manure} + N_{fert} + N_{irri} - N_{crop} - N_{min\ harvest} \quad (2)$$

where N_{loss} = apparent N loss, $N_{min\ initial}$ = soil N_{min} content at 0-0.6 m before transplanting, N_{manure} = total N input from organic manure, N_{fert} = N from mineral fertilizer, N_{irri} = NO_3^--N from irrigation water, N_{crop} = total N uptake by tomato aboveground parts, and $N_{min\ harvest}$ = soil N_{min} content at 0-0.6 m at the end of the harvest.

The parameters were calculated as follows:

Three soil cores (3.5 cm in diameter) were collected from each plot at a depth of 0.6 m and then subdivided into 0-0.3 m and 0.3-0.6 m increments before transplanting and at the end of the harvest for each season. Fresh soil cores were mixed thoroughly to give a composite sample from each plot and then passed through a 2 mm sieve. Next, 12 g subsamples were weighed and extracted by shaking with 100 mL of 1 mol L^{-1} KCl for 1 h. The extract was stored at -18°C until an analysis of the NO_3^--N and NH_4^+-N concentrations could be carried out with a continuous flow analyzer (TRAACS Model 2000). The water content of the soil samples was also gravimetrically determined to calculate the soil N_{min} (NO_3^--N+NH_4^+-N) content on a dry matter basis. The soil bulk density was used to convert the mineral N in mg per kg of soil to kg per hectare.

The input of N from irrigation water was determined by recording the amounts applied, and water samples were collected during each irrigation event over the entire growing season. The samples were stored frozen until NO_3^--N and NH_4^+-N analysis.

Plant samples were collected from each plot at the end of the harvest; divided into leaves, fruit, stems and roots; and weighed before and after drying at 70°C for 48 h. The dried shoots were ground before determining the total N, which was conducted using a modified Kjeldahl method with salicylic acid. N uptake was calculated as the product of dry matter and total N concentration in different parts.

Data analysis. Analysis of variance (*ANOVA*) was used to determine the significance of treatment effects based on a randomized complete block design. Multiple comparisons of mean values were performed using either Duncan's multiple range tests or Fisher's protected least significant difference (*LSD*) test at the 0.05 level of probability. Statistical analysis was performed using version 6.12 of the SAS software package (SAS Institute Inc., Cary, NC).

Results

Soil organic carbon and total N concentrations

Soil organic carbon (SOC) and total soil N (TN) concentrations in soil profiles to a depth of 60 cm are shown in Figure 2. The highest SOC concentration was seen in the 0-10 cm layer, below which it decreased for all treatments. In April of 2004, the SOC of the soil profile was similar across all N treatments, except there were some variations in the 30-60 cm layer associating with uneven soil fertility. After 6 years, SOC in the CK treatment was significantly lower than that of the other treatments in the 0-10 cm soil layer. However, there were no significant differences among the other treatments. In the 10-30 cm layer, SOC differed significantly between the RN and CN treatments. In comparison to the initial value, SOC decreased significantly in all CK treatment layers after 6 years of cultivation, but the only significant change in SOC for the CN treatment was an increase in the 0-10 cm layer.

Patterns in total N in the soil profile followed a similar pattern to that of the SOC (Figure 1). In January 2010, soil TN in the CK treatment was significantly lower than that of the other treatments in the 0-10 cm layer. In addition, no significant differences were observed among the MN, RN and CN treatments. The TN values did not show significant differences across N treatments in the 10-60 cm layer in 2010.

Soil organic carbon and total N pool

Total SOC and TN pool in the profile above 60 cm were 46.6-55.1 t C ha^{-1} and 6.8-8.3 t N ha^{-1}, respectively (Table 3). After 6 years of cultivation, approximately 13.1 t C ha^{-1} and 2.2 t C ha^{-1} were lost from the CK and MN treatments, respectively. The decreased bulk density was mainly attributed to the decreased SOC in the MN treatment (Table 2). The SOC pool increased to 0.8 t C ha^{-1} and 1.7 t C ha^{-1} in response to the RN and CN treatments, respectively. For the N pool, the only reduction occurred in the CK treatment. Approximately 0.63 t N ha^{-1} was lost over the last 6 years. Average accumulation rates of SOC and TN in the 0-60 cm layer were -2.17, -0.37, 0.14 and 0.28 t C ha^{-1} a^{-1} and -0.10, 0.14, 0.00 and 0.06 t N ha^{-1} a^{-1} for the CK, MN,

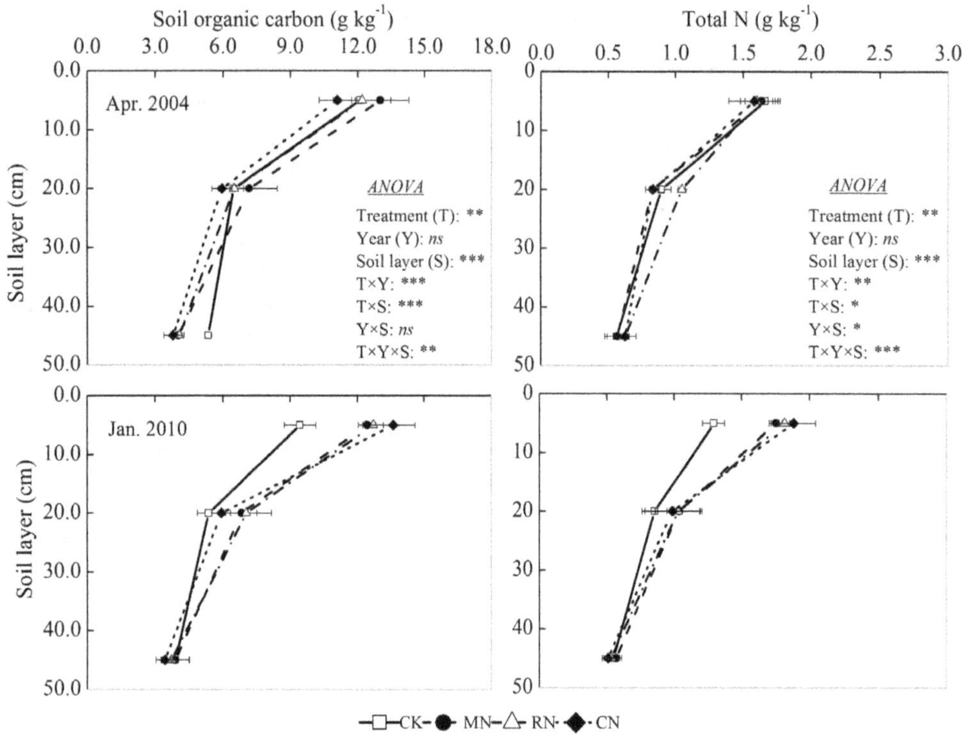

Figure 2. The distribution of soil organic carbon and total N concentrations in the soil profile with different N treatments in the year-round greenhouse tomato planting system in Shouguang, northern China. Note: *, ** and *** indicate significant differences at $P<0.05$, $P<0.01$ and $P<0.001$, ns denotes no significant difference.

RN and CN treatments, respectively. The addition of straw over 4 years made little difference to SOC and TN concentration

Soil organic carbon fractions

Figure 3 shows the distribution of labile and non-labile carbon in different soil layers. In April 2004, the labile C concentrations were similar across all N treatments in the same layer. However, in January 2010, the soil labile C concentration in the CK treatment was significantly lower than that of the other treatments in the 0-30 cm layer. Compared with the initial values in 2004, the soil labile C concentration in the CK treatment did not decrease significantly according to paired-sample T test; however, labile C increased significantly in the RN and CN treatments for the 0-10 cm layer.

Changes in the concentration of the non-labile C fraction in the soil profile were similar to those of the SOC. In January 2010, the soil non-labile C concentration in the 0-10 cm layer increased as the N application increased. When compared with April 2004, it decreased significantly in the CK treatment. Nevertheless, it increased significantly in the CN treatment. For all other treatments, no significant changes were observed in the soil profile.

Apparent N losses

Figure 4 shows the annual apparent N losses from 2004 to 2009 as obtained estimated from the nitrogen balance. For the CN treatment, the average annual mean apparent N losses were as high as 1529 kg N ha^{-1} a^{-1}, accounting for 77% of the annual exogenous N input. A significant decrease in the annual apparent N losses of 633 kg N ha^{-1} a^{-1} occurred in the RN treatment because the rate of N fertilization had been reduced to less than one-third of that in the CN treatment. For the CK treatment, soil N and nitrate in the irrigation water were the major sources of N, which were less input than N removed by plant uptake, resulting in a negative N balance. These changes could indirectly explain the decreased TN in the CK treatment. Although the straw treatments led to the addition of C, the apparent N losses from treatments with and without straw amendments did not differ significantly. The N surpluses were 490 and 1285 kg N ha^{-1} a^{-1} N for RN+S and CN+S, respectively.

Interactions between C and N in soil

Figure 5 shows the relationship between the changes in SOC and TN concentration after 6 years of cultivation, as well as annual apparent N losses and average N inputs. The SOC and TN concentrations increased in response to N addition but showed a negative response to excessive N application (Figure 5a, b). The total N in the soil did not increase linearly with the increase in applied mineral N, which might be explained by changes in the SOC and its fractions (Figure 5c). The minor changes in SOC in response to the current type and amount of organic manure application, limited N storage in soil and N was close to saturation. With the increase in fertilizer N application, the apparent N losses linearly increased and conventional N fertilizer management caused the highest apparent N losses (Figure 5d).

Discussion

Effect of N fertilizer on SOC

Greater root exudates and more crop residues in response to mineral N fertilizer application were the dominant reasons why N fertilizer application improved the SOC [6]· In this experiment, although 71.3% of mineral N fertilizer was cut down in the RN treatment compared with the CN treatment, there were no

Table 3. Distribution of the soil organic C and N pools in the soil profile of a greenhouse tomato production system in Shouguang, northern China.

		SOC pool (t ha^{-1})						N pool (t ha^{-1})					
		CK	MN (MN+S)[2]	RN	RN+S	CN	CN+S	CK	MN (MN+S)[2]	RN	RN+S	CN	CN+S
Apr-04	0-10 cm	17.2±0.2	18.5±1.8	17.3±1.8	-	15.7±1.2	-	2.36±0.11	2.32±0.17	2.27±0.17	-	2.25±0.27	-
	10-30 cm	18.8±1.3	20.8±3.7	18.8±2.0	-	17.3±1.3	-	2.61±0.21	2.41±0.15	3.04±0.07	-	2.43±0.05	-
	30-60 cm	23.7±0.6	17.8±0.5	18.1±0.9	-	16.8±1.8	-	2.50±0.39	2.52±0.33	2.79±0.37	-	2.78±0.14	-
Jan-10	0-10 cm	12.2±0.9	16.0±0.5	16.4±0.6	17.9±0.9	17.5±1.3	17.6±1.0	1.66±0.11	2.26±0.06	2.34±0.05	2.42±0.09	2.4±0.21	2.54±0.12
	10-30 cm	16.5±1.5	20.9±2.2	21.6±3.5	17.2±1.2	18.2±1.3	18.6±2.1	2.63±0.28	3.17±0.52	3.15±0.50	2.70±0.23	3.02±0.61	2.95±0.18
	30-60 cm	18.0±2.8	18.0±2.9	17.1±0.1	17.2±1.5	15.8±1.8	18.7±0.7	2.55±0.22	2.64±0.20	2.62±0.14	2.68±0.06	2.35±0.19	2.86±0.41
Δ(0-60 cm)[1]		-13.1±4.5	-2.2±1.0	0.8±6.1	-	1.7±3.6	-	-0.63±0.50	0.82±0.27	0.01±0.74	-	0.34±0.47	-
Accumulation rate (t ha^{-1} a^{-1})		-2.17	-0.37	0.14	-	0.28	-	-0.10	0.14	0.00	-	0.06	-

[1]Δ(0-60 cm) = SOC (N) pool$_{2004}$- SOC (N) pool $_{2010}$;
[2]Since 2006AW chopped wheat straw and dry chicken manure was broadcast as a basal fertilizer and the treatment was labeled as "MN+S" instead of "MN";

significant differences on fruit yields and plant biomass between the CN and RN treatment [24]; root exudates were presumed to be similar between these two treatments. Besides, in our greenhouse system crop residues were removed from the greenhouse at harvest because of the risks of disease carryover, so organic manure was the major C supplement. The C input from organic manure was the same for the CN and RN treatments in the same growing season (Table 1). Moreover, no significant changes in the root C/N ratio (data not shown), soil microbial community [36] and soil organic matter fractions (Figure 3) between the RN and CN treatments were found. Thus, no significant differences in SOC pool between the CN and the RN treatment were observed in this experiment, which was similar to other work [14], [17]. Apparently, N was not limiting factor in this cropping system and reduced mineral N input would not alter SOC and TN accumulation. All these findings indicate that when organic manure is used, optimizing N fertilizer input over a continuous 6-year period would not affect the SOC or TN contents in the greenhouse vegetable cropping system; but it was helpful in reducing the environmental risks without influencing fruit yields (Figure 4).

Effect of organic manure on SOC

Organic manure application brought lots of N, with the average of 252 kg N ha^{-1}. In contrast to the CK treatment, significant increments on fruit yields were achieved; yet there were no differences on fruit yields and plant biomass between MN+S treatment and RN, CN treatment in most growing seasons [24], demonstrating that N from organic manure was important N source for crop growth and excessive mineral N fertilizer application was wasteful without considering N from organic manure. In addition, Organic manure application is considered to be a consistent method for maintaining soil fertility over the long-term [10], [37]. In this unique production system, the major source of organic carbon is organic manure. If the input of organic manure is excluded, only 44-146 kg C ha^{-1} season^{-1} from residual roots is incorporated [38]. If no organic manure is applied, SOC concentration, especially for stable organic carbon, will decline significantly (Figure 2, 3). These results are similar to those of another long-term greenhouse tomato experiment [39].

Before the start of this experiment, about 210 t ha^{-1} chicken manure had been applied across the whole greenhouse during 1999-2003; and the high organic manure application might build higher soil organic carbon pool in a short time. Therefore, in contrast to the treatments with organic manure application, there was the lack of a big difference on SOC in the CK treatment for the next 6 years. As well, no observable changes in soil organic C was found over 6 years of successive chicken manure applications. Whether it implied the soil organic carbon pool in our study was saturated? Although the coarse-textured soil has lower capacity for C and N stabilization, the saturated soil organic carbon pool in 0-30cm soil layer could be high to 75.6-96.8 t ha^{-1} according to Hassink [40] and Six's C-saturation model [41]. These values were higher than it reported in our study, indicating that soil organic carbon could be improved with optimum management. Quantity and quality of input organic manure significantly influenced soil organic C dynamics. In contrast to farmers' normal manure application, the application rate of chicken manure in the experiment was lower, ranging from 8 t ha^{-1} season^{-1} to 12 t ha^{-1} season^{-1}. Whether SOC content would be enhanced with the organic manure application rate increase? Indeed, high rates of organic manure application were conducive to enhanced SOC and SOC fractions [42-43]. Ge et al [39] demonstrated that it would take 10-15y with 75 t ha^{-1} a^{-1} of horse manure to increase

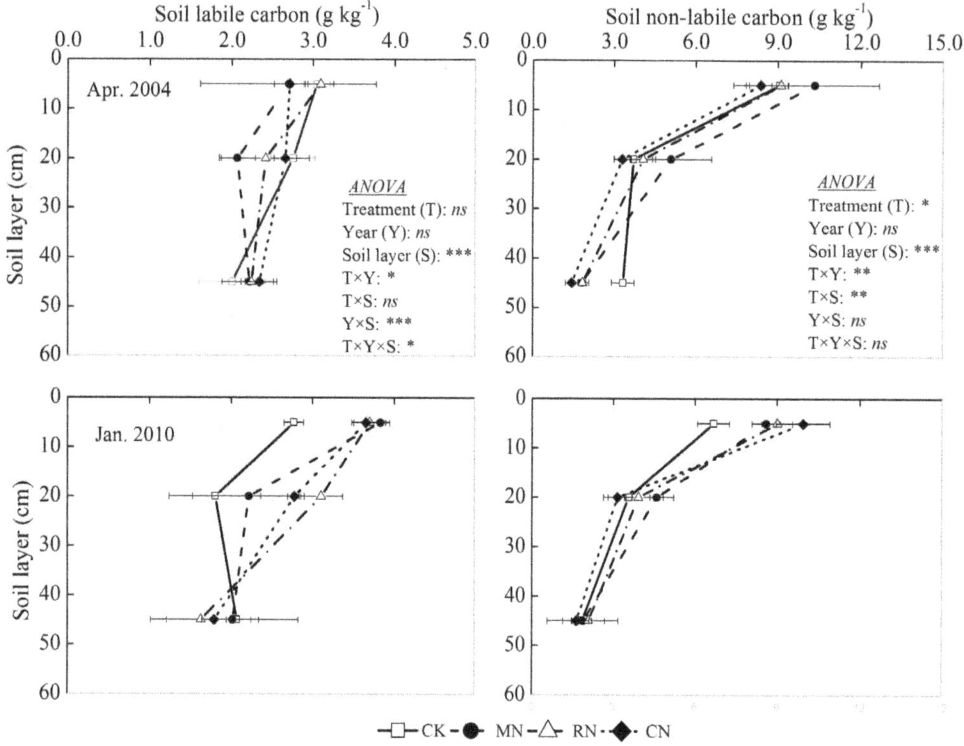

Figure 3. The distribution of the soil labile and non-labile carbon concentrations in the soil profile with different N treatments in the year-round greenhouse tomato planting system in Shouguang, northern China. Note: Note: *, ** and *** indicate significant differences at $P<0.05$, $P<0.01$ and $P<0.001$, ns denotes no significant difference.

the soil organic matter content from 24 g kg^{-1} to 30-40 g kg^{-1} in the greenhouse vegetable soil. However, environmental pressures associating with excessive application of organic manure were serious [44]; and in Europe the applications of organic manure are restricted to not exceed 170 kg N ha^{-1} y^{-1} by the legislation. In our experiment N input from organic manure averaged 252 kg N ha^{-1} $season^{-1}$, with 335 kg N ha^{-1} y^{-1} apparent N loss. Obviously it will not be an effective way to enhance SOC relying on the increments of organic manure application rate in greenhouse vegetable cropping system. Changing the type of organic manure input might be an important way to heighten SOC. For chicken and pig manure with high proportions of water-soluble C and easily biodegradable organic compounds, approximately 45-62% of C is evolved as CO_2-C within 30 days [45-46]. Plaza et al [47] reported a significant decrease in the total organic C in soils amended with pig manure slurry. However, manure with a greater ratio of C/N, or high content of recalcitrant C could reduce mineralization of bio-labile compounds, thereby enhancing soil organic matter [48-49]. Long term field experiments showed that the benefits in SOC content were higher from application of rice straw and compost than that from pig manure [42], [50].

To improve soil organic carbon content, wheat straw was added from in 2006AW. However, no significant increase in the SOC or the SOC fraction was observed after 4 years of cultivation. This result was similar to that of Antil's study [51], which found that the SOC in bulk soil decreased or was not affected by a slurry + straw treatment in both fallow and cropped plots, even after 28 and 38 yrs. According to the mechanisms of real and apparent priming effects [52], it is assumed that when chicken manure and wheat straw are applied together, microorganisms may first use the C from chicken manure to activate the microbial community; however, when the easily decomposed organic carbon is consumed, activated microorganisms will use the wheat straw carbon. Most of added straw carbon is then utilized by microorganisms, perhaps explaining why there was no effect on SOC when chicken manure and wheat straw were incorporated together. The short duration might be another important reason why no differences were seen. Additional long-term studies should be conducted to determine if the application of a mixture of wheat straw and chicken manure is an effective method of enhancing soil organic carbon in the greenhouse vegetable cropping systems over the long run. Overall, developing an optimum organic manure management system to enhance soil fertility is now one of the important issues in greenhouse vegetable cropping systems in China. In comparison to increasing the application rate, shifting the type of organic manure from pig and chicken manure to manure with a wider ratio of C/N or high content of recalcitrant C may be more practical for enhancing soil organic matter in greenhouse vegetable cropping system.

Effect of soil organic carbon content on the fate of N

Similar to the results from 2004 to 2007 [24], approximately 77% of the exogenous N input was surplus in the CN treatment. With the exception of immobilized N in soil clays, N leaching [53] and N_2O emissions [21] were the major N loss processes in these vegetable cropping systems. Furthermore, warm and moist conditions with sufficient available nitrate and labile carbon can lead to denitrification loss, which might also be an important N loss process [54]. In any case, an excessive N surplus would lead to high potentially N losses. Therefore, most studies on the prevention of N losses have focused on reducing the N input and on irrigation strategies [24], [55] and catch crops [56]. In the

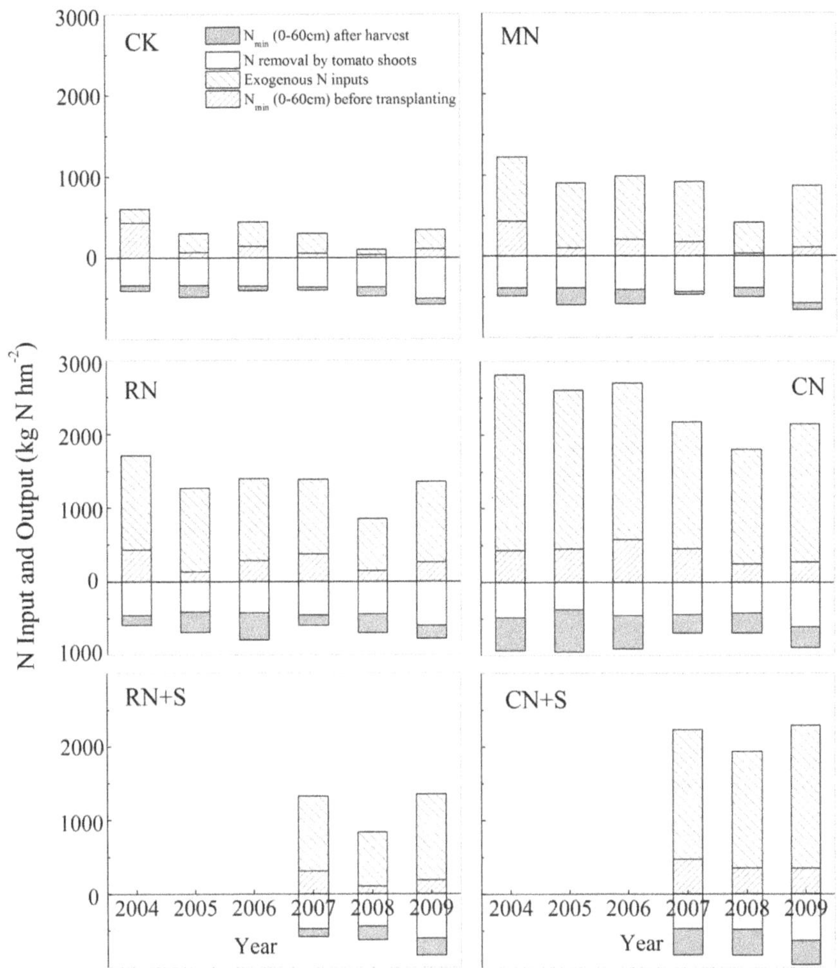

Figure 4. Apparent N loss with different treatments in the year-round greenhouse tomato planting system from 2004 to 2009 in Shouguang, northern China.

experiment, the rate of N fertilization in the RN treatment had been reduced to less than one-third of that in the CN treatment and apparent N losses were also decreased. Other than that, SOC plays an important role in regulating soil N turnover and the improvement of SOC is beneficial to increase potential rates of N immobilization and reduce N losses [57]. Yang et al [58] showed that the rate of absolute N change increased linearly with changes in the size of the C pool change and organic N capital was determined by long-term carbon sequestration. A similar tendency was observed in the present study (Figure 3c). In our experiment, SOC concentration was not increased even though high amounts of organic manure were applied during a 6 year period; Changes in organic N were similarly limited. Moreover, the mineralization rate of total organic N might be greater than the retention of exogenous N to soil organic N pool because of the great amount of organic manure applied within several years before the experiment started. Thus more exogenous N was lost to the environment. The low soil C/N ratio indirectly revealed that there was insufficient C in the cropping system would limiting N immobilization by soil microorganisms [2], [26] and leading to high N losses [28]. Therefore, improving soil organic matter content and enhancing potential rates of soil N immobilization according to optimal organic manure management was crucial, as well reducing N fertilizer input, to lower N losses in greenhouse vegetable cropping system.

Conclusion

Organic manure represents a major organic C source in conventional greenhouse vegetable cropping systems in China. Without additions of organic manure, SOC, particularly stable C is likely to decline. However, no significant increment in SOC was observed with the addition of high amounts of low C/N ratio organic manure or plant residues. Shifting the type of manure from chicken manure to manures with a wider ratio of C/N, or high content of recalcitrant may be more effective in enhancing soil fertility in greenhouse vegetable production. On the basis of organic manure application, optimized N fertilizer inputs according to root-zone N management did not influence the accumulation of SOC and TN in soil; but beneficial in reducing apparent N losses.

The SOC concentration was a dominant limiting factor for soil total N enhancement. Given the current type and quantity of organic manure application, the SOC concentration was unchanged and most applied N was in excess and was lost to the environment. Therefore, integrating nutrient management, including optimized N fertilizer input, as well as enhanced the soil

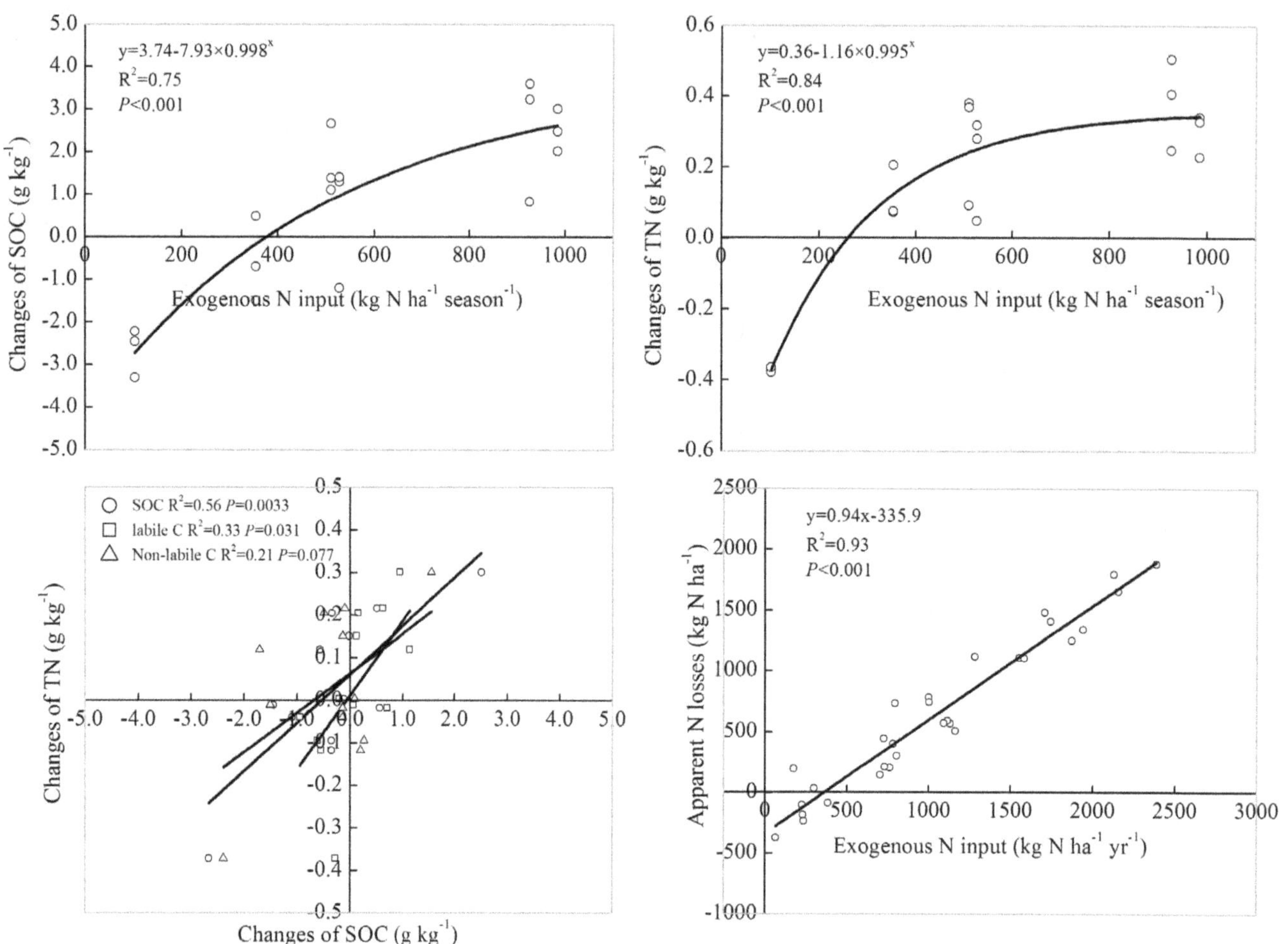

Figure 5. The relationship between the changes in SOC and TN concentration after 6 years of cultivation, as well as annual apparent N losses and the average N inputs in the year-round greenhouse tomato planting system in Shouguang, northern China.

organic matter content, should be considered to maintain soil fertility and productivity, minimize potential N losses and achieve sustainable development in greenhouse vegetable cropping systems.

Author Contributions

Conceived and designed the experiments: QC JW FZ. Performed the experiments: TR. Analyzed the data: TR SL. Contributed reagents/ materials/analysis tools: QC JW. Wrote the paper: TR QC.

References

1. Bird JA, van Kessel C, Horwath WR (2002) Nitrogen dynamics in humic fractions under alternative straw management in temperate rice. Soil Sci Soc Am J 66: 478-488.
2. Accoe F, Boeckx P, Busschaert J, Hofman G, Van Cleemput O (2004) Gross N transformation rates and net N mineralization rates related to the C and N contents of soil organic matter fractions in grassland soils of different age. Soil Biol Biochem 36: 2075-2087.
3. Paré MC, Bedard-Haughn A (2013) Soil organic matter quality influences mineralization and GHG emissions in cryosols: a field-based study of sub-to high Arctic. Global Change Biol 19: 1126-1140.
4. Compton JE, Boone RD (2002) Soil nitrogen transformation and the role of light fraction organic matter in forest soils. Soil Biol Biochem 34: 933-943.
5. Neff JC, Townsend AR, Gleixner G, Lehman SJ, Turnbull JT, et al. (2002) Variable effects of nitrogen additions on the stability and turnover of soil carbon. Nature 419: 915-917.
6. Christopher SF, Lal R (2007) Nitrogen management affects carbon sequestration in North American Cropland soils. Crit Rev Plant Sci 26: 45-64.
7. Swanston C, Homann PS, Caldwell BA, Myrold DD, Ganio L, et al. (2004) Long-term effects of elevated nitrogen on forest soil organic matter stability. Biogeochemistry 70: 227-250.
8. Hagedorn F, Spinnler D, Siegwolf R (2003) Increased N deposition retards mineralization of old soil organic matter. Soil Biol Biochem 35: 1683-1692.
9. Malhi SS, Harapiak JT, Nyborg M, Gill KS, Monreal CM, et al. (2003) Total and light fraction organic C in a thin Black Chernozemic grassland soil as affected by 27 annual application of six rates of fertilizer N. Nutr Cycl Agroecosyst 66: 33-41.
10. Blair N, Faulkner RD, Till AR, Poulton PR (2006) Long-term management impacts on soil C, N and physical fertility part I: Broadbalk experiment. Soil Till Res 91: 30-38.
11. Jagadamma S, Lal R, Hoeft RG, Nafziger ED, Adee EA (2007) Nitrogen fertilization and cropping systems effects on soil organic carbon and total nitrogen pools under chisel-plow tillage in Illinois. Soil Till Res 95: 348-356.
12. Lemke RL, VandenBygaart AJ, Campbell CA, Lafond GP, Grant B (2010) Crop residue removal and fertilizer N: effects on soil organic carbon in a long-term crop rotation experiment on a Udic Boroll. Agr Ecosyst Environ 135: 42-51.
13. Mack MC, Schuur EAG, Bret-Harte MS, Shaver GR, Chapin III FS (2004) Ecosystem carbon storage in arctic tundra reduced by long-term nutrient fertilization. Nature 431: 440-443.
14. Dolan MS, Clapp CE, Allmaras RR, Baker JM, Molina JAE (2006) Soil organic carbon and nitrogen in a Minnesoota soil as related to tillage, residue and nitrogen management. Soil Till Res 89: 221-231.
15. Fonte SJ, Yeboah E, Ofori P, Quansah GW, Vanlauwe B, et al. (2009) Fertilizer and residue quality effects on organic matter stabilization in soil aggregates. Soil Biol Biochem 73: 961-966.
16. Liu LL, Greaver TL (2010) A global perspective on belowground carbon dynamics under nitrogen enrichment. Ecol Lett 13: 819-828.

17. Lu M, Zhou XH, Luo YQ, Yang YH, Fang CM, et al. (2011) Minor stimulation of soil carbon storage by nitrogen addition: A meta-analysis. Agr Ecosys Environ 140: 234-244.
18. Chen Q, Zhang XS, Zhang HY, Christie P, Li XL (2004) Evaluation of current fertilizer practice and soil fertility in vegetable production in the Beijing region. Nutr Cycl Agroecosyst 69: 51-58.
19. He FF, Chen Q, Jiang RF, Chen XP, Zhang FS (2007) Yield and nitrogen balance of greenhouse tomato (*Lycopersicum esculentum* Mill.) with conventional and site-specific nitrogen management in Northern China. Nutr Cycl Agroecosyst 77: 1-14.
20. Song XZ, Zhao CX, Wang XL, Li J (2009) Study of nitrate leaching nitrogen fate under intensive vegetable production pattern in northern China. CR Biol 332: 385-392.
21. He FF, Jiang RF, Chen Q, Zhang FS, Su F (2009) Nitrous oxide emissions from an intensively managed greenhouse vegetable cropping system in Northern China. Environ Pollut 157(5): 1666-1672.
22. Mao XS, Liu MY, Wang XY, Liu CM, Hou ZM, et al. (2003) Effects of deficit irrigation on yield and water use of greenhouse grown cucumber in the North China Plain. Agr Water Manage 61(3): 219-228.
23. Guo RY, Li XL, Christie P, Chen Q, Jiang RF, et al. (2008) Influence of root zone nitrogen management and a summer catch crop on cucumber yield and soil mineral nitrogen dynamics in intensive production systems. Plant Soil 313: 55-70.
24. Ren T, Christie P, Wang JG, Chen Q, Zhang FS (2010) Root zone soil nitrogen management to maintain high tomato yields and minimum nitrogen losses to the environment. Sci Hortic 125: 25-33.
25. Lei BK, Fan MS, Chen Q, Six J, Zhang FS (2010) Conversion of wheat-maize to vegetable cropping systems changes soil organic matter characteristics. Soil Sci Soc Am J 74(4): 1320-1326.
26. Degens BP, Schipper LA, Sparling GP, Vojvodic-Vukovic M (2000) Decreased in organic C reserves in soils can reduce the catabolic diversity of soil microbial communities. Soil Biol Biochem 32: 189-196.
27. Cookson WR, Abaye DA, Marschner P, Murphy DV, Stockdale EA, et al. (2005) The contribution of soil organic matter fractions to carbon and nitrogen mineralization and microbial community size and structure. Soil Biol Biochem 37: 1726-1737.
28. Gundersen P, Callesen I, de Vries W (1998) Nitrate leaching in forest ecosystems is related to forest floor C/N ratios. Environ Pollut 102: 403-407.
29. Zeng XB, Bai LY, Li LF, Su SM (2009) The status and changes of organic matter, nitrogen, phosphorus and potassium under different soil using styles of Shouguang of Shangdong Province. Acta Ecol Sin 29(7): 3737-3746 (in Chinese).
30. Feller C, Fink M (2002) N_{min} target values for field vegetables. Acta Hort 571: 195-201.
31. Blair GJ, Lefroy RDB, Lisle L (1995) Soil carbon fractions based on their degree of oxidation, and the development of a carbon management index for agricultural systems. Aust J Agric Res 46: 1459–1466.
32. Six J, Paustian K, Elliott ET, Combrink C (2000) Soil structure and organic matter: I. Distribution of aggregate-size classes and aggregate-associated carbon. Soil Sci Soc Am J 64: 681-689.
33. McLauchlan KK, Hobbie S (2004) Comparison of labile soil organic matter fractionation techniques. Soil Sci Soc Am J 68: 1616-1625.
34. Öborn I, Edwards AC, Witter E, Oenema O, Ivarsson K, et al. (2003) Element balances as a tool for sustainable nutrient management: a critical appraisal of their merits and limitations within an agronomic and environmental context. Eur J Agron 20: 211-225.
35. Sieling K, Kage H (2006) N balance as an indicator of N leaching in an oilseed rape-winter wheat-winter barley rotation. Agric Ecosyst Environ 15: 261-269.
36. Zhao XC (2011) Effects of fertilization and crop rotation on soil microbial community structure of greenhouse tomato. Master Thesis, China Agricultural University, Beijing China (In Chinese).
37. Edmeades DC (2003) The long-term effects of manures and fertilizers on soil productivity and quality: a review. Nutr Cycl Agroecosyst 66: 165-180.
38. Lei BK, Chen Q, Fan MS, Zhang FS, Gan YD (2008) Changes of soil carbon and nitrogen in Shouguang intensive vegetable production fields and their impacts on soil properties. Plant Nutr Fert Sci 14(5): 914-922 (in Chinese).
39. Ge XG, Zhang EP, Zhang X, Wang XX, Gao H (2004) Studies on changes of filed-vegetable ecosystem under long-term fixed fertilizer experiment (I) changes of soil organic matter. Acta Hortic Sin 31(1): 34-38 (in Chinese).
40. Hassink J (1997) The capacity of soils to preserve organic C and N by their association with clay and silt particles. Plant Soil 191: 77-87.
41. Six J, Conant RT, Paul EA, Paustian K (2002) Stabilization mechanisms of soil organic matter: Implications for C-saturation of soils. Plant Soil 241: 155-176.
42. Liu J, Schulz H, Brandl S, Miehtke H, Huwe B, et al. (2012) Short-term effect of biochar and compost on soil fertility and water status of a Dystric Cambisol in NE Germany under field conditions. J Plant Nutr Soil Sci 175: 698-707.
43. Wang XJ, Jia ZK, Liang LY, Han QF, Ding RX, et al. (2012) Effects of organic manure application on dry land soil organic matter and water stable aggregates. Chin J Appl Ecol 23(1): 159-165 (in Chinese).
44. Ju XT, Kou CL, Zhang FS, Christie P (2006) Nitrogen balance and groundwater nitrate contamination: comparison among three intensive cropping systems on the North China Plain. Environ Pollut 143: 117-125.
45. Ajwa HA, Tabatabai MA (1994) Decomposition of different organic materials in soils. Biol Fert Soils 18: 175-182.
46. Cayuela ML, Velthof GL, Mondini C, Sinicco T, van Groenigen JW (2010) Nitrous oxide and carbon dioxide emissions during initial decomposition of animal by-products applied as fertilizers to soils. Geoderma 157: 235-242.
47. Plaza C, Garcia-Gil JC, Polo A (2005) Effects of pig slurry application on soil chemical properties under semiarid conditions. Agrochimica 49: 87-92.
48. Piccolo A, Spaccini R, Nieder R, Richter J (2004) Sequestration of a biologically labile organic carbon in soils by humified organic matter. Climatic Change 67: 329-343.
49. Adani F, Genevini P, Ricca G, Tambone F, Montoneri E (2007) Modification of soil humic matter after 4 years of compost application. Waste Manage 27: 319-324.
50. Li ZP, Liu M, Wu XC, Han FX, Zhang TL (2010) Effects of long-term chemical fertilization and organic amendments on dynamics of soil organic C and total N in paddy soil derived from barren land in subtropical China. Soil Till Res 106: 268-274.
51. Antil RS, Gerzabek MH, Haberhauer G, Eder G (2005) Long-term effects of cropped vs. fallow and fertilizer amendments on soil organic matter I. organic carbon. J Plant Nutr Soil Sci 168: 108-116.
52. Blagodatskaya E, Kuzyakov Y (2008) Mechanisms of real and apparent priming effects and their dependence on soil microbial biomass and community structure: critical review. Biol Fert Soils 45:115-131.
53. Lin Y (2010) Solute transportation and soil H^+ production budgets in a greenhouse vegetable production system. Master Thesis, China Agricultural University, Beijing China (In Chinese).
54. Ryden JC, Lund JL (1980) Nature and extent of directly measured denitrification losses from some irrigated vegetable crop production units. Soil Sci Soc Am J 44:505-511.
55. Zotareli L, Scholberg JM, Dukes MD, Mu~noz-Carpena R, Icerman J (2009) Tomato yield, biomass accumulation, root distribution and irrigation water use efficiency on a sandy soil, as affected by nitrogen rate and irrigation scheduling. Agr Water Manage 96: 23-34.
56. Constantin J, Mary B, Laurent F, Aubrion G, Fontaine A, et al. (2010) Effects of catch crops, no till and reduced nitrogen fertilization on nitrogen leaching and balance in three long-term experiments. Agr Ecosyst Environ 135: 268-278.
57. Schimel DS (1986) Carbon and nitrogen turnover in adjacent grassland and cropland ecosystems. Biogeochemistry 2: 345-357.
58. Yang YH, Luo YQ, Finzi AC (2011) Carbon and nitrogen dynamics during forest stand development: a global synthesis. New Phytol 190(4): 977-989.

2

Dynamics of Potassium Release and Adsorption on Rice Straw Residue

Jifu Li[1,2], Jianwei Lu[1,2]*, Xiaokun Li[1,2], Tao Ren[1,2], Rihuan Cong[1,2], Li Zhou[1,2]

1 College of Resources and Environment, Huazhong Agricultural University, Wuhan, China, **2** Key Laboratory of Arable Land Conservation (Middle and Lower Reaches of Yangtse River), Ministry of Agriculture, Wuhan, China

Abstract

Straw application can not only increase crop yields, improve soil structure and enrich soil fertility, but can also enhance water and nutrient retention. The aim of this study was to ascertain the relationships between straw decomposition and the release-adsorption processes of K^+. This study increases the understanding of the roles played by agricultural crop residues in the soil environment, informs more effective straw recycling and provides a method for reducing potassium loss. The influence of straw decomposition on the K^+ release rate in paddy soil under flooded condition was studied using incubation experiments, which indicated the decomposition process of rice straw could be divided into two main stages: (a) a rapid decomposition stage from 0 to 60 d and (b) a slow decomposition stage from 60 to 110 d. However, the characteristics of the straw potassium release were different from those of the overall straw decomposition, as 90% of total K was released by the third day of the study. The batches of the K sorption experiments showed that crop residues could adsorb K^+ from the ambient environment, which was subject to decomposition periods and extra K^+ concentration. In addition, a number of materials or binding sites were observed on straw residues using IR analysis, indicating possible coupling sites for K^+ ions. The aqueous solution experiments indicated that raw straw could absorb water at 3.88 g g^{-1}, and this rate rose to its maximum 15 d after incubation. All of the experiments demonstrated that crop residues could absorb large amount of aqueous solution to preserve K^+ indirectly during the initial decomposition period. These crop residues could also directly adsorb K^+ via physical and chemical adsorption in the later period, allowing part of this K^+ to be absorbed by plants for the next growing season.

Editor: Jörg Langowski, German Cancer Research Center, Germany

Funding: This work was supported by Non-profit Research Foundation for Agriculture (201103039), National Natural Science Foundation of China (41301319), Fundamental Research Funds for the Central Universities (2012BQ059) and International Potash Institute Co-operation Program. The funders had no role in study design, data collection and analysis, decision to publish, or preparation of the manuscript.

Competing Interests: The authors have declared that no competing interests exist.

* E-mail: lunm@mail.hzau.edu.cn

Introduction

The current Asian population of 4.3 billion is projected to increase by nearly 0.9 billion people, reaching roughly 5.2 billion, by 2050 [1], which will result in significantly increased regional food demand. Of this, population, 80% will be distributed in China, India and the southeast regions of Asia, posing a challenge to the economic development and social stability of these countries [2]. Rice-based cropping systems are the most productive agroecosystems in these areas and produce the most food for the most people [3]. To meet the food demand of the region, intensification and diversification have been applied as the two main strategies for rice-based cropping systems. In addition to the rise of multiple cropping indexes, fertilization consumption has played a very important role in production increases [4]. Compared with nitrogen and phosphorus fertilizer, potash fertilizer is often ignored by farmers, particularly in Asia [5–7]. Potash resources are comparatively limited [8,9], and in recent years, the higher price of potash on the international market has reduced the demand of potassium, as farmers in the area are unwilling to put more potash into the soil [10]. Soil K deficiency has become a major limiting factor in the modern agricultural process [11]. Therefore, it is of great importance to increase potash supplementation in these regions. K-bearing organic resources such as compost, green manure, farmyard manure and crop straws, particularly abundant crop residues, are again receiving attention from farmers [12].

Annually, the world production of straw is approximately 3.8 billion tons, 74% of which are cereal straws [13]; for rice-based land in Asia, 80% of straw production consists of rice residues [3]. Cereal straws usually have a higher potassium content than other straws (1.2%–1.7%). The results of Kaur and Benipal [14] have shown that returning straw to the field could improve soil available potassium to a significantly great extent than manure. A 30-years field trial conducted by Liao et al. indicated that straw management could increase exchangeable K by 26.4%, nonexchangeable K by 1.8% and SOC 21.0% in comparison to a CK treatment in reddish paddy soil [15]. As straw potassium is primarily present in the form of K^+ ions in the cell fluid [16], the release of K from stubble in field is influenced by rainfall [17]. Duong et al. [18] found that the distance of potassium migration from organic fertilizer is 10 mm. Excepting the K^+ adsorbed or fixed by soil clay particles, 50% of K^+ was retained in the soil solution [14]. Furthermore, farmers prefer to input potash fertilizers one time before sowing. The loss of K in the soil solution from such applications was 1.1- and 14.5- fold that of N and P, respectively

[19]; N and P losses were gradual, while leaching phenomena were observed for K [17]. However, Kozak et al. found that crop residues could intercept a maximum of 29% of the water loss for a given rainfall [20]. Soil water retention is also affected by the organic carbon content, as reported by Rawls et al [21]. These results indicate that crop residues have a positive effect on water absorption. Meanwhile, as a high-quality biological adsorbent, the biochars generated from crop straws can adsorb 0.48–1.40 mol kg^{-1} Cu(II) [22], and even unmodified rice straw can absorb 13.9 mg g^{-1} Cd(II) [23]. However, relatively few data exist regarding the adsorption of potassium by crop residues. Because straw decomposition is a slow and long-term process, a significant quality of plant residues can usually be found on farmland after one season of crop growth.

Since 1980, global warming has received increasing attention [24], bolstered by occasional extreme weather events such as the sustained hot temperature in Europe, North America and Asia in July 2013. Scorching weather causes a water shortage in rice farmland, thereby affecting the absorption of nutrients. The incorporation of residues into the soil may reserve water to slow down seasonal drought and could also adsorb cations. However, the capability of soil residues to preserve water and nutrients during different decomposition periods is unclear, especially, the fixation mechanism of residues for potassium.

Therefore, the aims of this research were as follow: (a) to investigate the characteristics of straw decomposition and K^+ release under flooded conditions; (b) to assess the retention capacity of straw for water and potassium during different decomposition periods; and (c) to ascertain the mechanism of K^+ adsorption on straw residues.

Materials and Methods

Ethics Statement

The authors of this study hereby confirm that no specific permissions were required for our experimental location and activities, as the experimental field belonged to our institute and is employed for scientific research only. The studies had negligible effects on the functioning of the broader ecosystem. The research did not involve measurements on humans or animals, and no endangered or protected species were involved.

Materials

Soil material. Paddy soil was collected from the plow layer (0–20 cm) after the rapeseed harvest in 2012. The sample was mixed, air-dried at room temperature and ground to pass through a 2 mm sieve. Soil physical and chemical properties were measured using conventional methods [25–27]. Soil pH was measured with a glass electrode in a 1:2.5 soil/water solution. Soil organic matter was measured using the dichromate oxidation method, and total nitrogen was measured using the Kjeldahl acid digestion method. Available phosphorus was determined using the Olsen method, and available potassium was measured by flame photometry after NH_4OAc neutral extraction. All parameters were measured three times and are presented as the mean±SD: pH 5.73±0.23, SOM 26.7±1.1 g kg^{-1}, TN 1.09±0.04 g kg^{-1}, Olsen-P 19.4±0.7 mg kg^{-1}, and NH_4OAc-K 107.8±4.3 mg kg^{-1}.

Rice straw material. Rice straw was cut into segments of approximately 2–3 cm and then conserved in a dryer until further use. The initial potassium (K) content of the rice straw was 21.87±0.12 mg g^{-1}.

Methods

Straw decomposition trial. The trial was performed at the experimental base of the College of Resources and Environment beginning on July 7, 2012. Three boxes, constructed of PVC (size 50×30×25 cm), were used for the incubation experiment. A total of 15 kg of dry bulk soil was packed in each box. Before the experiment, the rice straw was dried at 40°C for 3 h, and 10.0 g samples were then accurately weighed into 200 mesh (pore diameter 0.075 mm) nylon bag (size 25×20 cm), and sealed [28]. Each box contained 5 bags of straw, totaling 15 bags in the three boxes. According to the growth period of late rice, we removed nylon bags at 5, 15, 30, 60 and 110 d. During the incubation period, deionized water was added to the boxes to maintain a flooding layer of 1 cm. A schematic diagram of the trial is shown in Figure S1.

On each sampling date, one nylon bag was randomly removed from each box and rinsed with distilled water three times to remove any mud that had adhered to the bag. The residue of the rice straw was then dried at 40°C for 48 h until reaching a constant weight, weighed and ground to pass through a 1 mm sieve. Some of the residue powder was ground again using a mortar, until it passed through a 0.149 mm sieve, to obtain micron particles. The particles were dispersed in an aqueous solution of pH 6.0 to test the zeta potential using zeta potential and nanoparticle size analyzer (ZS90, Melvin British Company UK). Zeta potential is the potential difference between the dispersion medium and the stationary layer of fluid attached to the dispersed particle [29], which indicates the electric potential variation of the residue surface during the decomposition period [22]. All of the residue samples were digested with H_2SO_4-H_2O_2 to determine their potassium content using a flame photometer (M-410, Cole-Parmer USA) and to calculate their potassium release rates. The formulas employed in this study were as follow:

$$\text{Decomposition amount (g)} = \text{Dry matter at 0 d Remaining dry matter at } n \text{ d} \quad (1)$$

$$\text{Decomposition rate (\%)} = (\text{Decomposition amount/Dry matter at 0 d}) \times 100\% \quad (2)$$

$$\text{K release amount (K,mg)} = \text{K amount at 0 d} - \text{K remaining at } n \text{ d} \quad (3)$$

$$\text{K release rate (\%)} = \text{K release amount/K amount at 0 d} \times 100\% \quad (4)$$

Where n is the day of incubation.

Batches of K sorption experiments. Precise 0.30 g samples of the rice straw residue from different time points were added into 50 mL polythene bottles along with various concentrations of KCl solution (0, 10, 50 and 100 mg K L^{-1}). The bottles were shaken for 4 h at 160 r min^{-1} and filtered to test the K^+ concentration of

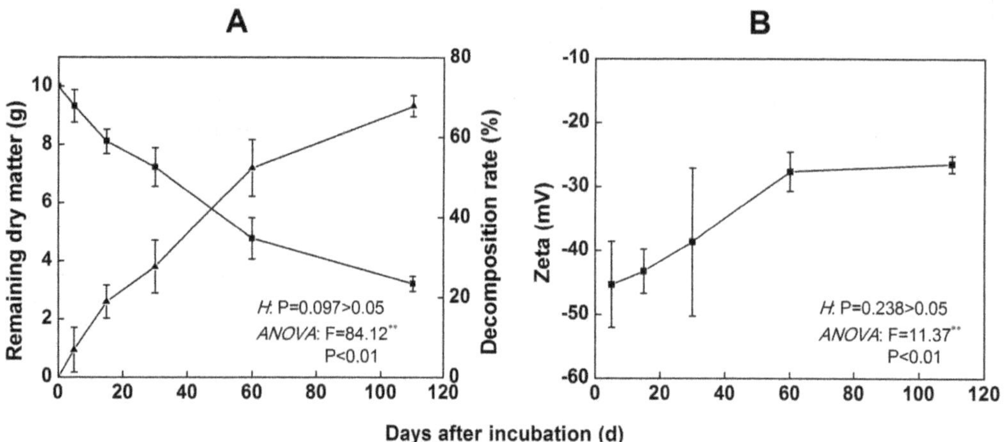

Figure 1. The characteristics of the decomposition and zeta potential of rice straw. The annotations in the panels indicate the homogeneity of variances (H) and ANOVA for the remaining dry matter and zeta potential. The H-test was performed using the Levene test. **indicates significant differences at P<0.01. The values are the means of 3 replicates (±standard deviation).

the equilibrium solution. The experiments were repeated 3 times, and the average values were used for analysis. The K adsorption on straw Q (mg g^{-1}) and the K removing rate R (%) were determined in the following manner [30]:

$$Q = (C_0 - C_t) \times V/m \quad (5)$$

$$R = (C_0 - C_t)/C_0 \times 100\% \quad (6)$$

Where C_0 and C_t are the initial and equilibrium K concentration, respectively, mg L^{-1}; V is the volume of the solution, mL; and m is the mass of the straw residue, g.

According to the K sorption experiment, the maximal Q residue was selected to test the changes of residue structure on K^+ fixation with the help of Fourier transform infrared spectroscopy. (Nexus, Thermo Nicolet USA).

Kinetics of water absorption on rice straw. The water absorption capacity of untreated dried straw [31] was determined by suspending 0.50 g of natural dried straw through a 1.0 mm sieve in a 50 mL beaker with 40 mL deionized water. three replicates were taken at each time interval and cleaned of gravitational water from the straw surface. The material was then weighed on an electronic balance, and the water absorption capacity was calculated.

The water absorption capacity of rice straw residues from different decomposition time points was also determined. A straw sample of 0.30 g was put into a 50 mL beaker with 40 mL of deionized water. After 300 min of immersion, the residues were removed and cleaned of gravitational water before the determination of water absorption capacity. The water absorption capacity of the straw was calculated as follows:

$$M = (M_t - M_0)/M_0 \quad (7)$$

Where M is the water absorption capacity of straw, g g−1; and Mt and M0 are the initial and final mass of straw, respectively, g.

Furthermore, to observe the influence of straw surface pore on the water absorption of rice straw, the surface morphology of residue from different time points was examined using a scanning electron microscope (JSM-6390LV, NTC Japan).

Data Analysis

All analyses were conducted on three replicates. Statistical analyses were performed using OrignPro8.0 software. The means were the average of three replicates, and analysis of variance (ANOVA) was performed according to standard procedures for factorial randomized block designs. Differences at $p<0.05$ level were considered statistically significant, as determined using the least significant difference (LSD) test.

Results

Characteristics of Straw Decomposition and Zeta Potential of Rice Straw Surface

The decomposition dynamics of rice straw during different stages of incubation under flooded conditions are shown in Figure 1A. The average rate of rice straw decomposition was 0.09 g d^{-1} from 0 to 60 d. The decomposition amount accounted for 52.3% (remaining mass 4.77 g) of the total mass after 60 d of incubation. The average rate of decomposition was relatively slow, 0.03 g d^{-1}, from 60 to 110 d. During this period, the decomposition amount accounted for 15.5% of the total mass. The remaining amounts of straw and the accumulation rate were 3.22 g and 67.8%, respectively, after the trial was ended at 110 d.

The zeta potential of the untreated rice straw surface was −56.36 mV in the pH 6.0 aqueous phase (Figure 1B). After 5 d of incubation, the zeta potential increased to −43.2 mV, and it continued to increase to −27.65 mV after 60 d of incubation. On the following days, the zeta potential was not significantly different, ranging from −30 to −20 mV.

Characteristics of Straw K Release

The potassium (K) content of untreated rice straw was 21.87 mg kg^{-1}. After 3 d of immersion, 90% of the total K had been released, and the K content of straw was only 1.47 mg g^{-1} after 5 d of immersion (Figure 2A); this value did not change significantly on the following days. These results indicated that 93.7% of the total potassium could be released into the ambient environment during the early stage of immersion (Figure 2B). After 30 d, the K content of the residues increased slightly and remained

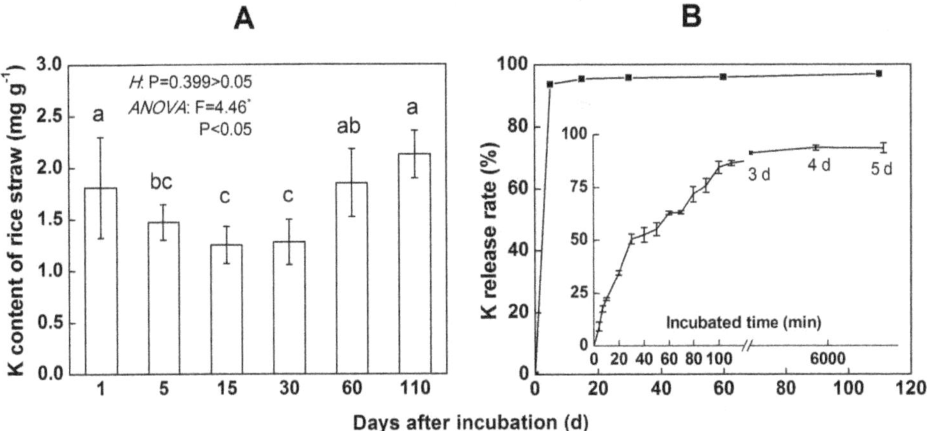

Figure 2. The characteristics of the potassium release of rice straw. The annotation in panel A indicates the homogeneity of variances (H) and ANOVA. The H-test was performed using the Levene test. *indicates significant differences at P<0.05. The values are the means of 3 replicates (±standard deviation). Means with the same letter are not significantly different. The insertion in panel B is the K release rate within 5 d.

at 1–2 mg g^{-1}. At the end of the trial, the K content of the residue and the K release rate were 2.13 mg g^{-1} and 96.9%, respectively.

K Adsorption in Different Decomposition Periods of Straw Residue

Figure 3 shows the adsorption-desorption equilibrium of exogenous potassium over different decomposition periods of straw residue. The results showed that over the entire decomposition period, K^+ ions were, on the whole, released in pure water. When the concentration of K^+ ions was 10 mg L^{-1} in solution, a positive effect of K^+ adsorption was observed after 60 d of decomposition, and the K^+ adsorption capacity of the residue reached 0.13 mg g^{-1} on the 110th d of the experiment. The residue at 15 d showed a positive adsorption effect when the concentration of external potassium increased to 50 mg L^{-1}. Meanwhile, the capacity also increased as decomposition continued. The amount of K^+ adsorption reached a maximum of 0.76 mg g^{-1} after 110 d of incubation, representing a considerable increase of 76.3% compared to its value on the 15th day. When additional increments of K were applied, up to 100 mg L^{-1}, the straw primarily released potassium into solution within 5 d. After 15 d, the adsorption capacity of the residue was higher than that in the trial with 50 mg L^{-1} K supplied over the same period.

Similarly, results of Figure 3B indicated that the K removing rate (R) increased with the decomposition of rice straw. When the concentration of extra K was 10 mg L^{-1}, for example, the value of R was −51.0% at 5 d after incubation, but reached 15.6% after 110 d. At the same time, when the concentration of external K was increased to 50 mg L^{-1}, the removing rate for each period of decomposition except for the first 5 d reached the maximum value. The values of R tended to decrease with increase external K concentration.

FT-IR Analysis for the K Adsorption on Rice Straw Residue

Figure 4 displays the infrared spectra of untreated dried straw both before and after the K adsorption of the rice straw residue on the 110th day after incubation. The results showed that the chemical structure of the straw had undergone a significant change after 110 d of decomposition (Figure 4A–B). New peaks emerged at 3698, 3619 and 779 cm^{-1}, and the existing peaks at 2851, 1086, 798, 695, 528 and 469 cm^{-1} were strengthened to various degrees.

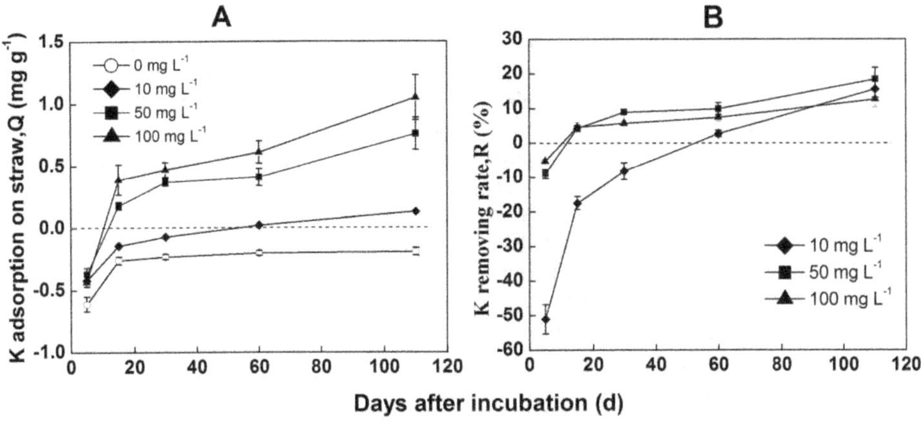

Figure 3. The adsorption of potassium by rice straw residues. The dashed line shows the zero point in the panels. The values are the means of 3 replicates (±standard deviation).

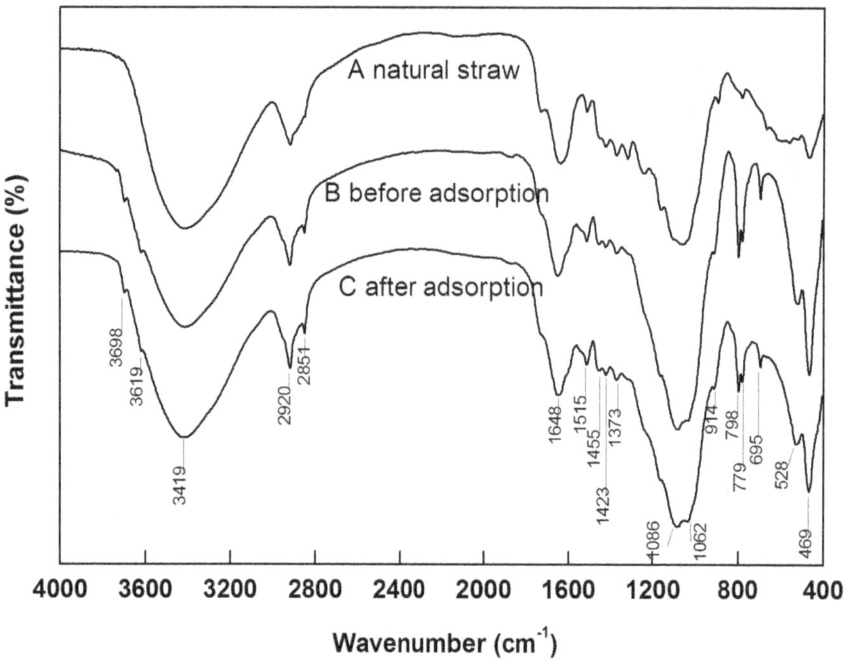

Figure 4. FT-IR spectra of natural dried rice straw and the rice straw residue at 110 d before and after potassium (K) adsorption. Solid line A represents the infrared spectrum of natural dried rice straw. Solid lines B and C represent the infrared spectra of the rice straw at 110 d before and after K adsorption in 50 mg L^{-1} KCl solution, respectively. The data in the panel are the peak wavenumbers.

Figure 4C illustrates the spectra after K adsorption for rice straw after 110 d. The main peaks of the functional groups at 4000–1300 cm^{-1} did not show significant changes. However, in the fingerprint region, the peaks at 798, 695, and 469 cm^{-1} decreased significantly compared to those before K adsorption on the residue.

Kinetics of Water Absorption on Rice Straw

The water absorption kinetic of untreated dried straw are shown in Figure 5A. The results indicated that the water storage capacity of straw reached 2.51 g g^{-1} after 5 min of immersions. From 5 to 90 min, the absorption rate continued to increase at an average rate of 0.01 g (g min) $^{-1}$. After 150 min, the rice straw could not absorb more water at the ambient temperature (23°C), approaching a saturated value of 3.88 g g^{-1}. The water retention also changed significantly during different periods of straw decomposition (Figure 5B). The results showed that the water absorption capacity of the residue continued to rise from 0 to 15 d, and the maximum water content was 5.17 g g^{-1}. Moreover, the water absorption capacity decreased gradually with the extension of straw decomposition. The water content of rice residue was the lowest at 110 d, with a value of 3.00 g g^{-1}.

Discussions

The Differences between Straw Decomposition and K Release

For field straw management, the rate of straw decomposition is influenced by many factors. These variables include internal factors, such as litter quality, material composition and structure [32,33], and external factors, such as temperature, moisture [34], methods of straw utilization (mulching or incorporation), time, length of straw [31] and the concentration of CO_2 in the soil. Litter quality and climate condition are considered to be key factors in the regulation of straw decomposition [35]. The present study was conducted from the beginning of July to the end of October. From 0 to 60 d, the soil temperature reached above 35°C, and this environment did significantly improve the decomposition rate of the straw. However, from 60 to 110 d, the soil temperature decreased to 20–24°C, and the decomposition rate consequently also decreased. After 110 d of degradation, the straw decomposition rate reached 60%. The results presented in Figure 1 illustrate that the decomposition of straw has two main periods: (1) a rapid decomposition stage from 0 to 60 d, during which the structure of the straw surface changed (Figure 6), and the zeta potential increased significantly; and (2) a slow decomposition stage from 60 to 110 d, during which the decomposition rate is significantly lower than before and the zeta potential is basically stable. The characteristic infrared peaks of cellulose were observed at 3412, 2900, 1425, 1370 and 895 cm^{-1} [30]. The characteristic peak of hemicellulose was observed at 1732 cm^{-1} [36], and the absorption peaks characteristic of lignin were seen at 1595 and 1516 cm^{-1}. All of these peaks were not changed dramatically over the study period, as seen in figure 4. Conversely, easily decomposable plant materials such as lipids, pectin, starch and carbohydrates were degraded over one season of crop growth. In addition, negligible amounts of cellulose and hemicellulose were also degraded. Lignin, along with the majority of the cellulose and hemicellulose, may require more time to be degraded by organisms [35].

The characteristics of K release are dramatically different from those of straw decomposition. As K mainly exists in ionic form in plants, it is able to move easily. After 5 d of immersion, more than 90% of potassium had been released from the straw. This result coincides with that of Rodríguez-Lizana et al. [17], who reported that the K of sunflower residue decreased by 98% over the study period while only 43% of the residue itself degraded over the same

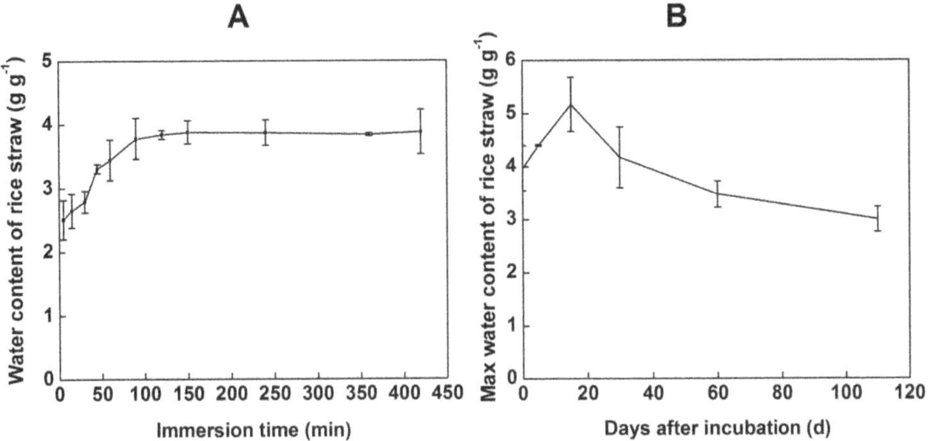

Figure 5. The kinetics of water absorption for rice straw. Panel A shows the changes of water absorption in dried rice straw. Panel B shows the changes of water absorption of straw residue for different decomposition periods. The values are the means of 3 replicates (±standard deviation).

time. Thus, the maintenance of residues under conservation tillage requires the application of large amount of K to the soil.

Mechanism of K Adsorption on Straw Residues

Currently, the ability of crop straw to serve as a bioadsorbent material has been widely studied and has received increasing attention, especially regarding the possibility of reducing the pollution of heavy metals such as Hg(II), Cd(II), and U(VI) [23,37]. The adsorption of metal ions mainly relies on physical adsorption and chemical adsorption. Usually, the latter process is dominated, as reported by Ke et al., who observed that modified rice straw coupled with metal ions to form organic compounds [38]. However, few report exist regarding the coupling interaction between heavy metals and residues during different decomposition period. In particular, alkali metal atoms (Li, Na, K) rarely form complexes with organic ligands because of their unique structures (which lack an empty orbital in the valence shell). When the active functional groups in rice straw bound Cd, nearly all the infrared bands from 3423 to 1000 cm^{-1} decreased in intensity [23]. In the residues after K adsorption, the functional group signatures at 4000 to 1300 cm^{-1} did not change (Figure 4). These results indicate that potassium chelated with C = C, C-O or carboxylic acids on straw is impossible. Moreover, greater focus has been placed in previous studies on the changes in the physical or chemical properties of soil after straw management [39], or on the conversion among diverse forms of nutrients [11]; the straw

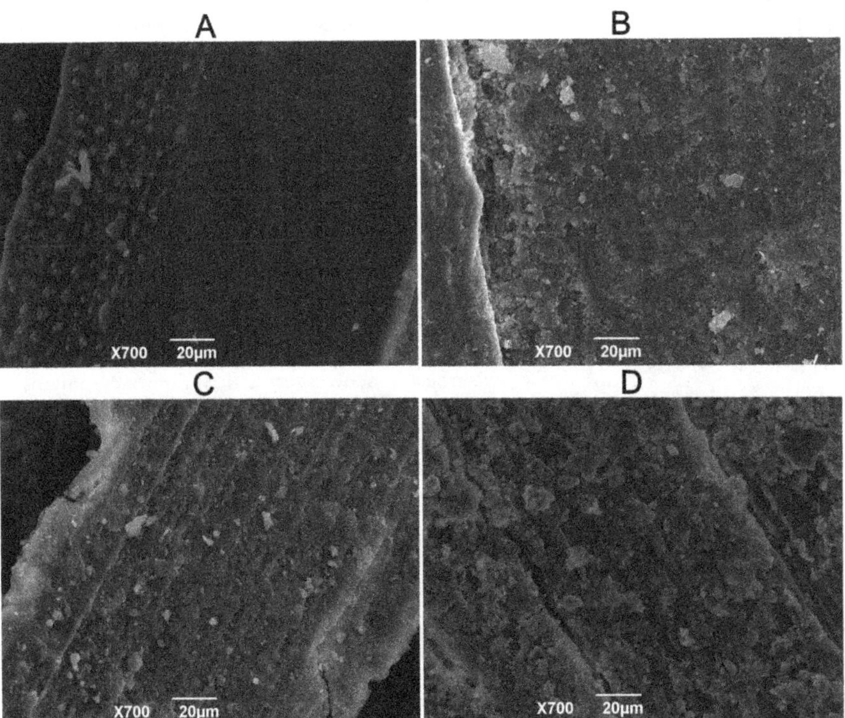

Figure 6. SEM micrographs of the rice straw residue surface on different days after incubation: A 0 d, B 15 d, C 60 d and D 110 d.

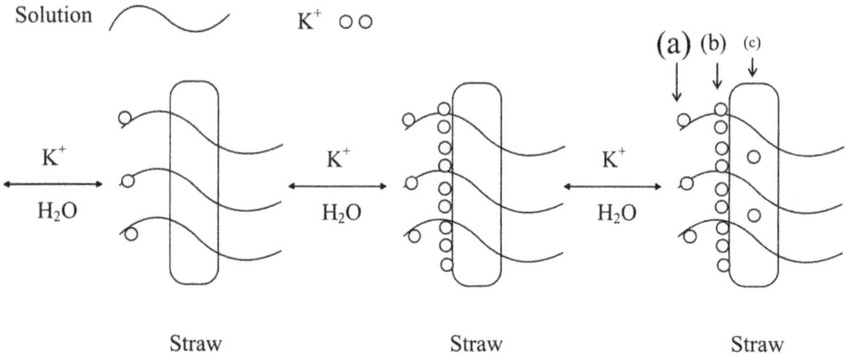

Figure 7. Schematic of the adsorption process of potassium (K^+) on the rice straw residue surface. (a) K^+ retained on the straw in the form of an aqueous solution absorbed by residue; (b) K^+ adsorbed via electrostatic interaction; and (c) K^+ fixed on the residue surface by chemically interfacial adsorption. The size of the three letters indicates the amount of K^+ reserved on the residue by the corresponding process.

residue itself has therefore often been ignored. The adequate prediction of K dynamics and bioavailability is further complicated by the limited knowledge of the reversibility of K adsorption in soil. However, it was noted in the present study that rice straw residue could adsorb potassium despite releasing most of its potassium at the beginning of the experiment. At the same time, rape (canola) and wheat straw at different points in decomposition could also adsorb a portion of the K^+ ions, indicating that the role of crop residues in potassium adsorption is widespread (Figure S2). The present study provides both direct and indirect observations demonstrating that the adsorption mechanism can be described by the following three aspects:

(1) Physically interfacial adsorption: electrostatic adsorption. Due to higher content of potassium in natural straw, the rate of K release is significantly greater than the rate of K adsorption capacity during the earlier stage of decomposition. As decomposition continues, the zeta potential of the straw surface increases. Straw residues could adsorb K^+ ions from the surrounding environment (Figure 3). The straw K adsorption capacity is a function of the decomposition period and the extra K concentration, indicating an equilibrium between the K pool adsorbed on the residue and that in the bulk solution. Residues had an especially strong adsorption capacity after long-term decomposition. Therefore, as vast majority of K fertilizer inputs to the soil are one-time events, a portion of these inputs could be replaced by accumulated debris (or organic matter). Considering the K uptake by plants when the K concentration in the soil solution is reduced, crop residues could release K to replenish the soil solution [18].

(2) Chemically interfacial adsorption. The infrared spectra of straw residues after K adsorption on the 110th day of decomposition showed decreased intensity of the peaks at 798, 695 and 468 cm^{-1} in the fingerprint region. This results suggested that binding sites existed on the residue surface that could couple with K^+ ions. For peak at 798 cm^{-1} indicates the presence of C-H stretching or Si-O stretching, and Lu et al has demonstrated that the peaks at 795 and 466 cm^{-1} represent symmetric Si-O stretching vibration and Si-O deformation vibration, respectively [40]. The peak at 695 cm^{-1} likely reflects the vibration of nitrogenous or boric compounds [41,42]. These results confirmed that the compounds containing Si, B and N on the residue surface were involved in coupling with K^+.

(3) Solution fixed on residue. A study of the water retention of the 11 main varieties of crop straw, as conducted by Iqbal et al., showed that the water retention of straw was not necessarily linked with content of its components (cellulose, hemicellulose, lignin) but was significant related with its porosity and the variation of this porosity during decomposition. These authors found that the water content of rice straw reached its maximum after 6 h of immersion [31]. According to the present research, the capacity of water absorbed on the straw reached saturation after 150 min of immersion, which was consistent with the pervious study. The water content of the decomposed residue reached its maximum of 5.17 g g^{-1} after 15 d of incubation. Due to the differences of porosity, the change of the water content of the decomposed rice residue differed from that of maize residue, which showed higher water retention as incubation proceeded [31]. On the fifth day, the rice straw displayed a regular and compact surface structure, showing no difference with the raw material [43], which likely relied on its inner surface to absorb water (Figure 6). After 15 d of incubation, pectin, starch and other biodegradable materials on the surface of straw were depredated and fell off; the resulting surface was rougher and retained more water. After 60 d, the easily decomposable materials on the straw surface were completely decomposed, and the lignin and cellulose were exposed to the surface. Due to the reduction of porosity, the water retention decreased.

Briefly, the K adsorption process can be described with the following steps (Figure 7): (1) K^+ is transferred from the solution to the rice straw surface; (2) fractional K^+ ions are transferred from the rice straw surface to active sites; and (3) the water content on the surface is maximized, and the K^+ on the residue surface reaches equilibrium with the bulk solution. Therefore, under the same conditions, the amount of K retention via water absorption in straw is greater than that from physical adsorption and chemical adsorption. The latter two methods may, however, play an additional roles in the potassium retention.

The loss of soil nutrients (especially N and K) occurs primarily through runoff and leakage [44,45]. In China and South Asia, rice-based cropping systems such as two- or three-cropping rotation modes are nearly universally deployed [3]. Because of the large population, food demand and soil nutrient depletion in these regions, potassium deficit has become a serious problem [46]. For these regions in particular, the combination of rain, hot and intensive farming increases the risk of nutrient losses and subsequent loss of soil fertility. The present results indicate that straw residues at different points in the decomposition process have different capacities for adsorbing potassium and retaining water. The accumulation of residues in the soil not only reduces water loss, mitigates soil erosion and provides nutrients [47,48], but can also maintain soil nutrients, enhance soil fertility and promote the development of modern agriculture.

Supporting Information

Figure S1 Schematic diagram of the experimental setup of the straw decomposition trial. The depth of the bulk soil in boxes was approximately 22 cm, and that of the flooding layer was 1 cm. The nylon bags were buried into the soil according to the orientation in the diagram.

Figure S2 The adsorption of potassium (K) on wheat and rape straws for different decomposition periods. The added K concentration is 50 mg L^{-1}. The annotations in the panels are the homogeneity of variances (H) and ANOVA. The H-test was performed using the Levene test. **indicates significant differences at P<0.01. The values are the means of 3 replicates (±standard deviation). The means with the same letter are not significantly different.

Acknowledgments

We gratefully acknowledge the reviewer and the editor for their constructive comments and suggestions. We also thank Wenjun Zhang for her comments on revising the manuscript.

Author Contributions

Conceived and designed the experiments: JWL JFL TR. Performed the experiments: JFL RHC. Analyzed the data: JFL JWL XKL. Contributed reagents/materials/analysis tools: XKL RHC LZ. Wrote the paper: JFL JWL LZ.

References

1. United Nations DoEaSA, Population Division (2013) World Population Prospects: The 2012 Revision, Highlights and Advance Tables. Working Paper No. ESA/P/WP.228.
2. Alexandratos N, Bruinsma J (2012) World agriculture towards 2030/2050: the 2012 revision. Rome: FAO. pp. ESA Working paper No.12–03.
3. Singh B, Shan YH, Johnson-Beebout SE, Singh Y, Buresh RJ (2008) Crop Residue Management for Lowland Rice-Based Cropping Systems in Asia. Advances in Agronomy 98: 117–199.
4. Stewart WM, Dibb DW, Johnston AE, Smyth TJ (2005) The contribution of commercia l fertilizer nutrients to food production. Agronomy Journal 97: 1–6.
5. Regmi AP, Ladha JK, Pasuquin E, Pathak H, Hobbs PR, et al. (2002) The role of potassium in sustaining yields in a long-term rice-wheat experiment in the Indo-Gangetic Plains of Nepal. Biology and Fertility of Soils 36: 240–247.
6. Panaullah GM, Timsina J, Saleque MA, Ishaque M, Pathan ABMBU, et al. (2006) Nutrient Uptake and Apparent Balances for Rice-Wheat Sequences. III. Potassium. Journal of Plant Nutrition 29: 173–187.
7. Wang XB, Hoogmoed WB, Cai DX, Perdok UD, Oenema O (2007) Crop residue, manure and fertilizer in dryland maize under reduced tillage in northern China: II nutrient balances and soil fertility. Nutrient Cycling in Agroecosystems 79: 17–34.
8. Sheldrick WF, Syers JK, Lingard J (2003) Soil nutrient audits for China to estimate nutrient balances and output input relationships. Agriculture Ecosystem & Environment 94: 341–354.
9. Pathak H, Mohanty S, Jain N, Bhatia A (2009) Nitrogen, phosphorus, and potassium budgets in Indian agriculture. Nutrient Cycling in Agroecosystems 86: 287–299.
10. Evenett SJ, Frédéric J (2012) Trade, competition, and the pricing of commodities. London: Centre for Economic Policy Research.
11. Li XK, Lu JW, Wu LS, Chen F (2009) The difference of potassium dynamics between yellowish red soil and yellow cinnamon soil under rapeseed (Brassica napus L.)–rice (Oryza sativa L.) rotation. Plant and Soil 320: 141–151.
12. Tejada M, Hernandez M, Garcia C (2009) Soil restoration using composted plant residues: Effects on soil properties. Soil and Tillage Research 102: 109–117.
13. Lal R (2005) World crop residues production and implications of its use as a biofuel. Environ Int 31: 575–584.
14. Kaur N, Benipal DS (2006) Effect of crop residue and farmyard manure on K forms on soils of long term fertility experiment. Indian Journal Crop Science 1: 161–164.
15. Liao YL, Zheng SX, Nie J, Xie J, Lu YH, et al. (2013) Long-Term Effect of Fertilizer and Rice Straw on Mineral Composition and Potassium Adsorption in a Reddish Paddy Soil. Journal of Integrative Agriculture 12: 694–710.
16. Jordan CF (1985) Nutrient cycling in tropical forest ecosystems. Principles and their application in management and conservation. New York: Wiley.
17. Rodríguez-Lizana A, Carbonell R, González P, Ordóñez R (2010) N, P and K released by the field decomposition of residues of a pea-wheat-sunflower rotation. Nutr Cycl Agroecosyst 87: 199–208.
18. Duong TTT, Verma SL, Penfold C, Marschner P (2013) Nutrient release from composts into the surrounding soil. Geoderma 195–196: 42–47.
19. Lin CW, Luo CY, Pang LY, Huang JJ, Tu SH (2011) Effect of Different Fertilization Methods and Rain Intensities on Soil Nutrient Loss from a Purple Soil. Scientia Agricultura Sinica 44: 1847–1854.
20. Kozak JA, Ahuja LR, Green TR, Ma LW (2007) Modelling crop canopy and residue rainfall interception effects on soil hydrological components for semi-arid agriculture. Hydrological Processes 21: 229–241.
21. Rawls WJ, Pachepsky YA, Ritchie JC, Sobecki TM, Bloodworth H (2003) Effect of soil organic carbon on soil water retention. Geoderma 116: 61–76.
22. Tong XJ, Li JY, Yuan JH, Xu RK (2011) Adsorption of Cu(II) by biochars generated from three crop straws. Chemical Engineering Journal 172: 828–834.
23. Ding Y, Jing DB, Gong HL, Zhou LB, Yang XS (2012) Biosorption of aquatic cadmium(II) by unmodified rice straw. Bioresource Technology 114: 20–25.
24. Vitousek PM (1994) Beyond Global Warming: Ecology and Global Change. Ecology 75: 1861–1876.
25. Walkley A (1947) A critical examination of a rapid method for determining organic carbon in soils: Effect of variations in digestion conditions and of inorganic soil constituents. Soil Science 63: 251–264.
26. Olsen SR, Cole CV, Watanabe FS, Dean LA (1954) Estimation of available phosphorus in soils by extraction with sodium bicarbonate. Washington, DC: USDA Circ. 939. U.S. Gov. Print. Office. 1–19.
27. van Reeuwijk LP (1992) Procedures for soil analysis 3rd edn. Wageningen, the Netherlands: ISRIC.
28. Daudu CK, Muchaonyerwa P, Mnkeni PNS (2009) Litterbag decomposition of genetically modified maize residues and their constituent Bacillus thuringiensis protein (Cry1Ab) under field conditions in the central region of the Eastern Cape, South Africa. Agriculture, Ecosystems & Environment 134: 153–158.
29. Honary S, Zahir F (2013) Effect of Zeta Potential on the Properties of Nano-Drug Delivery Systems -A Review (Part 1). Tropical Journal of Pharmaceutical Research 12: 255–264.
30. Cao W, Dang Z, Yi XY, Yang C, Lu GN, et al. (2013) Removal of chromium (VI) from electroplating wastewater using an anion exchanger derived from rice straw. Environmental Technology 34: 7–14.
31. Iqbal A, Beaugrand J, Garnier P, Recous S (2013) Tissue density determines the water storage characteristics of crop residues. Plant and Soil 367: 285–299.
32. Thomas MB, Spurway MI, Stewart DPC (1998) A review of factors influencing organic matter decomposition and nitrogen immobilisation in container media. In The International Plant Propagators' Society Combined Proceedings 48: 66–71.
33. Zhu JX, Yang WQ, He XH (2013) Temporal dynamics of abiotic and biotic factors on leaf litter of three plant species in relation to decomposition rate along a subalpine elevation gradient. PLoS ONE 8(4): e62073.
34. Pal D, Broadbent FE (1975) Influence of moisture on rice straw decomposition in soils. Soil Science Society of America Journal 39: 59–63.
35. Wang X, Sun B, Mao J, Sui Y, Cao X (2012) Structural Convergence of Maize and Wheat Straw during Two-Year Decomposition under Different Climate Conditions. Environmental Science & Technology 46: 7159–7165.
36. Chen XL, Yu J, Zhang ZB, Lu CH (2011) Study on structure and thermal stability properties of cellulose fibers from rice straw. Carbohydrate Polymers 85: 245–250.
37. Rocha CG, Zaia DAM, Alfaya RVS, Alfaya AAS (2009) Use of rice straw as biosorbent for removal of Cu(II), Zn(II), Cd(II) and Hg(II) ions in industrial effluents. J Hazard Mater 166: 383–388.
38. Ke X, Zhang Y, Li PJ, Li RD (2009) Study on mechanism of chestnut inner shells removal of heavy metals from acidic solutions. Journal of Shenzhen University (Science & Engineering) 26: 72–76 (In Chinese).
39. Karami A, Homaee M, Afzalinia S, Ruhipour H, Basirat S (2012) Organic resource management: Impacts on soil aggregate stability and other soil physico-chemical properties. Agriculture, Ecosystems & Environment 148: 22–28.
40. Lu P, Hsieh YL (2012) Highly pure amorphous silica nano-disks from rice straw. Powder Technology 225: 149–155.
41. Liao LB, Wang LJ, J.W Y, Fang QF (2010) Mineral material of modern testing technology. Beijing: Chemical industry press (In Chinese).
42. Peter L (2011) Infrared and raman spectroscopy: principles and spectral interpretation. Elsevier.

43. Yu G, Yano S, Inoue H, Inoue S, Endo T, et al. (2010) Pretreatment of Rice Straw by a Hot-Compressed Water Process for Enzymatic Hydrolysis. Applied Biochemistry and Biotechnology 160: 539–551.
44. Sims JT, Goggin N, McDermott J (1999) Nutrient management for water quality protection: Integrating research into environmental policy. Water Science and Technology 39: 291–298.
45. Lai FY, Yu G, Gui F (2006) Preliminary study on assessment of nutrient transport in the Taihu basin based on SWAT modeling. Science in China: Series D Earth Sciences 49: 135–145.
46. Cakmak I (2002) Plant nutrition research: priorities to meet human needs for food in sustainable ways. Plant Soil 247: 3–24.
47. Yang SM, Malhi SS, Li FM, Suo DR, Xu MG, et al. (2007) Long-term effects of manure and fertilization on soil organic matter and quality parameters of a calcareous soil in NW China. Journal of Plant Nutrition and Soil Science 170: 234–243.
48. Liu Y, Gao MS, Wu W, Tanveer SK, Wen XX, et al. (2013) The effects of conservation tillage practices on the soil water-holding capacity of a non-irrigated apple orchard in the Loess Plateau, China. Soil and Tillage Research 130: 7–12.

3

Miscanthus Establishment and Overwintering in the Midwest USA: A Regional Modeling Study of Crop Residue Management on Critical Minimum Soil Temperatures

Christopher J. Kucharik[1,2,3]*, Andy VanLoocke[4], John D. Lenters[5], Melissa M. Motew[2]

1 Department of Agronomy, University of Wisconsin-Madison, Madison, Wisconsin, United States of America, **2** Nelson Institute Center for Sustainability and the Global Environment, University of Wisconsin-Madison, Madison, Wisconsin, United States of America, **3** Department of Energy Great Lakes Bioenergy Research Center, University of Wisconsin-Madison, Madison, Wisconsin, United States of America, **4** Department of Atmospheric Sciences, University of Illinois, Urbana, Illinois, United States of America, **5** School of Natural Resources, University of Nebraska-Lincoln, Lincoln, Nebraska, United States of America

Abstract

Miscanthus is an intriguing cellulosic bioenergy feedstock because its aboveground productivity is high for low amounts of agrochemical inputs, but soil temperatures below −3.5°C could threaten successful cultivation in temperate regions. We used a combination of observed soil temperatures and the Agro-IBIS model to investigate how strategic residue management could reduce the risk of rhizome threatening soil temperatures. This objective was addressed using a historical (1978–2007) reconstruction of extreme minimum 10 cm soil temperatures experienced across the Midwest US and model sensitivity studies that quantified the impact of crop residue on soil temperatures. At observation sites and for simulations that had bare soil, two critical soil temperature thresholds (50% rhizome winterkill at −3.5°C and −6.0°C for different Miscanthus genotypes) were reached at rhizome planting depth (10 cm) over large geographic areas. The coldest average annual extreme 10 cm soil temperatures were between −8°C to −11°C across North Dakota, South Dakota, and Minnesota. Large portions of the region experienced 10 cm soil temperatures below −3.5°C in 75% or greater for all years, and portions of North and South Dakota, Minnesota, and Wisconsin experienced soil temperatures below −6.0°C in 50–60% of all years. For simulated management options that established varied thicknesses (1–5 cm) of miscanthus straw following harvest, extreme minimum soil temperatures increased by 2.5°C to 6°C compared to bare soil, with the greatest warming associated with thicker residue layers. While the likelihood of 10 cm soil temperatures reaching −3.5°C was greatly reduced with 2–5 cm of surface residue, portions of the Dakotas, Nebraska, Minnesota, and Wisconsin still experienced temperatures colder than −3.5°C in 50–80% of all years. Nonetheless, strategic residue management could help increase the likelihood of overwintering of miscanthus rhizomes in the first few years after establishment, although low productivity and biomass availability during these early stages could hamper such efforts.

Editor: Stephen S. Fong, Virginia Commonwealth University, United States of America

Funding: This research was partially supported by the U.S. Department of Energy's Office of Science through the Midwestern Regional Center for the National Institute for Climatic Change Research at Michigan Technological University, under Award Number DE-FC02-06ER64158. This work was also funded in part by the U.S. Department of Energy Great Lakes Bioenergy Research Center (Department of Energy Biological and Environmental Research Office of Science DE-FC02-07ER64494) and U.S. Department of Energy Biomass Program Office of Energy Efficiency and Renewable Energy (DE-AC05-76RL01830). The funders had no role in study design, data collection and analysis, decision to publish, or preparation of the manuscript.

Competing Interests: The authors have declared that no competing interests exist.

* E-mail: kucharik@wisc.edu

Introduction

The recent push towards developing new bioenergy cropping systems has focused on identifying highly productive plants – other than *Zea mays* (maize) – to provide biomass for lignocellulosic biorefineries [1]. Ideal bioenergy cropping systems should lead to improved soil, water, and air quality across agricultural regions, as well as reducing emissions of greenhouse gases without competing with food sources [2]. One of the plants of interest, miscanthus, is a highly productive C_4 perennial rhizomatous grass, which is not native to many temperate regions, but its bioenergy potential is now being studied extensively in Europe, the US, and Canada [3–5]. Specifically in the Midwest US, *Miscanthus* × *giganteus* is being studied as a model cellulosic feedstock because for low amounts of agrochemical inputs, its productivity is extremely high, ~60% higher than maize total aboveground biomass [5], and double that of another C_4 grass contender, switchgrass (*Panicum virgatum*), regardless of climate and nitrogen fertilizer applied [6].

Before cellulosic feedstocks can supplant maize grain as the dominant source there are significant technical obstacles to overcome [7]. Key barriers include developing an economically viable process to break down cellulose and establishing highly productive plants in environmental conditions that are more harsh than their native regions [8]. In the case of miscanthus, cultivation in the US has largely focused on *Miscanthus* × *giganteus*, which is a natural sterile triploid hybrid of *Miscanthus* × *sinensis* and *Miscanthus* × *sacchariflorus* [9]. Because the triploid *M.* × *giganteus* clones are sterile, establishment results from the planting of rhizomes at a typical depth of 5 to 10 cm [10], and is therefore more costly to establish on the basis of time and money [11].

A key factor that may affect the viability of rhizome propagation for establishing miscanthus in non-native regions is the susceptibility to damage at cold temperatures [11]. Temperatures below freezing can lead to significant miscanthus production losses in two forms. First, soil temperatures that fall below a critical threshold can damage rhizomes, where vulnerability appears elevated in the first winter after establishment [3]. Second, air temperatures below zero after the emergence of new leaves can damage young vegetation, and rhizomes may not sprout again [1,3]. There appears to be a wide variation in frost tolerance among different genotypes because miscanthus has been found to exist naturally from warm subtropical regions to more northern locations of the subarctic [12,13]. Previous research conducted as part of the European Union (EU) Miscanthus Productivity Network [3,14,15] suggested that at some sites in northern Europe including Denmark, Ireland, and Germany, rhizomes in newly established stands did not survive the first year (winter). This has led to subsequent research on the cold tolerance of different genotypes, including *M. sacchariflorus* and *M. sinesnsis* [13]. Results suggested that the lethal temperature at which 50% of shoots were killed was −8°C for *M. giganteus*, −7.5°C for *M. sacchariflorus*, and between −6 and −9°C for two hybrids of *M. sinensis* [1,16]. For rhizomes, the lethal temperature at which 50% of rhizomes were killed was −3.5°C for *M. giganteus* and *M. sacchariflorus*, and −4.5°C and −6°C, respectively for two different hybrids of *M. sinensis* [1,13].

In the US, research has focused predominantly on *M.* × *giganteus* in the Midwest, a region that typically sees extended periods of cold Arctic air outbreaks during the late fall and winter, and correspondingly is at risk to experience near-surface soil temperatures below 0°C as well as frequent freeze-thaw cycles. These cold temperature dynamics create a wide range of uncertainty concerning overwintering of miscanthus in the Midwest US [7]. Recent research has documented the impacts of rhizome size, planting depth, and cold storage on the success of establishment, and arrived at the conclusion that *M.* × *giganteus* rhizomes are best suited to be planted at a depth of 10 cm [8]. Heaton et al. [7] suggested that at this 10 cm depth in Illinois, mature stands of *M.* × *giganteus* have been able to consistently survive winter air temperatures as cold as −20°C and soil temperatures below −6°C. At the University of Wisconsin-Madison's agricultural research station near Arlington, Wisconsin, multiple plots of *M.* × *giganteus* that originated from hand planted rhizomes in spring of 2008 experienced 90% winterkill during the winter of 2008–09, which was representative of other plantings in the Midwest US that year [7].

Crop residue, the unharvested portion of above-ground biomass, can have a significant impact on the surface energy balance and soil properties including temperature. It has been well documented that surface residue management in agricultural systems can aide in conserving soil moisture by reducing maximum soil temperatures by up to 10°C and decreasing evapotranspiration due to a higher surface albedo [17–21]. Residue can also act as an effective insulating barrier to the mineral soil during winter, increasing soil temperatures by 5 to 8°C in the central US [21,22]. Therefore, improved or adapted agronomic management of *M.* × *giganteus* residue in the first several years may be key to increasing the likelihood of successful establishment.

Our primary objective was to investigate how the risk of *M.* × *giganteus* rhizome threatening temperatures could be reduced with strategic residue management. This objective was based on using observations and an agroecosystem model to examine historical spatial and temporal patterns of extreme minimum 10 cm soil temperatures across the Midwest US. Given previous research results, we focus on rhizome losses because this type of damage appears to be more devastating because frost damage to leaves appears to be survivable. Specifically, we created a simulated reconstruction of daily wintertime soil temperatures at high spatial resolution (0.08333°, 5 minute, or ~10 km) across the Midwest US (a region bounded by 36°N to 50°N lat and −79°W and −105°W lon) from 1948–2007, and quantified the frequency that lethal soil temperature thresholds, previously suggested for two miscanthus genotypes (−3.5°C and −6.0°C), were reached at 10 cm depth. Through model sensitivity studies, we investigate how varied thicknesses of prostrate layers of miscanthus straw and corn residue impact wintertime soil temperatures, and how management post harvest could reduce the risk of miscanthus establishment failure. We hypothesize that a prostrate thatch layer of miscanthus straw or corn residue of 1 cm or greater will increase wintertime soil temperatures compared to removal of the residue post harvest, and reduce the risk of soil temperatures at 10 cm reaching critical rhizome kill thresholds. To address our objectives, we employ a dynamic agroecosystem model, Agro-IBIS [21,23] that has been recently modified to include representation of miscanthus and switchgrass [24,25]. We conduct a validation of Agro-IBIS simulated snow depth and 10 cm soil temperatures using several Midwest observational datasets. We conclude with an analysis of trends in the coldest annual soil temperatures to determine whether climate change has led to decreased risk of winterkill of *M.* × *giganteus*.

Methods

Model description

Agro-IBIS is a process-based ecosystem model capable of simulating managed and natural ecosystem dynamics of North America, with coupled carbon, water, and energy cycles. Agro-IBIS was developed by adapting a Global Dynamic Vegetation Model (DGVM), called the Integrated Biosphere Simulator (IBIS) [26,27], to simulate corn, soybean, and wheat cropping systems across the continental US [28], and most recently miscanthus and switchgrass [24,25]. Agro-IBIS simulates the energy, water, carbon, and momentum balance of the soil-plant-atmosphere system at a 60-min time step. The model includes two vegetation layers with eight potential forest plant functional types (PFTs) in the upper canopy, and two grasses (cool and warm season) and two shrub PFTs in the lower canopy. Row crops and miscanthus are simulated as part of the lower canopy layer. The model version used in this study includes 11 soil layers of varying thicknesses to a 250 cm depth, which are parameterized with one of eleven soil textural categories and corresponding physical attributes [29]. A three-layer thermodynamic snow model simulates the energy balance of the snow surface and changes in snow cover in terms of temperature, fractional coverage, and total snow thickness [30]. Physiologically-based formulations of leaf-level photosynthesis, stomatal conductance [31–33] and respiration control canopy exchange processes, and parameters vary according to generalized vegetation categories (e.g., trees, shrubs, C_3 and C_4 grasses or crops). The reader is referred to Li et al. [34] for more details about root water uptake and hydrology and Soylu et al. [35] for description of one dimensional water movement through the soil profile.

Agro-IBIS simulates crop growth transitions through phenological stages of development using an accumulated thermal time approach, and characterizes seasonal changes in carbon (C) allocation to specific crop C pools (i.e. leaf, stem, root, and grain). Leaf area index (LAI) is calculated each timestep using the accumulated leaf tissue C multiplied by a crop specific leaf area value. Canopy and land surface processes in Agro-IBIS are based on the key differences in C_3 and C_4 physiology, daily phenology, and carbon allocation so that coupled carbon-water exchange is responsive to agricultural management and environmental stresses. For a complete description of the modeling approach, the reader is referred to several other publications [23,27,28,36]. IBIS and Agro-IBIS have been validated extensively from the individual farm scale to the global scale to improve model formulations and parameterizations [21,23–25,27,28,30,37–41].

Other modeling approaches and the selection of Agro-IBIS

We briefly review two other well validated models (HYDRUS and SHAW) that are often used to study the impacts of agricultural management on soil heat and water flow and offer reasoning why Agro-IBIS was selected to carry out the study objectives. HYDRUS is a variably saturated soil water flow model that numerically solves the mixed-based Richards' equation for saturated - unsaturated water flow and heat transport [42,43]. HYDRUS simulates water and heat movement in one-dimensional, variably saturated homogeneous media and represents infiltration, evaporation, root water uptake and transpiration, soil water storage, deep drainage, and groundwater recharge. The Simultaneous Heat And Water (SHAW) model is a one-dimensional physically based model of water and heat transport in soils [44], and is capable of simulating infiltration, evapotranspiration (ET), interception, and other hydrologic processes. SHAW has been used to address soil tillage and residue effects on soil freezing and soil water conservation [45] and the interaction between vegetation, soil properties, and other land surface characteristics on frozen soil processes [46]. The Agro-IBIS soil physics module uses Richard's equation to calculate the time rate of change of liquid soil moisture, and the vertical flux of water is modeled according to Darcy's Law. The water budget of soil is controlled by the rate of infiltration, evaporation of water from the soil surface, the transpiration stream originating from plants, and redistribution of water in the profile [27]. Each soil layer in Agro-IBIS is described in terms of soil temperature, volumetric water content, and ice content for any time step [27].

Therefore, while the three models discussed have similar capabilities in simulating soil heat and water flow, HYDRUS and SHAW have some limitations that make them difficult to apply to this study. First, they are designed as point models, and therefore operate at a limited spatial scale compared to Agro-IBIS, which makes it difficult to answer questions across broad regional scales. Second, HYDRUS and SHAW require temporal changes in leaf area index (LAI) and some other vegetation characteristics to be input by the user; therefore, they cannot explicitly simulate phenological development, growth (i.e. photosynthesis), or differences in carbon allocation among different plant species such as miscanthus and corn. Lastly, HYDRUS and SHAW do not explicitly represent carbon and nutrient cycling coupled to water and energy exchanges in the soil-plant-atmosphere system. Therefore, in the context of studying the impact of variable plant growth and development, litter decomposition, and residue management from 1978–2007 on the water cycle and the magnitude of minimum wintertime soil temperatures, HYDRUS and SHAW have limited abilities to generate a historical record for the entire Midwest USA. While SHAW and HYDRUS are exceptional modeling tools in their own right, the expanded capabilities of Agro-IBIS made it a better choice for this particular study.

Agro-IBIS inputs

Climate inputs required at each model time step for each grid cell include solar radiation, temperature, precipitation, relative humidity, and wind speed. ZedX Inc. (Bellefonte, PA) developed a daily gridded weather dataset at 10 km resolution (0.08333°) for a sixty-year period from 1948 through 2007, which included all six variables needed as model input. ZedX Inc. (Bellefonte, PA) generated the gridded weather data using statistical interpolation of observational data that was subject to a rigorous quality control procedure. Before the data were input into the interpolation algorithm, quality control checks were performed which included assessments of plausibility, checks against observational extremes, and checks against neighboring stations using a quality control threshold based on standard deviations. The spatial extent of this dataset spanned from 24°N to 52°N latitude, and 50°W to 130°W longitude. Three data sets were used to generate the gridded maximum and minimum daily temperatures and precipitation. Input station data for Canada and Mexico were obtained from the Global Historical Daily Climatology (GHCND) database, and the National Climatic Data Center (NCDC) TD3200 and TD3210 station data were used for the United States. Relative humidity and wind speed were generated using the Global Summary of the Day (GSOD) daily gridded data. The 10-km gridded data of solar radiation were produced using coarser resolution NCEP/NCAR reanalysis 1 data [47] and the NCEP/DOE AMIP 2 reanalysis data [48]. Hourly variations in climatic variables are simulated through the use of empirical formulations of temperature, specific humidity, precipitation, and radiation variability [29].

Land surface inputs at model initialization include soil textural class at each soil layer to a depth of 250 cm. The dominant soil texture for each soil layer in each 5-minute grid cell was derived from the USDA State Soil Geographic Database (STATSGO) 1 km resolution dataset [49]. The standard thicknesses of the 11 soil layers are 5 cm (layers 1 and 2), 10 cm (layers 3–5), 20 cm (layers 6–8), and 50 cm (layers 9–11), coinciding with the CONUS-Soil dataset. From the assignment of a textural category in each grid cell and each soil layer, the porosity, field capacity, wilting point, saturated air-entry potential and hydraulic conductivity, and moisture release curve "b" (Campbell) coefficient are obtained from a look-up table [29]. Soil moisture is used in combination with snow and vegetative properties to determine the land surface albedo in the absence of surface residues.

Implementation

The geographic region delineated for this study was between −79W and −105W longitude, and 35N to 50N latitude, excluding portions of Canada. While we carry out simulations of miscanthus growing everywhere across this region, this is done purely as a scientific exercise and not as a specific recommendation. We performed multiple simulations to (1) validate simulations of snow depth and 10 cm soil temperature against historical observations, and (2) examine the effects of changing land cover and management on annual minimum wintertime soil temperatures deemed critical to miscanthus rhizome overwintering, and how those temperatures are impacted by differing soil surface residue thicknesses, laying prostrate and evenly distributed (Table 1). Simulations represented potential (natural) vegetation (POTVEG); maize managed with conventional tillage (MAIZE+TILL) leaving no surface residue or stubble post harvest; maize managed with no-tillage (MAIZE+NOTILL) that left a 5 cm thick surface residue layer post harvest, but no standing stubble; *M.* × *giganteus* with an annual fall harvest leaving varying thicknesses of surface residue (1.0, 2.5, and 5.0 cm), with the intent to have a thatch layer present during each subsequent winter and spring, (MISCAN+R; Table 1). The prescribed range of residue thicknesses for miscanthus used in regional simulations are consistent with observations [50].

Given the focus of this study on regional soil temperatures and the important connection between snow depth and wintertime minimum soil temperatures [22,51], we reassessed Agro-IBIS simulations of soil temperature and monthly and seasonal snow depth across the Midwest US. Numerous studies have discussed the insulating properties of snowpack, thereby decreasing the depth of frost penetration, as well as the difficulty in modeling the transition from snowpack to a existence of a bare soil surface during spring in temperate latitudes [52,53]. We first used snow depth and soil temperature datasets previously constructed by Lenters et al. [30] and Hu and Feng [54], respectively. Iowa soil temperatures from the Hu and Feng [54] dataset were selected for model validation because of the high number of soil temperature observations available from 1982–2002. The majority of stations in the Hu and Feng [54] dataset had soil temperature readings made beneath soils void of vegetative or residue cover but could not always be confirmed.

We also used additional sources of soil temperature data at agricultural research sites to further validate Agro-IBIS. First, simulated 10 cm soil temperatures for the 2009–2011 period were compared to 10 cm soil temperature data (chromel-constantan thermocouples) collected in fields managed for conventionally tilled maize and switchgrass with a sparse thatch layer at the University of Wisconsin-Madison Arlington Agricultural Research Station, near Arlington, WI (43.31°N Lat, −89.38°W Lon). Next, we used observed 10 cm soil temperatures (Hydra Probe II, Stevens, Portland, OR USA) for the 2009–2011 period collected in five *M.* × *giganteus* plots at the University of Illinois Energy Farm (40.06°N Lat, −88.20°W Lon; for full site and plot description see Zeri et al. [55] and Smith et al. [56]). For this comparison, the model was driven with a climate data set developed specifically for the University of Illinois Energy Farm [24,25]. Lastly, we compiled 10 cm soil temperature data from 125 observation sites in the Midwest US – across eleven states – that have collected continuous 10 cm daily soil temperature data during a portion of the years 1981 through 2011 (SI, Table S1). Given our inability to exactly replicate specific site management history, soil profiles, and hourly meteorological conditions at each of the 125 observation sites in this comparison, we used a more conservative 3-day running mean of 10 cm soil temperatures in Agro-IBIS to estimate the simulated extreme minimum soil temperatures, and to compare with the daily extreme minimums measured at each site. These station data were used to assess the ability of Agro-IBIS to simulate the average annual extreme minimum 10 cm soil temperatures across a large geographic extent, and the frequency that the 10 cm soil temperature falls below −3.5°C and −6°C, respectively. The majority of stations had soil temperature readings made beneath soils void of vegetative or residue cover with the exception of a subset of the 125 sites located in Illinois, which had grass cover (Carl Bernacchi, personal communication).

The POTVEG, MAIZE+TILL, and MAIZE+NOTILL model simulations were used to validate monthly and seasonal (Nov-Apr) averages of snow depth for the 1963–1995 time period [30] and monthly 10 cm soil temperatures from 1980–2002 across Iowa [54]. For validation of simulated monthly and seasonal snow depth, we used the same observational snow depth dataset from the National Weather Service Summary of the Day that was used in a previous validation of the IBIS-2 model by Lenters et al. [30]. Agro-IBIS spatial averages were formed for all grid cells in northern quadrants (grid cells within 43.5° to 47.5°N lat and −94.0 and −83.0°W lon) and southern quadrants (from 39.5° to 43.49°N lat and −94.0° and −83.0°W lon) to compare with averages for 34 station observations. The simulation of extreme minimum soil temperatures was validated using the daily average 10 cm soil temperature output from the MAIZE+TILL simulation was compared with minimum temperature data obtained from the 125 sites (SI, Table S1).

Regional simulations were conducted from 1940 to 2007. However, because of changes in climate across the region [57,58], we limited our historical analyses of minimum 10 cm soil temperatures and frequency of occurrences to a shortened 30-year period from 1978–2007. The first eight years of the 1940–2007 simulations were discarded as spin-up, needed to bring the soil water balance into equilibrium. We drove those eight years with randomly selected years of climate data. Simulations assumed that nitrogen (N) was not a limiting factor to plant growth. For simulations of maize, yearly changes in optimal planting dates and cultivar selection (total growing degree days required to physiological maturity) were simulated. All simulations were performed with a static atmospheric CO_2 concentration of 370 ppm.

The Midwest US has experienced significant warming temperatures during the past several decades [57,58], particularly during winter and springtime. Given these changes, we also investigated whether soil temperatures have experienced similar warming,

Table 1. Description of Agro-IBIS model runs.

Model Simulation	Description	Residue Layer?
POTVEG	Potential vegetation representing natural vegetation types that could grow in each grid cell based on bioclimatic limits of each plant functional type; dynamic vegetation modeling	No
MAIZE+TILL	Continuous maize managed with conventional tillage; fall harvest removes all aboveground vegetation, leaving a bare soil surface	No
MAIZE+NOTILL	Continuous maize managed with no-tillage; fall harvest management leaves dry plant matter on field with an assumed thickness of 5 cm.	Yes
MISCAN+$R_{1\ cm}$	Miscanthus grown each year; fall harvest management leaves dry plant matter on field with an average thickness of 1 cm.	Yes
MISCAN+$R_{2.5\ cm}$	Miscanthus grown each year; fall harvest management leaves dry plant matter on field with an average thickness of 2.5 cm.	Yes
MISCAN+$R_{5\ cm}$	Miscanthus grown each year; fall harvest management leaves dry plant matter on field with an average thickness of 5 cm.	Yes

thereby lowering the risk of unsuccessful overwintering of miscanthus. We used Agro-IBIS and our daily climate dataset from 1948–2007 to analyze trends in annual average soil temperatures, as well as changes in the extreme minimum values at a 10 cm depth. We also analyzed trends in the annual average soil temperatures at three depths in the MAIZE+TILL simulation from 1967–2002 to coincide with the Hu and Feng [54] study for comparison. All statistics were performed off-line using a commercial software package [59]. To be considered statistically significant, trends had to differ from zero at $P<0.05$.

Simulating the effects of residue in Agro-IBIS

For the MAIZE+NOTILL and the three MISCAN+R regional simulations, the top model soil layer (0–5 cm for MAIZE+NO-TILL, and varying thicknesses for MISCAN+R; Table 1) was modified to represent an organic residue layer, lying prostrate, that persisted throughout the year. The variables modified to represent a thatch layer and the values for key maize and miscanthus residue properties are presented in Table 2. We also used several additional sensitivity analyses at eight geographic locations that experience a wide range in average wintertime soil temperatures and snowfall. These additional model runs were used to further investigate the impact of (1) a wider range of ten different miscanthus straw residue thicknesses (from 1 cm to 20 cm), (2) varying residue albedo (from 0.15 to 0.50), and (3) porosity (bulk density) of residue material (from 0.5 to 0.99) on annual average extreme minimum soil temperatures. The ten additional miscanthus residue thickness simulations were performed to develop more easily interpreted response curves (e.g., soil temperature warming response to residue thickness), as well as to investigate how changing residue thicknesses impacts interannual variability in minimum soil temperatures. These types of responses would be difficult to illustrate succinctly with a series of spatial maps. We note that crop residue thicknesses greater than about 10 cm should be considered extreme scenarios that were used solely for the purpose of building response curves, and are not an easily implemented or recommended residue management option. In the case of miscanthus residue albedo and porosity, there is currently very little published data concerning values for these variables. In order to understand whether our simulation results could be biased due to choosing a mean value for miscanthus residue albedo and bulk density, we further investigated whether large changes in these quantities can have a significant impact on soil temperature responses associated with residue management.

Results

Validation of simulated snow depth

Observed monthly mean snow depth in the northern region illustrated gradual increases from November through February, with an average maximum of approximately 30.3 cm occurring in February, declining to 19.6 cm in March and 3.4 cm in April (Fig. 1a). Compared with a previous version of IBIS that was executed over the Midwest with different climate and soils datasets at coarser spatial resolution [30], Agro-IBIS simulations exhibited increased snow depth in all months from December through March for the POTVEG scenario, which is also how land cover in Lenters et al. [30] was parameterized. In the POTVEG scenario, simulated snow depths were within ±5–10% of observations in all months from November through April, and the model captured the timing of the observed seasonal maximum snow depth in February. Model simulations for MAIZE+TILL scenario also showed higher simulated monthly mean snow depth from December through February compared to previous IBIS-2 simulations, but were approximately 15% and 75% lower than observed averages for February and March, respectively, and simulated maximum values occurred in January (Fig. 1a).

Observations of mean snow depth in the southern region illustrated gradual increases in monthly values from November through January, with an average maximum of approximately 8.0 cm occurring in January, declining to 7.4 cm in February and 2.0 cm in March (Fig. 1b). Compared to the previous IBIS-2 model results for a POTVEG scenario [30], simulations exhibited increased snow depth in all months for a similar vegetation parameterization and were 10–50% (i.e. 1–4 cm) higher than observed values in all months but matched the observed January maximum (Fig. 1b). Model simulations for MAIZE+TILL scenario also suggested improved simulated monthly mean snow depth from December through February compared to previous IBIS-2 simulations, were within 2 cm of observed values from December through March, and correctly simulated the timing of the observed snow depth maximum in January (Fig. 1b).

In the current study, all three scenarios showed significant improvement over the previous IBIS-2 model validation (slope = 0.61; $r^2 = 0.78$) across the northern region (Fig. 2a). Both cropping system scenarios resulted in a negative bias and underestimated mean annual snow depth (slope = 0.73; $r^2 = 0.81$ for MAIZE+TILL), while the POTVEG scenario closely captured the observed mean annual snow depth for each year (slope = 1.02, $r^2 = 0.82$). In the southern region (Fig. 2b), all three scenarios

Table 2. Plant residue biophysical values used to modify Agro-IBIS to simulate the effects of crop residue on soil surface energy balance and heat transfer.

Quantity	Maize residue	Reference	Miscanthus straw	Reference
Residue layer thickness (m)	0.05	[18]	0.022–0.042	[50]
Roughness length (m)	0.012	[76]	0.0065	[77]
Bulk density (kg m^{-3})	36.4	[78]	22.0	[50]
Cellulose density (kg m^{-3})	1450	[18]	1350	[79]
Thermal conductivity (W m^{-1} K^{-1})	0.126	[80]	0.08	[81]
Specific heat (J kg^{-1} K^{-1})	1900	[82]	1335	[81]
Porosity	0.975	[78]	0.96	[83]
Albedo	0.25	[84]	0.32	[85]
Fractional cover	0.95	[18]	0.90	Kucharik (unpublished data)

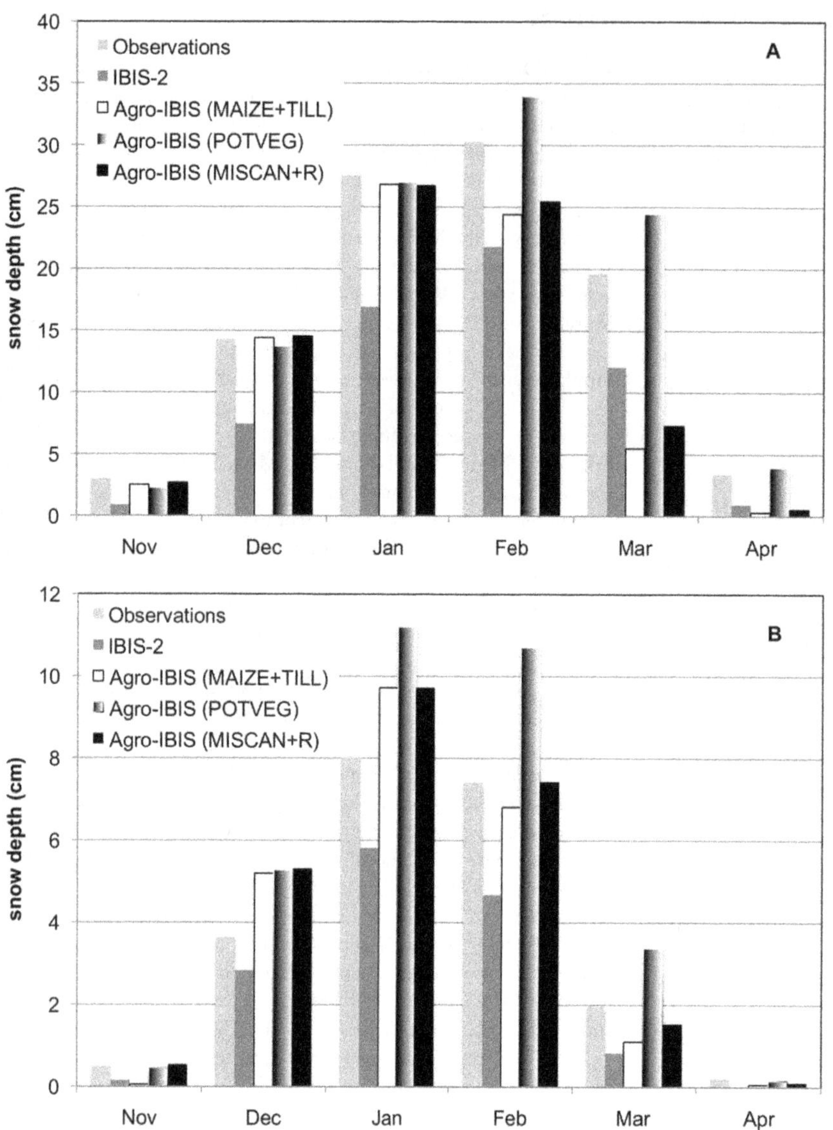

Figure 1. Comparison of observed and simulated monthly mean snow depths. Long-term monthly mean snow depths (1963–1995) for 34 Midwest station observations compared to previous IBIS and current Agro-IBIS simulations in (a) northern (43.5° to 47.5°N lat; −94.0 and −83.0°W lon) and (b) southern (39.5° to 43.49°N lat; −94.0 and −83.0°W lon) areas where observation stations were located. Observations and IBIS-2 results were obtained directly from previous statistical analyses for comparison here (Lenters et al., 2000).

plotted showed improvement over the previous IBIS-2 model validation (slope = 0.61; r^2 = 0.79). Both cropping system scenarios resulted in close approximations to mean annual snow depth and simulated interannual variability well (slope = 0.93; r^2 = 0.88 for MAIZE+TILL and slope = 0.99), while the POTVEG scenario generally overestimated the observed mean annual snow depth (slope = 1.23), but captured interannual variability in snow depth as well as the cropping system simulations (r^2 = 0.88).

Validation of simulated monthly soil temperatures

The MAIZE+TILL simulations had a warm bias compared to the Iowa observations of about 3 to 8°C from March to June (Fig. 3a), which corresponds with the bias of an early spring snowmelt across the northern regions of the study area (Figs. 1, 2). The MAIZE+NOTILL and POTVEG simulations were in much better agreement with observed values across Iowa, although the MAIZE+NOTILL model runs showed a warm bias from September to January, and January is the month when the coldest monthly average temperatures occur (Fig. 3a). The MAIZE+TILL simulation exhibited a small cold bias in the December-February time period of about 0.25 to 3°C.

Simulated soil temperatures at the University of Wisconsin-Madison site showed a similar pattern to the findings in the state of Iowa for the two maize simulations (Fig. 3b). There was a general model warm bias in early spring to early summer of 2–7°C for the MAIZE+TILL simulation (bare soil), very similar to observations from June through December, and a slight cold bias of about 1°C in January. However, the presence of the 5 cm thick residue layer in both the MAIZE+NOTILL and MISCAN+$R_{5\ cm}$ simulations caused soil temperatures to have a cool bias of about 1–6°C during the peak of the summer and then remain warmer than typical observed values from October through February (Fig. 3b).

Simulated 3-day running mean 10 cm soil temperatures at the University of Illinois Energy Farm site for miscanthus without a

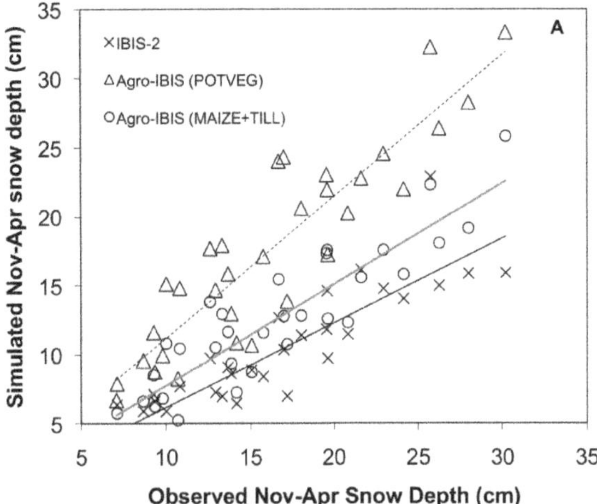

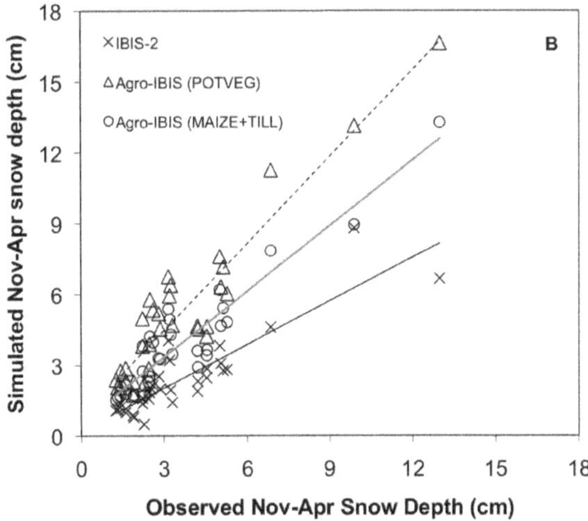

Figure 2. Observed versus simulated interannual variability of mean Nov-Apr snow depth. Comparison of observed versus Agro-IBIS model simulated mean Nov-Apr snow depth for each year from 1963–1995 in (a) northern (43.5° to 47.5°N lat; −94.0 and −83.0°W lon) and (b) southern (39.5° to 43.49°N lat; −94.0 and −83.0°W lon) areas where observation stations were located. Previous IBIS-2 model results (Lenters et al., 2000) are plotted for comparison. Linear regression fits between observations and model results are denoted by the following lines: thin black (IBIS-2); gray solid (MAIZE+TILL); black dashed (POTVEG).

residue layer present compare well with observations (Fig. 4a; $r^2 = 0.92$, P<0.0001) with a slope of 0.99 (S.E. 0.008) and intercept of 0.66 (S.E. 0.126). There were relatively large model errors at the freezing point, however, simulated values are typically within 2.5°C when observed temperatures were less than −1°C. At the monthly time scale, simulated values were ±1.7°C of the observed values for all months except April, and within 1.6°C during the winter months (December, January, February; Fig. 4b). Overall simulated values showed no consistent bias relative to observations, however, there was a slight underestimate (<0.8°C) of monthly mean 10 cm soil temperature during winter months.

Validation of simulated annual extreme soil temperatures

Overall, Agro-IBIS performed exceedingly well for extreme minimum temperatures greater than approximately −6°C, but there was an increasing cold bias in model simulations as average annual observed extreme minimum temperatures decreased from about −6°C to −12°C (Fig. 5a; $r^2 = 0.78$, P<0.0001). For example, the simulated cold bias was about −0.5°C at −6°C (observed temperature), and −2°C at −8°C. The validation exercise suggested that a 2nd degree polynomial model fit best approximated the relationship between simulated and observed average annual extreme minimum 10 cm temperatures (Fig. 5a).

Before comparing simulations to the observed fraction of years below the −3.5°C and −6°C thresholds, simulated annual extreme minimum soil temperatures were adjusted based on the regression analysis from the validation exercise (Fig. 5a). The simulated annual average minimum 10 cm soil temperatures had the mean bias removed by adjusting simulated values using the mean response relationship (2nd degree polynomial) between observed and simulated quantities and its numerical deviation from a linear relationship with a slope = 1.0. The model corrected data, also shown in Fig. 5a, exhibited greatly improved agreement, particularly at the coldest soil temperatures ($r^2 = 0.77$, P<0.0001, slope = 0.947). This statistical adjustment process did not significantly affect our ability to capture the soil temperature variability among observational sites, denoted by the similar r^2 values (0.78 vs. 0.77) for the original and corrected model fits (Fig. 5a). The linear model fit through the observed data and model simulated output for frequency of occurrence of −3.5°C temperatures (Fig. 5b) resulted in an $r^2 = 0.76$ (P<0.0001), with a slope of 0.86 (S.E. 0.043) and intercept of 0.046 (S.E. 0.029). For the frequency of occurrence of −6.0°C temperatures across the region (Fig. 5c), the linear model fit through the observed data and model simulated output resulted in an $r^2 = 0.76$ (P<0.0001), with a slope of 0.72 (S.E. 0.036) and intercept of 0.067 (S.E. 0.018).

Average annual extreme minimum 10 cm soil temperatures: 1978–2007

Analysis of the average annual extreme minimum soil temperatures at a 10 cm depth for the control bare soil case (MAIZE+TILL) model simulation suggest that an absence of any residue layer after fall crop harvest would result in the majority of the Midwest region commonly experiencing temperatures below 0.0°C each year (Fig. 6a). The −3.5°C and −6°C thresholds are generally reached over smaller regions of the Upper Midwest, but are still present in core areas of the Corn Belt. The coldest average annual extreme soil temperatures at 10 cm are in the −8°C to −11°C range confined to large portions of North Dakota and Minnesota (Fig. 6a). However, the MAIZE+NOTILL simulations (Fig. 6b) suggest that a 5 cm thick, continuous cover of maize residue helps to provide a widespread insulating effect on minimum soil temperatures. Most of the region experienced a warming of the extreme minimum temperatures from 2.5°C to 6°C for MAIZE+NOTILL compared to the bare soil (MAIZE+–TILL) simulation. However, this analysis suggests regions that typically experience greater snowfall in the far northern portions (e.g., North Dakota and Minnesota), the upper peninsula of Michigan, and on the eastern side of lake Michigan, would not warm as much during the winter with a persistent 5 cm thick residue layer. The areas that saw the greatest warming impact of residue were located in the central portions of the Midwest, across northern Iowa, eastern South Dakota, southern and central Minnesota, and much of Wisconsin.

Analysis of the average annual extreme minimum soil temperatures at a 10 cm depth for the series of miscanthus simulations with varied residue thicknesses (1, 2.5, and 5 cm) suggest that as residue thicknesses increase, the magnitude of the insulating effect on annual minimum soil temperatures also increases (Figs. 6c–6e). The residue simulations for the 5 cm thick residue layer for

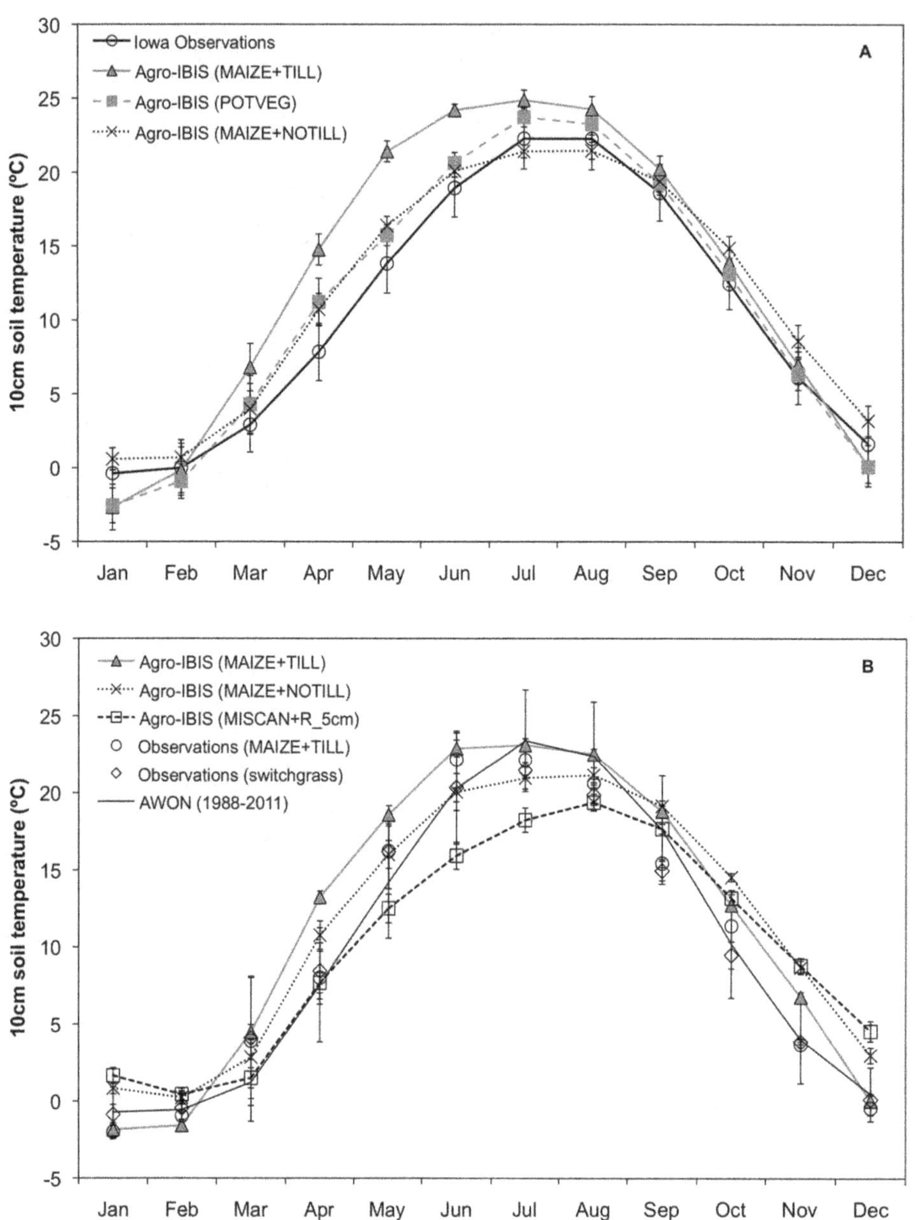

Figure 3. Comparison of observed and simulated soil temperatures in Iowa and Wisconsin. (a) Comparison of monthly average 10 cm soil temperatures for three Agro-IBIS model simulations with observational station data from Hu and Feng (2003) averaged over the state of Iowa for 1982–2002. Error bars are ±1 S.E. for both simulated and observed values. (b) Comparison of monthly average 10 cm soil temperatures for three Agro-IBIS models simulations compared with observational data at the University of Wisconsin-Madison Arlington Agricultural Research Station in tilled maize and switchgrass study plots for July 2009- June 2012. Observed data (maize and switchgrass) represent the monthly mean among 3 replicate plots (n = 3). Long-term averages (1988–2011) of monthly mean 10 cm soil temperature collected at the UW-Madison Automated Weather Observing Network (AWON) site at Arlington are plotted for comparison. Error bars are ±1 S.E. for all simulated and observed values.

miscanthus (MISCAN+$R_{5\ cm}$; Fig. 6e) suggest that miscanthus straw has a slightly increased insulating effect than maize stover, given different biophysical properties (Table 2). Regionally, the largest difference between miscanthus straw and maize stover was in southern portions, with a ~2°C increase in the extreme minimum temperatures while differences in the far northwest portion were typically less than 0.5°C (Fig. 6f). The impacts of a 1 cm thick miscanthus straw layer, compared to the bare soil scenario, were minimal over northwest portions of the region, where no warming of minimum soil temperatures occurred (Fig. 7a). More widespread warming of minimum soil temperatures of 0.5°C to as much as 6°C occurred with miscanthus residue thicknesses of 2.5 cm (Fig. 7b) and 5 cm (Fig. 7c), respectively, compared to the bare soil case. However, even in the 2.5 cm scenario, the magnitude of warming was minimal across far northwestern portions of the region (0°C to only 0.5°C), as well as areas in central Wisconsin and northcentral Nebraska that had soils with higher sand content (Fig. 7b). Thus, many areas still had annual average minimum 10 cm soil temperatures that were below −3.5°C and −6.0°C when miscanthus residue was less than or equal 2.5 cm thick (Figs. 6c,d). The largest magnitude of warming associated with both 1 cm and 2.5 cm miscanthus

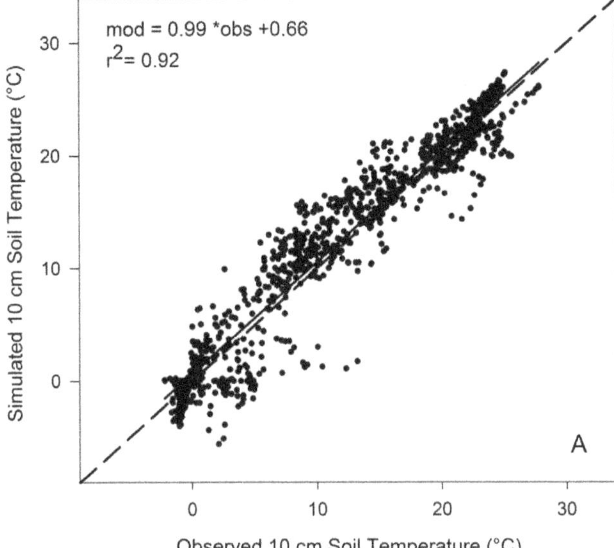

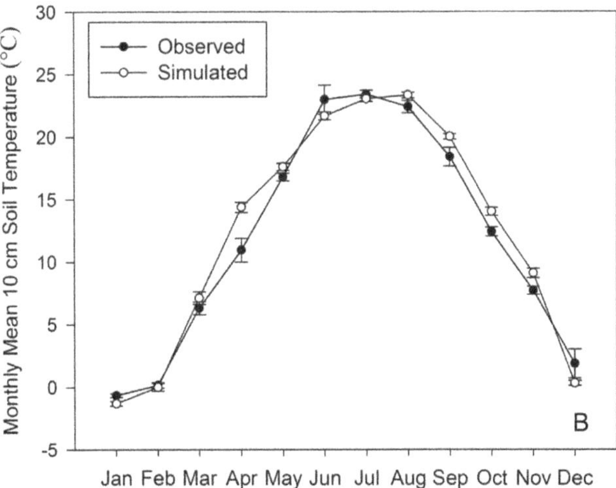

Figure 4. Comparison of observed and simulated soil temperatures in Illinois. (a) Comparison of 3-day running mean 10 cm soil temperatures for Agro-IBIS model simulations of miscanthus with no residue layer with observed values at the University of Illinois Energy Farm from 2009 to 2011. Observed values are the daily mean (n = 1 to 5) of observation for the miscanthus plots, n was less than five for periods when sensors were damaged, with n at least 3 for 80% of the days. (b) Comparison of monthly mean 10 cm soil temperatures for the same site and simulation. Data points are the mean (n = 3) of the 2009–20011 monthly values, error bars are ±1 S.E. for both simulated and observed values.

residue thicknesses was focused in a corridor from eastern Nebraska, through Iowa, southern Minnesota, northern Illinois and southern Wisconsin.

Expanded investigation of varied miscanthus residue depth on minimum soil temperatures

The warming of annual average minimum temperatures with 20 cm of miscanthus straw residue could be as little as approximately 4°C, to as great as 12°C depending on geographic location (Fig. 8). Although the soil warming effects were maximized for a residue thickness of around 20 cm, the shape of the response curves suggests that even greater warming could occur with greater thicknesses regardless of location throughout the Midwest US (Fig. 8). However, we reiterate that residue thicknesses greater than approximately 10 cm are extreme scenarios and not realistic management options in the field. The 5 cm and 10 cm thick residue layers for miscanthus produced about 60% and 84%, respectively, of the warming benefit associated with the thickest residue cover simulated. Additionally, the interannual variability of the coldest 10 cm soil temperatures was reduced as residue thickness increased (Fig. 8).

Likelihood of reaching critical minimum soil temperatures: 1978–2007

In the bare soil control simulation (MAIZE+TILL), large portions of the upper Midwest would have experienced 10 cm soil temperatures below −3.5°C in 75–95% of the years, and the risk of those temperatures being reached is not completely eliminated unless fields are located in far southern regions (Fig. 9a). The risk for these cold temperatures is reduced for the three scenarios of varied miscanthus straw thicknesses, with a corresponding relationship between the magnitude of reduced probabilities and residue thickness (Figs. 9b–d). In these simulations, the likelihood of reaching −3.5°C was considerably lower over southern and eastern portions of the Midwest US, but across the Dakotas as well as portions of Nebraska, Minnesota, and Wisconsin, about 40–80% of years had 10 cm soil temperatures reaching −3.5°C even with a 5 cm residue layer (Fig. 8d). While the effectiveness of a 1 cm thick miscanthus residue layer was much lower, even this small amount greatly reduced the probability of reaching −3.5°C across large portions of Kansas, Missouri, Illinois, Indiana, and Ohio (Fig. 9b).

In the bare soil control simulation (MAIZE+TILL), smaller portions of the upper Midwest would have experienced 10 cm soil temperatures below −6.0°C in about 75% of the years and large sections of the Dakotas, Minnesota, and Wisconsin, reached this threshold in 50 to 60% of the years (Fig. 10a). The risk for these very cold temperatures was greatly reduced, but not completely eliminated, in the three miscanthus simulations (Figs. 10b–d). Similar to the results presented in Figure 9, the likelihood of reaching the −6.0°C threshold is reduced the most with a 5 cm residue layer (Fig. 10c), and the least in the 1 cm thickness scenario (Fig. 10b). However, even in the 5 cm simulation, the −6.0°C threshold was reached in 60–90% of all years in far northern regions of North Dakota and Minnesota (Fig. 10d).

Impact of varied miscanthus residue albedo and bulk density on minimum soil temperatures

In general, variations in residue albedo or bulk density had small impacts on the simulated annual minimum soil temperatures. For residue thicknesses of 1 cm and 5 cm, a change in residue albedo from 0.15 to 0.5 contributed to minimum soil temperatures at 10 cm that were 0.3°C to 0.6°C colder. Over a more realistic range of likely albedo values (0.2 to 0.4), the contribution was only 0.1 to 0.2°C, or about 5–10% of the mean annual minimum 10 cm soil temperatures for the two residue thicknesses simulated in this sensitivity study. Bulk density values were varied, with particle (cellulose) density held fixed, to generate a range in porosity of the 1 cm and 5 cm layers from 0.5 to 0.99. Here, the net effect on minimum soil temperatures was even less than for albedo, contributing to a net change of 0.01–0.05°C.

Soil temperature trends

Our regionally averaged results for soil temperature trends (e.g., a spatial average for the entire study region) produced values equal

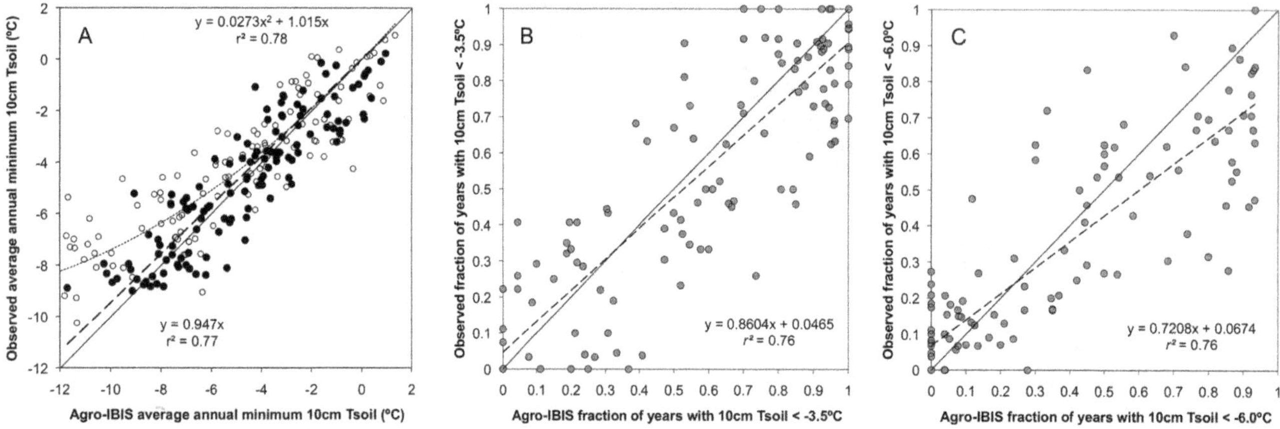

Figure 5. Comparison of observed and simulated annual extreme minimum 10 cm soil temperatures. (a) Agro-IBIS average annual extreme minimum 10 cm soil temperature (based on 3-day running mean temperatures) compared with observations from 125 observation sites (SI, Table S1) from the Midwest USA for original model values (open circles and dotted regression line), and statistically adjusted model values (filled circles and dashed regression line); (b) Agro-IBIS simulated frequency of occurrence of 10 cm soil temperatures (3-day running mean) reaching −3.5°C and (c) −6.0°C, respectively, compared with results from 125 observation sites in the Midwest USA. In these comparisons, Agro-IBIS was only simulated for a period from the beginning year that data was available for each observation station through 2007, which denotes the last year that gridded daily climate data was available.

to 0.250°C $(10\ \text{yr})^{-1}$ at 10 cm, 0.252°C $(10\ \text{yr})^{-1}$ at 60 cm, and 0.253°C $(10\ \text{yr})^{-1}$ at 100 cm during 1967–2002 for the MAIZE+TILL simulations (data not shown). We also compared observed trends in the annual extreme minimum 10 cm soil temperatures for station observations in our study (SI, Table S1) with the average simulated response at a subset of those sites that had continuous records of at least 27 years. The observed trend in annual extreme minimum 10 cm soil temperature, averaged across all 36 stations, was 0.88°C $(10\ \text{yr})^{-1}$ (S.D. 0.097) compared to 0.78°C $(10\ \text{yr})^{-1}$ (S.D. 0.030) for Agro-IBIS simulations.

The overall trend (°C per 60 years) in 10 cm annual average soil temperature varied widely from 1948–2007 (Fig. 11a). Some regions of the Midwest experienced significant cooling in contrast to warming of about 1–3°C. For the entire 1948–2007 period our analysis produced regionally averaged trends that were much lower than those for the 1967–2002 period; 0.059°C $(10\ \text{yr})^{-1}$ at 10 cm; 0.052°C $(10\ \text{yr})^{-1}$ at 60 cm, and 0.049°C $(10\ \text{yr})^{-1}$ at 100 cm. There are similar spatial patterns in the trends in the annual extreme minimum 10 cm soil temperatures from 1948–2007, with the most significant warming of about 2–3°C having taken place in the northern Plains (Fig 11b). However, a dramatically different spatial pattern emerges when the analysis of trends is limited to the 1981–2007 time period (Fig 11c). Here, southern regions appear to have experienced the most significant warming up to 3–4°C of change in the annual extreme minimum temperatures, with a smaller magnitude of change across more northern states. The regionally averaged annual average 10 cm soil temperature trend was 0.33°C $(10\ \text{yr})^{-1}$ for 1981–2007, and the spatial patterns of change across the Midwest were similar to trends in extreme minimum temperatures.

Discussion

This study has described a dataset for annual extreme minimum temperatures across the Midwest US for both soils that are managed without leaving residue on the soil surface after harvest of crops as well as for soils that have a full cover, 5 cm thick prostrate residue layer for maize in place, and for varying thicknesses of miscanthus straw. Our results indicated that strategic residue management in the region has potential to help increase extreme minimum soil temperatures that can threaten overwintering of miscanthus rhizomes in the first year of establishment and potentially in later years. Based on previous research that has investigated the ability of winter wheat to survive extremely cold winters in the central US [60], it appears that a combination of leaving behind standing stubble after harvest to preferentially trap snow, coupled with a prostate residue layer, offers the highest likelihood of insulating soils and to increase the odds that miscanthus rhizomes can survive the first winter after establishment. However, as our results have shown, the thickness of that residue layer has significant bearing on the insulating effect.

Influence of residue, soils, snowpack, and management on minimum extreme soil temperatures

The Midwest US is subjected each year to rapid and extreme temperature changes. However, due to the large presence of human management of agricultural lands, as well as differences in the timing of snowfall and the buildup of a consistent snow cover each year, air temperatures should not be perceived as a guide for estimating extreme minimum soil temperatures near the surface. For example, while air temperatures across regions of the Midwest can drop to −25°C to −40°C during the winter for extended periods of time, the absolute coldest soil temperatures at 10 cm were about −12° to −16°C, based on both observed data and simulated results for individual years in our study. According to observations in the region over the past 30 years, the average 10 cm minimum soil temperatures are in the range of −8°C to −10°C (Fig. 5a), which is much warmer than the typical average low air temperatures. This is attributed to the insulating effect of snow cover. The timing and duration of a consistent snow pack is highly influential in determining minimum soil temperatures in mid-January to early March, which is historically when soils reach their lowest temperatures [22,51,61].

Agro-IBIS simulations suggested that miscanthus straw could function better as an insulator than maize residue for a comparable 5 cm thick layer on the soil surface, which is largely attributed to differences in biophysical properties (Table 2). However, both plant residue types keep soils warmer during

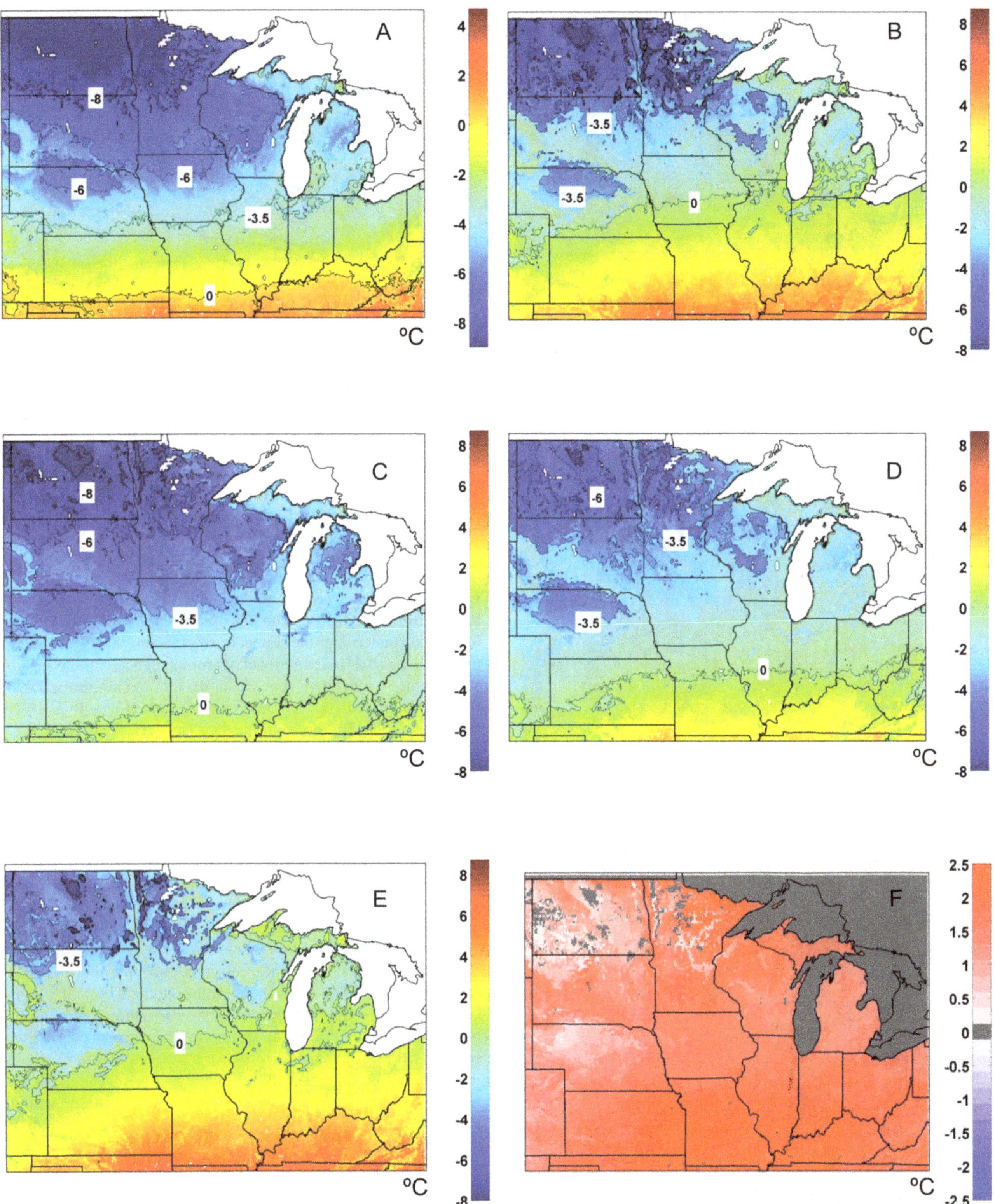

Figure 6. Impacts of soil surface residue management on annual average extreme minimum 10 cm soil temperatures. Average annual (1978–2007) extreme minimum 10 cm soil temperatures (based on 3-day running mean) for (a) MAIZE+TILL (bare soil post harvest), (b) MAIZE+NOTILL, (c) MISCAN+$R_{1\ cm}$, (d) MISCAN+$R_{2.5\ cm}$, and (e) MISCAN+$R_{5\ cm}$ simulations. The position of the 0.0°C, −3.5°C, −6.0°C, and −8.0°C 10 cm soil temperature isopleths are highlighted by labels on solid black lines; (f) differences in average annual extreme minimum 10 cm soil temperatures for MISCAN+$R_{5\ cm}$ – MAIZE+NOTILL (Fig. 6e–Fig. 6b results).

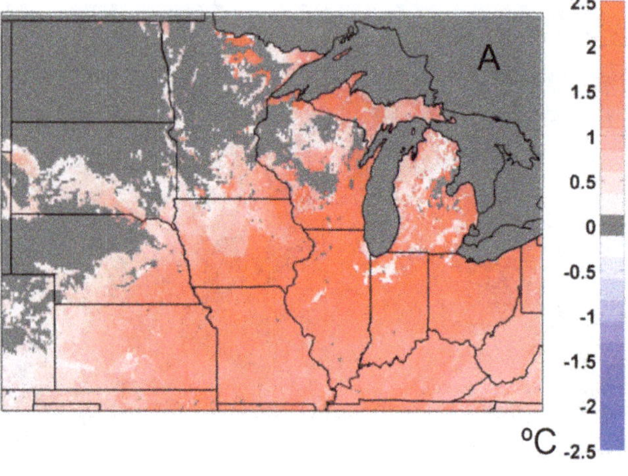

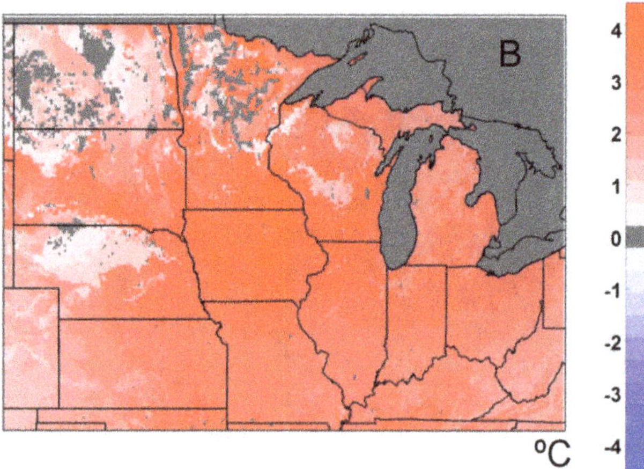

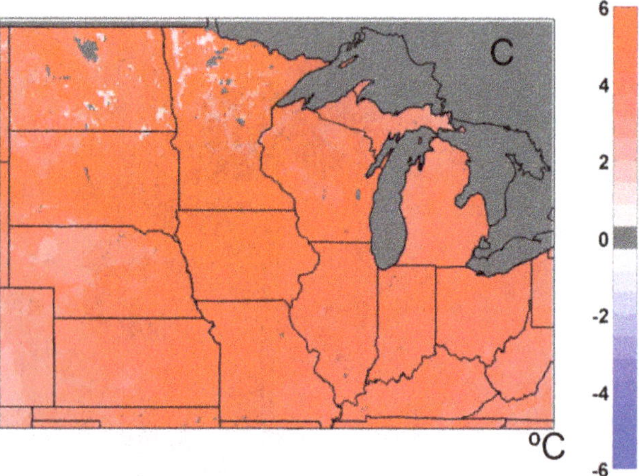

Figure 7. Soil temperature differences between different residue layer thicknesses. Average annual (1978–2007) extreme minimum 10 cm soil temperature differences, based on a 3-day running mean, for the following paired simulations: (a) MISCAN+$R_{1\ cm}$ minus MAIZE+TILL (bare soil post harvest), (b) MISCAN+$R_{2.5\ cm}$ minus MAIZE+TILL, (c) MISCAN+$R_{5\ cm}$ minus MAIZE+TILL.

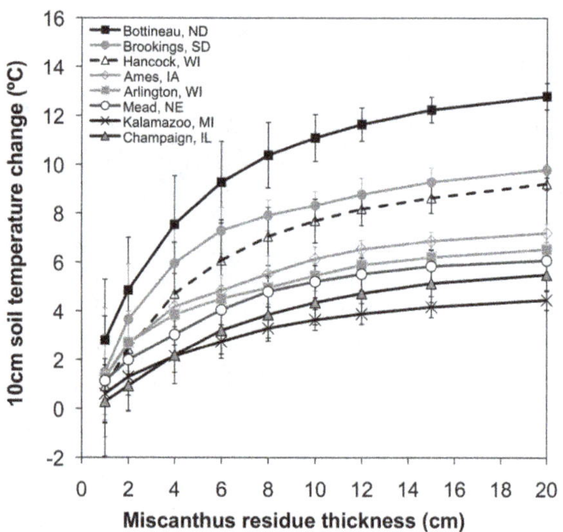

Figure 8. Impacts of miscanthus residue thickness on annual average extreme 10 cm soil temperatures. Average annual (1978–2007) extreme minimum 10 cm soil temperature changes at eight locations in the Midwest, based on a 3-day running mean, for MISCAN+R simulations with varied residue thicknesses relative to the MAIZE+TILL (bare soil) simulation.

winter, thereby decreasing the risk of 10 cm soil temperatures going below −3.5°C over a large part of the Midwest. However, for decreased thicknesses of miscanthus residue (1 and 2.5 cm), the warming effect is considerably less, and in some locations across the Midwest, negligible. Our expanded sensitivity tests at eight locations concerning the impacts of a wide range of residue thickness on the magnitude of warming of the minimum soil temperatures suggested that 60% of the maximum warming benefit occurs with a 5 cm thick layer, and 84% with a 10 cm thick layer (Fig. 8). For 1 cm and 2.5 cm thick layers, only 15% and 40% of the maximum warming benefit occurs. While thicker residue layers clearly increase the warming benefit, data from field experiments (Table 2) suggest that a nominal thickness of only a few centimeters is likely easier to sustain across large fields given issues related to wind blowing loose residue around. We emphasize that the 20 cm residue thickness is an extreme scenario that was created to understand system behavior, and is not suggested as a recommended management practice.

The simulated soil warming attributed to residue in this study is in agreement with previous studies for an experiment in Minnesota that examined the effect of different maize residue management options on soil temperatures in the 0.05 to 0.3 m soil layer. Sharratt et al. [22] reported that wintertime minimum soil temperatures at a 1 cm depth were 5 to 8°C warmer over a three year period for a management scenario that left 60 cm of stubble standing in combination with residue laying prostrate on the soil surface compared to a treatment that had 0 cm stubble with all residue removed from the soil surface. While the magnitude of the warming effect associated with residue was greater in Sharratt et al. [22] then in our study, their temperatures were reported for a 1 cm depth. They also found that snow depth was influenced by the residue treatments, whereby fields that had stalks cut closer to the ground (e.g., 30 cm compared to 60 cm) had a lower average snow depth during the winter.

However, not all Midwest regions may see a significant soil warming from straw or maize residue due to the confounding influence of snow cover on winter soil temperatures. Model results illustrated that even with the addition of a 5 cm thick residue layer of either miscanthus straw or maize residue, this had minimal effects on the extreme minimum temperatures in small portions of

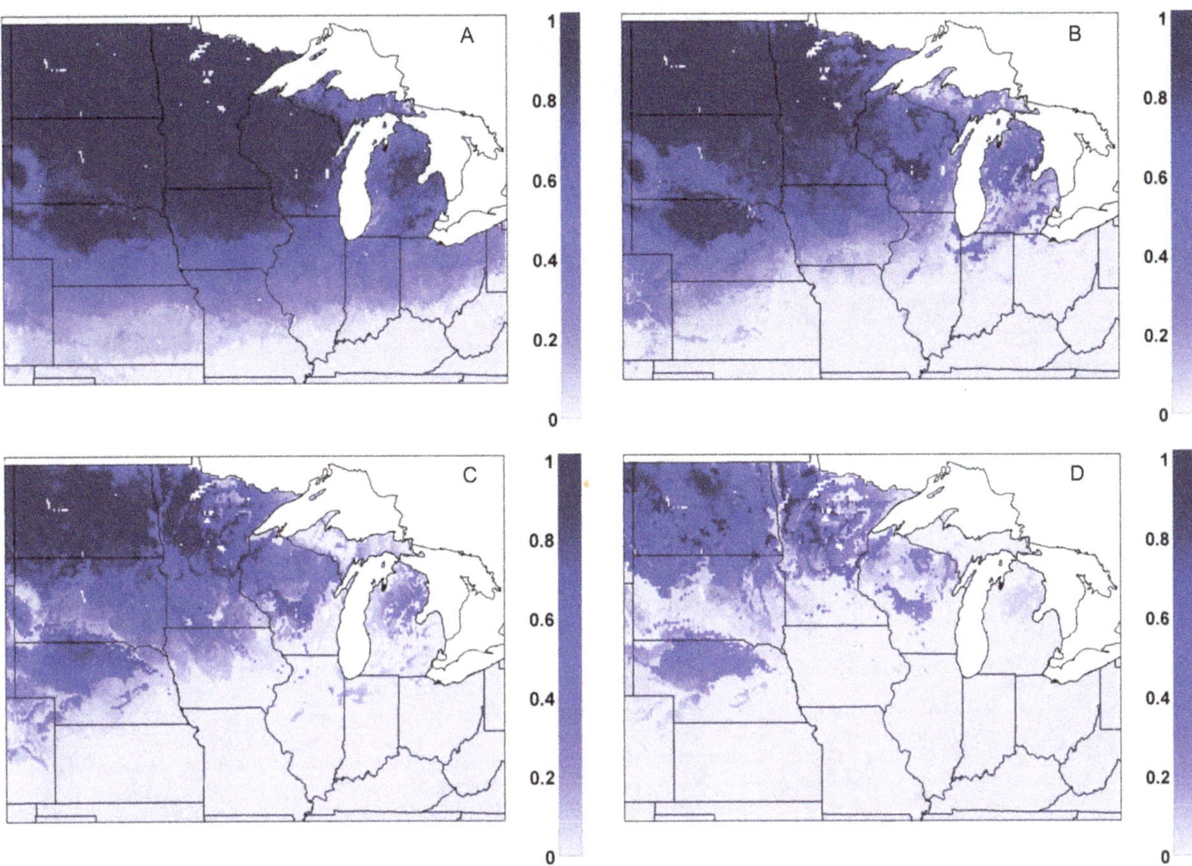

Figure 9. Frequency of 10 cm soil temperatures reaching −3.5°C or colder for varied miscanthus residue thicknesses. Fraction of total years during the 1978–2007 time period that simulated annual 10 cm soil temperatures were at or below a −3.5°C threshold (based on a 3-day running mean) for (a) MAIZE+TILL (bare soil), (b) MISCAN+$R_{1\ cm}$, (c) MISCAN+$R_{2.5\ cm}$, and (d) MISCAN+$R_{5\ cm}$ simulations.

North Dakota and Minnesota, and to a lesser extent the sandier soils of north and central Nebraska and the central sands region of Wisconsin (Figs. 6–7). This may be attributed to two reasons: first, across the most northern regions of the Midwest, snowfall comes earlier, and due to colder temperatures, snowpack depth and duration is generally greater and longer, respectively, than in more southern locations [62]. Sharratt et al. [63] suggested that approximately 15–42 cm of snow cover is needed to insulate the top portion of the soil profile from cold wintertime temperatures, resulting in near steady-state soil temperatures. Thus, the potential impact of a residue layer may be minimized in regions that historically have deeper and more consistent snow cover (e.g., North Dakota and northern Minnesota, as well as lake effect snow regions in Michigan), which might explain the simulated spatial patterns and results in our study. However, we note that in northern locations that typically have a significant, consistent snowpack during winter (e.g., >15 cm), observed and simulated 10 cm soil temperatures still reached well below 0.0°C; thus the timing of a building and retreating snowpack is also crucial, besides the thickness, in determining extreme minimum soil temperatures. Second, some of the spatial patterns of soil temperature change attributed to residue management suggest that soil texture plays an important role also in determining the magnitude of extreme minimum soil temperatures. In sandy soil regions, we hypothesize that a thicker residue layer may add a more prominent insulating effect on these soils that lose heat more rapidly due to their lower average volumetric water content in fall and inherent mineral properties [29].

These results suggest that in the first years of establishment of miscanthus, a soil surface residue layer could increase the probability of successful overwintering of the plant rhizomes, but the thickness of that layer is highly deterministic to the overall soil warming. This management option, coupled with leaving standing stubble that could preferentially trap snow, would likely provide the greatest likelihood of maximizing soil warming [22,51]. However, during the first year or two of establishment when *M. × giganteus* might be the most susceptible to winterkill or damage to rhizomes, the amount of biomass produced may not be sufficient to support a residue layer thickness that significantly reduces the risk of lethal soil temperatures. Several studies from the literature suggest that *M. × giganteus* will take at least three years to reach the expected yield ceilings, and during the first year, annual productivity can typically be in the 1–4 Mg ha^{-1} range [4,15,64–66]. With a typical residue bulk density of 22 kg m^{-3} (Table 2), 2.2 Mg ha^{-1} of aboveground biomass is needed for each 1 cm of thatch depth. Thus, to achieve a 5 cm thatch thickness, approximately 11 Mg ha^{-1} of aboveground biomass would be required; these values may not be observed until year three and beyond [50,55,67]. Given these results, a producer might be faced with a new dilemma in the context of residue management. For example, a farmer may not have enough miscanthus biomass to sell to make a profit after the first year, so they would probably mow the crop. In the next two years, they will have to hedge the

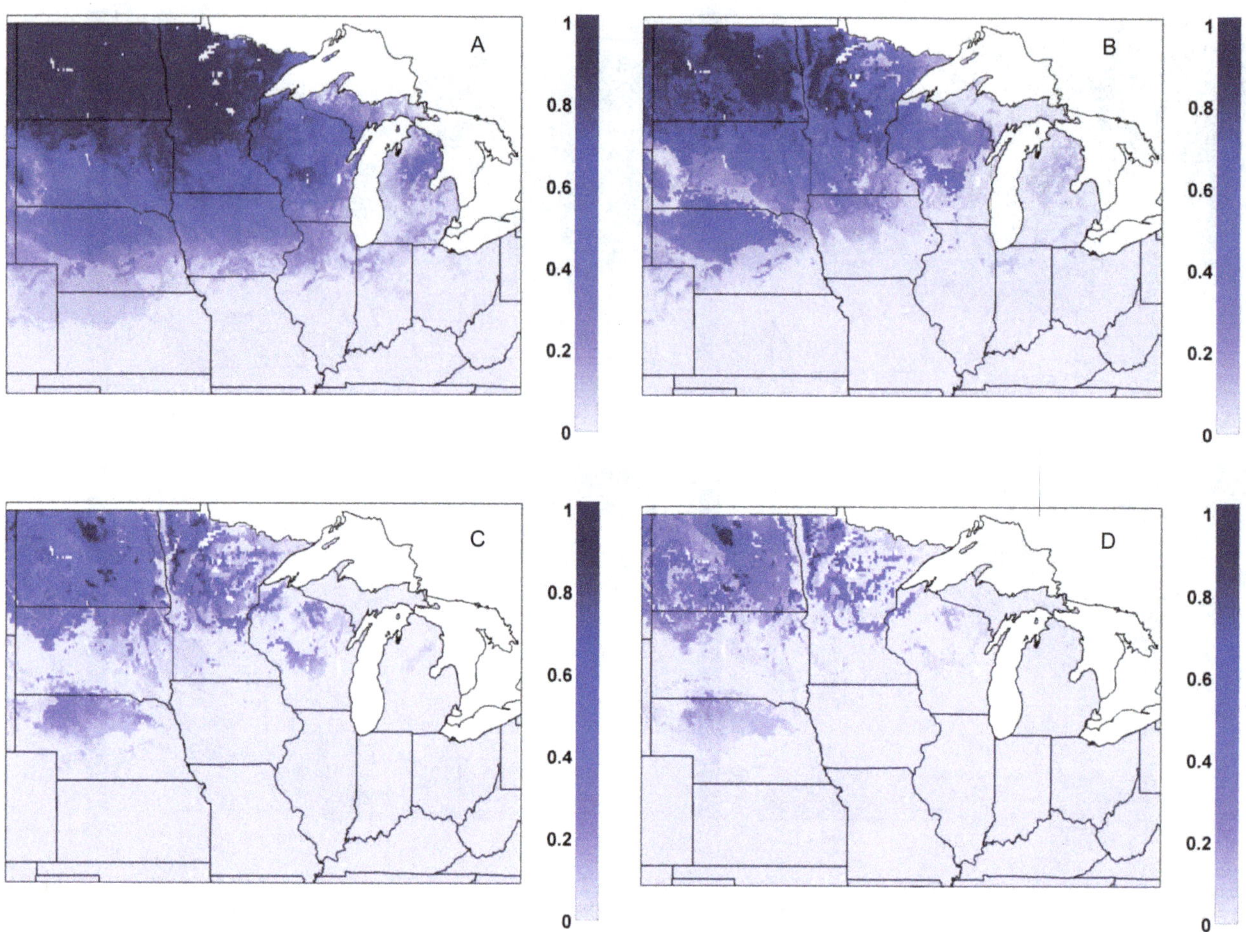

Figure 10. Frequency of 10 cm soil temperatures reaching −6.0°C or colder for varied miscanthus residue thicknesses. Fraction of total years during the 1978–2007 time period that simulated annual 10 cm soil temperatures were at or below a −6.0°C threshold (based on a 3-day running mean) for (a) MAIZE+TILL (bare soil), (b) MISCAN+$R_{1\ cm}$, (c) MISCAN+$R_{2.5\ cm}$, and (d) MISCAN+$R_{5\ cm}$ simulations.

risk of maximizing profits vs. minimizing the cold temperature threat. Furthermore, they may face a more complicated decision on what to do with available residue from nearby corn fields; should it be used to build a solid residue layer to protect rhizomes?

Additional studies that quantify the relationship between miscanthus productivity, residue bulk density, and thatch thickness during early establishment years will help determine whether additional sources of crop residue from nearby fields (e.g., maize) are needed to build a more substantial protective layer to reduce the odds of winterkill. The research presented here showed that very small differences in thatch thickness lead to significant differences in minimum winter soil temperatures. Therefore, field trials should continue to be established that strategically manage residue in varied amounts on fields as well as investigate how stubble height and snow depth variations impact soil temperatures over several years to account for interannual variability. These data would also prove useful to help identify an optimal growing region in the US and Canada for miscanthus. Currently, there is also a lack of biophysical information on the properties of miscanthus straw (e.g., bulk density, albedo, heat capacity), which is needed to better constrain the parameterization of agroecosystem models.

Soil temperature trends: are soils warming or cooling?

This study, as well as three other studies, have documented long-term changes in soil temperatures but arrived at different conclusions [54,68,69]. Using a limited number (38) of observation stations in the US that had a period of record from 1967–2002, Hu and Feng [54] reported that annual average 10 cm and 100 cm soil temperatures across these sites were increasing at a rate of 0.3°C $(10\ yr)^{-1}$, and the sites that had the greatest rate of warming were across northern regions. Sinha et al. [69] also reported warming soil temperatures at 10 cm during 1967–2006 in regions of Minnesota, Illinois, and Indiana, as well as a reduction in the number of days with soil frost in the Midwest. In contrast, Isard et al. [68] used a biophysical modeling approach and reported that even though wintertime air temperatures from 1951–2000 were increasing, wintertime soil temperatures at 50 cm depth were decreasing across the Great Lakes region, likely due to thinning and more variable snowpacks. However, a study by Dyer and Mote [62] suggested that minimal changes in North American snow depth has occurred in the November through January period from 1960–2000, but noted an earlier onset and acceleration of spring snowmelt in the March and April timeframe. While the long-term observational data presented in Hu and Feng [54] do not corroborate a reported trend of decreasing soil temperatures across Wisconsin and Michigan by Isard et al. [68], there were no

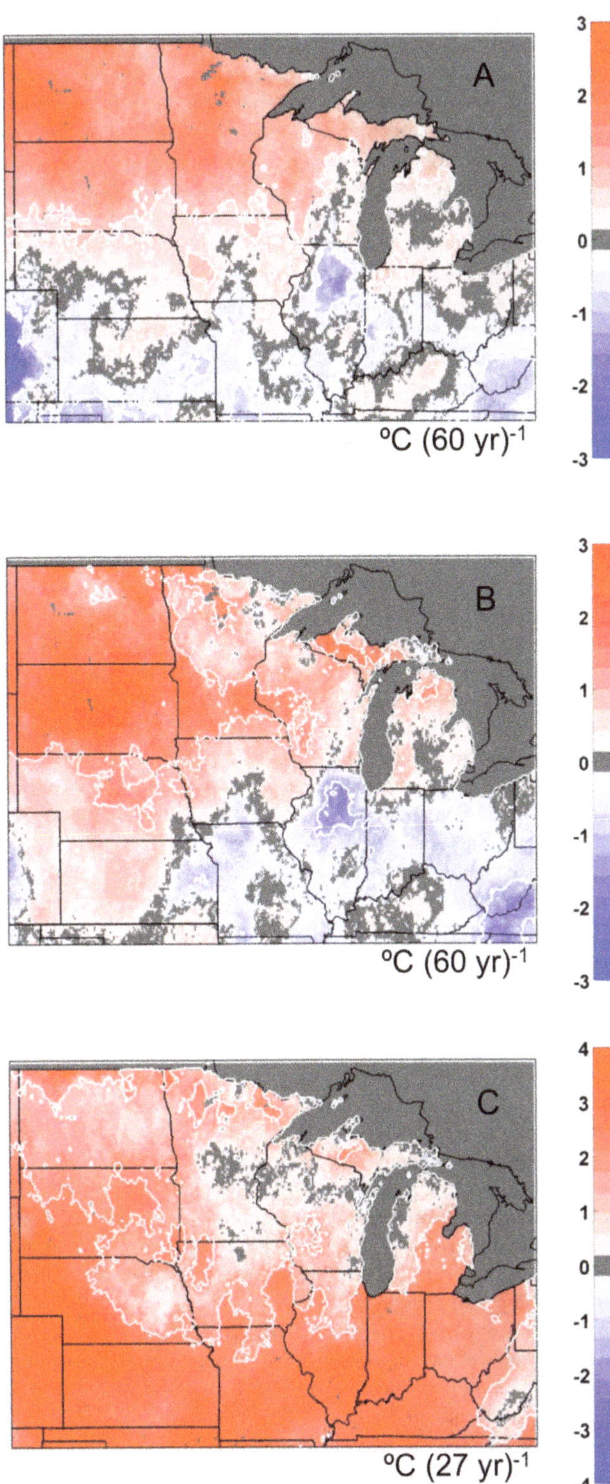

Figure 11. Simulated soil temperature trends across the Midwest US. (a) Total change (from linear regression) in annual average 10 cm soil temperatures for 1948–2007 for the MAIZE+TILL (bare soil) simulation; (b) total change (from linear regression) in annual extreme minimum 10 cm soil temperature for the MAIZE+TILL (bare soil) simulation from 1948–2007; (c) total change (from linear regression) in annual extreme minimum 10 cm soil temperature for the MAIZE+TILL (bare soil) simulation from 1981–2007. Regions bounded by solid white lines indicate trends with $P<0.05$.

observation stations across Wisconsin and Michigan included in the aforementioned analysis.

In our study, depending on the initial year for the calculation of linear trends (e.g., Fig. 11b compared to Fig. 11c), the magnitude of 10 cm soil temperature trends varied significantly across the Midwest. As with any type of linear trend analysis over time, the time period of choice can have a significant influence on the results. Agro-IBIS results for the 1948–2007 time period (Fig. 11b) illustrated a reduced warming signal, or no change in extreme minimum soil temperatures, across eastern Wisconsin and central lower Michigan, but are not necessarily indicative of a widespread cooling trend as suggested by Isard et al. [68]. We also analyzed Agro-IBIS extreme minimum soil temperature trends for the 50–60 cm soil depth, and did not find a significant difference from trends occurring at 10 cm (Fig. 11b). However, Agro-IBIS depicted several large areas of Missouri, Illinois, Indiana, and Ohio that exhibiting a cooling trend of annual average 10 cm soil temperatures (Fig. 11a), as well as the extreme minimum values (Fig. 11b) over a longer timeframe. Based on Agro-IBIS and soil temperature observation trends that agree, on average, for the 1981–2007 period for 36 station locations (SI, Table S1), we conclude that the coldest wintertime soil temperatures at 10 cm have been warming at a rate of approximately 0.8°C $(10\ \mathrm{yr})^{-1}$ to 0.9°C $(10\ \mathrm{yr})^{-1}$ over the last several decades. There is evidence to suggest that significant warming of soils and a reduction in soil frost would continue in the future, based on the modeling study of Sinha et al. [70]. They used future climate scenarios for the mid (2040–2069) and late (2070–2099) 21st century to drive a macroscale land surface model and found that increased wintertime soil temperatures, increased frequency of freeze-thaw cycles, and a reduction in soil frost days across the region would occur with continued climate change.

Challenges for ecosystem models

The model showed an overall good ability to simulate the dynamics of 10 cm soil temperatures observed at a number of locations in the Midwest (Figures 3–4), supporting the accuracy of our model estimates. The differences between observed soil temperatures in the tilled maize and switchgrass plots at the Arlington, WI site were attributed to a lack of an established residue layer in switchgrass experimental plots attributed to harvest of the majority of aboveground vegetation for three straight years. The simulation of interannual variability of annual extreme minimum temperatures at 10 cm proved more difficult than reproducing the average annual extreme soil temperatures at 10 cm. When we compared the frequency of occurrence of −3.5°C and −6.0°C soil temperatures with observations across the region (Figs. 5b,c), the model fit was not as good as we found when comparing the average annual extreme minimum temperatures (Fig. 5a). This result may be attributed to difficulties in simulating snow cover dynamics and the ability to capture the physical properties (i.e. density, compaction, water content) of snowpack on any particular day. Many ecosystem models currently do not have complicated dynamics such as the ability of standing vegetation or crop stubble to preferentially capture snow and lead to a deeper and longer duration of snow depth, which could lead to an accentuated insulating effect, higher soil temperatures, and decreased frost penetration [51,71] and alter the surface albedo [72]. However, this is an area of great potential for future model improvement if more data are collected in a variety of land management settings. We also understand that physical processes in soils such as freeze-thaw cycles, and soil ice and frost formation can influence soil structure and infiltration [52,53,73], which ultimately affect heat transfer, and make soil temperature prediction particularly challenging at short timescales. Incorporation of these factors or refinement of current modeling approaches as more data become available should also improve ecosystem model realism.

While simulated snow depth agreed quite well with observations from November through March for the POTVEG simulation in the northern portion of the study region (Figs. 1a, 2a), snowmelt occurred more rapidly in March compared to observations when the model was parameterized with maize (row crops) or miscanthus (grasses). Additionally, the seasonal timing of the maximum snow depth occurred one month earlier than observations in those scenarios. Across the southern region (Fig. 1b, 2b), simulated snow depth agreed quite well with observations across all months and years for maize as prescribed vegetation, whereas the POTVEG yielded the poorest comparison with observations.

The connection between snowpack, the timing of snowmelt, and simulated soil temperatures (Fig. 3) was obvious. When comparing simulated soil temperatures to observations, the scenarios that yielded lower than observed snow depth and an earlier occurrence of snowmelt in March-April produced soil temperatures that were 5–10°C warmer than observed values during the late winter and spring (Fig. 3a). Thus, we conclude that accurate simulation of snowpack and snowmelt in ecosystem models is a crucial, but potentially overlooked, step towards simulating an accurate portrayal of changing energy balance at the soil surface when transitioning from winter to spring. This transition period typically coincides with rapid changes in ecosystem processes (e.g., net ecosystem exchange, ET), and the timing of warming and drying of soils impacts farmer management across the Corn Belt [74]. Subsequently, land cover and management choices further influence plant phenology, leaf area index, and canopy architecture, which all play important roles on changing radiation interception and energy balance [21,38,52,61].

However, not all of the simulated error should probably be attributed to the model. When taking a closer examination of where and when the model performed well, we note that in the northern regions simulations of snow depth compared best with observations for the POTVEG simulations (Fig. 1a, 2a). Across the south, the opposite was true; the best comparisons with observations were found with simulations of maize (Fig. 1b, 2b). This is potentially a result of snow depth observations across the northern study having a higher likelihood of being collected in land cover/land use settings that are more reflective of natural vegetation (trees, shrubs, grasses), and therefore the snow depth in POTVEG simulations would likely agree better in those areas given closer agreement in plant phenology and LAI. Across the south, a higher proportion of land cover is in crops and might be the more likely land cover type where observations are collected, and therefore it might not be surprising that snow depth from the crop simulations across the southern region (Fig. 1b, 2b) compares the best across the south. Furthermore, accurate measurements of snow depth are known to be difficult to attain, attributed to a variety of factors, but recent improvements in technology may improve accuracy [75].

As discussed, there are several factors, concerning both modeling and observations, that contribute to perceived simulation error, and could call into question the statistical significance of the results. While using observational data across a wide range of research sites in the Midwest to validate the model would be considered a positive attribute of this study, we do not have a clear picture of the land-use history at those sites that could influence soil temperatures and we are focusing on the ability of the model to simulate the coldest (extreme) soil temperatures each year, which may not be sustained for more than a few days. We also know how difficult it is to simulate snow depth and density at any given site without having hourly weather data, and how influential hourly air temperature and other atmospheric conditions are in determining the ratio of liquid precipitation to accumulated snow depth. In Agro-IBIS, an air temperature threshold of 1.1°C is used to determine whether precipitation falls in liquid (rain) or frozen (snow) form, and a constant snow density is assumed. While the gridded climate data that we employed in this study are of extremely high quality and 5 min spatial resolution, some of the model bias for extremely cold soil temperatures (<-6.0°C) is likely induced by uncertainty in simulation of snowfall, snowmelt, and snow density, which is attributed to differences between the gridded climate data used to drive the model and what actually occurred at each site. For example, when additional model validation was performed at two specific locations (experimental stations in Wisconsin and Illinois) and implemented site-specific meteorological (hourly) and management data as drivers, the model agreed with observational data quite well, especially at the monthly time scale. Therefore, when the model had the best available land management and meteorological data to drive simulations, the comparisons were very strong.

The statistical significance of the results should be perceived as strong across large spatial regions like the Midwest US, and a good representation of how varied residue management impacts the typical average minimum soil temperatures. However, subtle changes in land management that effect soil structure, surface roughness, and the ability of the landscape to preferentially trap snow, as well as the short timescale temporal weather patterns, particularly those producing rapid changes in air temperature without snowpack, can lead to widely varying results. These temperature responses may be even greater than the magnitude of changes associated with varied residue management. The results presented here are robust as a broad generalization of what could happen at any particular Midwest US location in the context of residue management on soil temperatures based on the mean climate, but there are a range of other factors in any single winter season that could lead to a significant departure from the simulated mean responses.

Conclusions

There are numerous factors that will influence the ability of miscanthus to overwinter in the first year or two after establishment, including rhizome size, end of season harvest management, planting depth, soil water content, rhizome moisture content at the end of the season, and soil characteristics [4,7,10]. However, there is no denying that residue management plays a significant role in soil thermal dynamics, particularly on wintertime soil temperatures, and therefore on miscanthus rhizome survival during the early stages of establishment. The results presented here illustrated that very small differences in thatch thickness, on the order of a few centimeters, lead to significant differences in minimum winter soil temperatures. Continued field research across a wide climate gradient, and assessing varied management scenarios, will help to fill the gaps in our understanding of rhizome winter survival. A potential wild card might be how long-term climate change impacts snowpack variability and annual minimum soil temperatures, the latter of which have been shown to be warming over the last several decades. In the case of miscanthus, producers could be presented with a new cropping system to support the production of renewable fuels in the future. However, they may be faced with dilemmas on how to best manage miscanthus in its establishment phase to ensure long-term survival while simultaneously remaining profitable.

Supporting Information

Table S1 Soil temperature observation sites used in model validation and minimum soil temperature assessment. List of observation sites that were used to validate Agro-IBIS and for further assessment of annual minimum extreme 10 cm soil temperatures across the Midwest US. Soil surface refers to ground cover present; initial year is the beginning of the observation record used (starting January 1) and end year denotes the last year of the observation record used in this study (last day of record is December 31). Italicized and bolded station names denote the 36 locations that were used in an assessment of soil temperature trends simulated by Agro-IBIS.

Acknowledgments

We thank Tom Voigt, Emily Heaton, Daryl Herzmann, Kent Berns, and Gregg Sanford for helpful discussions. We extend gratitude to Nathan Healy, Michael Cruse, and James Tesmer for collecting and processing soil temperature data.

Author Contributions

Conceived and designed the experiments: CJK AV. Performed the experiments: CJK AV. Analyzed the data: CJK AV JDL MMM. Wrote the paper: CJK AV.

References

1. Zub HW, Brancourt-Hulmel M (2010) Agronomic and physiological performances of different species of Miscanthus, a major energy crop. A review. Agron Sustain Dev 30: 201–214.
2. Robertson GP, Dale VH, Doering OC, Hamburg SP, Melillo JM, et al. (2008) Agriculture - Sustainable biofuels Redux. Science 322: 49–50.
3. Lewandowski I, Clifton-Brown JC, Scurlock JMO, Huisman W (2000) Miscanthus: European experience with a novel energy crop. Biomass Bioenerg 19: 209–227.
4. Heaton EA, Dohleman FG, Long SP (2008) Meeting US biofuel goals with less land: the potential of Miscanthus. Global Change Biol 14: 2000–2014.
5. Dohleman FG, Long SP (2009) More Productive Than Maize in the Midwest: How Does Miscanthus Do It? Plant Physiol 150: 2104–2115.
6. Heaton E, Voigt T, Long SP (2004) A quantitative review comparing the yields of two candidate C-4 perennial biomass crops in relation to nitrogen, temperature and water. Biomass Bioenerg 27: 21–30.
7. Heaton EA, Dohleman FG, Miguez AF, Juvik JA, Lozovaya V, et al. (2010) Miscanthus: A Promising Biomass Crop. Adv Bot Res 56: 75–137.
8. Pyter RJ, Dohleman FG, Voigt TB (2010) Effects of rhizome size, depth of planting and cold storage on Miscanthus × giganteus establishment in the Midwestern USA. Biomass Bioenerg 34: 1466–1470.
9. Greef JM, Deuter M (1993) Syntaxonomy of Miscanthus-X-Giganteus Greef-Et-Deu. Angew Bot 67: 87–90.
10. Pyter R, Heaton E, Dohleman F, Voigt T, Long S (2009) Agronomic Experiences with Miscanthus × giganteus in Illinois, USA. Biofuels: Methods and Protocols 581: 41–52.
11. Khanna M, Dhungana B, Clifton-Brown J (2008) Costs of producing miscanthus and switchgrass for bioenergy in Illinois. Biomass Bioenerg 32: 482–493.
12. Numata M (1974) Grassland vegetation. In: Numata M, The flora and vegetation of Japan. Tokyo: Elsevier. 125–147.
13. Clifton-Brown JC, Lewandowski I (2000) Overwintering problems of newly established Miscanthus plantations can be overcome by identifying genotypes with improved rhizome cold tolerance. New Phytol 148: 287–294.
14. Clifton-Brown JC, Lewandowski I (2000) Water use efficiency and biomass partitioning of three different Miscanthus genotypes with limited and unlimited water supply. Ann Bot 86: 191–200.
15. Clifton-Brown JC, Lewandowski I, Andersson B, Basch G, Christian DG, et al. (2001) Performance of 15 Miscanthus genotypes at five sites in Europe. Agron J 93: 1013–1019.
16. Farrell AD, Clifton-Brown JC, Lewandowski I, Jones MB (2006) Genotypic variation in cold tolerance influences the yield of Miscanthus. Ann Appl Biol 149: 337–345.
17. Horton R, Bristow KL, Kluitenberg GJ, Sauer TJ (1996) Crop residue effects on surface radiation and energy balance - review. Theor Appl Climatol 54: 27–37.
18. Sauer TJ, Hatfield JL, Prueger JH, Norman JM (1998) Surface energy balance of a corn residue-covered field. Agr Forest Meteorol 89: 155–168.
19. Steiner JL, Schomberg HH (1996) Impacts of crop residue at the Earth-atmosphere interface: introduction. Theor Appl Climatol 54: 1–4.
20. Wilhelm WW, Johnson JMF, Hatfield JL, Voorhees WB, Linden DR (2004) Crop and soil productivity response to corn residue removal: A literature review. Agron J 96: 1–17.
21. Kucharik CJ, Twine TE (2007) Residue, respiration, and residuals: Evaluation of a dynamic agroecosystem model using eddy flux measurements and biometric data. Agr Forest Meteorol 146: 134–158.
22. Sharratt BS (2002) Corn stubble height and residue placement in the northern US Corn Belt Part I. Soil physical environment during winter. Soil Till Res 64: 243–252.
23. Kucharik CJ (2003) Evaluation of a Process-Based Agro-Ecosystem Model (Agro-IBIS) across the US Corn Belt: Simulations of the Interannual Variability in Maize Yield. Earth Interact 7.
24. Vanloocke A, Bernacchi CJ, Twine TE (2010) The impacts of Miscanthus × giganteus production on the Midwest US hydrologic cycle. GCB Bioenergy 2: 180–191.
25. VanLoocke A, Twine TE, Zeri M, Bernacchi CJ (2012) A regional comparison of water use efficiency for miscanthus, switchgrass and maize. Agr Forest Meteorol 164: 82–95.
26. Foley JA, Prentice IC, Ramankutty N, Levis S, Pollard D, et al. (1996) An integrated biosphere model of land surface processes, terrestrial carbon balance, and vegetation dynamics. Global Biogeochem Cycles 10: 603–628.
27. Kucharik CJ, Foley JA, Delire C, Fisher VA, Coe MT, et al. (2000) Testing the performance of a Dynamic Global Ecosystem Model: Water balance, carbon balance, and vegetation structure. Global Biogeochem Cycles 14: 795–825.
28. Kucharik CJ, Brye KR (2003) Integrated BIosphere Simulator (IBIS) yield and nitrate loss predictions for Wisconsin maize receiving varied amounts of Nitrogen fertilizer. J Environ Qual 32: 247–268.
29. Campbell GS, Norman JM (1998) An introduction to Environmental biophysics. New York, NY: Springer-Verlag.
30. Lenters JD, Coe MT, Foley JA (2000) Surface water balance of the continental United States, 1963–1995: Regional evaluation of a terrestrial biosphere model and the NCEP/NCAR reanalysis. J Geophys Res-Atmos 105: 22393–22425.
31. Collatz JG, Ball JT, Grivet C, Berry JA (1991) Physiological and environmental regulation of stomatal conductance, photosynthesis and transpiration: a model that includes a laminar boundary layer. Agr Forest Meteorol 53: 107–136.
32. Collatz GJ, Ribas-Carbo M, Berry JA (1992) Coupled photosynthesis-stomatal conductance model for leaves of C_4 plants. Aust J Plant Physiol 19: 519–538.
33. Farquhar GD, von Caemmerer S, Berry JA (1980) A biochemical model of photosynthetic CO_2 assimilation in leaves of C_3 species. Planta 149: 78–90.
34. Li KY, Coe MT, Ramankutty N (2005) Investigation of hydrological variability in west Africa using land surface models. J Clim 18: 3173–3188.
35. Soylu ME, Istanbulluoglu E, Lenters JD, Wang T (2011) Quantifying the impact of groundwater depth on evapotranspiration in a semi-arid grassland region. Hydrol Earth Syst Sci 15: 787–806.
36. Donner SD, Kucharik CJ (2003) Evaluating the impacts of land management and climate variability on crop production and nitrate export across the Upper Mississippi Basin. Global Biogeochem Cycles Doi 10.1029/2001gb001808.
37. Delire C, Foley JA (1999) Evaluating the performance of a land Surface/ecosystem model with biophysical measurements from contrasting environments. J Geophys Res-Atmos 104: 16895–16909.
38. Kucharik CJ, Barford CC, El Maayar M, Wofsy SC, Monson RK, et al. (2006) A multiyear evaluation of a Dynamic Global Vegetation Model at three AmeriFlux forest sites: Vegetation structure, phenology, soil temperature, and CO2 and H2O vapor exchange. Ecol Model 196: 1–31.
39. Twine TE, Kucharik CJ (2008) Evaluating a terrestrial ecosystem model with satellite information of greenness. J Geophys Res-Biogeo 113.
40. El Maayar M, Price DT, Delire C, Foley JA, Back AT, et al. (2001) Validation of the Integrated BIosphere Simulator over Canadian deciduous and coniferous boreal forest stands. J Geophys Res-Atmos 106: 14339–14355.
41. Molling CC, Strikwerda JC, Norman JM, Rodgers CA, Wayne R, et al. (2005) Distributed runoff formulation designed for a precision agricultural landscape modeling system. J Am Water Resour Assoc 41: 1289–1313.
42. Simunek J, Sejna H, Saito H, Sakai M, Van Genuchthen Mt (2008) The HYDRUS-1D Software Package for Simulating the Movement of Water, Heat, and Multiple Solutes in Variably Saturated Media, Version 4.08, HYDRUS Software Series 3. Riverside, CA: University of California Riverside. 330.
43. Simunek J, van Genuchten MT (2008) Modeling nonequilibrium flow and transport processes using HYDRUS. Vadose Zone J 7: 782–797.
44. Flerchinger GN (1991) Sensitivity of Soil Freezing Simulated by the Shaw Model. Trans ASAE 34: 2381–2389.
45. Flerchinger GN, Sauer TJ, Aiken RA (2003) Effects of crop residue cover and architecture on heat and water transfer at the soil surface. Geoderma 116: 217–233.
46. Flerchinger GN, Caldwell TG, Cho J, Hardegree SP (2012) Simultaneous Heat and Water (Shaw) Model: Model Use, Calibration, and Validation. Trans ASAE 55: 1395–1411.
47. Kalnay E, Kanamitsu M, Kistler R, Collins W, Deaven D, et al. (1996) The NCEP/NCAR 40-year reanalysis project. Bull Am Meteorol Soc 77: 437–471.

48. Kanamitsu M, Ebisuzaki W, Woollen J, Yang SK, Hnilo JJ, et al. (2002) NCEP-DOE AMIP-II Reanalysis (R-2). Bull Am Meteorol Soc 83: 1631–1643.
49. Miller DA, White RA (1998) A conterminous United States multilayer soil characteristics dataset for regional climate and hydrology modeling. Earth Interact 5: 42.
50. Amougou N, Bertrand I, Cadoux S, Reous S (2012) Miscanthus × giganteus leaf senescence, decomposition and C and N inputs to soil. GCB Bioenergy 10:1111/j.1757–1707.2012.01192.x.
51. Sharratt BS, Benoit GR, Voorhees WB (1998) Winter soil microclimate altered by corn residue management in the northern Corn Belt of the USA. Soil Till Res 49: 243–248.
52. Cherkauer KA, Lettenmaier DP (2003) Simulation of spatial variability in snow and frozen soil. J Geophys Res-Atmos 108.
53. Sinha T, Cherkauer KA (2008) Time Series Analysis of Soil Freeze and Thaw Processes in Indiana. J Hydrometeorol 9: 936–950.
54. Hu Q, Feng S (2003) A daily soil temperature dataset and soil temperature climatology of the contiguous United States. J Appl Meteorol 42: 1139–1156.
55. Zeri M, Anderson-Teixeira K, Hickman G, Masters M, DeLucia E, et al. (2011) Carbon exchange by establishing biofuel crops in Central Illinois. Agric Ecosyst Environ 144: 319–329.
56. Smith CM, David MB, Mitchell CA, Masters MD, Anderson-Teixeira KJ, et al. (2012) Reduced nitrogen losses following conversion of row crop agriculture to perennial biofuel crops. J Environ Qual 10.2134/jeq2012.0210.
57. Portmann RW, Solomon S, Hegerl GC (2009) Spatial and seasonal patterns in climate change, temperatures, and precipitation across the United States. Proc Nat Acad Sci USA 106: 7324–7329.
58. Kucharik CJ, Serbin SP, Vavrus S, Hopkins EJ, Motew MM (2010) Patterns of Climate Change across Wisconsin from 1950 to 2006. Phys Geogr 31: 1–28.
59. MathWorks (2010) MATLAB 7.11. Natick, MA.
60. Aase JK, Siddoway FH (1979) Crown-Depth Soil Temperatures and Winter Protection for Winter-Wheat Survival. Soil Sci Soc Am J 43: 1229–1233.
61. Zheng D, Hunt Jr., R.H Running, S.W. (1993) A daily soil temperature model based on air temperature and precipitation for continental applications. Climate Res 2.
62. Dyer JL, Mote TL (2006) Spatial variability and trends in observed snow depth over North America. Geophys Res Lett 33.
63. Sharratt BS, Baker DG, Wall DB, Skaggs RH, Ruschy DL (1992) Snow Depth Required for near Steady-State Soil Temperatures. Agr Forest Meteorol 57: 243–251.
64. Jørgensen U, Muhs H.J. (2001) Micanthus breeding and improvement. In: Jones MB, Walsh, M., Miscanthus for energy and fibre. London: Earthscan. 68–85.
65. Price L, Bullard M, Lyons H, Anthony S, Nixon P (2004) Identifying the yield potential of Miscanthus × giganteus: an assessment of the spatial and temporal variability of M-x giganteus biomass productivity across England and Wales. Biomass Bioenerg 26: 3–13.
66. Gauder M, Graeff-Honninger S, Lewandowski I, Claupein W (2012) Long-term yield and performance of 15 different Miscanthus genotypes in southwest Germany. Ann Appl Biol 160: 126–136.
67. Heaton EA, Flavell RB, Mascia PN, Thomas SR, Dohleman FG, et al. (2008) Herbaceous energy crop development: recent progress and future prospects. Curr Opin Biotechnol 19: 202–209.
68. Isard SA, Schaetzl RJ, Andresen JA (2007) Soils cool as climate warms in the great lakes region: 1951–2000. Ann Assoc Am Geogr 97: 467–476.
69. Sinha T, Cherkauer KA, Mishra V (2010) Impacts of Historic Climate Variability on Seasonal Soil Frost in the Midwestern United States. J Hydrometeorol 11: 229–252.
70. Sinha T, Cherkauer KA (2010) Impacts of future climate change on soil frost in the midwestern United States. J Geophys Res-Atmos 115.
71. Cherkauer KA, Lettenmaier DP (1999) Hydrologic effects of frozen soils in the upper Mississippi River basin. J Geophys Res-Atmos 104: 19599–19610.
72. Kunkel KE, Isard SA, Hollinger SE, Gleason H, Belding M (1999) Spatial heterogeneity of albedo over a snow-covered agricultural landscape. J Geophys Res-Atmos 104: 19551–19557.
73. Hillel D (2004) An introduction to environmental soil physics. San Diego: Academic Press.
74. Kucharik CJ (2006) A multidecadal trend of earlier corn planting in the central USA. Agron J 98: 1544–1550.
75. Kunkel KE, Palecki MA, Hubbard KG, Robinson DA, Redmond KT, et al. (2007) Trend identification in twentieth-century US snowfall: The challenges. J Atmos Oceanic Technol 24: 64–73.
76. Sauer TJ, Hatfield JL, Prueger JH (1996) Aerodynamic characteristics of standing corn stubble. Agron J 88: 733–739.
77. Bristow KL, Campbell GS, Papendick RI, Elliott LF (1986) Simulation of Heat and Moisture Transfer through a Surface Residue Soil System. Agr Forest Meteorol 36: 193–214.
78. Shen Y, Tanner CB (1990) Radiative and Conductive Transport of Heat through Flail-Chopped Corn Residue. Soil Sci Soc Am J 54: 653–658.
79. Miao Z, Grift TE, Hansen AC, Ting KC (2011) Energy requirement for comminution of biomass in relation to particle physical properties. Ind Crops Prod 33: 504–513.
80. Chung SO, Horton R (1987) Soil Heat and Water-Flow with a Partial Surface Mulch. Water Resour Res 23: 2175–2186.
81. Wagenaar BM, VandenHeuvel EJMT (1997) Co-combustion of Miscanthus in a pulverised coal combustor: Experiments in a droptube furnace. Biomass Bioenerg 12: 185–197.
82. Van Wijk WR, DeVries DA (1963) Periodic temperature variations in a homogeneous soil. In: Van Wijk WR, Physics of the plant environment. Amsterdam: North-Holland. 102–143.
83. Clemmensen AW (2004) Physical characteristics of Miscanthus composts compared to peat and wood fiber growth substrates. Compost Sci Util 12: 219–224.
84. Tanner CB, Shen Y (1990) Water-Vapor Transport through a Flail-Chopped Corn Residue. Soil Sci Soc Am J 54: 945–951.
85. Sui HJ, Zeng DC, Chen FZ (1992) A Numerical-Model for Simulating the Temperature and Moisture Regimes of Soil under Various Mulches. Agr Forest Meteorol 61: 281–299.

4

Straw Mulching Reduces the Harmful Effects of Extreme Hydrological and Temperature Conditions in Citrus Orchards

Yi Liu[1], Jing Wang[1], Dongbi Liu[2], Zhiguo Li[1], Guoshi Zhang[1,3], Yong Tao[1], Juan Xie[1], Junfeng Pan[1], Fang Chen[1,3]*

1 Laboratory of Aquatic Botany and Watershed Ecology, Wuhan Botanical Garden, Chinese Academy of Sciences China, Wuhan, China, **2** Institute of Plant Protection and Soil Fertilizer, Hubei Academy of Agricultural Sciences, Wuhan, China, **3** China Program, International Plant Nutrition Institute (IPNI), Wuhan, China

Abstract

Extreme weather conditions with negative impacts can strongly affect agricultural production. In the Danjiangkou reservoir area, citrus yields were greatly influenced by cold weather conditions and drought stress in 2011. Soil straw mulching (SM) practices have a major effect on soil water and thermal regimes. A two-year field experiment was conducted to evaluate whether the SM practices can help achieve favorable citrus fruit yields. Results showed that the annual total runoff was significantly ($P<0.05$) reduced with SM as compared to the control (CK). Correspondingly, mean soil water storage in the top 100 cm of the soil profile was increased in the SM as compared to the CK treatment. However, this result was significant only in the dry season (Jan to Mar), and not in the wet season (Jul to Sep) for both years. Interestingly, the SM treatment did not significantly increase citrus fruit yield in 2010 but did so in 2011, when the citrus crop was completely destroyed (zero fruit yield) in the CK treatment plot due to extremely low temperatures during the citrus overwintering stage. The mulch probably acted as an insulator, resulting in smaller fluctuations in soil temperature in the SM than in the CK treatment. The results suggested that the small effects on soil water and temperature changes created by surface mulch had limited impact on citrus fruit yield in a normal year (e.g., in 2010). However, SM practices can positively impact citrus fruit yield in extreme weather conditions.

Editor: Yi Liu, Wuhan Botanical Garden, Chinese Academy of Sciences China, China

Funding: This work was financially supported by the National Key Technology R&D Program (2012BAD15B01), the NSFC (31100386), the Open Funding Project (Y152741s04) of the Key Laboratory of Aquatic Botany and Watershed Ecology, Chinese Academy of Sciences and the Cooperated Program with the International Plant Nutrition Institute (IPNI-HB-33). The funders had no role in study design, data collection and analysis, decision to publish, or preparation of the manuscript.

Competing Interests: The authors have declared that no competing interests exist.

* E-mail: fchenipni@126.com

Introduction

One of the major challenges of climate change to agriculture and food security involves ever increasing extreme weather events world widely, such as droughts, heat waves, excessive cold, heavy and prolonged rainfalls, hailstorms, and so on [1]. They usually cause negative effects and sometimes fatal damages to crops, physiologically and/or physically [2]. However, the stresses could be relieved and neutralized by certain positive effects from field microclimate, beneficial soil water and thermal conditions under some farmland management practices [3]. Hence, there is an increasing need for understanding the response of soil water and temperature dynamics to changes in extreme weather conditions [4].

Soil water is considered to be one of the most important factors affecting plant growth and development [5,6]. Even a small change in soil water storage could greatly affect crop productivity [7]. Soil surface mulching, such as with plastic film [8], crop residue [9], or gravel and sand [10], has a large impact on many of the hydrological and biological processes of soil ecosystems, and the most prominent of these changes is the modification of the soil–plant–atmosphere continuum (SPAC) water cycling. For example, numerous reports indicate that soil straw mulching (SM) favorably influences the soil moisture regime by reducing evaporation from the soil surface [11,12], improving infiltration [13], and soil water retention [14]. SM has also led to improvements in crop yields in arid and semi-arid environments [5,12] and economized the use of irrigation water [14]. Thus, a better understanding of the impact of SM practices on soil hydrological processes is becoming critical, especially from the crop production perspective, because of the increasing shortage of water resources worldwide.

Soil temperature controls the rate of crop development, particularly when the meristem is within the soil [15]. Higher soil temperature accelerated the rates of leaf tip appearance and full leaf expansion, enabling the crop to attain maximum green leaf area index more rapidly [10]. SM has been reported to cause either a decrease, an increase or a negligible effect on soil temperature. For instance, SM during over-wintering period can improve soil thermal regime according to several studies [12,16,17]. However, Sarkar et al. [18] reported that SM could reduce soil temperature, while effective soil water conservation with SM may result in higher production. While Ghosh et al. [19] argued SM had little or no effect on soil temperature, and that its effects were almost entirely due to increased organic matter. These inconsistent results may depend on multiple factors including soil properties, climate, and species planted [15]. Lower soil temperature under SM has mostly been attributed to the reduced solar energy reaching the soil during hot periods, while increased soil temperature under SM has been attributed to the reduction in outgoing heat radiation from the soil during cold periods.

The Danjiangkou reservoir, established in the 1970s with a drainage area of 95,200 km^2, is a water source area for China's Middle Route of the South-to-North Water Transfer Project [20]. The staple crops of the region (wheat and rice) are generally grown in the flat land part of the Danjiangkou reservoir area. Citrus is one of the main types of fruit growing on sloping lands, and its high yields, averaging about 40 t hm^{-2}, are assumed to be a result of the beneficial thermal effects of the great lakes. After the wheat (or rice) harvest, farmers would generally burn the stalks, but this practice is now prohibited to restore and protect the Danjiangkou reservoir riparian ecosystem. Therefore, farmers have been using wheat (or rice) straw for mulching in citrus orchards. Notably, SM practices can effectively contribute to water conservation and decreased nutrient losses on sloping lands [21]. Although several studies have reported changes in water quality [22] and the role of surface mulching in soil nutrient losses [23], little is known about the impact of SM practices on reducing the harmful effects of extreme weather conditions. In the Danjiangkou reservoir area, citrus yields were strongly affected by cold and drought stress in 2011. The average temperature measured at the Danjiangkuo city meteorological station was 1.5°C during December 2010 to February 2011, which broke the record set in year 2000 and lowered the long-term (10 years) average temperature by 3°C for the corresponding period.

In the present study, we analyzed runoff, soil water content and storage, and seasonal variations in soil temperature under SM and non-mulching or control (CK) treatments. We specifically focused on the role of SM on soil water and temperature dynamics by comparing fruit yield under mulching and no mulching practices. This was done to test the hypothesis that the small effects on soil water conservation and thermoregulation created by surface SM practices can greatly impact citrus fruit yields in extreme weather conditions. The objectives of this study were: (i) to determine how productivity of citrus fruit was affected by soil water and temperature, (ii) to evaluate the influence of SM on the soil water and temperature in sloping citrus orchards in the Danjiangkou Reservoir area.

Materials and Methods

Site description

This study was conducted in 2010 and 2011 at the Xiaofuling experimental station (32°45′46″N, 111°9′26″E) in the Danjiangkou reservoir area (Fig. 1). The area has a subtropical zone climate with mean annual temperature of 15.7°C, and a monthly average temperature of 27.3°C in July and 4.2°C in January. Mean annual rainfall is approximately 834 mm, 80% of which concentrates between May and October. The study was carried out in a 10-year-old citrus orchard. The experimental site is owned by Wuhan Botanical Garden, Chinese Academy of Sciences China. The field studies did not involve endangered or protected species and no specific permits were required for the described field studies. The site is located about 300 m asl, with an average slope of 15°. The soil at the site is a cinnamon yellow soil as defined by the Chinese soil classification system [24] and textural composition is 14% clay, 23% silt, and 64% sand. At start of the field experiment in 2009, the soil had a pH of 6.5 and a bulk density of 1.45 g cm^{-3}. The amounts of organic matter, total nitrogen, available phosphorus, available potassium, and inorganic nitrogen were 9.1 g kg^{-1}, 0.88 g kg^{-1}, 16.0 mg kg^{-1}, 106.3 mg kg^{-1} and 101.8 mg kg^{-1}, respectively. These nutrient contents using routine analytical methods [25].

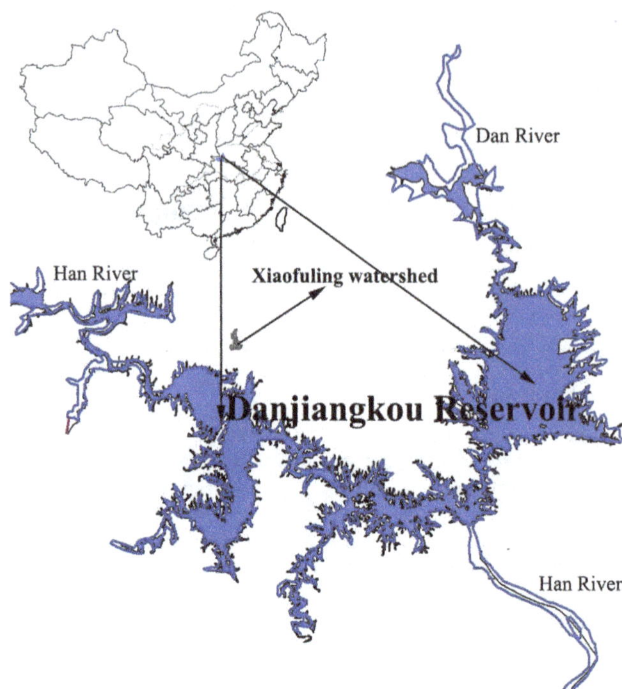

Figure 1. Location of the study area.

Experimental treatments and field management

The experiment was designed with two treatments, CK or without mulching and SM, (Fig. 2) with three replicates for each treatment. All plots, separated by concrete borders, were set up with a tank at the base of each plot to collect the runoff. The size of each experimental plot was 40.5 m^2 (4.5 m×9 m). The plantation consisted of trained citrus trees at 1.5 m×3 m spacing, with rows perpendicular to the slope. Hence, there are eight trees at each plot. In the plots mulched with straw, rice (or wheat) straw was uniformly applied at a rate of 6,000 kg hm^{-2}. Each citrus tree was fertilized in April by hand, with 0.5 kg N (urea), 0.3 kg P_2O_5 (superphosphate), and 0.4 kg K_2O (potassium chloride) in both years (2010 and 2011). Manual weeding was undertaken as required during the citrus growing season.

Sampling measurements and data calculation

To measure the runoff caused by rainfall, a standard recording rain gauge was sited about 100 m from the experimental plots. Rainfall was calculated at 1-day intervals. Runoff was collected at each plot after each rainfall event, and the depth of water in the runoff collection tank was recorded to calculate runoff volumes. The runoff were determined by the following formula: $R = S \times h \times 10/Pa$, where R is runoff (mm), S is collection tank floorage (m^2), h is water depth (cm), Pa is plot area (m^2). The monthly soil water content was determined gravimetrically by oven drying (105°C for 24 h) the core samples that were taken at depth intervals of 20 cm down the 0–100 cm profile in each plot on the 26th (or 27th) day of each month (January 2010–December 2011). The soil water storage (W) in the profile was considered to be the total storage in all of the sampled layers in the plot, as was

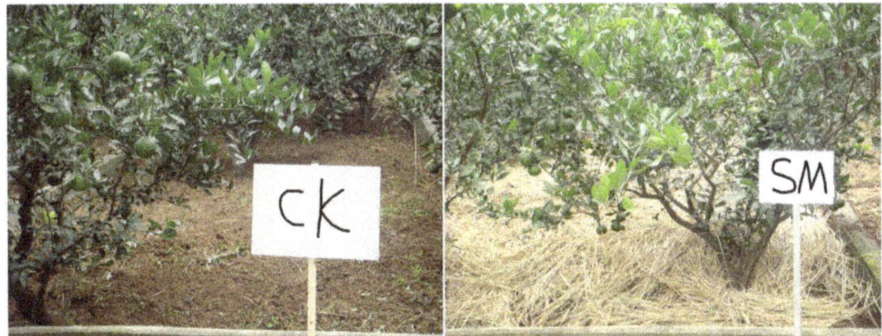

Figure 2. Photographs showing plots under control or without mulching (CK) and straw mulching (SM) treatments.

calculated using the formula: $W = h \times \rho \times \theta \times 1000$, where h is soil depth (cm), ρ is soil bulk density (g cm^{-3}), θ is soil gravimetric water content (g g^{-1}). Soil bulk density was determined from the inner diameter of the core sampler, segment depth, and the oven-dry weight of the core samples at start of the field experiment in 2009 (see Table S1). Air temperature within the canopy (CA, height = 1 m) and soil temperature at 5 cm depth were also recorded at 0.5 h intervals by the StowAway TidbiT temperature recorder (Range: −20° to 70°C in air; −20° to 30°C in water; Accuracy: ±0.4° at 20°C). During the 24-h period, the values were averaged to calculate mean daily temperature.

At the time of commercial harvest, the citrus fruits were harvested gradually when they were ripe. The yield per plot (t hm^{-2}) was obtained by weighing the harvested fruit. A random sample of 25 fruits from each plot was collected to determine the average fruit weight. In addition, fruit size was measured by measuring the equatorial diameter with the help of Vernier caliper from each experimental plot. Using a standard juicer, 25 fruits were juiced. The juice was weighed and expressed as a percentage of the total fruit weight.

Statistical analyses

Statistically significant differences in mean runoff, soil water storage, and soil temperature between the CK and SM treatments were determined utilizing the Wilcoxon signed rank test (WSRT). A P value of <0.05 was considered statistically significant. Statistical analysis for fruit yield and quality was performed using Analysis of Variance (ANOVA). One-way analysis of variance was

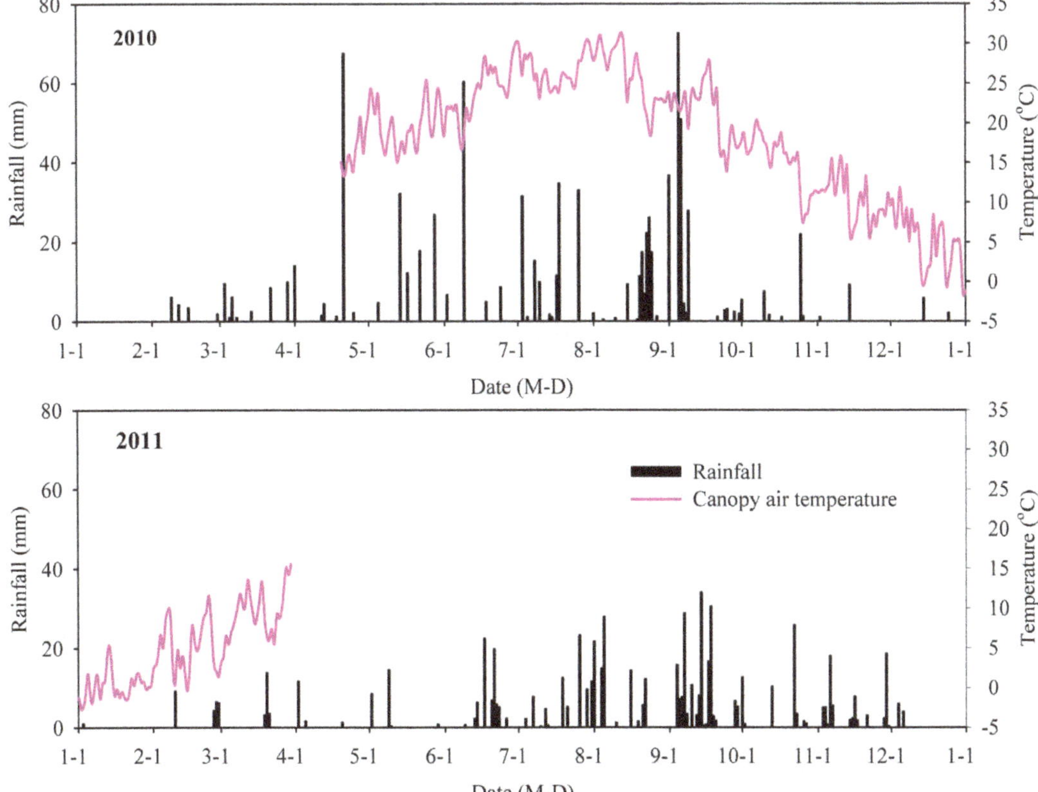

Figure 3. Canopy air temperature and rainfall data for 2010 and 2011 at Xiaofuling experimental station in Danjiangkou reservoir area in Hubei, China. Please note that the record of the air temperatures ended on March 31, 2011 due to the loss of the logger.

carried out to determine the differences between the measured parameters for different treatments. Least significant difference (LSD) at P = 0.05 was used to elucidate any significant differences.

Results

Rainfall and air temperature

The air temperature and rainfall of experimental field are shown in Fig. 3. Field data as well as summary statistics of air temperature and rainfall are provided as Table S2 and S3 in supporting information. The annual precipitation recorded at the experimental site was 849.3 mm in 2010 and 655.5 mm in 2011. 2011 was a dry year, which lowered precipitation by 21.5% compared with the long-term average. Fig. 3 also shows the air temperature within the canopy from April 20, 2010 to March 31, 2011. It should be noted that the record of the air temperatures ended on March 31, 2011 due to the loss of the logger. The daily air temperature varied from −2.5 to 31.2°C, averaging 15.0°C. The average minimum air temperature in the winter of 2010 (i.e., from December 2010 to February 2011) was −2.0°C. There were 66 days with minimum temperature <0°C during the study period, and not only the mean but also the maximum/minimum air temperatures were lower compared to the average of the last 10 years for the corresponding period.

Runoff

The surface mean runoff ranged from 1.0 to 37.7 mm in 2010 and from 0.4 to 15.8 mm in 2011 (see Fig. 4 or Table S4), and had three peak values with time (on April 21, June 9, and September 5) in 2010 and one peak value with time (on September 20) in 2011 indicating extreme rainfall events. Three rainstorm events were registered in 2010 as 67.7 mm (April 21, 2010), 60.4 mm (June 9, 2010) and 72.5 mm (September 5, 2010). While one continuous 7 day (From September 13 to 20) rainfall was registered in 2011 as 86.8 mm for accumulative precipitation (see Table S3). Annual total runoff volumes, calculated by adding the readings taken at the sampled points throughout the whole year, were much lower in the SM (107 mm in 2010, 78 mm in 2011) than in the CK (145 mm in 2010, 97 mm in 2011) plots (see Table S4). Lower surface runoff values observed in the SM plots were probably due to good ground coverage and slightly higher water infiltration than in the CK plots.

Soil water

The seasonal variations in water storage in the soil profile (0–100 cm) under CK and SM treatments are shown in Fig. 5. Field data as well as summary statistics of soil water storage are provided as Table S5 in supporting information. Mean soil water storage values, calculated by averaging the readings taken at the sampled points over one year, were much higher in the SM (ranged from 245 to 303 mm in 2010, and 254 to 291 mm in 2011, see Table S5) treatment than in the CK (ranged from 231 to 303 mm in 2010, and 237 to 290 mm in 2011, see Table S5) treatment. Soil water storage exhibited pronounced seasonal variations with minimum values at 231±2 mm during the dry season and maximum values at 303±26 mm during the wet season (Fig. 5). Largest differences in soil water storage between CK and SM occurred during the dry season (from January to March in 2011),

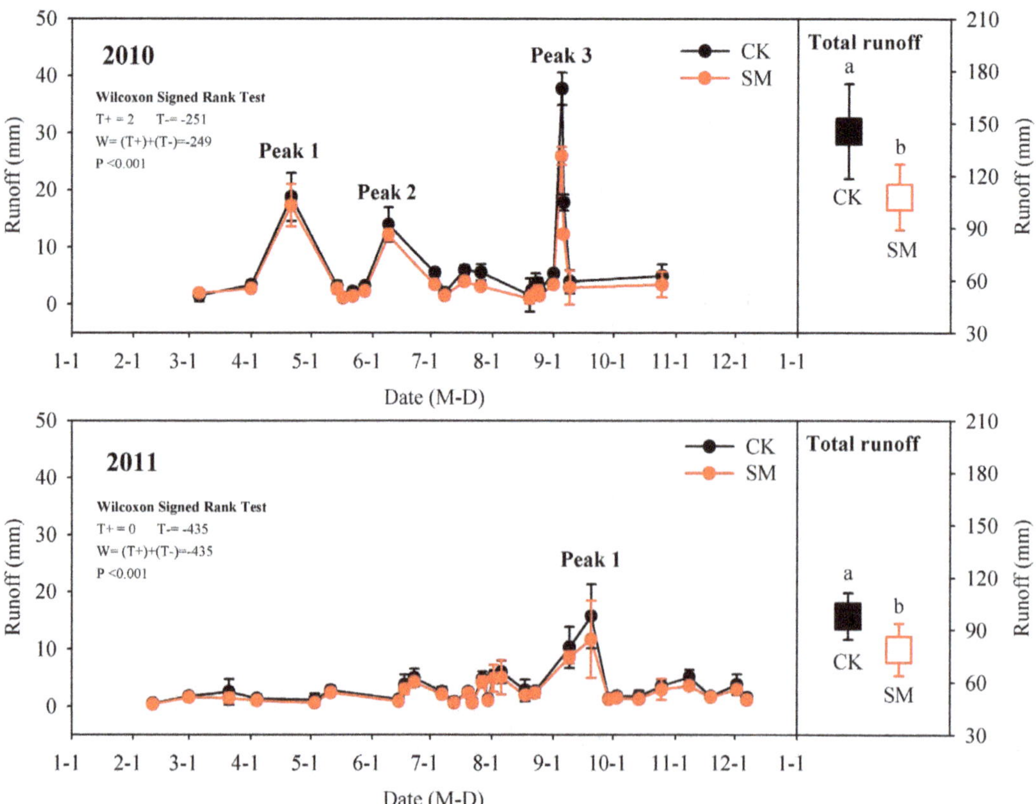

Figure 4. Seasonal variation in and annual total runoff under control (CK) and straw mulching (SM) treatments during 2010 and 2011. Error bars are twice the standard error of the mean (n = 3). Statistically significant differences are given after Wilcoxon signed rank test; Notations a and b indicate statistical significance at P<0.05 between CK and SM.

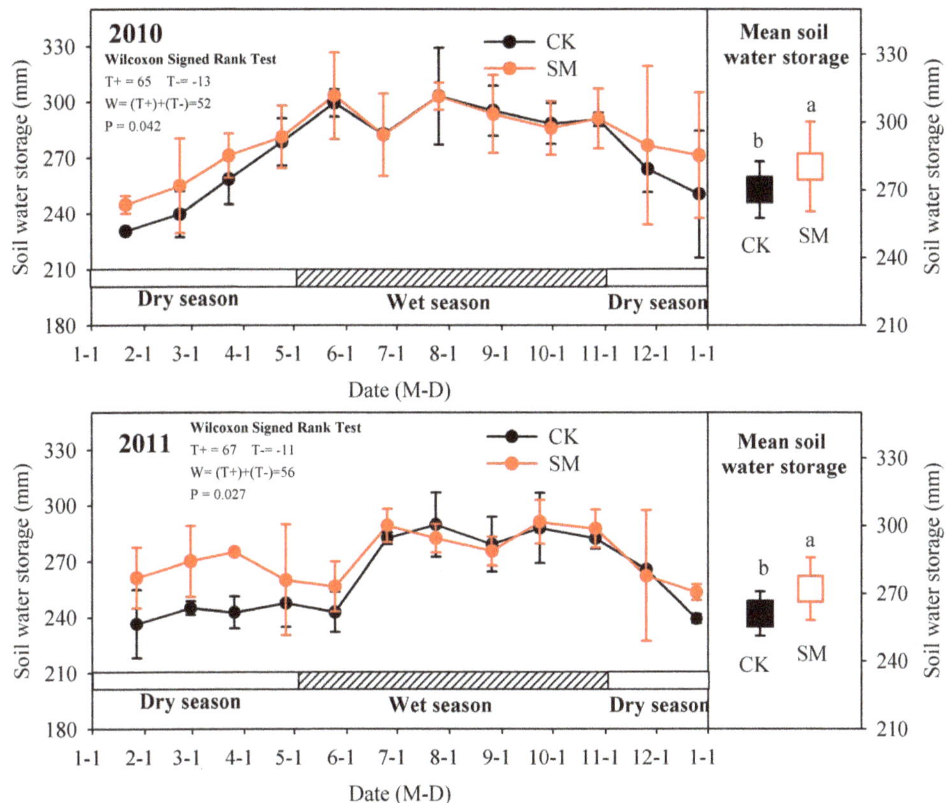

Figure 5. Seasonal variation in and mean soil water storage in the 0–100 cm soil profile under control (CK) and straw mulching (SM) treatments during 2010 and 2011. Error bars are twice the standard error of the mean (n = 3). Statistically significant differences are given after Wilcoxon signed rank test; Notations a and b indicate statistical significance at P<0.05 between CK and SM.

when soil under the SM treatment had about 10% (ranged from 5 to 13%) greater water storage than under the CK treatment. During the rainy season in the Danjiangkou reservoir area (July to September), citrus plants take up a great deal of water to maintain their luxuriant growth; the variation in seasonal soil water storage was therefore mainly affected by the amount of precipitation and citrus growth. As a result, no significant difference in soil water storage was observed between CK and SM treatments during the wet season.

The vertical distribution of soil water in a profile under both dry (26th February 2010 and 26th March 2011) and wet (27th August 2010 and 26th August 2011) seasons is shown in Fig. 6. Soil water distribution within the profile results from the combined effects of precipitation amount and movement of soil water. The soil water content was significantly higher in the SM treatment than in the CK treatment in the 0–40 cm soil layer during the dry season, indicating that SM reduced evaporation during the dry period because of the increased surface residue cover and/or the lack of

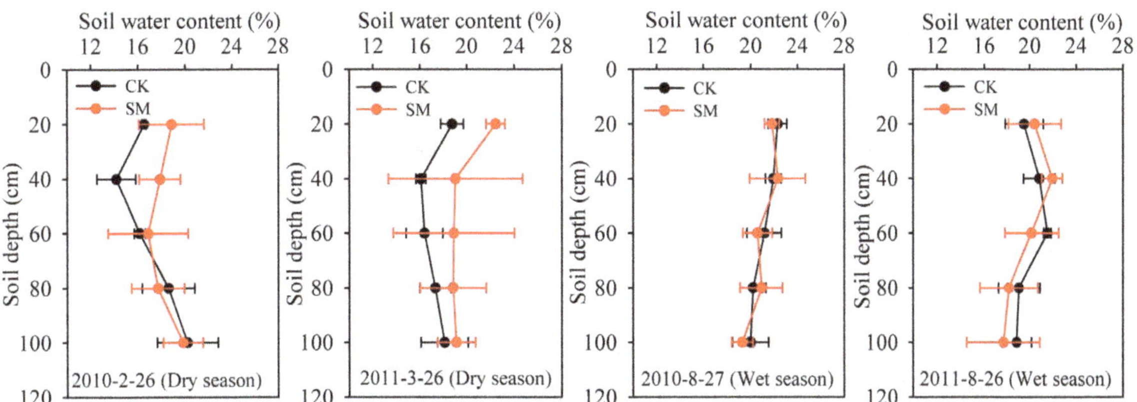

Figure 6. Differences in soil water content down the profile (0–100 cm) in dry (26th February 2010 and 26th March 2010) and wet (27th August 2010 and 26th August 2011) seasons between control (CK) and straw mulching (SM) treatment plots. Error bars are twice the standard error of the mean (n = 3).

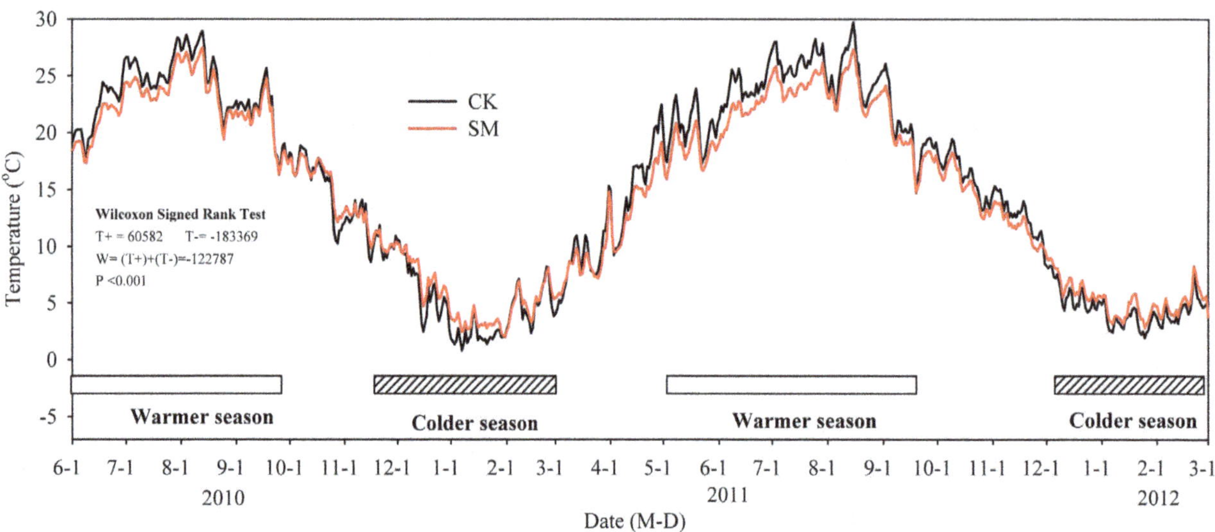

Figure 7. Seasonal variations in soil temperature under control (CK) and straw mulching (SM) treatment plots from June 2010 to March 2012. Statistically significant differences are given after Wilcoxon signed rank test; a P value of <0.05 was considered statistically significant.

soil disturbance. However, there was no significant difference in the soil water content in all soil layers between the SM and CK treatments during the wet season.

Soil temperature

The presence of SM altered soil temperature. Fig.7 shows the seasonal variations in soil temperature under CK and SM treatments. Monthly mean soil temperature as well as P values of WSRT are provided as Table 1. Our study demonstrated that soil temperatures under SM plots were higher during the colder seasons and lower during the warmer seasons when compared with the soil temperatures under CK plots. During the warmer period, reductions in soil temperature under the SM treatment were as high as 1.5°C (ranged from 0.3 to 2.9°C) as compared to the CK treatment (see Table S6). Because straw, that covered the soil surface, has a higher albedo and lower thermal conductivity than the bare soil, it helps to reduce the solar energy reaching the soil and, as a result, reduces temperature increases during warm conditions. Conversely, during the colder seasons, the presence of SM on the soil surface insulates the soil from the colder air temperatures. Therefore, heat loss from the soil is somewhat lower and soil temperatures are consequently higher under SM than under CK.

Four sets of typical diurnal trends (for summer, autumn, winter and spring) of CA and soil temperatures under both SM and CK conditions are presented in Fig. 8. In spring and autumn, when diurnal temperature range was large, soil temperature under mulch was lower during daytime, but higher at night. As daily radiation increased in summer, soil temperature under SM was always lower than that without mulch (CK). However, soil temperature was always higher in SM than in CK plots when the air temperature reached its minimum value in winter.

Table 1. Monthly mean soil temperature under the control (CK) and straw mulching (SM) treatments from June 2010 to March 2012.

Month-year	Soil temperature (°C)		WSRTP	Month-year	Soil temperature (°C)		WSRTP
	CK	SM			CK	SM	
Jun-2010	21.86±0.41	20.59±0.34	< 0.001	May-2011	20.33±0.32	18.55±0.24	< 0.001
Jul-2010	25.48±0.23	24.18±0.20	< 0.001	Jun-2011	23.95±0.24	22.02±0.24	< 0.001
Aug-2010	25.37±0.48	24.30±0.42	< 0.001	Jul-2011	26.32±0.19	24.40±0.16	< 0.001
Sep-2010	21.45±0.45	20.85±0.41	< 0.001	Aug-2011	25.10±0.37	23.82±0.30	< 0.001
Oct-2010	15.85±0.46	16.17±0.33	0.090	Sep-2011	20.08±0.47	19.04±0.41	< 0.001
Nov-2010	11.28±0.31	11.54±0.26	0.029	Oct-2011	16.32±0.33	15.42±0.31	< 0.001
Dec-2010	6.33±0.43	7.48±0.32	< 0.001	Nov-2011	12.75±0.30	11.82±0.25	< 0.001
Jan-2011	2.10±0.13	3.18±0.11	< 0.001	Dec-2011	5.80±0.24	6.69±0.23	< 0.001
Feb-2011	5.00±0.30	5.32±0.27	0.012	Jan-2012	3.47±0.16	4.27±0.16	< 0.001
Mar-2011	8.61±0.45	8.20±0.33	0.016	Feb-2012	4.41±0.20	5.14±0.21	< 0.001
Apr-2011	15.54±0.68	13.99±0.50	< 0.001	Mar-2012	8.05±0.57	6.37±0.52	< 0.001

Values are given as means ± standard error of means. WSRT: Wilcoxon Signed Rank Test, a P value of <0.05 was considered statistically significant.

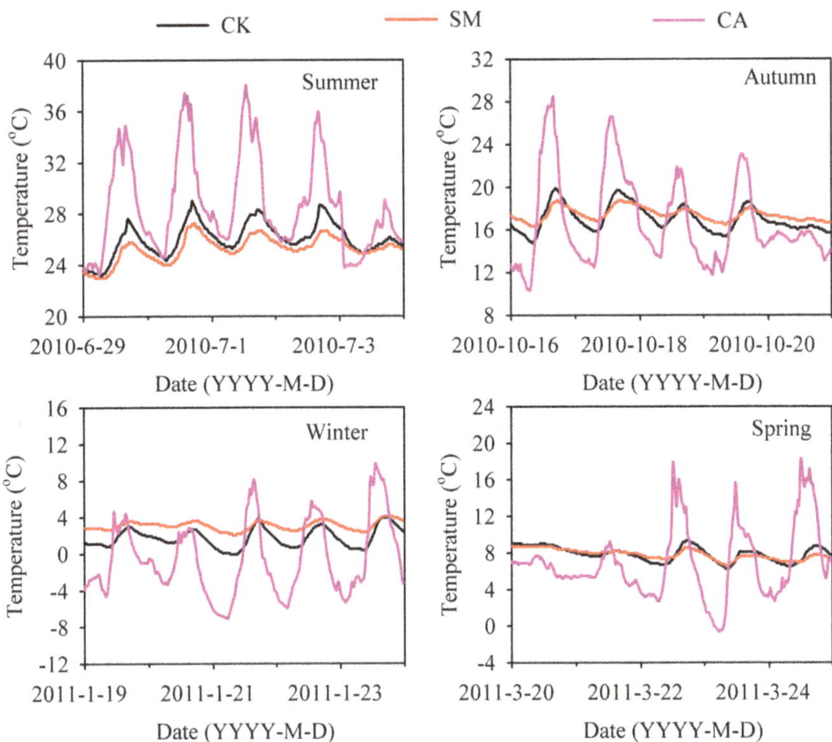

Figure 8. Typical diurnal air temperature trends for Summer, Autumn, Winter and Spring seasons within the canopy (CA) and soil temperatures under control (CK) and straw mulching (SM) treatment plots.

Fruit yield

Differences in citrus fruit yields were recorded in the SM and CK plots in different years (Table 2). There was no significant difference in fruit yield and fruit quality between CK and SM treatments in 2010, while in 2011, this difference in fruit yield was significant. Interestingly, the citrus yield was completely destroyed (zero fruit yield) in the CK treatment plot in 2011 due to extremely low temperatures during the citrus overwintering stage. According to visual observation, almost 50 to 60% of the leaves of citrus trees died or etiolated as a result of a cold and dry weather conditions that prevailed. In SM plots, despite the large reduction in fruit production, some fruit yield was recorded. The citrus fruit yield was 10.3±9.3 t hm^{-2} for SM in 2011.

Discussion

Field observations indicated that the SM treatment significantly decreased the annual runoff as compared to the CK treatment. For example, Adekalu, et al. [13] found that elephant grass (*Pennisetum purpureum*) can be a good alternative for rice straw to effectively reduce runoff and increase infiltration on sloping lands in southwestern Nigeria. Lal [26] found mulching tilled soil with 4–6 t hm^{-2} of rice straw to be effective in reducing soil loss and runoff on slopes ranging from 1% to 15%. Earlier studies have extensively discussed how soil surface mulching reduces the runoff by buffering the ground from raindrop action [21,27] and by modifying soil physical properties through the addition of litter and organic soil matter [28]. Furthermore, the absorption capability of straw provides additional pathways for water infiltration [21]. This trapped water in the straw is gradually released over several days, resulting in decreasing the velocity of surface flow and increasing infiltration. In general, surface runoff decreases with an increase of SM [13]. During the rainy season, a straw cover of 50% is necessary to significantly reduce runoff [29]. Results from the present study also provide indirect support for this conclusion, since the straw cover reached 100% in the SM plots, which was effective in controlling runoff water loss.

Table 2. Yield and fruit quality under the control (CK) and straw mulching (SM) treatments in 2010 and 2011.

Treatment	Yield (t hm^{-2})		Single fruit weight (g)		Equatorial diameter (mm)		Juice content (%)	
	2010	2011	2010	2011	2010	2011	2010	2011
CK	39.5±6.7 a	- -	172±32 a	- -	72.3±12.5 a	- -	42.5±4.0 a	- -
SM	43.2±5.6 a	10.3±9.3	168±7 a	179±42	70.8±3.8 a	76.0±10.9	41.2±5.1 a	40.9±3.6

Values are given as means ± standard error of means (n = 3). Values followed by different letters within a column are significantly different ($P<0.05$). Please note that the citrus crop was completely destroyed (zero citrus fruit yield) in the CK treatment plot in 2011 due to extremely low temperatures during the citrus overwintering stage.

Reduced runoff means an improvement in the soil water status in the root zone and a reduction in soil loss, which in turn leads to reduced land degradation and crop water stress [30]. Moreover, due to the evaporation reducing property of the surface-placed straw layer, mulching (SM treatment) increased soil water storage on an average by 10 mm as compared to the CK treatment (Fig 5). The effect was particularly pronounced during dry periods, when no rain occurred. Similar results have also been reported earlier [13]. Adekalu, et al. [13] stated that the large pores of crop residues permit rapid infiltration of water into the soil but retard evaporation. Water moves back to the atmosphere across the straw pores almost entirely in the vapor phase. The straw tends to act as a one-way water valve for the soil, thus water remains in the soil longer and benefits growing plants. Consequently, more soil water content in the SM treatment plot mainly resulted from higher infiltration, lesser runoff, and lower evaporation than in the CK treatment plot.

Under SM, soil temperature has been reported to have increased [31] as well as decreased [18]. And this can be mainly attributed to differences in climatic conditions. This effect can be explained with two basic mechanisms as observed in our field experiment. In the SM treatment, the mulch layer reduced soil radiation absorption during daytime, while at nighttime it reduced the outgoing heat radiation from the soil. Moreover, the mulch layer contains a significant amount of pore space. The majority of this pore space is likely to be filled with air, and air is known to be a very good insulator. The air space in the mulched layer prevents energy conduction. Therefore, in our study the SM treatment had lower thermal conductivity than the non-mulch control (CK), and acted as an insulator during the warmer period and helped to retain soil heat during the colder period, resulting in smaller fluctuations in soil temperatures (Fig. 7 and 8). Chen, et al. [12] and Olasantan [15] also observed similar results.

Previous research has shown that SM is an effective method to improve crop yield and soil water utilization [32]. Li, et al. [32] found that SM increased soil water content and maize (*Zea mays L.*) yields. Kar and Kumar [33] reported that potato tuber yield was higher in the SM treatments than with bare flat planting in eastern India. However, application of SM is restricted in some places because it is liable to lower the soil surface temperature and lead to a reduction in yield [34,35]. As pointed out by Doring, et al. [34], higher yields under mulch have mostly been attributed to increase soil water under arid and semiarid conditions [17,35], while reduced yields under SM have also been reported and have been attributed to below-optimum soil temperature, reduced soil nitrate levels, and mulching too early [35]. In our study, the results indicate that the effect of SM on citrus fruit yield can be positive in extreme weather conditions. This was possibly because SM reduced the outgoing heat radiation from the soil and, thus, increased the soil temperature compared to the no mulching or CK treatment. Higher soil water content during the dry season in the SM treatment may also have attributed to some fruit yield vis-à-vis the complete fruiting failure in the CK treatment.

Conclusions

From the comparison of runoff, soil water and temperature under both SM and CK conditions we conclude: (1) the surface runoff from the sloping citrus orchards were lower when the soil was mulching with straw than when it was unmulching. (2) Soil water storage in the top 100 cm of the soil profile was increased in the SM as compared to the CK treatment. However, this result was significant only in the dry season (Jan to Mar), and not in the wet season (Jul to Sep) for both years. (3) The mulch probably acted as an insulator, resulting in smaller fluctuations in soil temperature in the SM than in the CK treatment. The results of our study suggested that the small effects on soil water and temperature changes created by surface mulch had limited impact on citrus fruit yield in a normal year (e.g., in 2010). However, SM practices can positively impact citrus fruit yield in extreme weather conditions.

Supporting Information

Table S1 Soil bulk density at start of the field experiment in 2009.

Table S2 Field data and summary statistics of canopy air temperature for 2010 and 2011 at Xiaofuling experimental station.

Table S3 Field data and summary statistics of rainfall for 2010 and 2011 at Xiaofuling experimental station.

Table S4 Field data and summary statistics of runoff under control (CK) and straw mulching (SM) treatments during 2010 and 2011.

Table S5 Field data and summary statistics of soil water storage in the 0–100 cm soil profile under control (CK) and straw mulching (SM) treatments during 2010 and 2011.

Table S6 Field data and summary statistics of soil temperature under control (CK) and straw mulching (SM) treatment plots from May 2010 to March 2012.

Author Contributions

Conceived and designed the experiments: YL FC. Performed the experiments: JW ZGL. Analyzed the data: DBL GSZ JFP. Contributed reagents/materials/analysis tools: DBL YT JX. Wrote the paper: YL.

References

1. van der Velde M, Tubiello FN, Vrieling A, Bouraoui F (2012) Impacts of extreme weather on wheat and maize in France: evaluating regional crop simulations against observed data. Climatic Change 113: 751–765.
2. Porter JR, Semenov MA (2005) Crop responses to climatic variation. Philosophical Transactions of the Royal Society B-Biological Sciences 360: 2021–2035.
3. Farkas C, Birkas M, Varallyay G (2009) Soil tillage systems to reduce the harmful effect of extreme weather and hydrological situations. Biologia 64: 624–628.
4. Carrer M, Motta R, Nola P (2012) Significant Mean and Extreme Climate Sensitivity of Norway Spruce and Silver Fir at Mid-Elevation Mesic Sites in the Alps. Plos One 7: e50755.
5. Liu Y, Yang SJ, Li SQ, Chen XP, Chen F (2010) Growth and development of maize (*Zea mays L.*) in response to different field water management practices: Resource capture and use efficiency. Agricultural and Forest Meteorology 150: 606–613.
6. Silvente S, Sobolev AP, Lara M (2012) Metabolite Adjustments in Drought Tolerant and Sensitive Soybean Genotypes in Response to Water Stress. Plos One 7: e38554.

7. Liu Y, Li SQ, Chen F, Yang SJ, Chen XP (2010) Soil water dynamics and water use efficiency in spring maize (*Zea mays L.*) fields subjected to different water management practices on the Loess Plateau, China. Agricultural Water Management 97: 769–775.
8. Zhao H, Xiong YC, Li FM, Wang RY, Qiang SC, et al. (2012) Plastic film mulch for half growing-season maximized WUE and yield of potato via moisture-temperature improvement in a semi-arid.
9. Lou YL, Liang WJ, Xu MG, He XH, Wang YD, et al. (2011) Straw coverage alleviates seasonal variability of the topsoil microbial biomass and activity. Catena 86: 117–120.
10. Wang YJ, Xie ZK, Malhi SS, Vera CL, Zhang YB, et al. (2011) Effects of gravel-sand mulch, plastic mulch and ridge and furrow rainfall harvesting system combinations on water use efficiency, soil temperature and watermelon yield in a semi-arid Loess Plateau of northwestern China. Agricultural Water Management 101: 88–92.
11. Novak MD, Chen WJ, Orchansky AL, Ketler R (2000) Turbulent exchange processes within and above a straw mulch. Part II: Thermal and moisture regimes. Agricultural and Forest Meteorology 102: 155–171.
12. Chen SY, Zhang XY, Pei D, Sun HY, Chen SL (2007) Effects of straw mulching on soil temperature, evaporation and yield of winter wheat: field experiments on the North China Plain. Annals of Applied Biology 150: 261–268.
13. Adekalu KO, Olorunfemi IA, Osunbitan JA (2007) Grass mulching effect on infiltration, surface runoff and soil loss of three agricultural soils in Nigeria. Bioresource Technology 98: 912–917.
14. Balwinder-Singh, Humphreys E, Eberbach PL, Katupitiya A, Yadvinder-Singh, et al. (2011) Growth, yield and water productivity of zero till wheat as affected by rice straw mulch and irrigation schedule. Field Crops Research 121: 209–225.
15. Olasantan FO (1999) Effect of time of mulching on soil temperature and moisture regime and emergence, growth and yield of white yam in western Nigeria. Soil & Tillage Research 50: 215–221.
16. Huang YL, Chen LD, Fu BJ, Huang ZL, Gong E (2005) The wheat yields and water-use efficiency in the Loess Plateau: straw mulch and irrigation effects. Agricultural Water Management 72: 209–222.
17. Zhang SL, Lovdahl L, Grip H, Tong YN, Yang XY, et al. (2009) Effects of mulching and catch cropping on soil temperature, soil moisture and wheat yield on the Loess Plateau of China. Soil & Tillage Research 102: 78–86.
18. Sarkar S, Paramanick M, Goswami SB (2007) Soil temperature, water use and yield of yellow sarson (*Brassica napus* L. var. glauca) in relation to tillage intensity and mulch management under rainfed lowland ecosystem in eastern India. Soil & Tillage Research 93: 94–101.
19. Ghosh PK, Dayal D, Bandyopadhyay KK, Mohanty K (2006) Evaluation of straw and polythene mulch for enhancing productivity of irrigated summer groundnut. Field Crops Research 99: 76–86
20. Zhang QF (2009) The South-to-North Water Transfer Project of China: Environmental Implications and Monitoring Strategy1. Journal of the American Water Resources Association 45: 1238–1247.
21. Li XH, Zhang ZY, Yang J, Zhang GH, Wang B (2011) Effects of Bahia Grass Cover and Mulch on Runoff and Sediment Yield of Sloping Red Soil in Southern China. Pedosphere 21: 238–243.
22. Liu WZ, Liu GH, Zhang QF (2011) Influence of Vegetation Characteristics on Soil Denitrification in Shoreline Wetlands of the Danjiangkou Reservoir in China. Clean-Soil Air Water 39: 109–115.
23. Liu Y, Tao Y, Wan KY, Zhang GS, Liu DB, et al. (2012) Runoff and nutrient losses in citrus orchards on sloping land subjected to different surface mulching practices in the Danjiangkou Reservoir area of China. Agricultural Water Management 110: 34–40.
24. National Soil Survey Office (1992) Soil survey technique in China. Beijing: Agricultural Press (in Chinese).
25. Bao S.D (2000) Agro-chemical Analysis of Soil. Beijing: Agricultural Press (in Chinese).
26. Lal R (1979) Soil erosion on alfisols in Western Nigeria II: Effect of mulch rates. Geoderma 16: 377–382.
27. Wakindiki IIC, Danga BO (2011) Effect of straw mulch application on nutrient concentration in runoff and sediment in a humid region in Kenya. African Journal of Agricultural Research 6: 725–731.
28. She DL, Shao MA, Timm LC, Reichardt K (2009) Temporal changes of an alfalfa succession and related soil physical properties on the Loess Plateau, China. Pesquisa Agropecuaria Brasileira 44: 189–196.
29. Gutierrez J, Hernandez II (1996) Runoff and interrill erosion as affected by grass cover in a semi-arid rangeland of northern Mexico. Journal of Arid Environments 34: 287–295.
30. Fang Y, Xun F, Bai WM, Zhang WH, Li LH (2012) Long-Term Nitrogen Addition Leads to Loss of Species Richness Due to Litter Accumulation and Soil Acidification in a Temperate Steppe. Plos One 7: e47369.
31. Ramakrishna A, Tam HM, Wani SP, Long TD (2006) Effect of mulch on soil temperature, moisture, weed infestation and yield of groundnut in northern Vietnam. Field Crops Research 95: 115–125.
32. Li R, Hou XQ, Jia ZK, Han QF, Yang BP (2012) Effects of rainfall harvesting and mulching technologies on soil water, temperature, and maize yield in Loess Plateau region of China. Soil Research 50: 105–113.
33. Kar G, Kumar A (2007) Effects of irrigation and straw mulch on water use and tuber yield of potato in eastern India. Agricultural Water Management 94: 109–116.
34. Doring TF, Brandt M, Hess J, Finckh MR, Saucke H (2005) Effects of straw mulch on soil nitrate dynamics, weeds, yield and soil erosion in organically grown potatoes. Field Crops Research 94: 238–249.
35. Edwards L, Burney JR, Richter G, MacRae AH (2000) Evaluation of compost and straw mulching on soil-loss characteristics in erosion plots of potatoes in Prince Edward Island, Canada. Agriculture Ecosystems & Environment 81: 217–222.

5

Effects of Different Tillage and Straw Return on Soil Organic Carbon in a Rice-Wheat Rotation System

Liqun Zhu[1]*, Naijuan Hu[1], Minfang Yang[2], Xinhua Zhan[2], Zhengwen Zhang[1]

1 College of Agriculture, Nanjing Agricultural University, Nanjing, China, 2 College of Resources and Environmental Science, Nanjing Agricultural University, Nanjing, China

Abstract

Soil management practices, such as tillage method or straw return, could alter soil organic carbon (C) contents. However, the effects of tillage method or straw return on soil organic C (SOC) have showed inconsistent results in different soil/climate/cropping systems. The Yangtze River Delta of China is the main production region of rice and wheat, and rice-wheat rotation is the most important cropping system in this region. However, few studies in this region have been conducted to assess the effects of different tillage methods combined with straw return on soil labile C fractions in the rice-wheat rotation system. In this study, a field experiment was used to evaluate the effects of different tillage methods, straw return and their interaction on soil total organic C (TOC) and labile organic C fractions at three soil depths (0–7, 7–14 and 14–21 cm) for a rice-wheat rotation in Yangzhong of the Yangtze River Delta of China. Soil TOC, easily oxidizable C (EOC), dissolved organic C (DOC) and microbial biomass C (MBC) contents were measured in this study. Soil TOC and labile organic C fractions contents were significantly affected by straw returns, and were higher under straw return treatments than non-straw return at three depths. At 0–7 cm depth, soil MBC was significantly higher under plowing tillage than rotary tillage, but EOC was just opposite. Rotary tillage had significantly higher soil TOC than plowing tillage at 7–14 cm depth. However, at 14–21 cm depth, TOC, DOC and MBC were significantly higher under plowing tillage than rotary tillage except for EOC. Consequently, under short-term condition, rice and wheat straw both return in rice-wheat rotation system could increase SOC content and improve soil quality in the Yangtze River Delta.

Editor: Shuijin Hu, North Carolina State University, United States of America

Funding: This work was supported by Special Fund for Agro-scientific Research in the Public Interest (201103001). The funders had no role in study design, data collection and analysis, decision to publish, or preparation of the manuscript.

Competing Interests: The authors have declared that no competing interests exist.

* E-mail: zhulq@njau.edu.cn

Introduction

Soil organic carbon (C) has profound effects on soil physical, chemical and biological properties [1]. Maintenance of soil organic C (SOC) in cropland is important, not only for improvement of agricultural productivity but also for reduction in C emission [2]. However, short- and medium-term changes of SOC are difficult to detect because of its high temporal and spatial variability [3]. On the contrary, soil labile organic C fractions (i.e., microbial biomass C (MBC), dissolved organic C (DOC), and easily oxidizable C (EOC)) that turn over quickly can respond to soil disturbance more rapidly than total organic C (TOC) [1,3,4]. Therefore, these fractions have been suggested as early sensitive indicators of the effects of land use change on soil quality (e.g. [3,5,6]).

Agricultural practices such as tillage methods are conventionally used for loosening soils to grow crops. But long-term soil disturbance by tillage is believed to be one of the major factors reducing SOC in agriculture [7]. Frequent tillage may destroy soil organic matter (SOM) [8] and speed up the movement of SOM to deep soil layers [9]. As a consequence, agricultural practices that reduce soil degradation are essential to improve soil quality and agricultural sustainability. Crop residue plays an important role in SOC sequestration, increasing crop yield, improving soil organic matter, and reducing the greenhouse gas (e.g. [10–13]). As an important agricultural practice, straw return is often implemented with tillage in the production process. Although numerous studies have indicated that tillage methods combined with straw return had a significant effect on labile SOC fractions, the results varied under different soil/climate conditions. For example, both no-tillage and shallow tillage with residue cover had significantly higher SOC than conventional tillage without residue cover in Loess Plateau of China [14], while Wang et al. [15] reported that the difference between the treatments of plowing with straw return and no-tillage with straw return on TOC in central China was not significant. Rajan et al. [2] showed that in Chitwan Valley of Nepal, no-tillage with crop residue application at upper soil depth had distinctly higher SOC sequestration than conventional tillage with crop residue. The effects of tillage on soil labile organic C vary with regional climate [16], soil condition (e.g. [17–20]), residue management practice, and crop rotation (e.g. [21,22]). Therefore, the investigation on soil labile organic C for specific soil, climate, and cropping system is necessary to improve the soil quality.

The Yangtze River Delta of China is the main production region of rice and wheat, and rice-wheat rotation is the most important cropping system in this region [23]. The total sown area of rice and wheat in the Yangtze River Delta accounted for about 20.1% of that in China in 2011, and the total yield was 22.1% of the national yield for these two crops [24]. Many field experiments in this region (e.g. [25–27]) about the effects of tillage methods

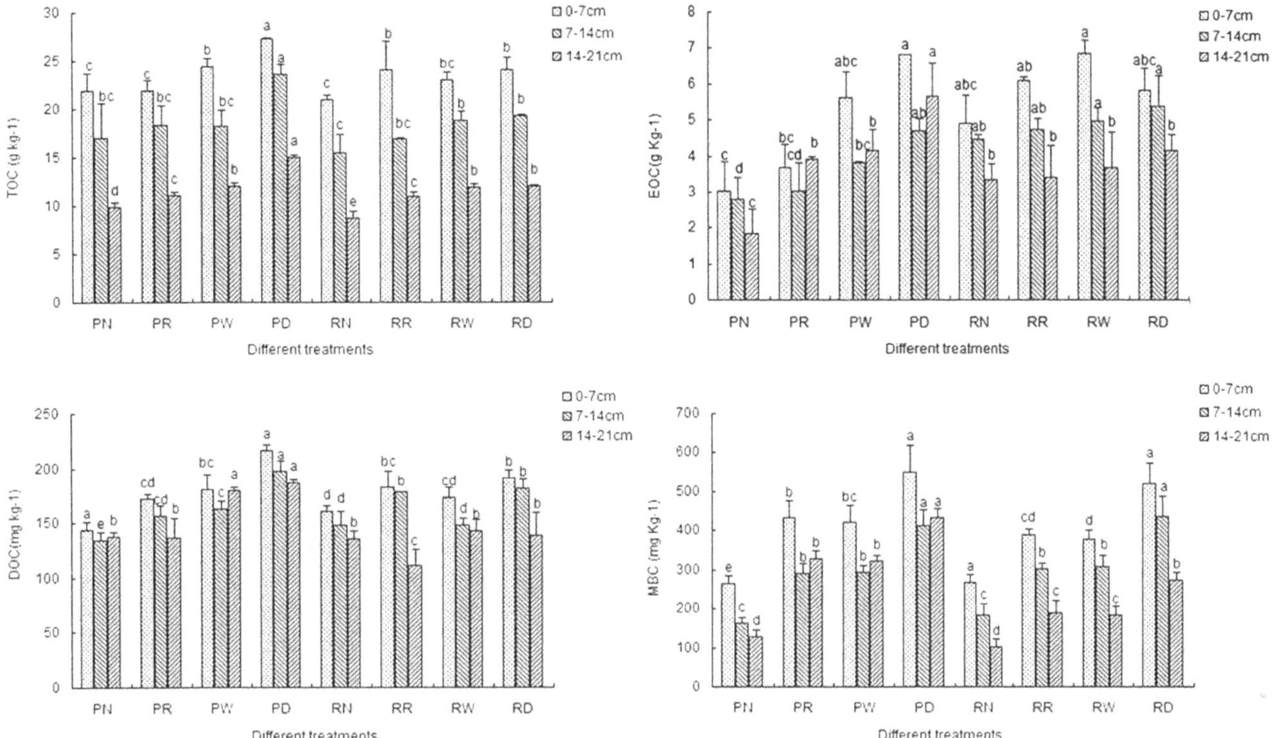

Figure 1. Effects of eight treatments on soil TOC, EOC, DOC and MBC contents at three depths.

combined with straw return on cropland ecosystem in rice-wheat rotation system have been studied during these years. However, most of them are focused on soil physical-chemical properties, soil nutrient and crop yield. To our knowledge, there is little information about the effects of different tillage and straw return on soil labile C fractions in rice-wheat rotation system. Thus, the objectives of this study were (1) to quantify the effects of tillage methods and straw return on soil TOC, MBC, DOC, EOC contents in the rice-wheat rotation system in the Yangtze River Delta, and (2) to explore an optimal management practice combination of tillage and straw return for improving the soil quality and increasing the local crop production.

Table 1. Linear correlations among soil TOC and labile organic C fractions at the 0–21 cm depth.

Index	TOC	EOC	DOC	MBC
TOC	1			
EOC	0.638**	1		
DOC	0.758**	0.684**	1	
MBC	0.741**	0.639**	0.908**	1

TOC: total organic carbon; MBC: microbial biomass carbon; DOC: dissolved organic carbon; EOC: easily oxidizable carbon.
* $P<0.05$.
**$P<0.01$.

Materials and Methods

Site Description

The experiment was conducted at Changwang Country, Youfang Town, Yangzhong City, Jiangsu Province, China (119°42′–119°58′E, 32°–2°19′N, 4–4.5 m above mean sea level) from November, 2009 to June, 2011. Access to the study site was obtained in the form of a rent contract, in which we had to confirm that our study did not involve endangered or protected species.

The experimental site had a subtropical monsoon climate with an average annual precipitation of 1000 mm, an average annual temperature of 15.1°C, and a mean annual sunshine hour of 2135 h. The soil of the experimental site was a loam and classified as an anthrosols. Rice-wheat double cropping system was the most important cropping system in the region. The main properties of soil (0–20 cm depth) sampled in November 2009 were as follows: soil organic matter 29.81 g kg^{-1}; alkali-hydrolyzale nitrogen 194.02 mg kg^{-1}; available phosphorus 13.60 mg kg^{-1}; available potassium 51.45 mg kg^{-1}; and pH 7.34.

The variety of wheat used in this study was Yangmai16 *(Triticum aestivum* L.*)* and rice was Nangeng47 (*Oryza sativa* L.).

Experimental Design and Field Managements

The experiment had a split-plot design with two tillage methods in the main plots and four straw return modes in subplots with three replications (6 m×5 m). Tillage methods included plowing tillage (P) and rotary tillage (R). Straw return modes were as follows: no straw return (N), only rice straw return (R), only wheat straw return (W), and rice and wheat straw both return (D). There were eight treatments in this study: (1) plowing tillage with no straw return (PN: rice with plowing tillage-wheat with plowing tillage); (2) plowing tillage with only rice straw return (PR: rice with

Table 2. Effects of different tillage factor and straw factor on soil TOC, EOC, DOC and MBC at 0–7 cm, 7–14 cm and 14–21 cm depth.

Soil dept(cm)	Factors	Treatments	TOC(g kg^{-1})	labile organic carbon fractions contents		
				EOC (g kg^{-1})	DOC(mg kg^{-1})	MBC(mg kg^{-1})
0–7	Tillage factor	P	23.87±2.67a	4.78±1.23a	178.36±33.74a	417.16±17.20a
		R	23.00±2.38a	5.92±1.48a	177.05±30.96a	389.22±36.56b
	Straw factor	N	21.40±1.87c	3.97±0.92b	152.25±10.76c	266.73±11.68c
		R	23.01±2.97bc	4.88±1.40ab	177.61±33.57b	411.67±27.32b
		W	23.65±1.52b	6.23±1.56a	176.92±30.87b	398.57±27.34b
		D	25.68±2.63a	6.32±1.68a	204.04±22.67a	535.79±20.31a
7–14	Tillage factor	P	19.30±1.23a	3.58±1.76b	163.05±37.30a	311.42±43.89a
		R	17.64±2.48b	4.89±1.52a	164.11±27.23a	306.50±36.09a
	Straw factor	N	16.28±3.20c	3.62±1.76c	141.49±18.49d	172.63±16.58c
		R	17.63±2.86bc	3.90±2.70bc	167.39±21.63b	315.19±20.00b
		W	18.50±1.75b	4.39±0.54ab	155.47±22.41c	313.93±14.31b
		D	21.47±2.02a	5.04±0.76a	189.98±17.71a	434.09±11.65a
14–21	Tillage factor	P	12.00±2.56a	3.89±1.42a	160.06±30.21a	280.68±58.67a
		R	10.90±2.61b	3.64±1.14a	132.00±25.31b	187.08±44.86b
	Straw factor	N	9.28±1.69d	2.58±1.18c	136.34±11.66b	114.99±18.05d
		R	10.99±1.96c	3.66±1.26b	123.67±29.32c	239.95±39.07b
		W	11.97±3.06b	3.91±1.39b	161.19±25.91a	238.92±35.06c
		D	13.56±1.75a	4.90±0.68a	162.93±26.49a	341.67±31.14a

Different letters in a line under a specific influence factor denote significant difference at the 5% level. Different capitals in a column at different soil depths and treatments present significant different at the 0.05 level. TOC: total organic carbon; MBC: microbial biomass carbon; DOC: dissolved organic carbon; EOC: easily oxidizable carbon.

plowing tillage - wheat with plowing tillage+rice straw return); (3) plowing tillage with only wheat straw return (PW: rice with plowing tillage+wheat straw return - wheat with plowing tillage); (4) plowing tillage with rice and wheat straw both return(PD: rice with plowing tillage+wheat straw return -wheat with plowing tillage+rice straw return); (5) rotary tillage with no straw return (RN: rice with rotary tillage - wheat with rotary tillage); (6) rotary tillage with only rice straw return (RR: rice with rotary tillage - wheat with rotary tillage+rice straw return); (7) rotary tillage with only wheat straw return (RW: rice with rotary tillage+wheat straw return - wheat with rotary tillage); (8) rotary tillage with rice and wheat straw both return (RD: rice with rotary tillage+wheat straw return - wheat with rotary tillage+rice straw return).

The experimental site was cultivated with a rice-wheat rotation prior to November 2009, where wheat was planted with plowing tillage from November to the following June, and rice was transplanted by plowing tillage from June to November. In this study, after wheat or rice was harvested, they were cultivated at a depth of 10–15 cm by rotary cultivation in rotary tillage plots while for the plowing tillage plots, cultivation was at a depth of 20–25 cm with a moldboard plough. Before the rice and wheat were sown, the plowing tillage plots were disked and moldboard plowed for weed control and bedding. This was followed by an application of fertilizer. For straw returned plots, the wheat and rice straw were cut into 8–10 cm after air-dried, and placed back on the surface of the soil in June or November of each year, with returned amount of 6000 kg·hm^{-2} for both wheat and rice straw.

In this study, wheat was sown on November 3, 2009 and November 24, 2010, respectively. The seed quantity was 150 kg hm^{-1} by machine. The base fertilizer applied before sowing was 135 kg·hm^{-2} pure N, 67.5 kg·hm^{-2} P_2O_5, and 67.5 kg·hm^{-2} K_2O, and topdressing was at the elongation stage with 135 kg hm^{-1} pure N. For all treatments, N was applied in the form of $CO(NH_2)_2$, and the fertilizer in wheat seasons was applied at the same rate. The wheat was harvested on June 3, 2010 and June 8, 2011, respectively. Rice was transplanted at about 3–4 seedlings per hole, 255,000 holes per hectare on June 15, 2010 and June 15, 2011, respectively. The rice was fertilized just before transplanting with a base fertilizer (120 kg hm^{-1} pure N; 60 kg hm^{-1} P_2O_5; 60 kg hm^{-1} K_2O), and at tillering stage and earing stage with a topdressing (180 kg hm^{-1} N (3:1 ratio)). The fertilizer applied in the two rice seasons were the same. The rice was harvested on November 12, 2010 and November 29, 2011, respectively. The pesticide management of both rice and wheat seasons was in accordance with the conventional, and all other management procedures were identical for the eight treatments.

Soil Sampling and Analytical Methods

Soil samples were collected by a geotome (5 cm diameter) on October 29, 2011 (just before the rice was harvested). Five random locations were chosen in each of the 24 observational plots and samples were taken from each location at three soil depths (0–7, 7–14, and 14–21 cm) separately. Soil samples from each depth were about 200 g, fully blended. The collected moist samples were ground and sieved through a 10 mesh screen. Sieved soil samples were divided into two sub-samples. One was air-dried and sieved again through 100 mesh screen for determining soil TOC and EOC. Another was immediately stored in 4°C refrigerators for determining DOC and MBC. During sieving, crop residues, root material and stones were removed.

Table 3. Affecting force analysis of different tillage and straw return and their interaction on soil TOC, EOC, DOC and MBC at 0-7 cm, 7-14 cm and 14-21 cm depth.

Soil depth(cm)	Difference source	Affecting force(%)			
		TOC	EOC	DOC	MBC
0-7	Block	10.35	8.53	6.55	0.26
	Tillage	0.34	28.06**	0.64	2.06**
	Straw return	19.69*	24.79**	38.58**	95.95**
	Straw return × tillage	10.71	4.34	4.86	0.93
	Error	58.92	34.28	49.37	0.79
7-14	Block	19.97**	5.36	11.53	0.29
	Tillage	0.02	13.03*	0.85	0.07
	Straw return	21.41**	13.16	39.21**	97.83**
	Straw return × tillage	26.30**	3.89	11.12*	0.96
	Error	32.32	64.56	37.28	0.86
14-21	Block	10.22	1.61	16.66**	0.27
	Tillage	1.05	3.21	11.77**	23.81**
	Straw return	32.35**	25.06**	35.65**	70.17**
	Straw return × tillage	14.72*	13.95*	9.60*	4.93**
	Error	41.66	56.17	26.33	0.82

The affecting force of tillage = tillage variables (square)/total variables (sum of total squares) ×100%; the affecting force of straw = straw variables (square)/ total variable (sum of total squares) ×100%; the affecting force of the interaction = interaction variable (square)/total variables ((total squares of sum) ×100%. TOC: total organic carbon; MBC: microbial biomass carbon; DOC: dissolved organic carbon; EOC: easily oxidizable carbon.
* $P<0.05$.
** $P<0.01$.

Total organic C (TOC) concentration was determined by oxidation with potassium dichromate and titration with ferrous ammonium sulphate [28].

Dissolved organic C (DOC) was extracted from 10 g of moist soil with 1:2.5 ratio of soil to water at 25.8°C [29]. After shaking for 1 h and centrifuging for 10 min at 4500 r min^{-1}, the supernatant was filtered with a 0.45 mm membrane filter. The filtrate was measured by oxidation with potassium dichromate and titration with ferrous ammonium sulphate.

Microbial biomass C (MBC) was analyzed by the fumigation extraction method [30]. Each sample was weighed into two equivalent portions, one was fumigated for 24 h with ethanol-free chloroform and the other was the unfumigated control. Both fumigated and unfumigated soils were shaken for 1 h with 0.5 M K_2SO_4 (2:5 soil: extraction ratio), centrifuged and filtered.

Easily oxidizable C (EOC) was measured as described by Blair et al. [3]. Finely ground air-dried soil samples were reacted with 333 mmol L^{-1} $KMnO_4$ by shaking at 60 r min^{-1} for 1 h. The suspension was then centrifuged at 2000 r min^{-1} for 5 min. The supernatant was diluted and measured spectrophotometrically at 565 nm. All soil samples were analyzed in triplicate.

Data Analysis

The SPSS 16.0 analytical software package was used for all statistical analyses. A 2-factor analysis of variance (ANOVA) was employed for difference test among eight treatments at $P<0.05$, with separation of means by least significant difference (LSD). Correlation analysis were performed to determine correlations among soil labile organic C fractions in the 0–21 cm soil depth, and the significant probability levels of the results were given at the $P<0.05$ (*) and $P<0.01$ (**), respectively. Moreover, the affecting force analysis of tillage factor, straw factor and their interaction influence on labile organic C fractions was calculated based on the method of Leng [31]: the affecting force of tillage = tillage variables (square)/total variables (sum of total squares) × 100%; the affecting force of straw return = straw return variables (square)/total variable (sum of total squares) × 100%; the affecting force of interaction = interaction variable (square)/total variables ((total squares of sum) × 100%.

Results

Soil TOC, DOC, MBC and EOC Contents in Different Treatments

As shown in Fig. 1, the different treatments significantly affected the contents of soil TOC and labile organic C fractions, where PD generally had the highest contents of TOC, DOC, MBC and EOC at the three soil depths. Crop straw return treatments (PR, PW, PD, RR, RW, RD) had consistently higher amount of TOC and labile organic C fractions at the three soil depths than without crop straw return treatments (PN, RN). Moreover, PN had significantly lower TOC, DOC, MBC and EOC at 0–7 cm and 7–14 cm, and RN had the lowest TOC and MBC at 14–21 cm compared to other treatments (Fig. 1). Soil TOC and labile organic C fractions generally decreased with an increase in soil depth under all treatments. As expected, soil TOC and labile organic C fractions were significantly and positively correlated with each other (Table 1).

Effects of Different Tillage Methods on Soil TOC and Labile Organic C Fractions

Tillage had a significant effect on MBC at 0–7 cm soil depth, but seldom on soil TOC, DOC and EOC. Soil TOC, DOC and MBC contents were all higher under plowing tillage (P) than rotary tillage (R), while EOC was opposite at 0–7 cm soil depth (Table 2). At 7–14 cm, soil EOC under rotary tillage (R) was significantly higher than plowing tillage (P), but TOC had the contrary results, and there were no significant differences on DOC and MBC (Table 2). At 14–21 cm, soil TOC, DOC and MBC were significantly higher under plowing tillage (P) than rotary tillage (R), except EOC (Table 2).

Effects of Different Straw Return on Soil TOC and Labile Organic C Fractions

Straw return had significant effects on soil TOC and labile organic C at the three depths as shown in Table 2. In general, soil TOC and three labile organic C ranged in the following order: rice and wheat straw both return>only wheat or rice straw return>no straw return at three depths (Table 2). At 7–14 cm depth, only rice straw return in the wheat season had significantly higher DOC than only wheat straw return in the rice season (Table 2). However, at 14–21 cm depth, except for MBC, soil TOC, EOC, DOC under only rice straw return in the wheat season were lower than only wheat straw return in the rice season (Table 2). Moreover, there were significant differences in TOC and MBC among the four straw return at 14–21 cm depth (Table 2).

Affecting Force Analysis of Different Tillage, Straw Return and their Interaction on Soil TOC and Labile Organic C Fractions

Affecting force of different tillage, straw return and their interaction on soil TOC and labile organic C were different with increasing soil depth (Table 3). The affecting force of tillage increased with the increase of soil depth (Table 3). Tillage had significant affecting force on EOC and MBC at 0–7 cm depth, but seldom on TOC and DOC (Table 3). At 7–14 cm depth, the affecting force of tillage on EOC was lower than at 0–7 cm, and there was no significant affecting force on soil TOC and other labile organic C fractions (Table 3). At 14–21 cm depth, tillage had significant affecting force on DOC and MBC. However, there was no significant affecting force on TOC and EOC (Table 3).

Straw return had significant affecting force on soil TOC, DOC and MBC at the three depths, but there was no significant affecting force on EOC at 7–14 cm (Table 3). Among the four indictors, the affecting force of straw return on MBC was the greatest at the three depths, which reached 95.95%, 97.83% and 70.17%, respectively (Table 3).

The affecting force of the interaction generally increased with an increase in soil depth (Table 3). At 0–7 cm soil depth, the interaction had no significant affecting force on soil TOC and labile organic C (Table 3). Soil TOC and DOC were mainly dominated by the interaction at 7–14 cm depth, but there was no significant affecting force on MBC and EOC. At 14–21 cm depth, the interaction had significant affecting force on soil TOC and all labile organic C (Table 3).

Discussion

Suitable soil tillage practice can increase the SOC content, and improve SOC density of the plough layer [32]. The effect size of tillage methods on SOC dynamics depends on the tillage intensity [33]. Compared to conventional tillage (CT), no-tillage and reduced tillage could significantly improve the SOC content in cropland. Frequent tillage under CT easily exacerbate C-rich macroaggregates in soils broken down due to the increase of tillage intensity, then forming a large number of small aggregates with relatively low organic carbon content and free organic matter particles. Free organic matter particles have poor stability and are easy to degradation, thereby causing the loss of SOC [33,34]. In our study, at 0–7 cm soil depth, soil EOC under plowing tillage was lower than rotary tillage (Table 2). The reason could be attributed to the tillage method. Tillage increases the effect of drying–rewetting and freezing-thawing on soil, which increases macroaggregate susceptibility to disruption [21,35,36], and accelerates the labile organic C mineralization and SOM degradation, thus increasing the loss of EOC [14,37]. At 7–14 cm, rotary tillage had higher soil EOC and DOC than plowing tillage, but lower at 14–21 cm soil depth, indicating that tillage affected the vertical distribution of EOC and DOC (Table 2). The difference in soil condition after plowing tillage or rotary tillage affects the rate of straw decomposition, thereby resulting in a difference in the soil nutrient accumulation [38]. Similarly, Liu et al. [39] have found that SOM content under plowing and rotary tillage at deeper soil both were higher than that of the upper soil. The reason might be that rotary tillage and plowing tillage mixed crop straw into the deeper soil layer, making SOM well-distributed at different depths [40].

Carbon input can be increased by adopting straw return in cropland [14]. Fresh residues are C source for microbial activity and nucleation centers for aggregation when returned to cropland. The enhanced microbial activity induces the binding of residue and soil particles into macroaggregates [34,41], which could increase aggregates stability, fix the unstable C, thus improving the concentration of SOC [42] and increasing C sequestration [14]. In our research, straw return had significantly higher soil TOC and labile organic carbon fractions contents at the three soil depths than no crop straw return (Table 2). Soil TOC and labile organic C fractions in both rice and wheat straw return treatments were higher than only wheat or rice straw return (Table 2), indicating that straw return plays an very important role in increasing soil TOC and labile organic C fractions. Similar observations have been reported by other researchers [43–45]. At the three depths, soil TOC in the treatment of only wheat straw return in rice season was higher than only rice straw return in wheat season, moreover, the difference was significant ($p<0.05$) at 7–14 cm (Table 2). This was related to the relatively near-surface higher water content and favorable soil temperature during the rice growing season, resulting in relatively fast straw decomposition [46]. The decomposition of wheat straw provides enough energy and carbon source for soil microorganisms, thus increases the microorganisms' activities. Alternatively, after wheat straw return, the high temperature and humid conditions accelerates the reduction of the C/N ratio of the straw, allowing for sufficient decomposition. More nutrients are released and utilized by the crops, which therefore lightens the pressure of burning straw and improves the soil quality [47]. According to table 1, the study showed that MBC was affected by the straw return factor with an affecting force of 95.95% at 0–7 cm depth and 97.83% at 7–14 cm depth. The probable explanation maybe that crop residue might enter the labile C pool, provide substrate for the soil microorganisms, and contribute to the accumulation of labile C [48].

In our study, PD had the highest content of soil TOC at all the three soil depths (Fig. 1). The reason might be that plowing tillage made the soil and straw in the plow layer turned over quarterly, which increased the stability of the TOC content at each soil layer [49]. In addition, the rice and wheat straw were both returned under PD treatment from 2009, plowing tillage made much SOM enter into the soil and accumulate [45]. However, Tian et al. [50] found that rotary tillage with straw return had higher SOC than plowing tillage with straw return at 0–10 cm soil depth in wheat field. The diverse results might be due to the different regional climate, soil type, crop rotation and the length of study [22]. In this study, at upper soil layer, the interaction effect between tillage and straw return was not significant, but generally increased with an increase in soil depth (Table 3). Rajan et al [2] also found that single effect of residue application was not significant but its significance became apparent after its interaction with tillage system.

In our study, soil labile organic C fractions were significantly and positively correlated with TOC concentrations at 0–21 cm soil depth (Table 1). Such correlations suggested that TOC was a major determinant of soil labile organic C fractions. MBC, DOC and EOC were also significantly and positively correlated with each other in this study (Table 1). The results were consistent with Chen et al. [14], who reported similar correlations between soil TOC, labile organic C fractions (MBC, DOC, particulate organic C, EOC and hot-water extractable C), and macroaggregate C within 0–15 cm depth. Dou et al. [51] also observed the same results. MBC is the living part of SOM, which plays an important role in maintenance of soil fertility [52]. It serves as a sensitive indicator of change and future trends in organic matter level [53]. Dissolved organic C consists of organic compounds present in soil solution, acts as a substrate for microbial activity, and is the primary energy source for soil microorganisms [1]. Easily oxidizable C partly reflects enzymatic decomposition of labile SOC [54]. Therefore, it is not surprising to find the positive correlations among the labile C pools as they have a close association with each other.

Conclusions

In this study, after 2 years of a rice-wheat rotation, soil TOC and labile organic C fractions in PR, PW, PD, and RR, RW, RD were all higher than PN and RN. PD and RD had more significant effects on EOC, DOC and MBC compared to other treatments at 0–21 cm depth. Soil TOC and labile organic C fractions were highly correlated with each other. Under short-term conditions, rice and wheat straw both return in rice-wheat rotation system can increase SOC content and improve soil quality in the Yangtze River Delta, which is a suitable agricultural practice in this region under rice-wheat cropping system.

Acknowledgments

We thank Mr. Kejun Gu (Jiangsu Province Academy of Agricultural Sciences, China) for supporting the field experiment, Dr. Changqing Chen (Nanjing Agricultural University, China) for his kind help in data analysis,and Mr. Michael Rickaille (University of the West Indies) for his critical reading.

Author Contributions

Conceived and designed the experiments: LQZ. Performed the experiments: MFY NJH ZWZ. Analyzed the data: MFY. Contributed reagents/ materials/analysis tools: LQZ. Wrote the paper: LQZ NJH XHZ.

References

1. Haynes RJ (2005) Labile organic matter fractions as central components of the quality of agricultural soils: an overview. Adv. Agron 85: 221–268.
2. Rajan G, Keshav RA, Zueng-Sang C, Shree CS, Khem RD (2012) Soil organic carbon sequestration as affected by tillage, crop residue, and nitrogen application in rice–wheat rotation system. Paddy Water Environ 10: 95–102.
3. Blair GJ, Lefory RDB, Lise L (1995) Soil carbon fractions based on their degree of oxidation and the development of a carbon management index for agricultural system. Aust. J. Agric. Res 46: 1459–1466.
4. Ghani A, Dexter M, Perrott WK (2003) Hot-water extractable carbon in soils: a sensitive measurement for determining impacts of fertilization, grazing and cultivation. Soil Biol. Biochem 35: 1231–1243.
5. Rudrappa L, Purakayastha TJ, Singh D, Bhadraray S (2006) Long-term manuring and fertilization effects on soil organic carbon pools in a Typic Haplustept of semi-arid sub-tropical India. Soil Till. Res 88: 180–192.
6. Yang CM, Yang LZ, Zhu OY (2005) Organic carbon and its fractions in paddy soil as affected by different nutrient and water regimes. Geoderma 124: 133–142.
7. Baker JM, Ochsner TE, Venterea RT, Griffis TJ (2007) Tillage and soil carbon sequestration–what do we really know? Agric Ecosyst Environ 118, 1–5.
8. Hernanz JL, L'opez R, Navarrete L, S'anchez-Gir'on V (2002) Long-term effects of tillage systems and rotations on soil structural stability and organic carbon stratification in semiarid central Spain. Soil Till. Res 66 (2): 129–141.
9. Shan YH, Yang LZ, Yan TM, Wang JG (2005) Downward movement of phosphorus in paddy soil installed in large-scale monolith lysimeters. Agr. Ecosyst. Environ 111 (1–4): 270–278.
10. Zhang ZJ (1998) Effects of long-term wheat-straw returning on yield of crop and soil fertility. Chinese Journal of Soil Science 29(4): 154−155.
11. Sun X, Liu Q, Wang DJ, Zhang B (2007) Effect of long-term application of straw on soil fertility. Chinese Journal of Eco-Agriculture 16(3): 587−592.
12. West TO, Post WM (2002) Soil organic carbon sequestration rates by tillage and crop rotation. Soil Science Society of American Journal 66: 1930–1946.
13. Liu SP, Nie XT, Zhang HC, Dai QG, Huo ZY, et al. (2006) Effects of tillage and straw returning on soil fertility and grain yield in a wheat-rice double cropping system. Transactions of the CSAE 22(7): 48–51.
14. Chen HQ, Hou RX, Gong YS, Li HW, Fan MS, et al. (2009) Effects of 11 years of conservation tillage on soil organic matter fractions in wheat monoculture in Loess Plateau of China. Soil Till. Res 106: 85–94.
15. Wang DD, Zhou L, Huang SQ, Li CF, Cao CG (2013) Short-term Effects of Tillage Practices and Wheat-straw Returned to the Field on Topsoil Labile Organic Carbon Fractions and Yields in Central China. Journal of Agro-Environment Science 32(4): 735–740.
16. Miller AJ, Amundson R, Burke IC, Yonker C (2004) The effect of climate and cultivation on soil organic C and N. Biogeoche mistry 67: 57–72.
17. Diekow J, Mielniczuk J, Knicker H, Bayer C, Dick DP, et al. (2005) Soil C and N stocks as affected by cropping systems and nitrogen fertilisation in a southern Brazil Acrisol managed under no-tillage for 17 years. Soil and Tillage Research 81, 87–95.
18. Galantini JA, Senesi N, Brunetti G, Rosell R (2004) In fluence of texture on organic matter distribution and quality and nitrogen and sulphur status in semiarid Pampean grassland soils of Argentina. Geoderma 123, 143–152.
19. Ouédraogo E, Mando A, Stroosnijder L (2006) Effects of tillage, organic resources and nitrogen fertiliser on soil carbon dynamics and crop nitrogen uptake in semi-arid West Africa. Soil and Tillage Res 91: 57–67.
20. Yamashita T, Feiner H, Bettina J, Helfrich M, Ludwig B (2006) Organic matter in density fractions of water-stable aggregates in silty soils: effect of land use. Soil Biology and Biochemistry 38: 3222–3234.
21. Paustian K, Collins HP, Paul EA (1997) Management controls in soil carbon. In: Paul, E.A., Paustian, K.A., Elliott, E.T., Cole, C.V. (Eds). Soil Organic Matter in Temperate Ecosystems: Long Term Experiments in North America. RC, Boca Raton, FL 15–49.
22. Puget P, Lal R (2005) Soil organic carbon and nitrogen in a Mollisol in central Ohio as affected by tillage and land use. Soil Till. Res 80, 201–213.
23. Ding LL, Cheng H, Liu ZF, Ren WW (2013) Experimental warming on the rice-wheat rotation agro-ecosystem. Plant Science Journal 31(1): 49–56.
24. Editorial Board of China Agriculture Yearbook (2012) China Agriculture Yearbook 2009, Electronic Edition. China Agriculture Press, Beijing, China.
25. Hao JH, Ding YF, Wang QS, Liu ZH, Li GH, et al. (2010) Effect of wheat crop straw application on the quality of rice population and soil properties. Journal of Nanjing Agricultural University 33(3): 13–18.
26. Zhu LQ, Zhang DW, Bian XM (2011) Effects of continuous returning straws to field and shifting different tillage methods on changes of physical-chemical properties of soil and yield components of rice. Chinese Journal of Soil Science 42(1): 81–85.
27. Liu SP, Nie XT, Zhang HC, Dai QG, Huo ZY, et al. (2006) Effects of tillage and straw returning on soil fertility and grain yield in a wheat-rice double cropping system. Transactions of the CSAE 22(7): 48–51.
28. Lu RK (1999) Soil agricultural chemistry analysis. China's agricultural science and technology press, 106–110.
29. Jiang PK, Xu QF, Xu ZH, Cao ZH (2006) Seasonal changes in soil labile organic carbon pools within a Phyllostachys praecox stand under high rate fertilization and winter mulch in subtropical China. For. Ecol. Manage 236: 30–36.
30. Vance F, Brookes P, Jenkinson D (1987) Microbial biomass measurements in forest soils: the use of the chloroform fumigation-incubation method in strongly acid soils. Soil Biochem 19: 697–702.
31. Leng SC (1992) Biological Statistic and Field Experimental Design. Beijing: China Radio& Television Press (in chinese).
32. Duan HP, Niu YZ, Bian XM (2012) Effects of tillage mode and straw return on soil organic carbon and rice yield in direct seeding rice field. Bulletin of Soil and Water Conservation 32(3): 23–27.
33. Yang JC, Han XG, Huang JH, Pan QM (2003) The dynamics of soil organic matter in cropland responding to agricultural practices. Acta Ecologica Sinica 23(4): 787–796.
34. Six J, Elliott ET, Paustian K (1999) Aggregate and soil organic matter dynamics under conventional and no-tillage systems. Soil Sci. Soc. Am. J. 63, 1350–1358.
35. Beare MH, Hendrix PF, Coleman DC (1994) Water-stable aggregates and organic matter fractions in conventional-tillage and no-tillage soils. Soil Sci. Soc. Am. J. 58, 777–786.
36. Mikha MM, Rice CW (2004) Tillage and manure effects on soil and aggregate-associated carbon and nitrogen. Soil Sci. Soc, AM. J 68: 809–816.
37. Wang J, Zhang RZ, Li AZ (2008) Effect on soil active carbon and C cool management index of different tillages. Agricultural Research in the Arid Areas 26(6): 8–12.
38. Li XJ, Zhang ZG (1999) Influence on soil floods properties of mulching straws and soil-returning straw. Territory and Natural Resources Study 1: 43−45.
39. Liu DY (2009) Physiological and ecological mechanism of stable and high yield of broadcasted rice in paddy field with high standing-stubbles under no-tillage condition. PhD: Sichuan Agricultural University.
40. Gao YJ, Zhu PL, Huang DM, Wang ZM, Li SX (2000) Long-term impact of different soil management on organic matter and total nitrogen in rice-based cropping system. Soil and Environmental Sciences 9(1): 27−30.
41. Jastrow JD (1996) Soil aggregate formation and the accrual of particulate and mineral associated organic matter. Soil Biol. Biochem. 28, 656–676.
42. Govaerts B, Sayre KD, Lichter K, Dendooven L, Deckers J (2007) Influence of permanent raised bed planting and residue management on physical and chemical soil quality in rain fed maize/wheat systems. Plant Soil 291, 39–54.
43. Stockfisch N, Forstreuter T, Ehlers W (1999) Ploughing effects on soil organic matter after twenty years of conservation tillage in Lower Saxony, Germany. Soil Till. Res 52: 91−101.

44. Chen SH, Zhu ZL, Liu DH, Shu L, Wang CQ (2008) Influence of straw mulching with no-till on soil nutrients and carbon pool management index. Plant Nutrition and Fertilizer Science 14(4): 806−809.
45. Song MW, Li AZ, Cai LQ, Zhang RS (2008) Effects of different tillage methods on soil organic carbon pool. Journal of Agro-Environment Science 27(2): 622−626.
46. Zuo YP, Jia ZK (2004) Effect of soil moisture content o n straw decomposing and its dynamic changes. Joural of Northwest Sci-Tech University of Agri. and For. (Nat. Sci. Ed.) 32(5): 61–63.
47. Dai ZG (2009) Study on nutrient release characteristics of crop residue and effect of crop residue returning on crop yield and soil fertility. PhD: Huazhong Agricultural University (in Chinese).
48. Li CF, Yue LX, Kou ZK, Zhang ZS, Wang JP, et al. (2012) Short-term effects of conservation management practices on soil labile organic carbon fractions under a rape–rice rotation in central China. Soil Till, Res 119: 31–37.
49. Zhang P, Li H, Jia ZK, Wang W, Lu WT, et al. (2011) Effects of straw returning on soil organic carbon and carbon mineralization in Semi-arid areas of southern Ningxia, China. Journal of Agro-Environment Science 30(12): 2518−2525.
50. Tian SZ, Ning YT, Wang Y, Li HJ, Zhong WL, et al. (2010) Effects of different tillage methods and straw-returning on soil organic carbon content in a winter wheat field. Chinese Journal of Applied Ecology 21(2): 373–378.
51. Dou FG, Wright AL, Hons FM (2008) Sensitivity of labile soil organic carbon to tillage in wheat-based cropping systems. Soil Sci. Soc. Am. J. 72: 1445–1453.
52. Wu TY, Schoenau JJ, Li FM, Qian PY, Malhi SS, et al. (2004) Influence of cultivation and fertilization on total organic carbon and carbon fractions in soils from the Loess Plateau of China. Soil & Till, Res 77: 59–68.
53. Gregorich EG, Ellert BH, Gregorich EG, Carter MR, Monreal CM, et al. (1994) Towards a minimum data set to assess soil organic matter quality in agricultural soils. Can. J. Soil Sci 74: 367–385.
54. Loginow W, Wisniewski W, Gonet SS, Ciescinska B (1987) Fractionation of organic carbon based on susceptibility to oxidation. Pol. J. Soil Sci 20: 47–52.

6

Swedish Spring Wheat Varieties with the Rare High Grain Protein Allele of *NAM-B1* Differ in Leaf Senescence and Grain Mineral Content

Linnéa Asplund[1]*, Göran Bergkvist[1], Matti W. Leino[2,3], Anna Westerbergh[4], Martin Weih[1]

1 Department of Crop Production Ecology, Swedish University of Agricultural Sciences, Uppsala, Sweden, **2** Swedish Museum of Cultural History, Julita, Sweden, **3** IFM - Biology, Linköping University, Linköping, Sweden, **4** Department of Plant Biology and Forest Genetics, BioCenter, Swedish University of Agricultural Sciences, Uppsala, Sweden

Abstract

Some Swedish spring wheat varieties have recently been shown to carry a rare wildtype (wt) allele of the gene *NAM-B1*, known to affect leaf senescence and nutrient retranslocation to the grain. The wt allele is believed to increase grain protein concentration and has attracted interest from breeders since it could contribute to higher grain quality and more nitrogen-efficient varieties. This study investigated whether Swedish varieties with the wt allele differ from varieties with one of the more common, non-functional alleles in order to examine the effect of the gene in a wide genetic background, and possibly explain why the allele has been retained in Swedish varieties. Forty varieties of spring wheat differing in *NAM-B1* allele type were cultivated under controlled conditions. Senescence was monitored and grains were harvested and analyzed for mineral nutrient concentration. Varieties with the wt allele reached anthesis earlier and completed senescence faster than varieties with the non-functional allele. The wt varieties also had more ears, lighter grains and higher yields of P and K. Contrary to previous information on effects of the wt allele, our wt varieties did not have increased grain N concentration or grain N yield. In addition, temporal studies showed that straw length has decreased but grain N yield has remained unaffected over a century of Swedish spring wheat breeding. The faster development of wt varieties supports the hypothesis of *NAM-B1* being preserved in Fennoscandia, with its short growing season, because of accelerated development conferred by the *NAM-B1* wt allele. Although the possible effects of other gene actions were impossible to distinguish, the genetic resource of Fennoscandian spring wheats with the wt *NAM-B1* allele is interesting to investigate further for breeding purposes.

Editor: Dorian Q. Fuller, University College London, United Kingdom

Funding: This study was supported by the Swedish University of Agricultural Sciences. The funders had no role in study design, data collection and analysis, decision to publish, or preparation of the manuscript.

Competing Interests: The authors have declared that no competing interests exist.

* E-mail: linnea.asplund@slu.se

Introduction

Knowledge on the efficiency of agricultural crops in using N and other nutrients can help reduce e.g., nitrogen (N) fertilization and increase the value of the crops produced. There are many factors which affect nitrogen use, ranging from field management to soil, weather, and genotype of the crop, and their interactions [1]. For example, the genotypic traits affecting N uptake, N conversion to harvested product, and N retranslocation into plant parts that survive until the next growing season (e.g., grain) all influence N use efficiency [2]. Breeding could therefore be one method for improving N use efficiency.

Effective translocation of nitrogen to the grain during grain filling has been identified as a candidate trait for improving N use efficiency in bread wheat (*Triticum aestivum* L. ssp *aestivum*) [3,4]. One gene possibly involved in nutrient translocation, and thereby nitrogen use efficiency, in wheat is *NAM-B1 (Gpc-B1)*. The locus on chromosome 6B was originally identified in tetraploid wild emmer wheat (*T. turgidum* L. ssp. *dicoccoides*) using durum wheat (*T. turgidum* L. ssp. *durum*) - wild emmer wheat substitution lines [5,6] and later the gene has been mapped more precisely [7]. The effects of the locus on grain protein concentration have been shown in several tetraploid wheat backgrounds [8–10] and in hexaploid wheats [10–13]. The wildtype (wt) allele of *NAM-B1* originating in emmer wheat codes for a NAC-domain protein, a group of proteins known to be transcription factors involved in plant development processes. There are at least two additional alleles of the gene, both of which are believed to be non-functional. One has a +1 bp insertion likely causing a frame shift and a loss of the NAC-domain, the other one probably has a large deletion [14,15].

Studies indicate that the functional wt *NAM-B1* allele increases the rate of senescence and allows more effective translocation of nutrients to the grain, resulting in shorter grain filling and higher concentrations of protein, Fe, Zn, and Mn in the grain [10,14,16–18]. The hexaploid wheat genome has several *NAM* homologues, which have been down-regulated with RNAi in a hexaploid line (carrying a non-functional *NAM-B1* allele). At 12 days after anthesis there are changes in the expression pattern of 691 genes [19], indicating involvement of *NAM* genes in complex mechanisms governing the senescence of leaves and the associated retranslocation of nutrients. Wildtype varieties have down-regulation of genes likely related to functions no longer needed during senescence, such as signaling components and photosynthetic machinery components, and up-regulation of genes possibly involved in the onset of senescence, such as proteins induced by the hormones jasmonic acid and abscissic acid. Effects of the wt allele on weight per grain and grain yield differ with genetic background and environment. When six hexaploid and three tetraploid near isogenic lines (NILs) with and without the wt allele were grown in the field, the wt allele had a negative effect on weight per grain and no significant effect on grain yield [12]. Mainly negative but sometimes positive effects on weight per grain in different genotype×environment combinations of three tetraploid recombinant substitution lines have also been reported [10]. In the previously mentioned RNAi line, grain protein concentration decreased but there was no significant change in weight per grain in a greenhouse experiment [14].

It has been hypothesized that the reduction in weight per grain in the wt genotype in many environments led to fixation of the non-functional alleles in durum and bread wheat during domestication [20]. However, the wt allele was found in four hexaploid wheat varieties (two spelt wheats, one spring and one winter bread wheat) out of 63 wheat varieties in a museum collection of wheat varieties from 1865 [21]. It was subsequently found in five (four spring wheats and one winter wheat) out of 367 varieties in a core collection of bread wheat chosen to maximize the world's collected genetic diversity in bread wheat [15,22]. Since many of the wt varieties in the museum and in the core collection had a northern origin and were spring wheats, a larger set of 138 northern spring wheat cultivars was screened. The wt allele was found only in Fennoscandian varieties, where 46 out of 104 varieties investigated carried the wt allele [15]. To our knowledge, the possible effects of the wt allele in Fennoscandian varieties have not previously been studied.

Besides yield, fast maturation and good baking quality have been important traits in Swedish spring wheat breeding [23]. Both these traits are positively affected by the wt *NAM-B1* allele [14,24]. Even though the presence of the wt allele has declined during 20th century breeding [15], these positive influences could explain the preservation of the allele during breeding despite its possible negative influence on yield.

Although effects of *NAM-B1* have been identified in NILs, these effects are not necessarily so large and consistent in different genotypic backgrounds that they would be visible when studying the allele in diverse varieties. The recently identified Swedish varieties with the wt allele therefore offer a chance to investigate whether varieties with the wt allele of *NAM-B1* show a different phenotype than varieties with a non-functional allele when present in different genetic backgrounds. In the present pot experiment, we compared flag leaf senescence, yield, weight per grain, and nutrient content in a set of Swedish varieties with and without the wt allele under controlled conditions in a climate chamber. Our starting hypothesis was that varieties with the wt allele have faster senescence and higher concentrations of the minerals N, Fe, Zn, and Mn in the grain than varieties with a non-functional allele. We also sought to explain why the wt allele is relatively common in Swedish varieties and to give an understanding of how modern breeding has affected N-retranslocation and senescence. Such information could be useful for breeders interested in possible effects of the wt allele of *NAM-B1* in different genetic backgrounds.

Materials and Methods

Plant Material

Seeds of 41 varieties of spring wheat were used (Table 1), of which 38 were donated by NordGen and three by a local farmer. Forty of the varieties were hexaploid spring wheat (*Triticum aestivum* L. ssp *aestivum*) and one was tetraploid emmer wheat (*T. turgidum* L. ssp. *dicoccoides*), which was included for comparison. The selection included three landraces with uncertain background cultivated for 10 years by a farmer in Uppsala, Sweden, six landraces with Swedish origin preserved in NordGen, and 29 Swedish cultivars released during the 20th century. Three varieties from other countries which were used in early Swedish breeding were also included. Some measurements were only performed on a subset of 12 varieties with release years spanning the time period of all varieties (Table 1). All varieties were genotyped as described in [15]. The wt allele was present in 12 of the varieties, and the deletion allele (del) in 29 varieties. In the subset of 12 varieties, five had the wt allele.

Growth Conditions

The experiment was laid out in a complete randomized block design with four replicates and with pot as the experimental unit. The experiment was conducted in a climate chamber with 16 h light, a 9/18°C night/day temperature regime, PAR about 230 μmol m^{-2} s^{-1} at the top of the pots, and 60% relative humidity. The pots were 13 cm×13 cm×13 cm in size, with a volume of 1.5 L, and were filled with a fertilized soil mixture containing (% of volume) 72% peat, 20% perlite, 5.6% silica clay, and 2.4% gravel. The soil initially contained 260 mg N per pot and essential macro- and micronutrients (g m^{-3}, N 180, P 110, K 195, Mg 260, S 100, Ca 2000, Fe 6.0, Mn 3.5, Cu 2.5, Zn 1.5, B 0.6, Mo 3.0). Five seeds were sown per pot at a depth of 2.5 cm. Most of the plants had emerged after seven days, and day seven after sowing was considered day 1 of the experiment. The plants were thinned down to two plants per pot within the following week.

The pots were placed on trolleys, with about nine pots per trolley. The trolleys were moved within the blocks three times per week and the order of the blocks was rotated once a week until day 64, when moving the plants became impractical. The pots were initially placed on nets but were moved to trays on day 13 to allow nutrient solution to be soaked up when added later on. Nets were mounted above the trolleys to provide support when the plants grew taller. On day 69 the plants were moved to lower trolleys to allow further vertical growth of the plants. The plants were irrigated on the trays, initially only with deionized water, but from day 27 to 122 the plants were also supplied two times a week with 0.2 L nutrient solution per pot containing (mg L^{-1}) 102 N, 20 P, 86 K, 8 S, 6 Ca, 8 Mg, 0.34 Fe, 0.4 Mn, 0.2 B, 0.06 Zn, 0.03 Cu, 0.0008 Mo. From day 123 the dose was reduced to 0.1 L nutrient solution applied twice a week. The plants were watered with deionized water as needed to keep the soil moist, with up to two days between watering occasions. The beginning of anthesis of the main stem in each pot (BBCH 61, [25]) was recorded as the day when one of the two plants had reached this stage.

Measurements

Leaf chlorophyll content (SPAD index) was assessed with a portable chlorophyll meter (SPAD-502, Konica Minolta Sensing Inc., Japan). Two values were taken per pot, from the newest fully developed leaf on the main shoot of both plants. The value for each leaf was the mean of three measurements taken from along

Table 1. Varieties used in the study, their allele types and year of release, with the subset of 12 varieties indicated in bold style.

Deletion (non-functional) allele of *NAM-B1*			Wildtype (functional) allele of *NAM-B1*		
Acc. No.	Name	Year of release	Acc. No.	Name	Year of release
NGB6675	**Vårpärl**	**1901**	**NGB6678**	**Rubin**	**1921**
NGB6676	**Kolben**	**1892**	**NGB6679**	**Diamant**	**1928**
NGB6677	Extra Kolben	1919	**NGB6684**	**Rival**	**1952**
NGB6680	Fylgia I	1933	**NGB6688**	**Prins**	**1962**
NGB6682	**Progress**	**1941**	NGB6689	Amy	1971
NGB6683	Ella	1950	NGB13346	Sopu (Finland)	1935
NGB6685	Fylgia II	1952	NGB4499	Emmer wheat from Gotland **	landrace
NGB6686	**Drott**	**1955**	NGB6409	Halland	landrace
NGB6687	Safir	1955	**NGB6410**	**Dalarna**	**landrace**
NGB7455	Atle	1936	NGB6673	Landrace from Dalarna	landrace
NGB7456	Brons	1945	NGB13441	Västergötland	landrace
NGB7457	Kärn	1946	*	origin Öland	landrace
NGB7461	Svenno	1953			
NGB7462	Ring	1957			
NGB7464	Pompe	1967			
NGB7465	Snabbe	1968			
NGB7475	Kadett	1981			
NGB7479	Tjalve	1990			
NGB8923	Extra Kolben II	1926			
NGB9691	Blanka	1950			
NGB9954	**Dragon**	**1988**			
NGB9955	**Dacke**	**1990**			
NGB9956	Sport	1991			
NGB11010	Heines Kolben (Germany)	1900***			
NGB11280	Marquis (Canada)	1909			
NGB13917	**Vinjett**	**1998**			
NGB9708	Dalarna	landrace			
*	origin Dalarna	landrace			
*	origin Halland	landrace			

*Landrace obtained from farmer, all other accessions were obtained from NordGen.
***Triticum turgidum* ssp. *dicoccon,* was not included in the comparison of allele types.
***Exact year of origin unclear, 1900 was used in calculations.

the middle of the leaves. Starting from a few days before anthesis, SPAD measurements of the main stem flag leaf of all plants were taken initially three, then two times a week. Each of the two main shoots in each pot was individually labeled, using two different colors of cotton thread. When it was no longer possible to get a value (on dying leaves), it was recorded as zero.

The harvest was performed block-wise on days 138 and 139, when all plants were considered completely mature by visual inspection. At harvest, the number of ears was counted in each pot and the height was measured from the soil surface to the tip of the ears, excluding awns on the two main stems in each pot. The ears were cut off and stored at room temperature in open paper bags. The remaining straw was cut at ground level and dried for 48 hours at 60°C. The ears were threshed by hand and the grains and chaff were thereafter dried. The number of grains was counted by hand in each sample and weight per grain was calculated by dividing the dry weight by the number of grains. Straw (including chaff) and half of each grain sample were milled using a knife mill (Grindomix GM 200, Retsch GmbH) with titanium blade to minimize micronutrient contamination [26]. All grain samples and a subset (12 varieties) of straw samples were analyzed for concentration of N using dry combustion (CNS2000, LECO Corporation, Saint Joseph, Michigan, USA), and Ca, K, Mg, P, S, Fe, Zn, Mn, Cu, Na, and Al using inductively coupled plasma optical emission spectrometry (ICP Optima 7300 DV, PerkinElmer, Waltham, Massachusetts, USA) at the Department of Soil and Environment, Swedish University of Agricultural Sciences, Uppsala, Sweden. For occasional ears that were malformed, very late maturing, or where the stem had broken, the grains and straw were not analyzed for nutrient content due to the increased risk that retranslocation was affected. The not analyzed material was weighed and the total amount of nutrients in the biomass was calculated by assuming the same concentrations in the damaged as in the undamaged plant parts within the same pot. The yield of minerals, i.e., the weight, was calculated by multiplying the dry weight of biomass by the concentration of the mineral, adjusted for water content. All weights refer to the dry weight. Harvest index (HI) was calculated by dividing the grain biomass by the total aboveground biomass, and N harvest index (NHI) was calculated by dividing the grain N yield by the yield of N in the aboveground biomass.

Statistics

The SAS version 9.3 procedure Mixed [27] was used for analyzing the variables, with variety as fixed effect and block as random effect. The varieties of different alleles were compared with contrasts to test for significant differences. Only accessions of *Triticum aestivum* ssp *aestivum* L. were included in the comparison of allele types. Normality and heterogeneity of residuals were examined by residual plots, and some variables were transformed to meet the assumptions of the test. The varieties were released during a long period of time and there is a possibility that gradual genetic changes have taken place during this time, which could interfere with the analysis of the comparison of allele types. Therefore, the SAS procedure Reg was used to test for gradual changes in the varieties with time, by performing linear regressions with year of release as the independent variable. All landraces (from NordGen and from the farmer) were treated as if they had been released in 1890.

During the experiment, we observed and graded some leaf spots on many plants, but no relationships to occurrence of any pests/pathogens or nutrient deficiencies were found. To test for the possible effects of the leaf spots on the results of this experiment, the leaf spot grading data were subjected to statistical analysis. The SAS procedure Glimmix with the Cumulative Logit Link function was used to analyze the multinomial variable of the grading, and it was found that the spots were not connected either to allele type or variety. Furthermore, analysis of the variables was repeated with only the most common level of spots, and the results did not change in the majority of cases. Therefore, we did not consider the spots as a factor that affected the comparison of the allele groups in this study.

Senescence was analyzed by fitting curves to the SPAD data for each pot, and retrieved values were used in the SAS procedure Mixed model described above. Our different varieties had different curves, and it was not possible to fit one smooth curve to all pots. Therefore, we chose a model which could be fitted to most pots to identify interesting features of senescence, but not always with as perfect a fit as could have been achieved for a few of the pots if a smoother curve had been used. A two-segmented model with a left plateau [28] was fitted for each of the pots, merging the data series for both plants:

$$g(x_i;\ \beta)=\begin{cases}\beta_0 & \text{if} \quad x_i \leq \tau \\ \beta_0+\beta_2(x_i-\tau) & \text{if} \quad x_i > \tau\end{cases} \tag{1}$$

where β_0 represents the intercept and level of the left segment/plateau, β_2 is the slope of the right segment which is the senescence rate, and τ is the change point where the two segments meet in this continuous model (Figure 1). In this study, g = SPAD unit and x = days after anthesis. The SAS version 9.2 [27] procedure Nlin with the Marquardt minimization algorithm was used to fit the model. The value of τ was retrieved as the value at start of senescence, and the length of senescence was calculated as the difference in time between τ and the time of completely senesced flag leaf of the slowest plant in each pot. The statistical programming language R version 2.14.2 [29] was used to produce the plots.

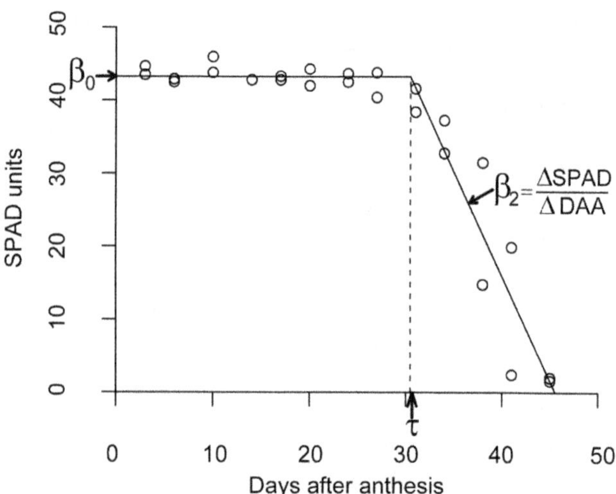

Figure 1. Example of model fit of senescence, variety 'Vårpärl' (NGB6675). Parameters are indicated in the graph; β_0 is the intercept on the y-axis, τ is the breakpoint on the x-axis and β_2 is the slope of the right segment.

Results

Leaf Senescence and Development

Varieties with the wt allele completed senescence faster than varieties with the del allele (p = 0.0004) (Figure 2). The range of mean values of the varieties with each type of allele was 11.8–27.3 days for the wt varieties and 15.0–47.0 days for the del varieties. The period between emergence and anthesis was also shorter in the wt varieties ($p<0.0001$). The total time from emergence to complete senescence of the flag leaf was shorter for the wt varieties than for the del varieties ($p<0.0001$). There was no difference in the time between anthesis and start of senescence (p = 0.84).

Biomass

Varieties with the wt allele had significantly lighter grains than varieties with the del allele (Table 2), ranging from 43.9 to 55.6 mg grain^{-1} among the wt varieties and from 39.4 to 57.1 mg grain^{-1} among the del varieties. However, the wt varieties had a larger total grain yield. The wt varieties also had more ears than the del varieties, but there was no difference in the number of grains per ear. The wt varieties had a larger total aboveground biomass than the del varieties, but the del varieties had a higher harvest index. The wt varieties were also taller than the del varieties.

Nutrient Content

There were significant differences between the groups of varieties with the different alleles in terms of both macro- and micronutrients in the grain (Table 3). The groups differed in grain macronutrient concentrations of Ca, Mg, P, and S. Of these, the wt group had a higher percentage of Mg and P, and a lower percentage of Ca and S. The yield of the macronutrients Mg and P was significantly higher in the wt group, but there were no significant differences between the groups in yield of the other macronutrients, including N. Regarding the micronutrient concentrations analyzed in the grains, there was only a significant difference in Mn, where the wt group had a lower concentration.

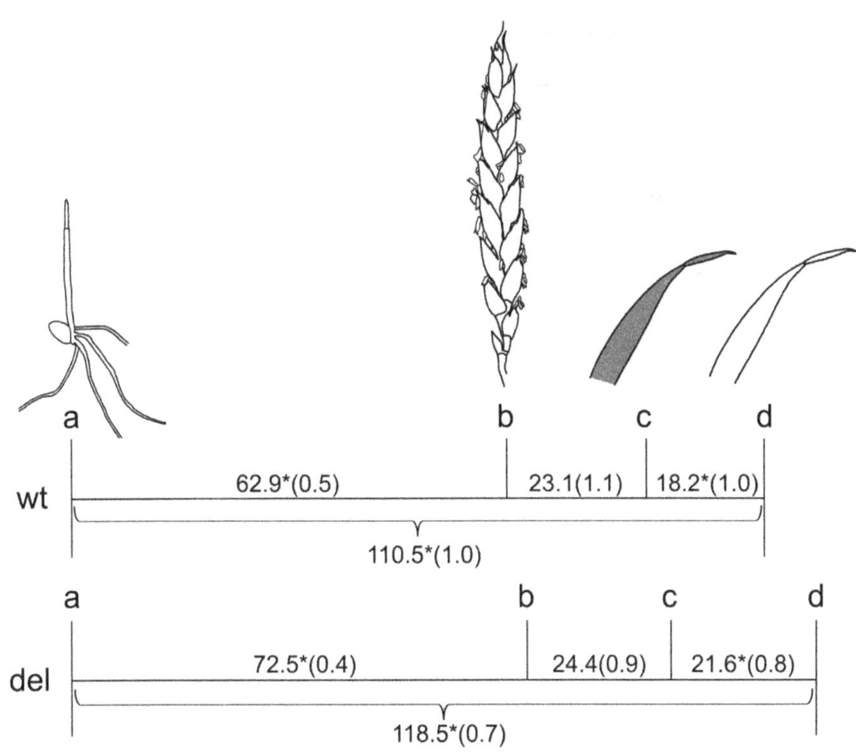

Figure 2. Timeline of development of varieties with the wt allele and varieties with the deletion. The scale is mean number of days (SE in brackets). Time spans that are significantly different between the allele types are indicated with * on both groups of varieties ($p<0.001$). a = emergence, b = anthesis, c = modeled start of senescence of flag leaf (τ), d = complete senescence of flag leaf.

There were no differences between the allele groups in the yield of any of the micronutrients in the grain.

For the straw, there were significant differences in macronutrient concentrations in the subset of 12 varieties (Table 4). The wt group had lower concentrations of N, K, Ca, Mg, and S. There were also lower yields of the macronutrients Ca and S in the wt group. The straw differed in micronutrients between allele groups, e.g., the concentration of Mn was lower in the wt group and the yield of Fe was higher in the wt group than in the del group. The yield of Al was significantly higher in the straw of the wt group. About 85% of the N was in the grain (NHI) in both allele groups.

Gradual Changes with Time

Grain yield did not change significantly with release year of the variety, but straw biomass decreased (Figure 3). The del varieties had significantly lower straw biomass, which made it difficult to separate the effect of gradual genetic changes from the effect of the specific *NAM-B1* allele in this trait. The absence of trend in grain biomass, on the other hand, supported the finding of a significant difference between the alleles in grain biomass. The plants became

Table 2. Means of some agronomic traits for the varieties grouped based on *NAM-B1* allele.

Variable	*NAM-B1* allele						F-value*	p-value*	Emmer mean value**
	wildtype			deletion					
	N	Mean	SE	N	Mean	SE			
Grain yield (g pot^{-1})	44	23.3	1.1	116	20.9	0.6	4.3	0.04	12.8
Weight per grain (mg grain^{-1})	44	48.9	0.6	116	51.2	0.4	20.6	<.0001	34
No. of ears	44	11.4	0.4	114	9.9	0.3	10.1	0.002	11.8
No. of grains per ear	44	40.0	1.6	114	39.2	1.2	0.4	0.56	30.9
Aboveground biomass (g)	44	55.5	2.6	116	47.1	1.3	10.9	0.001	46.3
Plant height (cm)	44	143	2	116	134	2	66.4	<.0001	157
HI (%)	44	42	0.8	116	44	0.5	44	<.0001	28

*F and p-values of contrasts from ANOVA tests with contrasts comparing the wt and del varieties.

**Mean value of the studied emmer wheat (with wildtype *NAM-B1* allele) included for comparison, but was not included in the statistical comparison of allele types.

Table 3. Means of concentrations and total amount per pot of some minerals in the grains for the varieties grouped based on *NAM-B1* allele.

Variable	*NAM-B1* allele						F-value*	p-value*	Emmer mean value**
	wildtype			deletion					
	N	Mean	SE	N	Mean	SE			
N conc. (g g^{-1})	44	0.0213	0.0003	116	0.0218	0.0002	2.5	0.12	2.37
N yield (g pot^{-1})	44	0.489	0.023	116	0.455	0.014	1.8	0.18	0.304
NHI (%)	20	84.8	1.1	28	84.8	0.5	0.0	0.95	-
K conc. (mg kg^{-1})	44	4661	61	116	4656	38	0.0	0.92	5517
K yield (g pot^{-1})	44	0.109	0.006	116	0.098	0.003	3.7	0.06	0.071
Ca conc. (mg kg^{-1})	44	257	6	116	278	4	18.3	<.0001	286
Ca yield (mg pot^{-1})	44	6.04	0.35	116	5.78	0.19	0.4	0.43	3.66
Mg conc. (mg kg^{-1})	44	1412	25	116	1371	12	4.3	0.04	1385
Mg yield (mg pot^{-1})	44	32.6	1.6	116	28.6	0.9	6.3	0.01	17.6
P conc. (mg kg^{-1})	44	3966	53	116	3810	31	9.7	0.002	4681
P yield (mg pot^{-1})	44	92.3	4.8	116	79.6	2.4	7.6	0.007	59.8
S conc. (mg kg^{-1})	44	1573	27	116	1642	13	9.7	0.002	1800
S yield (mg pot^{-1})	44	36.1	1.7	116	34.0	1.0	1.5	0.23	22.9
Fe conc. (mg kg^{-1})	44	32.2	1.2	116	33.3	0.6	1.7	0.19	43.5
Fe yield (mg pot^{-1})	44	0.719	0.033	116	0.675	0.017	2.2	0.14	0.548
Mn conc. (mg kg^{-1})	44	109	2	116	116	1	10.8	0.001	115
Mn yield (g pot^{-1})	44	2.49	0.11	116	2.37	0.06	1.4	0.23	1.46
Zn conc. (mg kg^{-1})	44	40.7	1.6	116	43.2	0.9	3.3	0.07	54.9
Zn yield (mg pot^{-1})	44	0.895	0.035	116	0.870	0.021	0.5	0.49	0.688
Cu conc. (mg kg^{-1})	44	2.7	0.1	116	2.6	0.1	1.2	0.28	2.42
Cu yield (mg pot^{-1})	44	0.064	0.005	116	0.056	0.003	3.2	0.07	0.0305
Al conc. (mg kg^{-1})	44	0.6	0.1	116	0.8	0.1	1.1	0.31	0.491
Al yield (mg pot^{-1})	44	0.015	0.002	116	0.016	0.001	0.1	0.82	0.00622
Na conc. (mg kg^{-1})	44	8.0	0.5	116	9.6	0.4	1.1	0.31	13
Na yield (mg pot^{-1})	44	0.19	0.02	116	0.21	0.01	0.1	0.82	0.169

*F and p-values of contrasts from ANOVA tests with contrasts comparing the wt and del varieties.
**Mean value of the studied emmer wheat (with wildtype *NAM-B1* allele) included for comparison, but was not included in the statistical comparison of allele types.

significantly shorter with release year, and again the del varieties had shorter straw than the wt varieties. The grain N concentration and N yield did not change significantly with release year, and neither of these traits differed between the allele groups. The concentration of several other grain minerals decreased (Figure 4), but since yield increased (even if not significantly), there was no significant trend in total yield for any of the minerals tested (data not shown). The modeled senescence parameters changed with release year: SPAD value at anthesis increased (with no difference between allele groups), the senescence rate was faster (wt varieties had faster senescence and these results are therefore supported), and senescence started later with release year (there was no significant difference in start of senescence between allele groups) (Figure 3). Although several of the variables changed significantly with year of release, the regressions still only explained relatively small parts of the variation in the variables based on the coefficients of determination (r^2). Plant height had the highest value (0.68), and we considered that regression to be strong. The other regressions had much lower r^2 values.

Discussion

Varieties with the wt allele of *NAM-B1* had faster development and completed leaf senescence faster. They also had lighter grains and higher yield of P and K in the grain compared with varieties with the del allele. In our set of 40 Swedish spring wheat varieties, the pleiotropic effects of *NAM-B1* genotype appeared consistent and large enough to be detectable when plants with different genetic backgrounds were compared.

To increase precision, we chose to study the effect of a single gene (*NAM-B1*) on a number of phenotypic traits under controlled conditions, in which additional variation caused by the actions of pests and diseases and other environmental factors was minimized. The values of yield and other traits assessed in controlled greenhouse environments frequently differ from results obtained in the field. For example, the number of ears per plant was much greater in our study than in the field. Therefore, field experiments could give additional information in evaluating the effect of the gene on e.g., yield components and nutrient retranslocation. However, we believe that the results to a large extent reflect the relative differences between the varieties. When two of the

Table 4. Means of concentrations and total amount per pot of some minerals in the straw for a subset of 12 spring wheat varieties, grouped based on *NAM-B1* allele.

Variable	*NAM-B1* allele						F-value*	p-value*
	wildtype			deletion				
	N	Mean	SE	N	Mean	SE		
N conc. (g kg^{-1})	20	2.80	0.18	28	3.24	0.10	7.3	0.01
N yield (g pot^{-1})	20	0.0764	0.0067	28	0.0835	0.0051	1.0	0.33
K conc. (mg kg^{-1})	20	14141	448	28	15635	450	8.7	0.006
K yield (g pot^{-1})	20	0.391	0.032	28	0.406	0.025	0.2	0.64
Ca conc. (mg kg^{-1})	20	1718	78	28	2366	148	27.1	<.0001
Ca yield (mg pot^{-1})	20	48.9	5.5	28	61.9	4.9	5.5	0.03
Mg conc. (mg kg^{-1})	20	734	35	28	925	50	14.6	0.0006
Mg yield (mg pot^{-1})	20	20.9	2.3	28	24.3	1.8	2.6	0.12
P conc. (mg kg^{-1})	20	1477	98	28	1506	72	0.1	0.78
P yield (mg pot^{-1})	20	42.8	4.9	28	40.0	3.1	0.5	0.50
S conc. (mg kg^{-1})	20	2336	154	28	2979	133	16.4	0.0003
S yield (mg pot^{-1})	20	59.7	2.4	28	73.9	3.2	14.5	0.0006
Fe conc. (mg kg^{-1})	20	19.7	2.3	28	16.0	1.6	2.8	0.11
Fe yield (mg pot^{-1})	20	0.565	0.100	28	0.375	0.030	5.3	0.03
Mn conc. (mg kg^{-1})	20	273	13	28	312	15	5.4	0.03
Mn yield (mg pot^{-1})	20	7.21	0.41	28	7.80	0.39	2.1	0.15
Zn conc. (mg kg^{-1})	20	14.7	1.2	28	14.4	0.7	0.1	0.78
Zn yield (mg pot^{-1})	20	0.400	0.035	28	0.366	0.022	1.7	0.20
Cu conc. (mg kg^{-1})	20	1.4	0.1	28	1.6	0.1	1.7	0.20
Cu yield (mg pot^{-1})	20	0.042	0.005	28	0.041	0.003	0.0	0.97
Al conc. (mg kg^{-1})	20	2.4	0.2	28	2.2	0.2	1.3	0.27
Al yield (mg pot^{-1})	20	0.064	0.005	28	0.054	0.003	6.0	0.02
Na conc. (mg kg^{-1})	20	41.8	4.7	28	27.8	1.6	15.5	0.0004
Na yield (mg pot^{-1})	20	1.26	0.22	28	0.73	0.06	11.7	0.002

*F and p-values of contrasts from ANOVA tests with contrasts comparing the wt and del varieties.

varieties, a landrace from Dalarna obtained from a farmer and Vinjett, were recently grown together in a field experiment at about 180 kg N ha^{-1}, the grain N concentrations were similar to those in the present study (2.1% and 1.9% for Vinjett in climate chamber vs field, 2.2% in both experiments for the Dalarna landrace; unpublished data), indicating that there is agreement between this climate chamber study and the field situation.

A relevant question is whether differences in phenotype between the varieties carrying different alleles are influenced by other genes than *NAM-B1*. A more precise method would be to compare the phenotype of NILs carrying different *NAM-B1* alleles in different Swedish variety backgrounds. Another approach would be to reduce the transcript level of the wt allele in different Swedish varieties [10] and study the resulting phenotype. Such lines are not currently available. However, our study exemplifies an approach of investigating effects of single alleles which can be used when NILS are not available in a given genetic background. The fact that many of the results in our study are in line with previously shown effects of the wt *NAM-B1* allele makes it likely that some of the phenotypic differences between the two variety groups studied are actually controlled by *NAM-B1*. The potential pleiotropic effect of the wt allele in different genetic backgrounds is of great interest in developing new bread wheat varieties with qualities controlled by the wt *NAM-B1* allele.

In parallel to historical developments in which the presence of the wt allele decreased during the 20th century in Fennoscandian varieties [15], some traits may have changed due to directional breeding. This poses a problem of separating the effects of the allele from the effects of other genes on the same traits. That is, the allele type may be causing a trend in a trait over years, or a trend can lead us to see a difference between allele types in the traits. The difference in plant height between the allele groups, for example, with the wt allele varieties being higher, could possibly be explained by directional breeding. Breeding has targeted shorter varieties, but there is no known effect of the wt allele on plant height. However, there are too few varieties with and without the wt allele originating from the same year to allow for a study without such trends. The study performed here can at least give an indication of the effect of the wt allele in these Swedish varieties, which are unique from an international perspective. The absence or opposite direction of gradual changes in many traits believed to be affected by *NAM-B1*, e.g., grain N concentration and senescence rate, supports our results for these traits.

Breeding has made a positive contribution to bread wheat yields in many countries during the 20th century [30], and a yield

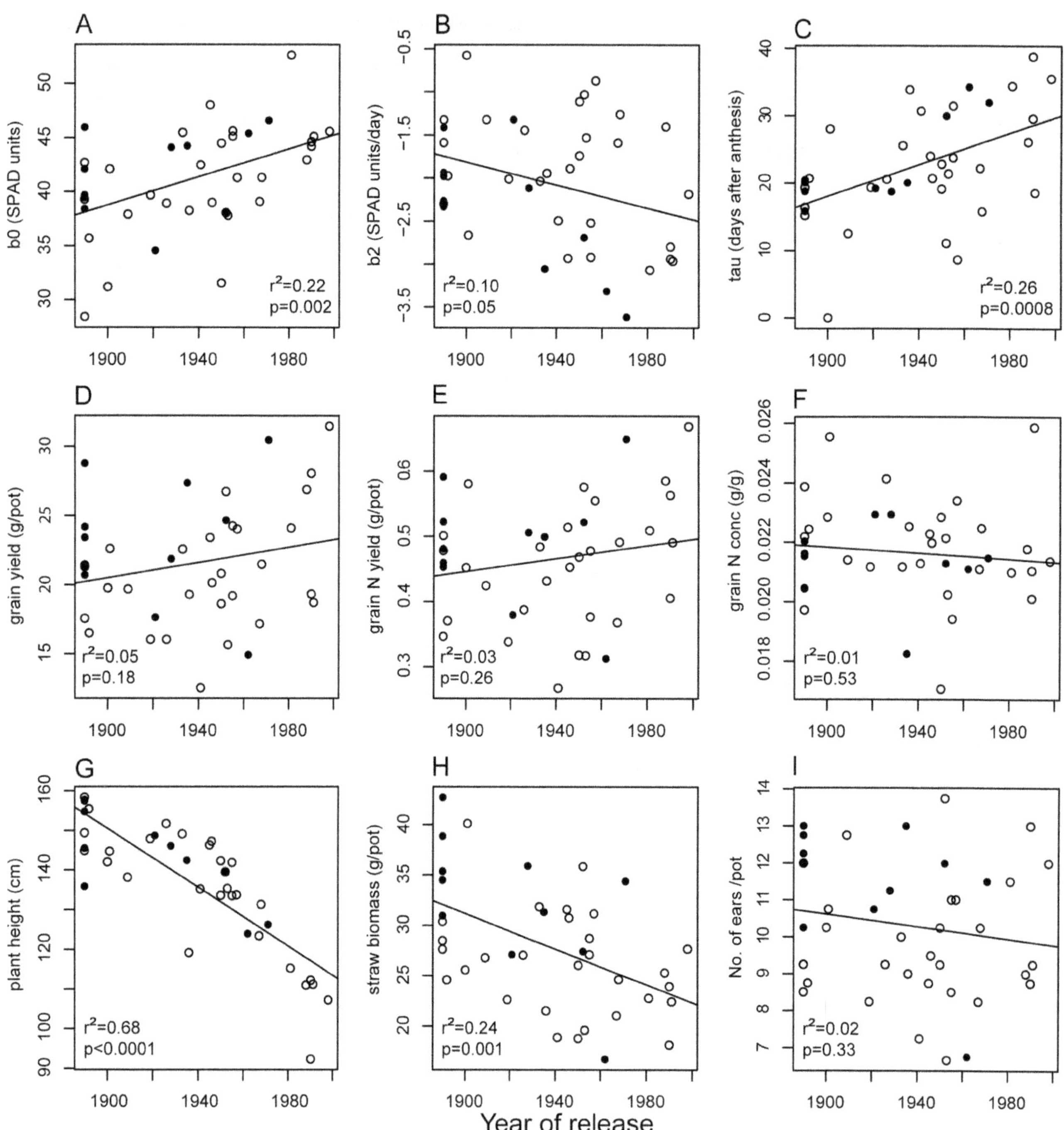

Figure 3. Changes in measured traits over time. Filled circles are varieties with the wildtype allele of *NAM-B1*, open circles are varieties with the deletion. A) modeled SPAD-units at anthesis, β_0, B) rate of senescence, β_2, C) modeled start of senescence in days after anthesis, τ, D) grain yield, E) grain N yield, F) grain N concentration, G) plant height, H) straw biomass, I) number of ears per pot.

increase has been reported for Nordic spring wheat varieties under field conditions, even if the r^2 was low in that study [31]. We did not find a significant yield increase with year of release in this study of spring wheat. Furthermore, we did not find a reduction in grain N concentration with time, as has been found in Argentinian [32,33] and Italian [34] bread wheat cultivars. It is known that there is often a negative relationship between grain yield and grain N concentration [35,36]. If previous breeding aimed for high N concentrations, it is possible that yield increases have been low as a consequence. The fact that the varieties were grown close together in this study may also have given an advantage to the higher (and in general older) varieties.

A novel finding was that varieties with the wt allele reached anthesis faster than the other varieties, which has not been observed previously for wheat. However, it has been noticed in a high grain protein concentration barley NIL [37]. Barley carries an ortholog of *NAM-B1* called *HvNAM-1*, which also has effects on senescence and grain protein concentration [37–39]. Wheat is sensitive to heat spells and droughts around anthesis, which reduce the number of grains per ear [40]. Earlier anthesis is one way to

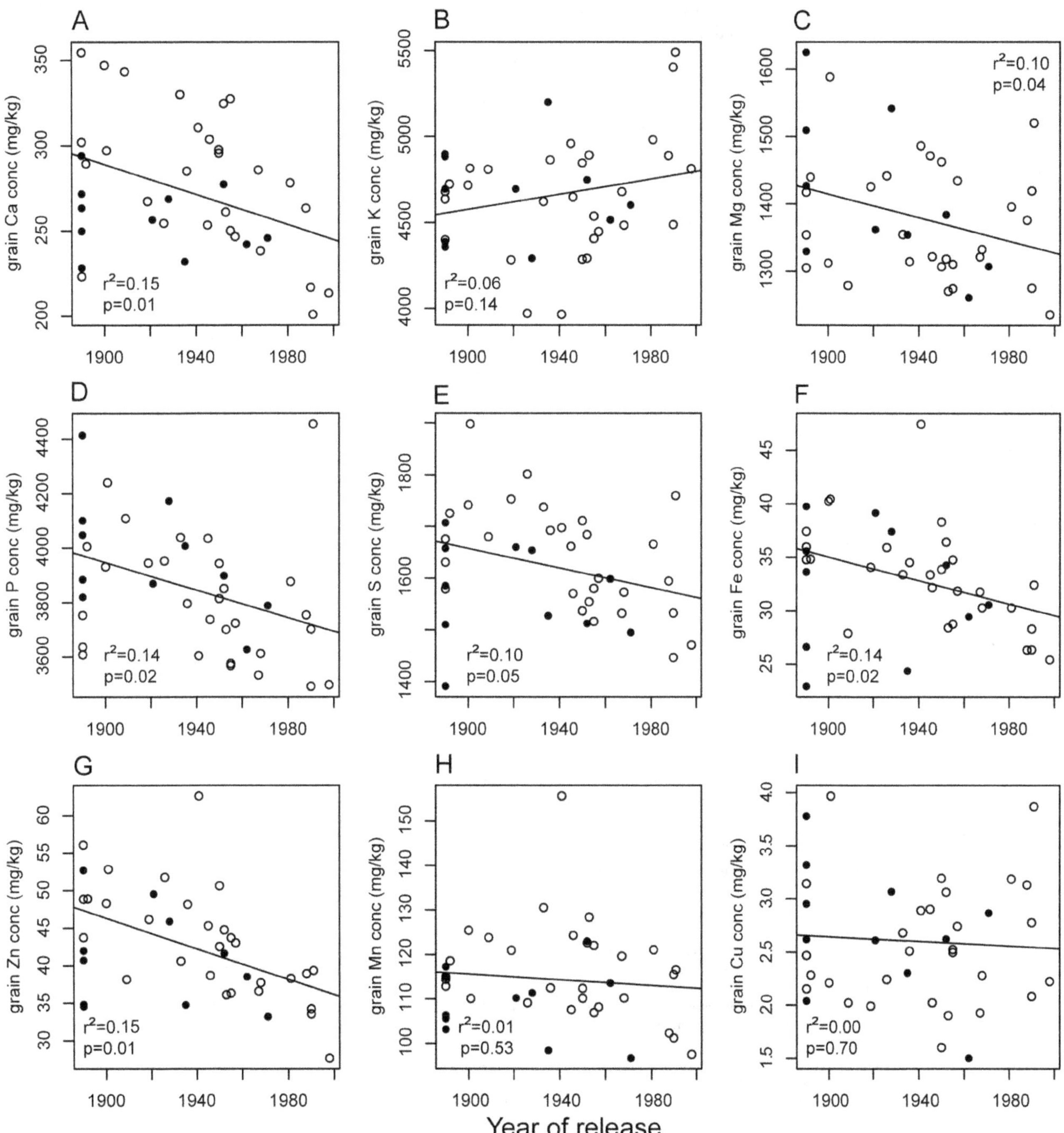

Figure 4. Changes in grain mineral concentrations over time. Filled circles are varieties with the wildtype allele of *NAM-B1*, open circles are varieties with the deletion. The graph indicates grain concentrations of A) calcium, B) potassium, C) magnesium, D) phosphorus, E) sulfur, F) iron, G) zinc, H) manganese, and I) copper.

escape this drought risk [41]. Fast maturation can be a desirable trait in cases of adverse conditions at the end of the growing season (e.g., cold and rain), which are common in the Fennoscandian climate. After the wt allele had been identified in Swedish germplasm, it was hypothesized that the faster maturation of the wt allele had led to its preservation also in other areas at high latitudes [15]. However, there was little evidence supporting this hypothesis, since the wt allele was found only in Fennoscandian varieties, and not in varieties grown in other areas at high latitude [15]. Nevertheless, if the wt varieties have faster maturation without a reduction in yield and possibly even a yield increase, this may still be one reason explaining why the wt allele has been retained in Fennoscandia. In the early Swedish breeding programs landraces were used in crosses with the deliberate goal of transferring their early ripening [42]. Some of the landraces used in the crosses (originating in Dalarna and Halland, Table 1) most likely carried the wt allele.

Varieties with the wt allele had lighter grains than varieties with the non-functional allele type, as hypothesized. However, they had higher total grain weight and more ears, which has not previously been reported as effects of the wt allele. In a study of the effect of the gene in some Indian bread wheat varieties into which the wt allele had been crossed, the lines with the wt allele all had more tillers than the lines from which they were derived [13]. This may be an actual effect of the wt allele that is only expressed in certain varieties or environments. It is also possible that the wt allele is linked to, or controlled by, other genes in the Swedish varieties studied here. More ears is a way of increasing the number of grains per unit area [43] and could be a way of increasing yield at a fixed N fertilization rate.

The concentrations and yields of N, Fe, Zn, and Mn in our Swedish varieties with the wt *NAM-B1* allele were similar to those of varieties with a non-functional allele, contrary to our expectations based on previous reports of *NAM-B1* effects on nutrient retranslocation [10,14,16–18]. This indicates that possible effects of the wt allele on N use efficiency through increased grain N concentration [2] are probably negligible. The only difference in yield was in the macronutrients P and Mg. It is possible that differences in micronutrients were too small to detect. Previous results indicate that the difference in minerals between allele types is smaller closer to complete senescence [18], and we harvested when all plants were mature, which would have decreased the differences.

An observation relevant for breeding is that we did not find indications of reduced yield in wt varieties in the environment studied here. Instead, the yield was higher in the wt varieties. This is positive if the allele is to be used in breeding, even if the effect on yield is known to be different in different environments. The absence of differences between the variety groups in grain N concentration and grain N yield indicates that any effects of the wt allele on these traits are small in some conditions, a fact which needs to be considered if the allele is to be used for increasing grain N. Since the varieties with the wt allele had lower amounts of several micro- and macronutrients in the straw, the question of retranslocation remains unclear. This issue is probably better examined by more detailed studies of nutrient allocation in NILs. Since varieties with the wt allele did not have higher grain N concentration or grain N yield, we have no indications that the good baking properties of Swedish spring wheat are related to the wt *NAM-B1* allele. However, it would be interesting to study this aspect under field conditions. Varieties with the wt allele reached anthesis faster and senesced faster than varieties with a non-functional allele, which is also interesting for breeding purposes.

Conclusions

Two groups of varieties, with the wt or the del allele of *NAM-B1*, differed in several traits. Varieties with the wt allele had lighter grains, more ears, and reached anthesis and completed senescence faster. The wt varieties had higher concentrations of Mg and P in the grain, while the del varieties had higher concentrations of Ca, S, and Mn in the grain. These findings suggest that the wt *NAM-B1* allele has multiple effects on the phenotype that are detectable when the different alleles are present in different genetic backgrounds. No difference between variety groups was found in grain N concentration or grain N yield, which does not support an association between the *NAM-B1* wt allele and the high baking quality of Swedish spring wheat. It also indicates that the relationship between nitrogen use efficiency and allele type might be small. The higher number of ears and faster development and senescence in *NAM-B1* wt varieties is interesting for breeding purposes and should be further investigated.

Acknowledgments

The authors thank NordGen for donating seed accessions, Lennart Karlsson for donating seed for the farmer's landraces, Jan-Eric Englund for statistical advice and Johan Gottfridsson for technical assistance.

Author Contributions

Conceived and designed the experiments: LA GB ML AW MW. Performed the experiments: LA. Analyzed the data: LA. Wrote the paper: LA GB ML AW MW.

References

1. Mosier A, Syers JK, Freney JR (2004) Agriculture and the nitrogen cycle assessing the impacts of fertilizer use on food production and the environment. Washington, D.C.: Island Press.
2. Weih M, Asplund L, Bergkvist G (2011) Assessment of nutrient use in annual and perennial crops: A functional concept for analyzing nitrogen use efficiency. Plant Soil 339: 513–520.
3. Foulkes MJ, Hawkesford MJ, Barraclough PB, Holdsworth MJ, Kerr S, et al. (2009) Identifying traits to improve the nitrogen economy of wheat: Recent advances and future prospects. Field Crop Res 114: 329–342.
4. Chardon F, Noël V, Masclaux-Daubresse C (2012) Exploring NUE in crops and in Arabidopsis ideotypes to improve yield and seed quality. J Exp Bot 63: 3401–3412.
5. Cantrell RG, Joppa LR (1991) Genetic Analysis of Quantitative Traits in Wild Emmer (Triticum turgidum L. var. dicoccoides). Crop Sci 31: 645–649.
6. Joppa LR, Du C, Hart GE, Hareland GA (1997) Mapping Gene(s) for Grain Protein in Tetraploid Wheat (Triticum turgidum L.) Using a Population of Recombinant Inbred Chromosome Lines. Crop Sci 37: 1586–1589.
7. Distelfeld A, Uauy C, Fahima T, Dubcovsky J (2006) Physical map of the wheat high-grain protein content gene Gpc-B1 and development of a high-throughput molecular marker. New Phytol 169: 753–763.
8. Steiger DK, Elias EM, Cantrell RG (1996) Evaluation of Lines Derived from Wild Emmer Chromosome Substitutions: I. Quality Traits. Crop Sci 36: 223–227.
9. Chee P, Elias E, Anderson J, Kianian S (2001) Evaluation of a high grain protein QTL from Triticum turgidum L. var. dicoccoides in an adapted durum wheat background. Crop Sci 41: 295–301.
10. Uauy C, Brevis JC, Dubcovsky J (2006) The high grain protein content gene Gpc-B1 accelerates senescence and has pleiotropic effects on protein content in wheat. J Exp Bot 57: 2785–2794.
11. Mesfin A, Frohberg RC, Anderson JA (1999) RFLP Markers Associated With High Grain Protein From Triticum Turgidum L. Var. Dicoccoides Introgressed Into Hard Red Spring Wheat. Crop Sci 39: 508–513.
12. Brevis JC, Dubcovsky J (2010) Effects of the Chromosome Region Including the Gpc-B1 Locus on Wheat Grain and Protein Yield. Crop Sci 50: 93–104.
13. Kumar J, Jaiswal V, Kumar A, Kumar N, Mir RR, et al. (2011) Introgression of a major gene for high grain protein content in some Indian bread wheat cultivars. Field Crop Res 123: 226–233.
14. Uauy C, Distelfeld A, Fahima T, Blechl A, Dubcovsky J (2006) A NAC gene regulating senescence improves grain protein, zinc, and iron content in wheat. Science 314: 1298–1301.
15. Hagenblad J, Asplund L, Balfourier F, Ravel C, Leino M (2012) Strong presence of the high grain protein content allele of NAM-B1 in Fennoscandian wheat. Theor Appl Genet 125: 1677–1686.
16. Kade M, Barneix AJ, Olmos S, Dubcovsky J (2005) Nitrogen uptake and remobilization in tetraploid "Langdon" durum wheat and a recombinant substitution line with the high grain protein gene Gpc-B1. Plant Breeding 124: 343–349.
17. Distelfeld A, Cakmak I, Peleg Z, Ozturk L, Yazici AM, et al. (2007) Multiple QTL-effects of wheat Gpc-B1 locus on grain protein and micronutrient concentrations. Physiol Plantarum 129: 635–643.
18. Waters BM, Uauy C, Dubcovsky J, Grusak MA (2009) Wheat (Triticum aestivum) NAM proteins regulate the translocation of iron, zinc, and nitrogen compounds from vegetative tissues to grain. J Exp Bot 60: 4263–4274.
19. Cantu D, Pearce SP, Distelfeld A, Christiansen MW, Uauy C, et al. (2011) Effect of the down-regulation of the high Grain Protein Content (GPC) genes on the wheat transcriptome during monocarpic senescence. BMC Genomics 12: 492.
20. Dubcovsky J, Dvorak J (2007) Genome Plasticity a Key Factor in the Success of Polyploid Wheat Under Domestication. Science 316: 1862–1866.

21. Asplund L, Hagenblad J, Leino MW (2010) Re-evaluating the history of the wheat domestication gene NAM-B1 using historical plant material. J Archaeol Sci 37: 2303–2307.
22. Balfourier F, Roussel V, Strelchenko P, Exbrayat-Vinson F, Sourdille P, et al. (2007) A worldwide bread wheat core collection arrayed in a 384-well plate. Theor Appl Genet 114: 1265–1275.
23. Olsson G (1997) Den svenska växtförädlingens historia: jordbruksväxternas utveckling sedan 1880-talet. Stockholm: Kungl. Skogs- och Lantbruksakad.
24. Brevis JC, Morris CF, Manthey F, Dubcovsky J (2010) Effect of the grain protein content locus Gpc-B1 on bread and pasta quality. J Cereal Sci 51: 357–365.
25. Lancashire P, Bleiholder H, Van Den Boom T, Langelüddeke P, Stauss R, et al. (1991) A uniform decimal code for growth stages of crops and weeds. Ann Appl Biol 119: 561–601.
26. Dahlin AS, Edwards AC, Lindström BEM, Ramezanian A, Shand CA, et al. (2012) Revisiting herbage sample collection and preparation procedures to minimise risks of trace element contamination. Eur J Agron 43: 33–39.
27. SAS software, Version 9.2 and 9.3 of the SAS System for Windows. Copyright © 2002–2008 and 2002–2010 SAS Institute Inc.
28. Piegorsch WW, Bailer AJ (2005) Analyzing environmental data. Hoboken, NJ: Wiley.
29. R Development Core Team (2012) R: A language and environment for statistical computing. R Foundation for Statistical Computing, Vienna, Austria. ISBN 3-900051-07-0. Available: http://www.R-project.org/.
30. Slafer G, Andrade F (1991) Changes in physiological attributes of the dry matter economy of bread wheat (Triticum aestivum) through genetic improvement of grain yield potential at different regions of the world. Euphytica 58: 37–49.
31. Ortiz R, Madsen S, Andersen SB (1998) Diversity in Nordic spring wheat cultivars (1901–93). Acta Agr Scand, Section B - Plant Soil Science 48: 229–238.
32. Slafer G, Andrade F, Feingold S (1990) Genetic improvement of bread wheat (Triticum aestivum L.) in Argentina: relationships between nitrogen and dry matter. Euphytica 50: 63–71.
33. Calderini DF, Torres-León S, Slafer GA (1995) Consequences of Wheat Breeding on Nitrogen and Phosphorus Yield, Grain Nitrogen and Phosphorus Concentration and Associated Traits. Ann Bot 76: 315–322.
34. Guarda G, Padovan S, Delogu G (2004) Grain yield, nitrogen-use efficiency and baking quality of old and modern Italian bread-wheat cultivars grown at different nitrogen levels. Eur J Agron 21: 181–192.
35. Kibite S, Evans LE (1984) Causes of negative correlations between grain yield and grain protein concentration in common wheat. Euphytica 33: 801–810.
36. Peltonen-Sainio P, Jauhiainen L, Nissilä E (2012) Improving cereal protein yields for high latitude conditions. Eur J Agron 39: 1–8.
37. Parrott DL, Downs EP, Fischer AM (2012) Control of barley (Hordeum vulgare L.) development and senescence by the interaction between a chromosome six grain protein content locus, day length, and vernalization. J Exp Bot 63: 1329–1339.
38. Distelfeld A, Korol A, Dubcovsky J, Uauy C, Blake T, et al. (2008) Colinearity between the barley grain protein content (GPC) QTL on chromosome arm 6HS and the wheat Gpc-B1 region. Mol Breeding 22: 25–38.
39. Heidlebaugh NM, Trethewey BR, Jukanti AK, Parrott DL, Martin JM, et al. (2008) Effects of a barley (Hordeum vulgare) chromosome 6 grain protein content locus on whole-plant nitrogen reallocation under two different fertilisation regimes. Funct Plant Biol 35: 619–632.
40. Ferris R, Ellis RH, Wheeler TR, Hadley P (1998) Effect of High Temperature Stress at Anthesis on Grain Yield and Biomass of Field-grown Crops of Wheat. Ann Bot 82: 631–639.
41. Semenov MA (2007) Development of high-resolution UKCIP02-based climate change scenarios in the UK. Agr Forest Meteorol 144: 127–138.
42. Åkerman Å (1951) Svensk växtförädling. Stockholm: Natur och Kultur.
43. Peltonen-Sainio P, Kangas A, Salo Y, Jauhiainen L (2007) Grain number dominates grain weight in temperate cereal yield determination: Evidence based on 30 years of multi-location trials. Field Crop Res 100: 179–188.

7

Effects of Straw Incorporation on Soil Organic Matter and Soil Water-Stable Aggregates Content in Semiarid Regions of Northwest China

Peng Zhang[1,2], Ting Wei[1,2]**, Zhikuan Jia[1,2]*, Qingfang Han[1,2], Xiaolong Ren[1,2], Yongping Li[3]**

1 The Chinese Institute of Water-Saving Agriculture, Northwest A&F University, Yangling, Shaanxi, China, 2 Key Laboratory of Crop Physi-Ecology and Tillage Science in Northwestern Loess Plateau, Ministry of Agriculture, Northwest A&F University, Yangling, Shaanxi, China, 3 Guyuan Institute of Agricultural Sciences, Guyuan, Ningxia, China

Abstract

The soil degradation caused by conventional tillage in rain-fed areas of northwest China is known to reduce the water-use efficiency and crop yield because of reduced soil porosity and the decreased availability of soil water and nutrients. Thus, we investigated the effects of straw incorporation on soil aggregates with different straw incorporation rates in semiarid areas of southern Ningxia for a three-year period (2008–2010). Four treatments were tested: (i) no straw incorporation (CK); (ii) incorporation of maize straw at a low rate of 4 500 kg ha^{-1} (L); (iii) incorporation of maize straw at a medium rate of 9000 kg ha^{-1} (M); (iv) incorporation of maize straw at a high rate of 13 500 kg ha^{-1} (H). The results in the final year of treatments (2010) showed that the mean soil organic carbon storage of the 0–60 cm soil layers were significantly ($P<0.05$) increased with H, M, and L, by 21.40%, 20.38% and 8.21% compared with CK, respectively. Straw incorporation increased >0.25 mm water-stable macroaggregates level, geometric mean diameter, mean weight diameter and the aggregate stability, which were ranked in order of increasing straw incorporation rates: H/M > L > CK. Straw incorporation significantly ($P<0.05$) reduced the fractal dimension in the 0–40 cm soil layers compared with CK. Our results suggest that straw incorporation is an effective practice for improving the soil aggregate structure and stability.

Editor: Raffaella Balestrini, Institute for Plant Protection (IPP), CNR, Italy

Funding: This work was supported by the China Support Program (2012BAD09B03, 2011AA100504, and 2011BAD29B09) for Dryland Farming in the 12th 5-year plan period, the Youth project of National Natural Science Fund (31201156), and the Basic Science Research Fund in Northwest A&F University (QN2013005). The funders had no role in study design, data collection and analysis, decision to publish, or preparation of the manuscript.

Competing Interests: The authors have declared that no competing interests exist.

* E-mail: jiazhk@126.com

These authors contributed equally to this work.

Introduction

Soil infertility [1], soil erosion, and water deficiency [2] are the major factors that limited crop growth in semiarid areas of northwest China. The rates of crop straw use for fuel and forage have declined significantly since the 1980s and crop straw is increasingly burned after the harvest, which leads to high losses of soil organic substances [3,4], and increased emission of CO_2 that pollute the environment [5]. Furthermore, this practice has led to the degradation of the agricultural ecological environment [6].

The soil organic matter (SOM) content is one of the major factors that affects soil properties and functions including a range of physical characteristics such as the water-holding capacity [7], water infiltration [8], and aggregate stability [9]. SOM is considered to be a major binding agent that stabilizes soil aggregates [10,11]. Soil aggregates are the basic units of the soil structure [12], which are composed of primary particles and binding agents that determine the microbial biomass and mineral nutrient reserves [13–15]. These soil properties are also affected by soil organic matter decomposition processes [16,17].

Many studies have shown that crop straw is rich in organic material and soil nutrients, so it is increasingly considered to be an important natural organic fertilizer [18–20]. Straw can be incorporated to soil either directly or indirectly, which can promote the production of a favorable soil environment. Straw also maintains the physicochemical condition of the soil and improves the overall ecological balance of the crop production system [20,21]. Nelson [22] and Wilhelm *et al.* [23] showed that the incorporation of crop residues into soil significantly prevented soil erosion and enhanced the soil quality. Sonnleitner *et al.* [24] found that straw incorporation also improved the aggregate stability and other soil properties compared with farmyard manure. Mulumba and Lal [25] also reported that the addition of crop residues to cultivated soil had positive effects on the soil porosity, available water content, soil aggregation, and bulk density. Bhagat and Verma [26] showed that the incorporation of crop straws for five years significantly increased the crop yield and improved the soil properties.

The soil improvement effect of straw incorporation has been recognized widely [23–25] but information is still limited on the responses of the SOC and water-stable aggregates under different rates of straw incorporation, particularly in the loessal soil in semiarid areas of northwest China. The theory and technique of straw incorporation in this region have also not been reported. Thus, the present study investigated the effects of different crop straw application rates combined with conventional planting on SOC, the >0.25 mm water-stable macroaggregate rate, and various soil properties in the southern Ningxia region of China.

Materials and Methods

Ethics Statements

The study was carried out on the private land, we rent the farmland from the local farmers, and contracts and deeds are signed. No specific permissions were required in this area to run the experiment as the study sites are farming area without any protection zone, and the farming activities won't hurt the local animals. And we only plant the grain crop in the field, so the field studies did not involve endangered or protected species.

Site description

The experiment was conducted between 2008 and 2010 at the Dryland Agricultural Research Station, Pengyang County, Ningxia, China (106°45'N, 35°79'E and 1800 m a.s.l.). The experimental area was in a hilly and gully region of the Loess Plateau, which was characterized by a semiarid, warm temperate, continental monsoon climate. The average annual rainfall was 435 mm, which fell mainly from June to September. The annual mean evaporation was 1050 mm and the annual temperature average was 8.1°C with a frost-free period of 155 days.

Rainfall during the experimental period was measured using an automatic weather station (WS-STD1, England) at the experimental site. Monthly precipitation distributions during the experimental period are shown in Fig. 1. The total precipitation for 2008, 2009, and 2010 was 390.9, 335.2, and 537 mm, while the precipitation during the maize-growing season was 362, 298.2, and 476.1 mm, respectively.

The soil at the experimental site was a loessal soil with a pH of 8.5. In the 0–40 cm soil layer, the organic matter, total N, P, and K were 8.32 g kg^{-1}, 0.61 g kg^{-1}, 0.58 g kg^{-1}, and 5.4 g kg^{-1}, respectively, while the available N, P, and K were 46.25 mg kg^{-1}, 10.41 mg kg^{-1}, and 104.82 mg kg^{-1}. In 2007, the site was planted with maize prior to the experiment

The experimental field was flat and, according to the FAO/UNESCO Soil Classification [27], the soil was a Calcic Cambisol (sand 14%, silt 26%, and clay 60%) with low fertility. The key physical properties of the soil layers (0–40 cm depth) are shown in Table 1.

Experimental design and field management

The experiment used a randomized block design with three replicates. Each plot was 3 m wide and 6 m long. The experiment included four straw incorporation rate treatments: (i) no straw incorporation (CK); (ii) incorporation of maize straw at a low rate of 4 500 kg ha^{-1} (L); (iii) incorporation of maize straw at a medium rate of 9000 kg ha^{-1} (M); (iv) incorporation of maize straw at a high rate of 13 500 kg ha^{-1} (H).

The maize straws were mixed manually with the top 25 cm of soil in the field. Before mixing with the soil, the maize straws were chopped into 5 cm pieces and then applied to the soil six months before the crop was planted to facilitate decomposition of the straw. The straw was incorporated into the soil layer on 15 October 2007 and after the crop harvests during 2008–2010.

Ten days before sowing, a basis fertilizer containing 102 kg N ha^{-1} and 90 kg P ha^{-1}, was spread evenly over the each plot and plowed into soil layer. Maize (cv. Shendan 16) was sown at a rate of 5.25 seed m^{-2} on 18 April 2008, 15 April 2009 and 20 April 2010 using a holesowing (3 cm in diameter) machine. An additional 102 kg N ha^{-1} was applied as a top dressing in late June. And on 7 October 2008, 5 October 2008, and 10 October 2010. No irrigation was provided during the experimental years. Manual weeding was performed throughout the experiment.

Sampling and measurement

Rainfall data were recorded using a standard weather station located at the experimental site. After the maize harvest in 2008 and 2010, soil samples were collected for the four incorporation treatments. A soil sample was collected from each plot at depths of 0–20 cm, 20–40 cm and 40–60 cm to determine the soil organic matter. A similar soil sample was collected at depths of 0–10 cm, 10–20 cm, 20–30 cm and 30–40 cm to determine the aggregate stability. The soil samples were collected from four points in each plot replicate and mixed to produce a composite sample. Each soil samples was passed through an 8 mm sieve by gently breaking the soil clods, whereas pebbles and stable clods >8 mm were discarded. Soil samples were air-dried for 24 h in the laboratory before analysis.

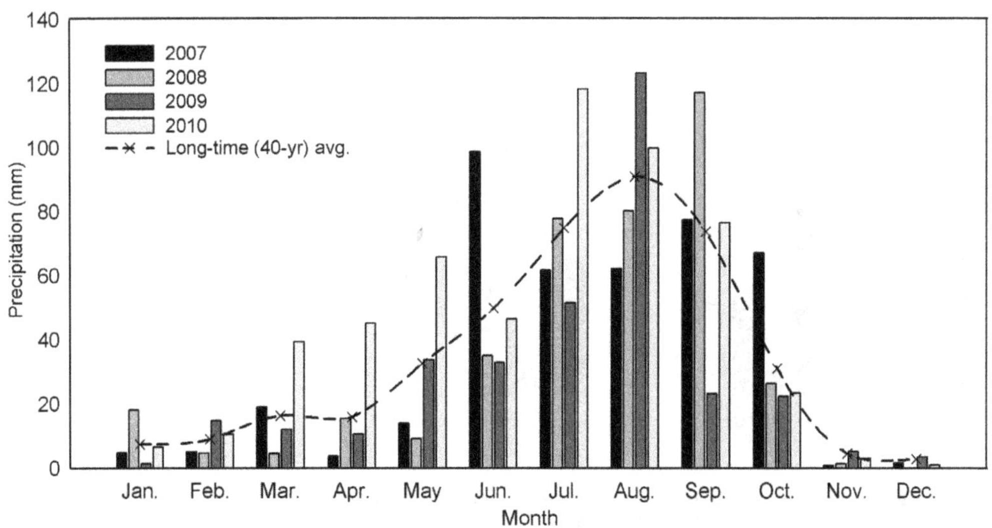

Figure 1. Distribution of mean monthly precipitation at the experimental site during 2007–2010.

Table 1. Physical properties of the tilth soil (0–40 cm depth) in the experimental site.

Depth (cm)	Bulk density (g cm^{-3})	Aggregate size (%)					
		>5 mm	5-2 mm	2-1 mm	1-0.5 mm	0.5-0.25 mm	<0.25 mm
0–10	1.33	0.1	0.27	2.25	4.5	4.47	88.41
10–20	1.33	0.15	0.19	1.38	4.2	4.15	89.93
20–30	1.36	0.1	0.28	1.41	4.02	3.36	90.83
30–40	1.38	0.1	0.22	1.07	3.29	3.8	91.52

Soil organic carbon was determined by the $K_2Cr_2O_7$–H_2SO_4 digestion method, and SOM content was calculated as a portion of SOC which has been described by Wang *et al.* [28].

$$OM_i = O_i \times A \quad (1)$$

$$C_i = \frac{d_i \times \rho_i \times O_i}{10} \quad (2)$$

where OM_i is the soil organic matter content (g kg^{-1}), $O_i$is the soil organic carbon content (g kg^{-1}), C_i is the soil organic carbon storage (Mg C ha^{-1}), $d_i$is the soil depth (cm), ρ_i is the soil bulk density (g cm^{-3}), A is Van Bemmelen coefficient (A = 2) [29].

The size distribution of water-stable aggregates was determined by placing a soil sample on a stack of sieves (5,2, 1, 0.5 and 0.25 mm) fitted with a soil aggregate analyzer (Japan, QD24–DIK–2001). The stacked sieves were immersed in water and moved up and down by 3.5 cm at a frequency of 30 cycles 60s^{-1} for 15 min. The proportions of aggregates that measured >5, 5–2, 2–0.5, 0.5–0.25 and <0.25 mm were calculated [30].

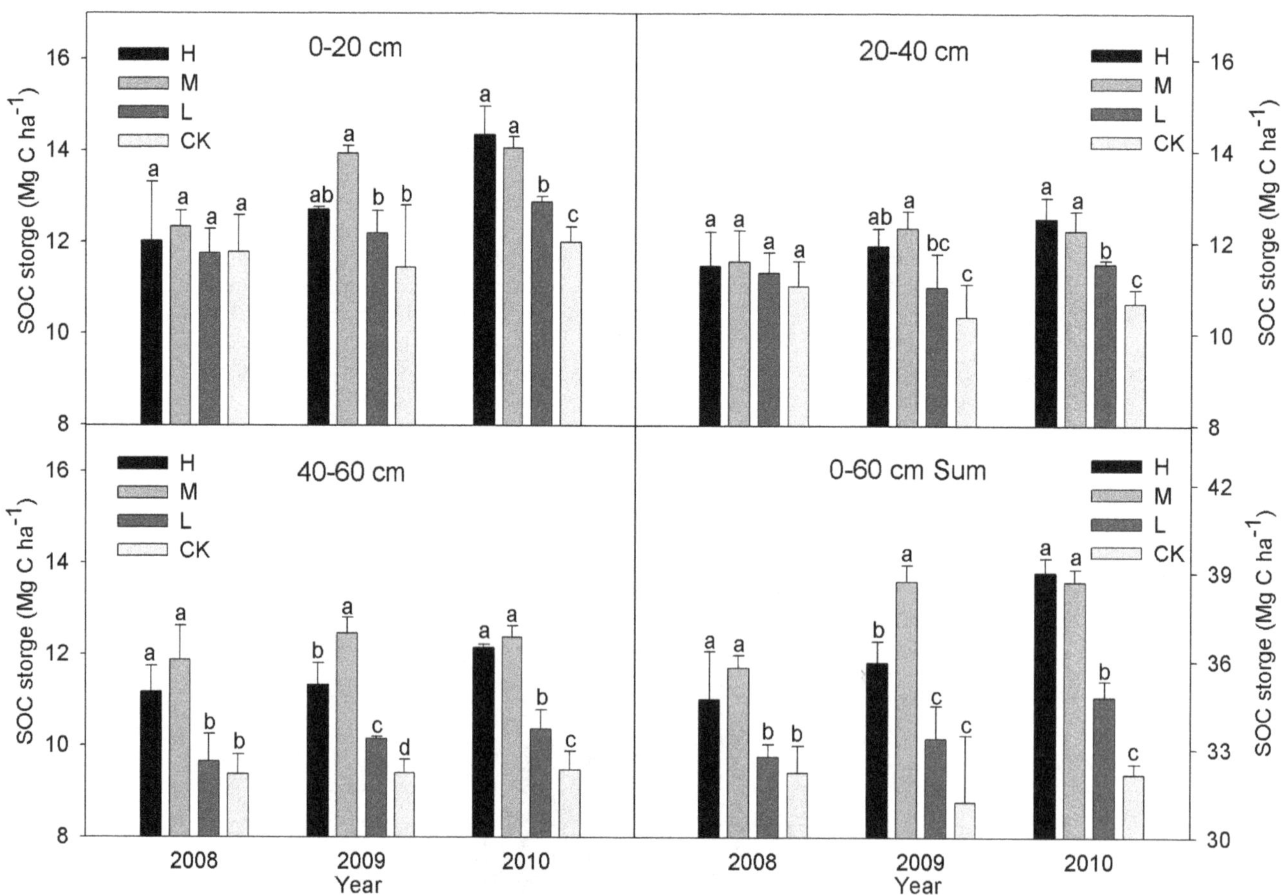

Figure 2. Soil organic carbon storage with different straw incorporation treatments (Mg C ha^{-1}). Note: CK, no straw incorporation; L, incorporation of straw at a low rate of 4 500 kg ha^{-1} maize straw; M, incorporation of straw at a medium rate of 9 000 kg ha^{-1} maize straw; H, incorporation of straw at a high rate of 13 500 kg ha^{-1} maize straw; Sum, the sum value of the 0–60 cm soil layers; Bars with different lower case letters indicate significant differences at $P<0.05$.

Table 2. $WR_{0.25}$ with different straw incorporation treatments (%).

Year	Treatment	Soil depth (cm)				0-40 AVG
		0—10	10—20	20—30	30—40	
2008	H	13.37a[1]±0.33B[2]	10.57a±1.28[3]A	7.72ab±0.27B	7.33a±0.51A	9.75a±0.31B
	M	11.79b±1.05A	9.28a±0.60B	7.97a±0.65A	6.62ab±0.85B	8.91b±0.26B
	L	10.10c±0.46A	7.39b±0.58A	6.53bc±1.08A	6.07ab±0.81A	7.52c±0.16A
	CK	9.12c±0.70A	6.93b±0.78A	5.89c±0.57A	5.68b±0.63A	6.91d±0.44A
2009	H	14.64a±0.49AB	11.46a±0.17A	9.51a±0.26AB	7.85a±1.00A	10.86a±0.19AB
	M	14.93a±0.89A	11.22a±0.14AB	9.34a±0.32A	7.06ab±1.46AB	10.64a±0.27AB
	L	10.39b±0.26A	9.48b±1.49A	7.02b±1.18A	6.48ab±1.10A	8.34b±0.32A
	CK	9.75b±0.26A	7.18c±0.73A	6.40b±1.00A	5.65b±0.35A	7.25c±0.10A
2010	H	15.81a±1.20A	12.76a±0.44A	11.50a±0.20A	8.57a±0.05A	12.16a±0.59A
	M	16.72a±1.36A	12.81a±0.05A	11.30a±1.10A	8.81a±0.44A	12.41a±0.72A
	L	11.00b±0.78A	9.06b±0.81A	8.03b±2.14A	6.96b±0.81A	8.76b±0.66A
	CK	10.23b±0.46A	7.79b±0.23A	6.07b±0.46A	5.71c±0.23A	7.45c±0.25A

Note: CK, no straw incorporation; L, incorporation of straw at a low rate of 4 500 kg ha^{-1} maize straw; M, incorporation of straw at a medium rate of 9 000 kg ha^{-1} maize straw; H, incorporation of straw at a high rate of 13 500 kg ha^{-1} maize straw; AVG, the mean value of the 0-40 cm soil layers.

[1]Values followed by the same lowercase letter in the same line are not significantly different according to Duncan's multiple range test ($P<0.05$) between the four straw incorporation treatments in the same year.

[2]Values followed by the same uppercase letter in the same line are not significantly different according to Duncan's multiple range test ($P<0.05$) between the different years of the same straw incorporation treatment.

[3]Means ± standard deviations.

And the proportions of aggregates were used to calculate the water-stable macroaggregates content with a diameter of >0.25 mm [31], mean weight diameter (MWD) [32,33], geometric mean diameter (GMD) [34], and the soil aggregate stability (WSAR) [35]. These parameters were calculated as follows:

$$macroaggregates(>0.25mm)=\sum_{i=1}^{n} W_i \quad (3)$$

$$MWD=\sum_{i=1}^{n} \overline{X_i} W_i \quad (4)$$

$$GMD=\exp\left(\frac{\sum_{i=1}^{n} W_i \log \overline{X_i}}{\sum_{i=1}^{n} W_i}\right) \quad (5)$$

$$WSAR=WSA/A\times 100\% \quad (6)$$

where $macroaggregates(>0.25mm)$ is the volume of soil particles with a diameter of >0.25 mm, $W_i$is weight of the aggregates in that size range as a fraction of total dry weight of the sample analysed, and n is number of sieves, $\overline{X_i}$ is the mean diameter of aggregates over each sieve size, WSA is the mass of $macroaggregates(>0.25mm)$, and A is the mass of the soil aggregates with a diameter of >0.25 mm.

Fractal dimension D were then obtained to describe the characteristics of soil aggregate size distribution. As suggested by Tyler and Wheatcraft [36] and Zhang *et al.*, [37], the volume of soil with particle diameter $>d_i(d_i>d_{i+1}, i=1,2,\ldots)$ is defined as:

$$V(\delta>d_i)=A\left[1-(d_i/k)^{3-D}\right)], \quad (7)$$

where δ is yard measure, k and A are the constants representing size and shape, respectively, and D is the fractal dimension. For a given soil, d_i represents the average value of soil particles between d_i and d_{i+1}.Generally, variations of particle density ρ among different soil particles could be ignored. And hence ρ is a constant. Therefore, another expression (Eq. (8)) is derived from Eq. (7):

$$W(\delta>d_i)=V(\delta>d_i)\rho=\rho A\left[1-(d_i/k)^{3-D}\right)], \quad (8)$$

where $W(\delta>d_i)$ is the cumulative mass of particles with sizes $\delta>d_i$, and W_0 is the total mass of any sizes of soil particles. The fractal equation, reflecting the relationship between the mass distributions of soil particles and average particle diameter, can be obtained as follows:

$$\frac{W(\delta>d_i)}{W_0}=1-\left(\frac{\overline{d_i}}{d_{max}}\right)^{3-D} \quad or \quad \frac{W(\delta>d_i)}{W_0}=\left(\frac{\overline{d_i}}{d_{max}}\right)^{3-D} \quad (9)$$

Then after regression analysis between $\log(\overline{d_i}/d_{max})$ and $\log((W(\delta<\overline{d_i}))/W_0)$, the fractal dimension D can be calculated.

Statistical analysis

Statistical analysis was carried out by using the SPSS 13.0 (Statistical Package for the Social Sciences) package. The effects of treatments on the measured parameters were evaluated using a one-way ANOVA. Duncan's new multiple range test was used to calculate the least significant difference (LSD) between means

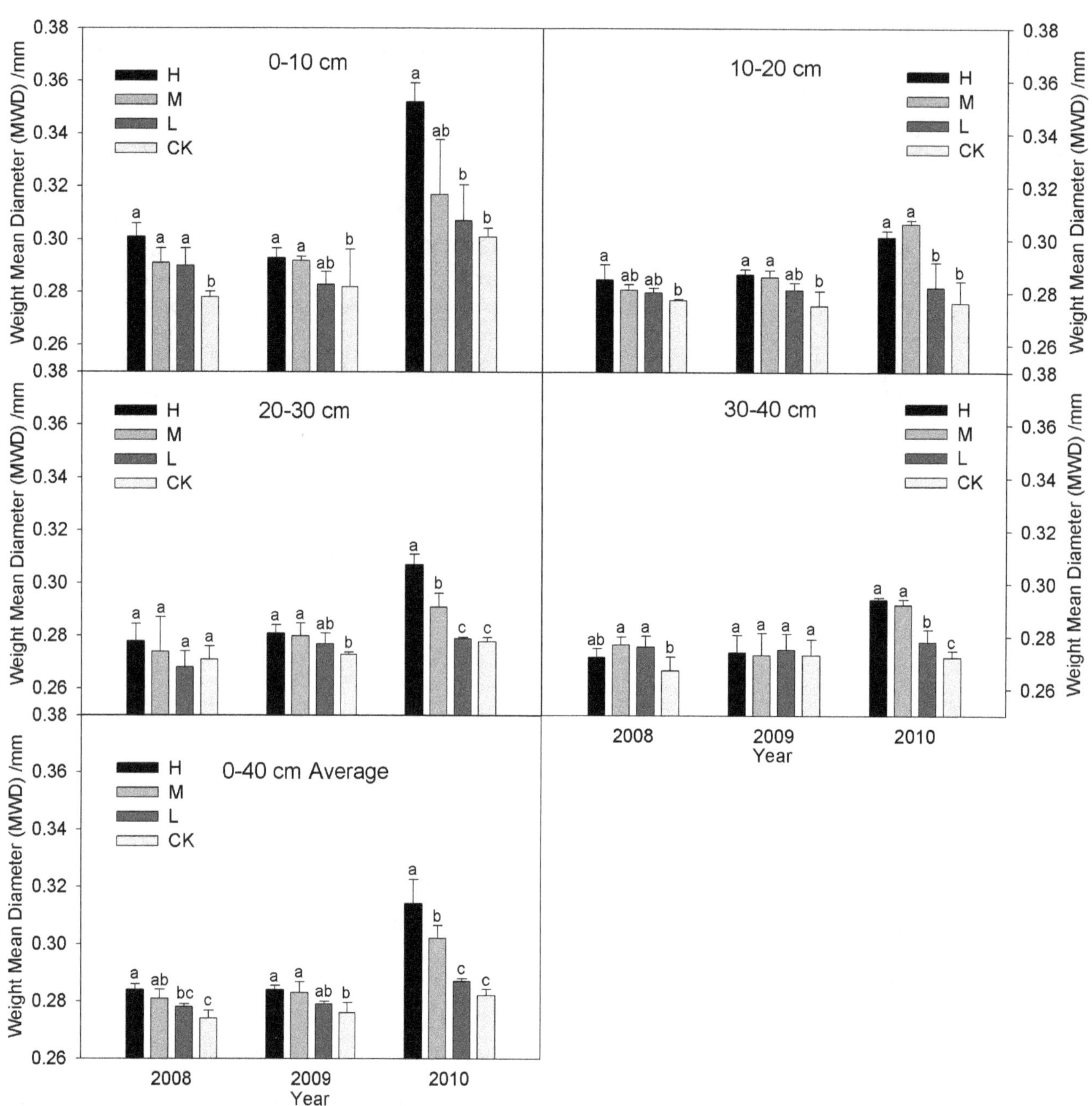

Figure 3. MWD values with wet sieving under the different straw incorporation treatments. Note: MWD, weight mean diameter; CK, no straw incorporation; L, incorporation of straw at a low rate of 4 500 kg ha^{-1} maize straw; M, incorporation of straw at a medium rate of 9 000 kg ha^{-1} maize straw; H, incorporation of straw at a high rate of 13 500 kg ha^{-1} maize straw; Average, the mean value of the 0–40 cm soil layers. Bars with different lower case letters indicate significant differences at $P<0.05$.

when F-values were significant. In all cases, differences were deemed to be significant if $P<0.05$.

Results

Soil organic carbon storage (SOC)

The effect of straw incorporation on SOC storage is shown in Fig. 2, where the soil organic carbon storage increased with the straw incorporation. The sum of SOC storage in 0–60 cm layers with the three incorporation treatments were higher than CK, i.e., 7.71% ($P<0.05$), 11.14% ($P<0.05$) and 1.70% in 2008, 15.15% ($P<0.05$), 24.00% ($P<0.05$) and 6.86% in 2009, and 21.40% ($P<0.05$), 20.38% ($P<0.05$) and 8.21% ($P<0.05$) in 2010, respectively. The SOC storage increased with the number of years of incorporation, i.e., the SOC storage (0–60 cm depth) in 2010 had increased by 6.19–12.48% compared with 2008, and decreased with the soil layer depth, i.e., by 3.75–25.68% in 2008, 11.85–21.70% in 2009 and 13.51–26.64% in 2010. The SOC storage of H and M was slightly higher than CK in 2008, although the difference was significant in 40–60 cm layer only. In 2009, compared with CK, H and M significantly increased the SOC storage by 11.01% and 21.74% ($P<0.05$) in 0–20 cm layer,

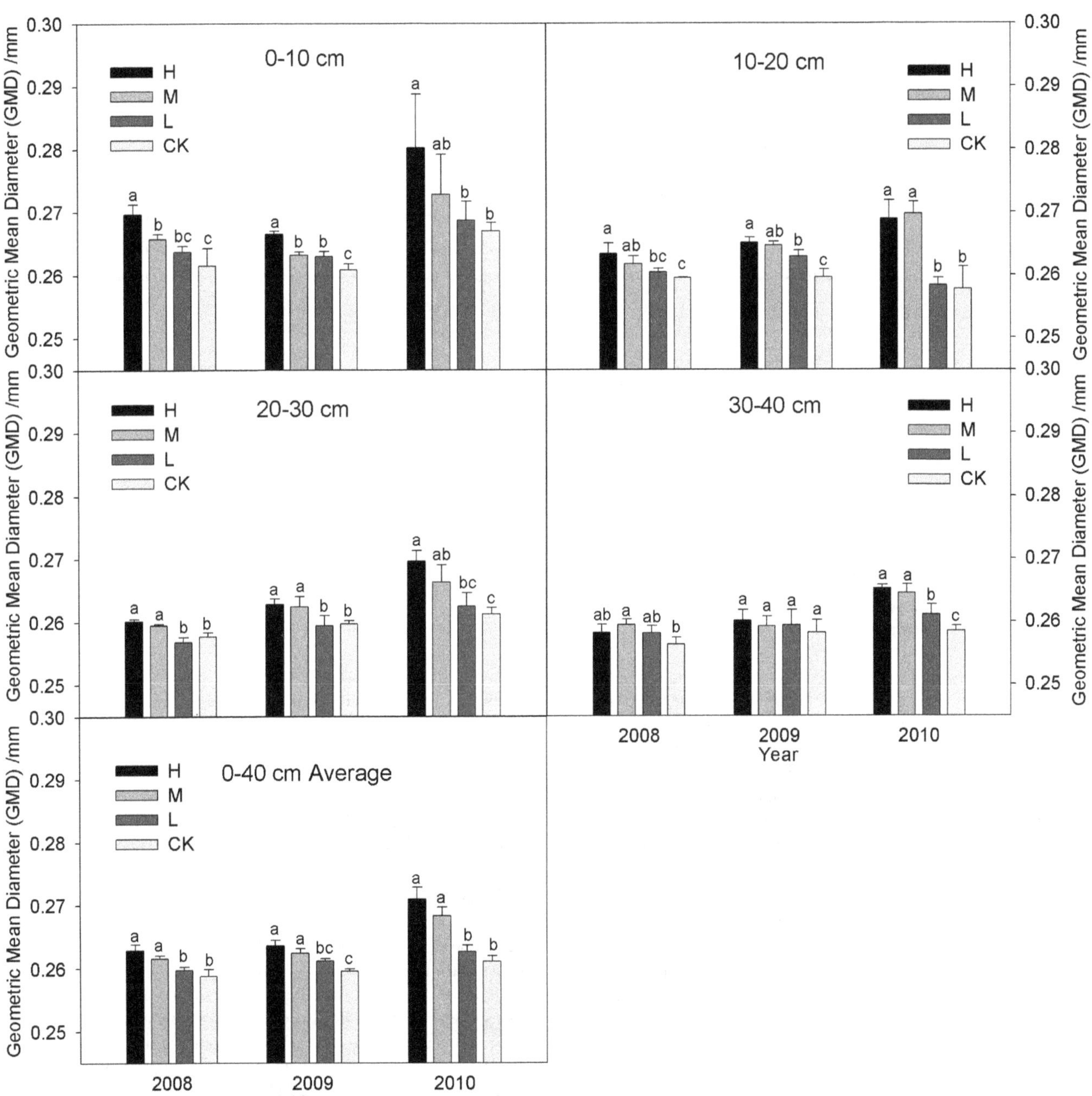

Figure 4. GMD values with wet sieving under the different straw incorporation treatments. Note: GMD, geometric mean diameter; CK, no straw incorporation; L, incorporation of straw at a low rate of 4 500 kg ha^{-1} maize straw; M, incorporation of straw at a medium rate of 9 000 kg ha^{-1} maize straw; H, incorporation of straw at a high rate of 13 500 kg ha^{-1} maize straw; Average, the mean value of the 0–40 cm soil layers. Bars with different lower case letters indicate significant differences at $P<0.05$.

14.94% and 18.81% ($P<0.05$) in 20–40 cm layer, and 20.40% and 32.47% ($P<0.05$) in 40–60 cm layer, respectively, and L was significantly increased by 7.89% ($P<0.05$) in 40–60 cm layer only. There was a significant difference between straw incorporation treatments and CK for each of the three soil layers in 2010, but there were no significant differences between H and M.

Water-stable macroaggregates (>0.25 mm)

The effects of straw incorporation on >0.25 mm macroaggregates are shown in Table 2. The mean >0.25 mm macroaggregates contents of the 0–40 cm layers of incorporation treatments were significantly ($P<0.05$) higher than CK during 2008–2010, i.e., 8.91–41.14%, 15.13–49.94%, and 17.64–66.58%, respectively. The >0.25 mm macroaggregates content also decreased with the soil layer depth, and increased with the number of years of incorporation,, i.e., the H, M, L and CK in 2010 was 24.72% ($P<0.05$), 39.28% ($P<0.05$), 16.49% and 7.81% higher than that in 2008. Compared with CK, the >0.25 mm macroaggregates content of three straw incorporation treatments in the 0–10 cm layer was increased by 10.75–46.60% in 2008, 6.56–53.13% in 2009, and 7.53–63.44% in 2010, respectively, while the differences between L and CK were not significant ($P>0.05$) during 2008–2010. The

Table 3. Effects on the soil aggregate stability rate with different straw incorporation treatments (%).

Year	Treatments	Soil Depth (cm)				0–40 AVG
		0–10	10–20	20–30	30–40	
2008	H	21.86a[1]±0.89[A2]	16.29a±0.41[3][A]	10.83a±0.54[B]	9.19a±0.65[A]	14.60a±0.33[A]
	M	19.62ab±1.09[A]	15.31b±0.62[A]	11.16a±0.82[A]	8.47b±0.21[A]	13.70b±0.62[A]
	L	17.83bc±1.56[A]	12.88c±0.24[A]	9.70b±0.64[A]	8.00bc±0.42[A]	11.83c±0.18[A]
	CK	16.45c±2.42[A]	12.52c±0.58[A]	9.39b±0.80[A]	7.53c±0.37[A]	11.79c±0.45[A]
2009	H	20.32a±0.27[A]	16.14a±0.54[A]	13.15a±0.34[AB]	9.74a±0.99[A]	14.84a±0.23[A]
	M	20.02a±0.46[A]	15.41a±0.50[A]	12.61a±0.41[A]	8.66ab±1.28[A]	14.18b±0.22[A]
	L	15.76b±0.49[A]	14.45a±1.78[A]	10.16b±1.36[A]	8.28ab±1.01[A]	12.16c±0.35[A]
	CK	15.47b±0.30[A]	11.81b±0.83[A]	9.59b±1.10[A]	7.50b±0.41[A]	11.09d±0.05[A]
2010	H	20.19a±1.93[A]	16.48a±1.73[A]	15.24a±0.12[A]	10.45a±0.11[A]	15.59a±0.26[A]
	M	20.87a±0.68[A]	15.78a±1.32[A]	13.99a±1.23[A]	10.08a±0.43[A]	15.18a±0.64[A]
	L	15.72b±0.72[A]	13.39a±1.81[A]	12.53b±0.46[A]	8.66b±0.77[A]	12.58b±1.00[A]
	CK	15.11b±0.12[A]	12.79a±1.70[A]	12.00b±0.76[A]	7.52c±0.26[A]	11.85b±1.32[A]

Note: CK, no straw incorporation; L, incorporation of straw at a low rate of 4 500 kg ha^{-1} maize straw; M, incorporation of straw at a medium rate of 9 000 kg ha^{-1} maize straw; H, incorporation of straw at a high rate of 13 500 kg ha^{-1} maize straw; AVG, the mean value of the 0–40 cm soil layers.
[1]Values followed by the same lowercase letter in the same line are not significantly different according to Duncan's multiple range test ($P<0.05$) between the four straw incorporation treatments in the same year.
[2]Values followed by the same uppercase letter in the same line are not significantly different according to Duncan's multiple range test ($P<0.05$) between the different years of the same straw incorporation treatment.
[3]Means ± standard deviations.

>0.25 mm macroaggregates content showed the same trends in the 10–20 and 20–30 cm layers, and the H and M treatments had the significant difference ($P<0.05$) with CK. The H, M and L increased by 29.05% ($P<0.05$), 16.55% and 6.87% in 2008, 38.94% ($P<0.05$), 24.96% and 14.69% in 2009, and 50.09% ($P<0.05$), 54.29% ($P<0.05$), 21.89% ($P<0.05$) in 2010, respectively. A linear correlation was found between the >0.25 mm macroaggregates and soil organic carbon ($R^2>0.64$, $P<0.01$).

Mean weight diameter (*MWD*) and Geometric mean diameter (*GMD*)

Fig. 3 & 4 show that the *MWD* and *GMD* values with the three incorporation treatments increased significantly throughout the three-year study. The average *MWD* and *GMD* values under the incorporation treatments were higher in the 0–40 cm layers than CK during 2008–2010, i.e., 1.46–3.65% and 0.39–1.54% in 2008, 1.09–2.90% and 0.62–1.55% in 2009, 1.77–11.35% and 0.77–3.83% in 2010, respectively. There was no significant difference ($P>0.05$) between L and CK during the study period. The *MWD* and *GMD* values decreased with the soil layer depth and increased with the number of years of incorporation, i.e., the H, M, L and CK in 2010 was 10.56% ($P<0.05$) and 3.04% ($P<0.05$), 7.47% ($P<0.05$) and 2.29% ($P<0.05$), 3.24% and 1.15%, and 2.92% and 0.77% higher than that in 2008. The *MWD* and *GMD* values of treatments in the 0–10 and 10–20 cm layers were ranked in the order: H > M > L > CK. The H and M levels were significantly higher than CK while H and M levels were similar during 2008–2010 (Fig. 3–4). The *MWD* and *GMD* exhibited the same trends in 20–30 cm layers and there were no significant ($P>0.05$) differences among incorporation treatments during 2008–2009, the *MWD* values with the H were significantly ($P<0.05$) higher than the other three treatments during 2010, i.e., by 5.50%, 10.04%, and 10.43%, respectively. The *GMD* did not differ significantly among the four treatments, with the exceptions of H and L, and H and CK in 2010. The *MWD* and *GMD* values in the 30–40 cm layers under the incorporation treatments were increased by 1.87%, 3.75% ($P<0.05$), 3.37% ($P<0.05$) and 0.76%, 1.76% ($P<0.05$), 0.78% in 2008, 0.37%, 0.01%, 0.73% and 0.73%, 0.45%, 0.45% in 2009, 8.09% ($P<0.05$), 7.35% ($P<0.05$), 2.21% ($P<0.05$) and 2.32% ($P<0.05$), 1.93% ($P<0.05$), 0.77% ($P<0.05$) in 2010, compared with CK, respectively. There was no significant difference ($P>0.05$) between H and M during the study period.

Soil aggregate stability (*WSAR*)

The straw incorporation significantly affected the *WSAR* after three years (Table 3). The *WSAR* values in the 0–40 cm layers of incorporation treatments was significantly higher than CK during 2008–2010, i.e., 0.34–19.80%, 9.65–33.79%, and 6.16–35.08%, respectively. The *WSAR* also decreased with the soil layer depth, while it increased with the number of years of incorporation. The *WSAR* in the 0–10 cm layers increased significantly with the amount of straw and the *WSAR* was highest with H throughout 2008–2010, i.e., the *WSAR* significantly ($P<0.05$) increased by 28.35%, 28.93% and 25.78% compared with L, respectively, and by 33.62%, 31.35% and 32.89% ($P<0.05$) compared with CK, respectively. Compared with L and CK, the WSAR values with M increased significantly by 20.08–24.67% and 26.87–29.78% during 2008–2010, respectively, but there were no significant ($P>0.05$) differences between H and M, and L and CK. With the four treatments, the *WSAR* values in the 10–20 and 20–30 cm layers were lower than the 0–10 cm layers, which were ranked in the order: H > M > L > CK. There were no significant differences ($P>0.05$) among the four treatments in the 30–40 cm layers throughout the three years, and the *WSAR* maintained at 7.50–10.45%.

Fractal dimension (D)

Fig. 5 shows that the fractal dimensions with the three incorporation treatments decreased significantly ($P<0.05$) after three years of straw incorporation (the R^2 of fitting curve is:

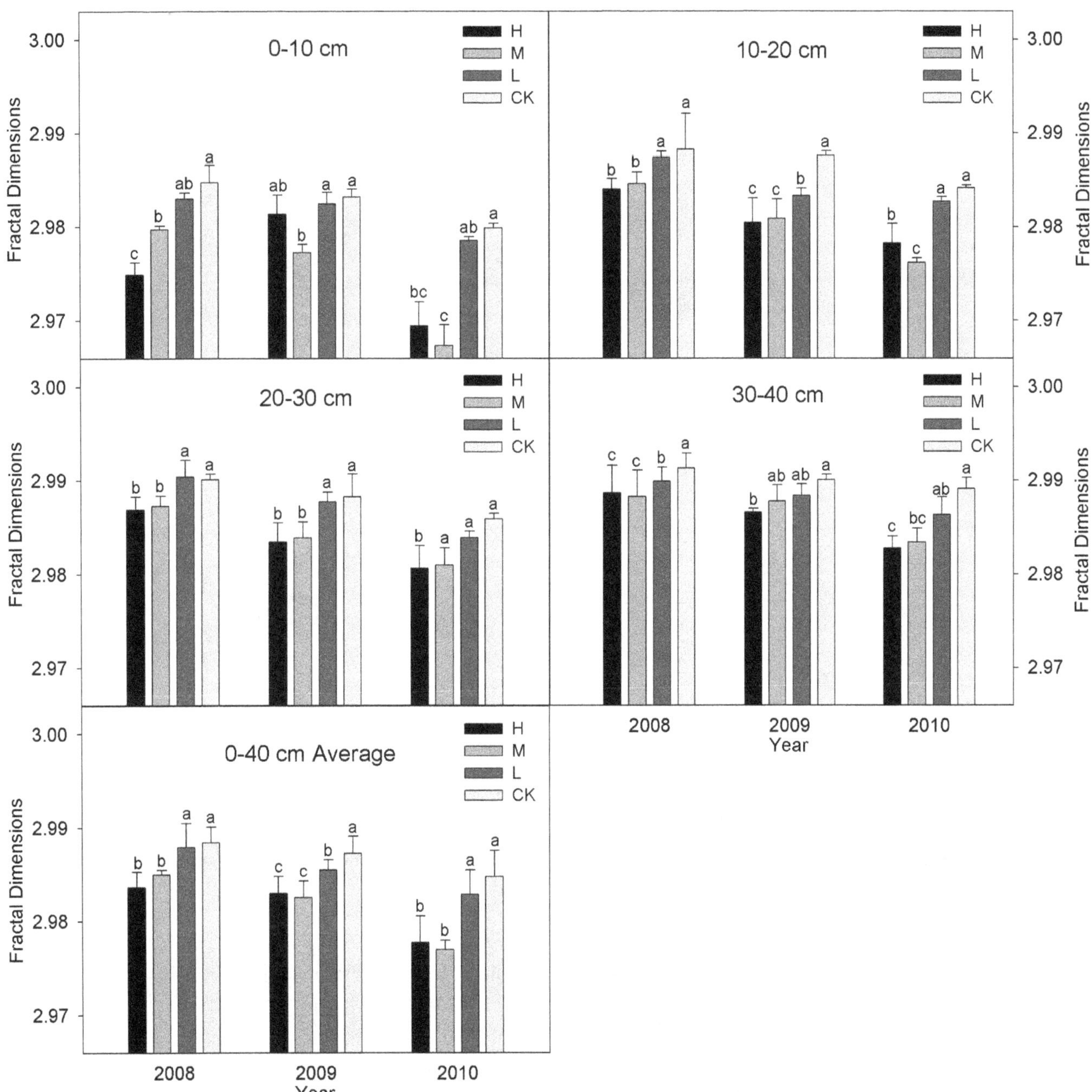

Figure 5. Fractal dimensions (D) of soil aggregates with different straw incorporation treatments. Note: CK, no straw incorporation; L, incorporation of straw at a low rate of 4 500 kg ha^{-1} maize straw; M, incorporation of straw at a medium rate of 9 000 kg ha^{-1} maize straw; H, incorporation of straw at a high rate of 13 500 kg ha^{-1} maize straw; Average, the mean value of the 0–60 cm soil layers; Bars with different lower case letters indicate significant differences at $P<0.05$; $0.92<R^2<0.99$.

0.92~0.99). The fractal dimensions (0–40 cm layer) with the incorporation treatments were lower than those of CK during 2008–2010, i.e., 0.16% ($P<0.05$), 0.11% ($P<0.05$) and 0.02% in 2008, 0.14% ($P<0.05$), 0.16% ($P<0.05$) and 0.06% ($P<0.05$) in 2009, and 0.24% ($P<0.05$), 0.26% ($P<0.05$) and 0.06% in 2010, respectively. The fractal dimensions also increased with the soil layer depth and decreased with the number years of incorporation, i.e., the H, M, L and CK in 2010 was 0.20% ($P<0.05$), 0.27% ($P<0.05$), 0.17% ($P<0.05$) and 0.12% higher than that in 2008. The fractal dimensions in the 0–10 cm layers with the four treatments were ranked in the order: H<M<L<CK in 2008. The fractal dimensions with H and M were significantly ($P<0.05$) lower than CK, i.e., 0.33%, 0.17% and 0.06%, respectively. However, the ranking was M<H<L<CK during 2009–2010, i.e., 0.06%, 0.20% ($P<0.05$), 0.02% in 2009, 0.35% ($P<0.05$), 0.42% ($P<0.05$) and 0.04% in 2010, respectively. While the differences between L and CK were not significant ($P>0.05$) during 2008–2010. Compared with L and CK, the fractal dimensions in the 10–20 cm layers under H was significantly ($P<0.05$) reduced the fractal dimensions by 0.11% and 0.14% in 2008, and 0.10% and 0.24% in 2009,, while M significantly ($P<0.05$) reduced by 0.09% and 0.12% in 2008, and 0.08% and 0.23% in 2009, respectively.

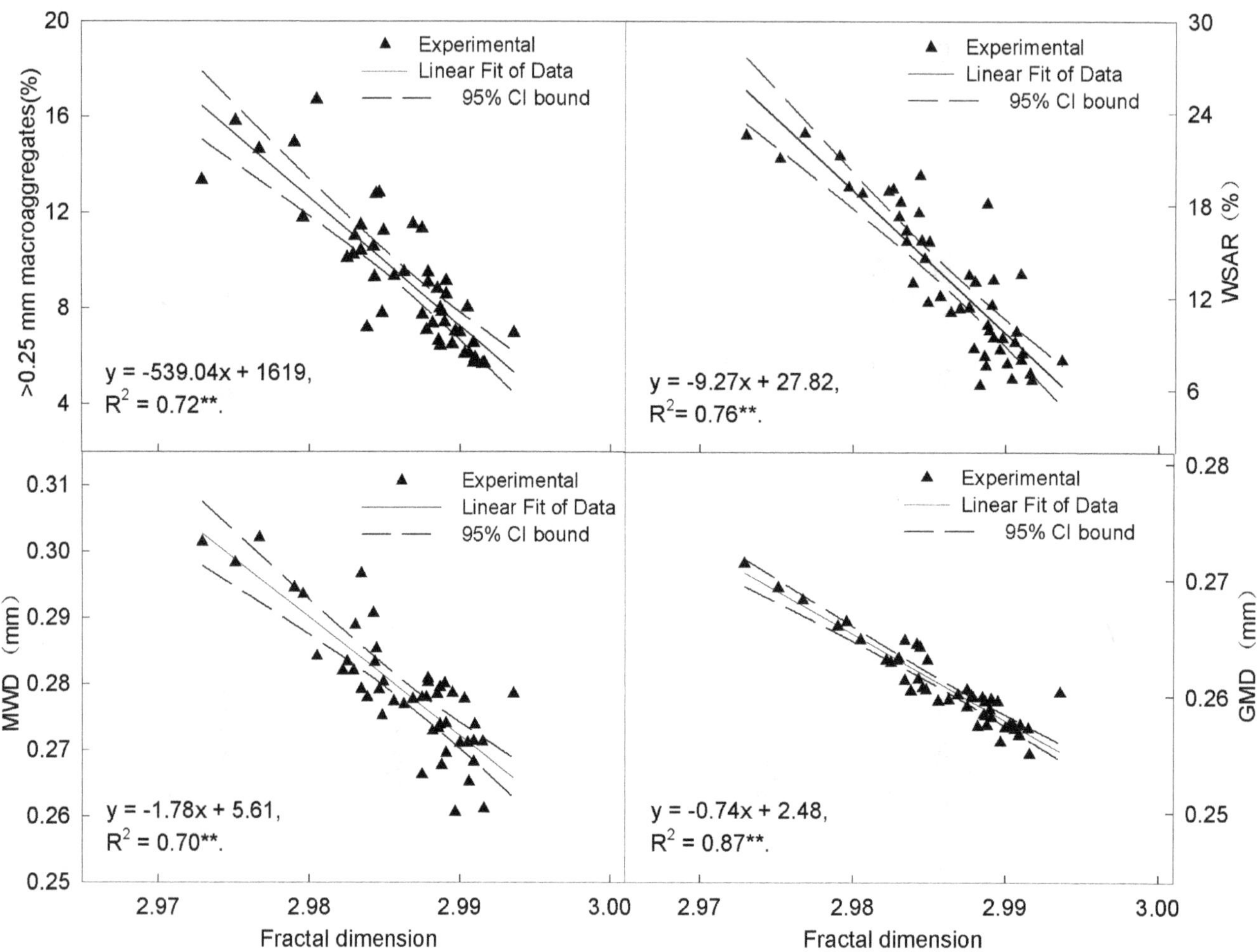

Figure 6. Correlations between D and >0.25 mm macroaggregates, MWD, GMD, WSAR of the soil aggregates. Note: MWD, weight mean diameter of soil aggregates; GMD, geometric mean diameter of soil aggregates; WSAR, soil aggregate stablility rate; D, Fractal dimension of soil aggregates. 95% CI, 95% confidence interval.** $P<0.01$.

The fractal dimension was lowest with M during 2010, i.e., it was decreased by 0.22% ($P<0.05$) compared with L, and by 0.27% ($P<0.05$) compared with CK. The differences among the four treatments decreased gradually in the 20–40 cm layer and the fractal dimension remained at 2.987~2.991.

Data fitting detected a linear correlation between the indexes of soil aggregates and the fractal dimension (Fig. 6).

Discussion

Results of this study demonstrated that application of crop straws had positive effects on the soil physico-chemical properties. On the other hand, there are Loess plateau regions at northwest China with intensive cultivation systems and poor soil management strategies. Therefore, crop residual management is very important for preserving natural ecosystems [38]. Another problem for this area is conventional tillage that reduces the soil water storage and destroys the aggregates and soil structure [39] which prompts deterioration of crop yield.

The results indicated that the SOC storage increased significantly in the 0–60 cm layers after three years of straw incorporation (Fig. 2). Rasmussen and Collins [40] reported that the soil organic matter content was strongly related to the amount of residues added and only weakly related to the type of residue applied. The H and M treatments were significantly different from the L throughout the three years and the soil organic matter level increased as the straw incorporated and decomposed [20], which effectively mitigated the loss of soil organic carbon from in the agroecosystem caused by intensive cropping [41]. The SOC storage decreased gradually in all the treatments with the soil layer depth because the degree of straw incorporation was lower in the deeper layers compared with the surface layers (0–20 cm) of the soil [42]. This was because the amount of straw incorporated in the topsoil was greater than that in the deeper layers [40,43].

Soil aggregates are the basic units of the soil structure and they are composed of primary particles and binding agents [12]. They are also necessary soil conditions for high crop yields [44]. Conventional tillage disturbs the soil and increase the effects of drying–rewetting and freezing–thawing, which increases the susceptibility of the macroaggregate (>0.25 mm) to disruption [45–47]. Pinheiro *et al.* [48] showed that soil exposure with tillage and the lack of residue inputs led to a decline in aggregation and organic carbon, both of which made the soil susceptible to erosion. Our study showed that the straw incorporation of straws determined significantly more and larger soil aggregates than CK, thus indicating an improvement of soil physical quality. This may have been attributable to the significant increase in the SOC storage (an average increase by 8.21–21.40% in 2010), the lower soil bulk density (an average decreased by 1.80–4.13% in 2010, data no shown) [49], and increased soil porosity (an average increased by 1.70–3.90% in 2010, data no shown) [50] after straw incorporation. It also stimulated the activity of soil microorganisms [51] and an abundance of polyose metabolites were produced during the straw decay process [52]. These soil physical and chemical conditions may have accelerated the incorporated SOM decomposing process and increased soil aggregation [24].

Soil aggregate structural stability is widely recognized as a key indicator of soil quality, which is closely related to a number of soil properties, processes, and functions, e.g., the quantity and composition of SOM [48], infiltration capacity [53], soil biotic activity [54] and the resistance to erosion [55,56]. Wei *et al.* [57] showed that the addition of crop residues was the most effective measure for increasing the rhizosphere aggregate stability. Sonnleitner *et al.* [24] and Karami *et al.* [41] also found that straw application improved the aggregate stability and other soil properties. In our study, the soil aggregate stability of the straw incorporation treatments were significantly higher than CK in 2010, and it was decreased with the soil layer depth. These results agreed with studies by Tripathy and Singh [42] and Karami *et al.* [41]. Our results also indicated that straw incorporation was positively related to the physical protection of organic matter [20] and an increased aggregate quantity [58], but it also improved the soil aggregate stability [20] and reduced soil degeneration [22,23].

Many studies have shown that soil is a porous medium with fractal characteristics [59–60]. Thus, fractal theory can be used to describe the complex characteristic of soil structure [61]. Castrignanò and Stelluti [62] reported that a higher fractal dimension indicated the heavier texture of a soil and its inferior permeation properties. This showed that fractal theory is an effective method to describing the soil aggregate distribution [37] and changed with different levels of straw incorporation [31]. The fractal dimensions of the 0–40 cm layers with the four treatments after three years were ranked in the order: H<M<L<CK and the three straw incorporation treatments were significantly different from CK. These results agreed with Zhang *et al.* [37] and Zhang *et al.* [31]. The low values of *D* indicated a size distribution dominated by a large number of macroaggregates (>0.25 mm) [59,63]. This improvement in the fractal dimension may have been accelerated by the incorporation of straw, which improved the soil structure, increased the SOM content and microbial activity [64], and significantly increased the mount and size of soil aggregates [37]. Our results indicated there were significant improvements in the soil macroaggregates and the aggregate structure after straw incorporation [20].

Conclusion

The incorporation of different amount of straw significant increased the *SOC* storage, >0.25 mm macroaggregates, *MWD* and *GMD* in a semiarid soil. The *SOC* storage, >0.25 mm macroaggregates, *MWD* and *GMD* also increased with higher straw incorporation rates. The fractal dimension decreased with increasing straw incorporation rates. Therefore, the incorporation of straw into the soil in semiarid areas is an effective practice for improving the soil aggregate content and stability.

Acknowledgments

We are grateful to Mr Li Yongping and Liu Shixin for managing the field experiments and professional English editor Jackson who is from UK, and kind help with the language of this manuscript.

Author Contributions

Conceived and designed the experiments: ZJ QH. Performed the experiments: PZ TW. Analyzed the data: PZ. Contributed reagents/materials/analysis tools: XR YL. Wrote the paper: PZ.

References

1. Hu KL, Li BG, Lin QM, Li GT, Chen DL (1999) Spatial variabi lity of soil nutrient in wheat field. Transactions of the Chinese Society of Agricultural Engineering 15:33–38. (in Chinese with English abstract)
2. Mupangwa W, Twomlow S, Walker S (2008) The influence of conservation tillage methods on soil water regimes in semi-arid southern Zimbabwe. Phys Chem Earth, Parts A/B/C 33: 762–767.
3. Biederbeck V, Campbell C, Bowren K, Schnitzer M, McIver R (1980) Effect of burning cereal straw on soil properties and grain yields in Saskatchewan. Soil Sci Soc Am J 44: 103–111.
4. Wuest SB, Caesar-TonThat T, Wright SF, Williams JD (2005) Organic matter addition, N, and residue burning effects on infiltration, biological, and physical properties of an intensively tilled silt-loam soil. Soil Tillage Res 84: 154–167.
5. Duan F, Liu X, Yu T, Cachier H (2004) Identification and estimate of biomass burning contribution to the urban aerosol organic carbon concentrations in Beijing. Atmos Environ 38: 1275–1282.
6. Mandal KG, Misra AK, Hati KM, Bandyopadhyay KK, Ghosh PK, et al. (2004) Rice residue-management options and effects on soil properties and crop productivity. J Food Agric Environ 2: 224–231.
7. Carter MR (2002) Soil quality for sustainable land management. Agron J 94: 38–47.
8. Hillel D (2004) Encyclopedia of Soils in the Environment: London, Elsevier, v. 4, p. 295–303.
9. Six J, Bossuyt H, Degryze S, Denef K (2004) A history of research on the link between (micro) aggregates, soil biota, and soil organic matter dynamics. Soil Tillage Res 79: 7–31.
10. Tisdall J, Oades J (1982) Organic matter and water-stable aggregates in soils. J Soil Sci 33:141–163. doi:10.1111/j.1365-2389.1982.tb01755.x
11. Haynes R, Beare M (1997) Influence of six crop species on aggregate stability and some labile organic matter fractions. Soil Biol Biochem 29: 1647–1653.
12. Scanlon BR, Andraski BJ, Bilskie J (2002) Methods of Soil Analysis: Part 4 Physical Methods: Soil Science Society of America. 643–668 p.
13. Hernández-Hernández R, López-Hernández D (2002) Microbial biomass, mineral nitrogen and carbon content in savanna soil aggregates under conventional and no-tillage. Soil Biol Biochem 34: 1563–1570.
14. Villar M, Petrikova V, Dıaz-Ravina M, Carballas T (2004) Changes in soil microbial biomass and aggregate stability following burning and soil rehabilitation. Geoderma 122: 73–82.
15. Ashagrie Y, Zech W, Guggenberger G (2005) Transformation of a *Podocarpus falcatu* dominated natural forest into a monoculture *Eucalyptus globulus* plantation at Munesa, Ethiopia: soil organic C, N and S dynamics in primary particle and aggregate-size fractions. Agric Ecosyst Environ 106: 89–98.
16. Jastrow J (1996) Soil aggregate formation and the accrual of particulate and mineral-associated organic matter. Soil Biol Biochem 28: 665–676.
17. Chevallier T, Blanchart E, Albrecht A, Feller C (2004) The physical protection of soil organic carbon in aggregates: a mechanism of carbon storage in a Vertisol under pasture and market gardening (Martinique, West Indies). Agric Ecosyst Environ 103: 375–387.
18. Duiker S, Lal R (1999) Crop residue and tillage effects on carbon sequestration in a Luvisol in central Ohio. Soil Tillage Res 52: 73–81.
19. Saroa G, Lal R (2003) Soil restorative effects of mulching on aggregation and carbon sequestration in a Miamian soil in central Ohio. Land Degrad Dev 14: 481–493.
20. Tan DS, Jin JY, H SW, Li ST, He P (2007) Effect of long-term application of K fertilizer and wheat straw to soil on crop yield and soil K under different planting systems. Agric Sci in China 6: 200–207.
21. Li Q, Chen X, Zhang F, Römheld V (2002) Study on balance of phosphorus and potassium in winter wheat and summer maize rotation system. Plant Nutr Fert Sci 8: 152–156. (in Chinese with English abstract)
22. Nelson RG (2002) Resource assessment and removal analysis for corn stover and wheat straw in the Eastern and Midwestern United States—rainfall and wind-induced soil erosion methodology. Biomass Bioenergy 22: 349–363.
23. Wilhelm W, Johnson JMF, Karlen DL, Lightle DT (2007) Corn stover to sustain soil organic carbon further constrains biomass supply. Agron J 99: 1665–1667.
24. Sonnleitner R, Lorbeer E, Schinner F (2003) Effects of straw, vegetable oil and whey on physical and microbiological properties of a chernozem. App Soil Ecol 22: 195–204.
25. Mulumba LN, Lal R (2008) Mulching effects on selected soil physical properties. Soil Tillage Res 98: 106–111.
26. Bhagat R, Verma T (1991) Impact of rice straw management on soil physical properties and wheat yield. Soil Sci 152: 108–115.

27. FAO/UNESCO (1993) 'World soil resources. An explanatory note on the FAO World Soil Resource Map at 1 : 25,000,000 scales.' (FAO: Rome)
28. Wang S, Tian H, Liu J, Pan S (2003) Pattern and change of soil organic carbon storage in China: 1960s–1980s. Tellus B 55: 416–427.
29. Pribyl DW (2010) A critical review of the conventional SOC to SOM conversion factor. Geoderma 156: 75–83.
30. Oades J, Waters A (1991) Aggregate hierarchy in soils. Soil Res 29: 815–828.
31. Zhang P, Jia ZK, Wang W, Lu WT, Gao F, et al. (2012) Effects of straw returning on characteristics of soil aggregates in semi-arid areas in southern Ningxia of China. Sci Agric Sin 45:1513-1520. (in Chinese with English abstract)
32. Van Bavel C (1949) Mean weight diameter of soil aggregates as a statistical index of aggregation. Soil Sci Soc Am Proc. pp. 20–23.
33. Youker R, McGuinness J (1957) A short method of obtaining mean weight-diameter values of aggregate analyses of soils. Soil Sci 83: 291–294.
34. Mazurak AP (1950) Effect of gaseous phase on water-stable synthetic aggregates. Soil Sci 69: 135–148.
35. Hou XQ, Li R, Jia ZK, Han QF, Wang W, et al. (2012) Effects of rotational tillage practices on soil properties, winter wheat yields and water-use efficiency in semi-arid areas of north-west China. Field Crops Res 129: 7–13.
36. Tyler SW, Wheatcraft SW (1992) Fractal scaling of soil particle-size distributions: analysis and limitations. Soil Sci Soc Am J 56: 362–369.
37. Zhang Z, Wei C, Xie D, Gao M, Zeng X (2008) Effects of land use patterns on soil aggregate stability in Sichuan Basin, China. Particuology 6: 157–166.
38. Zhu H, Wu J, Huang D, Zhu Q, Liu S, et al. (2010) Improving fertility and productivity of a highly-weathered upland soil in subtropical China by incorporating rice straw. Plant Soil 331: 427–437.
39. Mele PM, Crowley DE (2008) Application of self-organizing maps for assessing soil biological quality. Agric Ecosyst Environ 126: 139–152.
40. Rasmussen PE, Collins HP (1991) Long-term impacts of tillage, fertilizer, and crop residue on soil organic matter in temperate semiarid regions. Adv Agron 45: 93–134.
41. Karami A, Homaee M, Afzalinia S, Ruhipour H, Basirat S (2012) Organic resource management: Impacts on soil aggregate stability and other soil physico-chemical properties. Agric Ecosyst Environ 148: 22–28.
42. Tripathy R, Singh AK (2004) Effect of water and nitrogen management on aggregate size and carbon enrichment of soil in rice-wheat cropping system. J Plant Nutr Soil Sci 167: 216–228.
43. Prasad R, Power J (1991) Crop residue management. Adv Soil Sci 15: 205–251.
44. Limon-Ortega A, Govaerts B, Sayre KD (2009) Crop rotation, wheat straw management, and chicken manure effects on soil quality. Agron J 101: 600–606.
45. Beare M, Hendrix P, Cabrera M, Coleman D (1994) Aggregate-protected and unprotected organic matter pools in conventional-and no-tillage soils. Soil Sci Soc Am J 58: 787–795.
46. Paustian K, Levine E, Post WM, Ryzhova IM (1997) The use of models to integrate information and understanding of soil C at the regional scale. Geoderma 79: 227–260.
47. Mikha MM, Rice CW (2004) Tillage and manure effects on soil and aggregate-associated carbon and nitrogen. Soil Sci Soc Am J 68: 809–816.
48. Pinheiro E, Pereira M, Anjos L (2004) Aggregate distribution and soil organic matter under different tillage systems for vegetable crops in a Red Latosol from Brazil. Soil Tillage Res 77: 79–84.
49. Lal R (2000) Physical management of soils of the tropics: priorities for the 21st century. Soil Sci 165: 191–207.
50. Pagliai M, Vignozzi N, Pellegrini S (2004) Soil structure and the effect of management practices. Soil Tillage Res 79: 131–143.
51. Kasteel R, Garnier P, Vachier P, Coquet Y (2007) Dye tracer infiltration in the plough layer after straw incorporation. Geoderma 137: 360–369.
52. Pascual JA, García C, Hernandez T (1999) Comparison of fresh and composted organic waste in their efficacy for the improvement of arid soil quality. Bioresour Technol 68: 255–264.
53. Abu-Sharar T, Salameh A (1995) Reductions in hydraulic conductivity and infiltration rate in relation to aggregate stability and irrigation water turbidity. Agric Water Manage 29: 53–62.
54. Roldán A, Salinas-García J, Alguacil M, Caravaca F (2005) Changes in soil enzyme activity, fertility, aggregation and C sequestration mediated by conservation tillage practices and water regime in a maize field. Appl Soil Ecol 30: 11–20.
55. Barthes B, Roose E (2002) Aggregate stability as an indicator of soil susceptibility to runoff and erosion; validation at several levels. Catena 47: 133–149.
56. Ramos M, Nacci S, Pla I (2003) Effect of raindrop impact and its relationship with aggregate stability to different disaggregation forces. Catena 53: 365–376.
57. Wei CF, Gao M, Shao JG, Xie DT, Pan GX (2006) Soil aggregate and its response to land management practices. China Particuology 4: 211–219.
58. Tarafdar JC, Meena SC, Kathju S (2001) Influence of straw size on activity and biomass of soil microorganisms during decomposition. Eur J Soil Biol 37: 157–160.
59. Perfect E, Kay B (1991) Fractal theory applied to soil aggregation. Soil Sci Soc Am J 55: 1552–1558.
60. Rieu M, Sposito G (1991) Fractal fragmentation, soil porosity, and soil water properties: II. Applications. Soil Sci Soc Am J 55: 1239–1244.
61. Bird N, Bartoli F, Dexter A (2005) Water retention models for fractal soil structures. Eur J Soil Sci 47: 1–6.
62. Castrignanò A, Stelluti M (1999) Fractal geometry and geostatistics for describing the field variability of soil aggregation. J Agric Eng Res 73: 13–18.
63. Ding Q, Ding W (2007) Comparing stress wavelets with fragment fractals for soil structure quantification. Soil Tillage Res 93: 316–323.
64. Liu JF, Hong W, Wu CZ (2002) Fractal features of soil clusters under some precious hardwood stands in the central subtropical region, China. Acta Ecol. Sin. 22, 197. (in Chinese with English abstract).

8

Chopped or Long Roughage: What Do Calves Prefer? Using Cross Point Analysis of Double Demand Functions

Laura E. Webb[1]*, Margit Bak Jensen[2], Bas Engel[3], Cornelis G. van Reenen[4], Walter J. J. Gerrits[5], Imke J. M. de Boer[1], Eddie A. M. Bokkers[1]

1 Animal Production Systems Group, Wageningen University, Wageningen, Netherlands, **2** Department of Animal Sciences, Aarhus University, Tjele, Denmark, **3** Biometris, Wageningen University, Wageningen, Netherlands,, **4** Livestock Research, Wageningen University and Research Centre, Lelystad, Netherlands, **5** Animal Nutrition Group, Wageningen University, Wageningen, Netherlands

Abstract

The present study aimed to quantify calves'(*Bos taurus*) preference for long versus chopped hay and straw, and hay versus straw, using cross point analysis of double demand functions, in a context where energy intake was not a limiting factor. Nine calves, fed milk replacer and concentrate, were trained to work for roughage rewards from two simultaneously available panels. The cost (number of muzzle presses) required on the panels varied in each session (left panel/right panel): 7/35, 14/28, 21/21, 28/14, 35/7. Demand functions were estimated from the proportion of rewards achieved on one panel relative to the total number of rewards achieved in one session. Cross points (cp) were calculated as the cost at which an equal number of rewards was achieved from both panels. The deviation of the cp from the midpoint (here 21) indicates the strength of the preference. Calves showed a preference for long versus chopped hay (cp =14.5; P =0.004), and for hay versus straw (cp =38.9; P=0.004), both of which improve rumen function. Long hay may stimulate chewing more than chopped hay, and the preference for hay versus straw could be related to hedonic characteristics. No preference was found for chopped versus long straw (cp =20.8; P=0.910). These results could be used to improve the welfare of calves in production systems; for example, in systems where calves are fed hay along with high energy concentrate, providing long hay instead of chopped could promote roughage intake, rumen development, and rumination.

Editor: Johan J. Bolhuis, Utrecht University, Netherlands

Funding: This work was supported by the Division for Earth and Life Sciences (ALW) of the Netherlands Organisation for Scientific Research (NWO), and the Product Board Animal Feed. Travelling between The Netherlands and Denmark was supported by UFAW. The funders had no role in study design, data collection and analysis, decision to publish, or preparation of the manuscript.

Competing Interests: The authors have declared that no competing interests exist.

* E-mail: laura.webb@wur.nl

Introduction

Foraging animals gather information about available resources at the expense of optimising immediate rate of energy gain [1,2]. Ruminants have been found to trade-off between optimising rate of energy gain and minimising disadvantages to rumen function caused by the intake of high energy food, by including in their diets roughage high in fibre and low in energy [3,4]. This requires prior association between the sensory characteristics of feed and their post-ingestive consequences [5]. Ruminants spend extensive time feeding and ruminating. Mastication and rumination promote salivation, an important buffering agent in the rumen, and reduce feed particle size to enable passage of feed into the abomasum [6,7]. As a consequence, ruminants have a high incentive to chew and ruminate [8,9], and they may sometimes show a preference for roughages that require long chewing times [10]. The latter is especially relevant in farmed ruminants fed high energy diets with little fibre, as these animals develop abnormal oral behaviours due to limited opportunity to chew and ruminate [11–13]. Abnormal behaviours occur in sub-optimal environments and are a sign of poor welfare in captive animals [14].

A method for investigating foraging behaviour in ruminants is to quantify the preferences for two simultaneously available feeds. Manipulating the particle length of roughage is an easy way to control the rate of energy gain, without affecting taste and smell. Compared to longer ones, smaller particles of roughage are ingested at a higher rate [15–19], and pass faster/more easily through the reticulorumen [20], resulting in an increased rate of energy gain. However, feeding only small amounts of small particles of roughage, as opposed to longer roughage particles, on top of a high concentrate diet, may lower ruminal pH in the long term, increasing the chances of developing acidosis [7]. These diets may also lead to ruminal plaque formation, i.e. a sticky mass of hairs and small feed particles between the papillae [21], and ruminal hairball development [13]. In addition, small roughage particles often mean less chewing and rumination than longer particles. Less chewing and rumination increases energy intake rate by decreasing ingestion and digestion effort, but these behaviours also stimulate saliva secretion, which is an important buffering agent in the rumen [7]. Ruminants were capable of making foraging choices that favour good rumen function by selecting a large portion of chopped roughage particles (30%) in their total diet, when chopped and ground roughages were offered together [3,4]. In previous studies, however, animals had to balance energy intake and good rumen function, because no other feed was provided besides roughage. If energy intake was taken out of the equation, by, for example, feeding high energy concentrate, ruminants are expected to prefer longer particles of roughage, as the need for good rumen function would then become more important than rate of energy gain.

Previous research investigating preferences for different particle lengths of roughage in ruminants used short-term [18,22] or long-term [3,4] choice tests. Providing freely available alternative resources and imposing no cost on preference, however, does not reflect foraging environments in the wild and does not quantify the strength of a given preference. Cross point analysis of demand functions, where two substitutable resources are presented simultaneously and the workload for each resource is varied relative to the other, incorporates a 'cost' on the choice and is suggested as a more accurate and biologically relevant method for quantifying preferences [23,24]. In this method, demand function refers to the linear regression between rewards achieved and resource costs [25]. The cross point designates the combination of costs (one for each resource) at which an equal number of rewards is achieved for both resources. The cross point analysis of double demand functions enables quantification of preferences, and may be viewed as reflecting the natural foraging situation where food availability (cost) varies.

The present study aimed to quantify calves'preference for long versus chopped hay and straw, using double demand operant conditioning, in a context where energy intake was no limiting factor (i.e. feeding large quantities of milk replacer and concentrate). We hypothesised that calves would prefer long roughage particles over chopped because they value long chewing time and good rumen function. This presupposes that calves previously learnt post-ingestive consequences of different roughage types. Hay is associated with increased energy intake rate and better rumen function [26], but decreased chewing time [22], compared to straw. Moreover, sensory characteristics, such as smell, taste or texture, may also affect the relative preference of hay and straw. The preference for hay and straw was also quantified in the present study.

Materials and Methods

This study was carried out at Wageningen University's Animal Science Department experimental facilities, The Netherlands. The experiment ran from April to August 2012.

Ethics statement

All procedures met the terms of the Dutch law for animal experiments, which complies with the ETS123 (Council of Europe 1985 and the 86/609/EEC Directive), and were approved by Wageningen University's Committee on Animal Care and Use (DEC no. 2012006).

Animals and husbandry

Nine 7-week-old Holstein-Friesian bull calves (body weight mean ± SEM: 84.6±1.3 kg) were purchased from one Dutch veal farm. Calves were individually housed for the first 4 weeks after arrival at the veal farm (from 2 to 6 weeks of age), and thereafter, housed in a large group of 95 male calves. On the veal farm, calves had access to brushes (for grooming), bouncy balls (for head butting), and rubber teats (for sucking and chewing). The calves were fed milk replacer, concentrate (400 g per calf) and a small amount of chopped wheat straw (10 g per calf). The calves for the experiment were selected based on two criteria: similar size and no previous health treatment. At arrival at the experimental facilities, the nine calves were housed together in one 9.40 m×2.45 m home pen with a wooden slatted floor. The home pen was fitted with two brushes (for grooming) and one plastic ball hanging from a chain for enrichment. The calves received commercial milk replacer (18% crude protein and 18% crude fat) twice a day at 07:30 and 16:30 h in buckets with floating teats. Calves were also fed pelleted concentrate (17.5% crude protein, 37% starch, 24% NDF, based on 71% cereal and cereal by-products and 25% lupins as the main ingredients), which were provided once a day in the milk buckets immediately after the milk was consumed during the afternoon feeding. All calves finished their milk meal within 10 min. Calves were restrained during milk feeding to prevent them from ingesting other calves' milk. The daily allowance of milk replacer and concentrate corresponded to ad libitum intakes of these feeds in similar age calves in a previous study, where milk replacer, concentrate, maize silage, hay and barley straw were offered ad libitum (unpublished data). The allowance of milk replacer ranged from 10.0 L/d at 7 weeks of age to 15.6 L/d (122 g DM/L) at 5 months of age, while the allowance of concentrate ranged from 0.3 kg/d at 7 weeks of age to 2.7 kg/d at 5 months of age (Figure 1). The choice of the feeding strategy (milk fed twice a day and concentrate fed only at night) enabled control of intake before testing. After arrival, calves were offered five roughages: chopped barley straw, long barley straw, chopped grass hay, long grass hay (straw: 3.1% crude protein and 79% NDF; hay: 9.2% crude protein and 59% NDF), and chopped Lucerne hay mixed with 8% cane molasses and linseed oil (molashine, Gedizo Trading Int.). Chopped roughage particles were 2–3 cm, while long particles were unprocessed and around 20–30 cm. These particle lengths were chosen as providing the largest possible variation in length, with the smaller length reflecting what is commonly fed to fattening calves. The five roughages were offered one after the other in order to familiarise the calves with sensory and post-ingestive information associated with each roughage type. This familiarisation was done for three consecutive days per roughage type (i.e. 15 days of familiarization in total starting the day after arrival), offered ad libitum. After this initial familiarisation period, calves only received roughage (i.e. long and chopped hay and straw) in the home pen during days with no training or days with no testing. During the training period, which lasted a total of 6 weeks, calves were not brought into the operant pen during the weekend, i.e. there were 2 d/wk without training. During the testing period, which also lasted 6 weeks, the Sundays were used for habituation to the new roughage types on a low workload, i.e. there was 1 d/wk without testing (see subsection "Testing calves" below). All test-roughages (i.e. all roughage types except Lucerne hay) were offered in the home pen each weekend. Roughage intake in the home pen during familiarisation and during days without training or testing was recorded.

Milk and concentrate refusals in the home pen were weighed daily. Milk refusals only occurred once (on the day of arrival at the experimental facilities). Concentrate refusals were less than 5% of provision, on average, throughout the study. The calves received water ad libitum via two drinking nipples. Lights were on between 07:00 and 22:00 h. Temperature was regulated with a heater and mechanical ventilation, and ranged from 14.4 to 26.1°C. Relative humidity ranged from 50.6 to 97.1%. A radio was turned on during the day in an attempt to maintain constant ambient background noise. In the week after arrival, calves were blood sampled for haemoglobin (Hb) and serum iron (SeFe) analysis in order to ensure that they were not anaemic: (mean ± SEM) Hb = 6.8±0.1 mmol/L and SeFe = 36.3±3.2 µmol/L. Given these values, calves were not given extra iron.

In order to test the equipment and develop a training protocol for the calves in this study, a pilot study was conducted using three calves prior to the present study.

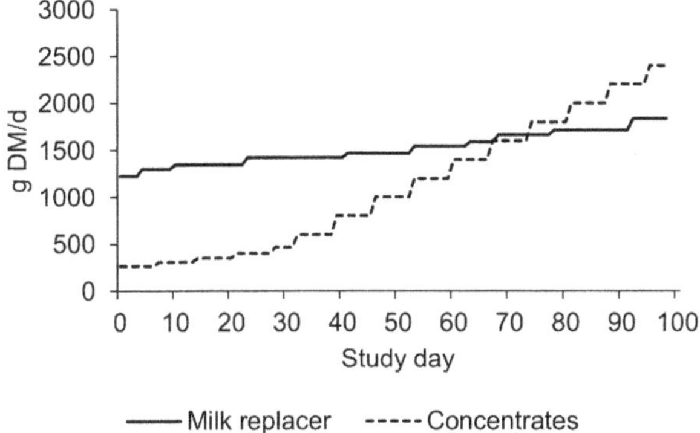

Figure 1. Milk replacer and concentrate feeding. Feeding schedule for milk replacer and concentrate in g DM per day per calf. Milk replacer was fed in two meals per day at 07:30 and 16:30 h, whereas concentrate were fed only at 16:30 h. Milk replacer and concentrate were fed in buckets, with floating teats for the milk.

Training calves on double demand operant conditioning

The test pen (2.35 m×2.45 m) was immediately adjacent to the home pen, fitted with a wooden slatted floor and black plastic walls (1.45 m high), and accessible from the home pen through a door. Calves could, therefore, be walked from the home pen, through the door, into the test pen. On the wall opposite the door were two panels (24 cm×20 cm) and two buckets (33 cm diameter). The two buckets were located between the two panels. Each bucket was 17 cm away from the corresponding panel, and the distance between the two buckets was 53 cm. The panels were raised 60 cm above the floor and the bottom of the buckets were raised 46 cm above the floor. Above the buckets were cylindrical automated feed delivery systems with a clap that opened to release roughage rewards into the buckets, via a computer that recorded the number of successful presses made on the panels. The left panel and bucket were associated to each other, in such a way that the correct number of presses on the left panel would result in the delivery of a roughage reward into the left bucket. The same applied to the right panel and right bucket. When panels were active, that is when the computer system was switched on, panels were lit with white led lamps. Each successful press made to an active panel was rewarded with a bell sound. When a reward was delivered, an alarm sound was played and the lights in both panels went off for 500 ms.

The nine calves were randomly assigned to groups of three, and randomly assigned to a working order within each group. During the entire experiment, including habituation, shaping, training and testing, calves were always placed in the test pen in the same order so that they could form expectations as to when they would be given the opportunity to work for roughage. One section of the home pen, adjacent to the test pen, could be closed off and formed a "waiting room" (2.35 m×2.45 m). To avoid disturbing all calves every time a new calf was collected for testing, calves were placed in the waiting room in their groups of three and remained there until all three calves had visited the test pen. Calves were first habituated to the test pen in their groups of three for 10 and 30 min. They were then habituated to the test pen individually for 10 and 20 min. Each calf visited the test pen once per day. During all habituation sessions, except the last two, the panels were inactive, meaning that the lights in the panels were off and a muzzle press resulted in neither sound nor reward. In the last two habituation sessions, the panels were active in order for calves to habituate to the lights in the panels. One muzzle press resulted in reward delivery.

During shaping and training, the reward on both panels was 10 g of Lucerne hay. During shaping, one panel and its corresponding bucket were blocked off with a barrier, and calves could only access the other panel and its corresponding bucket. During shaping, calves were rewarded for the following behaviours in the following sequence: approach the panel, sniff the panel from any angle, sniff the top of the panel, touch the top of the panel with the muzzle, and press the panel. When calves successfully learnt to press the panel to gain access to a reward, they were shaped on the other side. The side made accessible first was balanced for each group of calves.

Once calves were shaped on both panels, the fixed ratio (FR), i.e. number of presses required for one reward, was increased to two (FR2). After this, the barrier was removed and calves were trained on both panels, which were accessible simultaneously, on FR2. Subsequently, the FR on both panels was gradually increased while maintaining the same FR on both panels until FR10. Finally, the difference in FR between the two panels was gradually increased until calves could be trained on the five FR pairs used during testing: (Left-right panel) 7/35, 14/28, 21/21, 28/14, 35/7. Training ended when all calves worked economically, i.e. accessed over 60% of rewards from the panel with the lowest FR. At this stage, calves were 15 weeks old. Training sessions lasted a minimum of 30 min, but no maximum duration was imposed on the calves. This was done to enable all calves to work at their own individual speed and to access the number of rewards that they were motivated to get. Training sessions were ended when the calves had received no rewards for 3 min, after the initial 30 min. Training sessions lasted 39 min on average. For testing sessions, the minimum session time was reduced to 20 min, but again no maximum session time was imposed. When calves did not receive a reward for 3 min between 20 and 40 min in the test pen, or when calves walked away from the panels after 40 min in the test pen, the session was ended. Testing sessions lasted 39 min on average. Therefore, changing the criteria used during training did not affect average session duration.

Testing calves

Calves' preference for three combinations of roughage types was tested, and each combination was tested for 2 weeks: 1) chopped hay versus long hay, 2) chopped straw versus long straw, 3) chopped hay versus chopped straw. Each week comprised of one day of habituation with FR7 on both panels (to allow calves to familiarise themselves with the two roughage types and the location of each type) and five testing days; i.e. one day per FR pair: (Left-right panel) 7/35, 14/28, 21/21, 28/14, 35/7 presented in a random order). The two weeks with the same combination were repetitions of each other, but the location of the two roughage types was switched in order to control for any pre-existing side bias. The first two combinations of roughage types, which both investigated preference for different particle lengths, were presented in a cross-over design, with half the calves starting with chopped versus long hay and the other half starting with chopped versus long straw. After this, calves' preference for hay versus straw (both chopped) was tested. During testing of chopped versus long roughage, the reward size was 5 g, whereas during the testing of hay versus straw, the reward size was 8 g. The reward size was increased in an attempt to reduce test session duration and to take into account the older age of the calves. If calves did not consume all rewards, refusals were weighed at the end of the session and noted for each roughage type. The number of rewards used in the analysis was based on consumed rewards (number of rewards delivered minus number of rewards not consumed).

Post-mortem measurements

In order to check for any underlying health issues that may have affected the preferences of calves for different types of roughage, post-mortem health measurements were collected. At 6 months, all calves were slaughtered in a small slaughter house and routine Welfare Quality® post-mortem measurements were carried out [27]. Respiratory and gastrointestinal health measurements were made on all calves. Pneumonia was scored from 0 to 3 based on damaged area on the lungs, and presence of pleuritis was noted. Plaque and hyperkeratosis in the rumen, as well as lesions in the torus pylorus and pylorus areas of the abomasum were noted as present or absent. Rumen development was scored from 1: low to 4: full. A rumen score was calculated as the median of the rumen scores on the 9 rumens. Damage from abomasal lesions of <0.5 cm^2 (category 1), 0.5–1.0 cm^2 (category 2), and >1.0 cm^2 (category 3), were scored from 0 (absent) to 4 based on the number present. An abomasal lesion score was calculated for each calf as the sum of the lesion number, multiplied by the lesion category. The median of these scores was then calculated.

Data analysis

The response variable was the proportion of rewards of one resource over the total number of rewards for both resources within a session. This choice for a response variable differs from previous studies using cross point analysis of double demand functions, which generally used (logarithms of) reward counts [23,24,28–30]. We suggest that using proportions is more appropriate, as it takes into account the dependence between two simultaneously presented resources. A two-step approach was followed where (1) a model was fitted to the data of each individual animal and individual cross points were estimated, and (2) these individual cross points were compared to the midpoint. The midpoint in the present study was 21, i.e. the point where the FR values for the two resources were the same.

The two-step approach circumvented the need for modelling a dependence structure between proportions of the same animal over different sessions (resulting from repeated measures design). The model fitted to the data per animal was a generalised linear model (GLM) [31] with a logit link, the variance was specified as a multiple of the binomial variance function, and FR (of the chopped reward or of the hay reward, depending on whether particle lengths or roughage sources were compared) was introduced as an explanatory variable. Individual cross points corresponded to the values of FR where the expected proportion $p = 0.5$ and differed across animals. Individual cross points were calculated as: $cp = -\alpha/\beta$, where α and β are an animal's estimated intercept and slope on the logit scale. The overall cross point was defined as the median of the cross points of all animals in the target population and estimated by the median of the individual cross point of the animals in the experiment. The overall cross point was compared to the midpoint (i.e. 21) using Wilcoxon's signed rank test, applied to the differences between the individual cross points and the midpoint, and an associated 0.95-confidence interval for the overall cross point was constructed.

In order to demonstrate the meaning of "cross point" when using proportions instead of counts, a graphical representation, plotting predicted proportions of chopped hay rewards against FR for chopped hay, is shown for calf no. 2 (Figure 2). The curves fitted by proportions are sigmoid, and the curve for long hay is the opposite (*1-p*) of the curve for chopped hay (*p*). The cross point

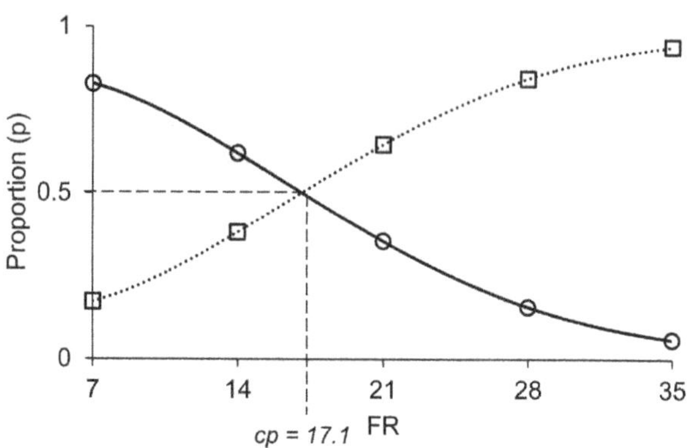

Figure 2. Cross point analysis illustrated. Graphical representation of the cross point (cp) of calf no. 2 for the comparison chopped hay (circles) versus long hay (squares) using proportions (p) of chopped hay rewards over total number of rewards. The proportions for long hay rewards were calculated as 1 - p. The x axis shows fixed ratio (FR) values for the chopped hay (the long hay fixed ratio values are 42 - FR). The lines connecting the points are 4th order polynomials.

corresponds to the point where $p = 0.5$, which in this figure is illustrated by the intersection between the two curves (Figure 2).

P-values lower than 0.05 were considered significant. Calculations were conducted using SAS version 9.2 [32] and Genstat version 15 [33].

Results

At the end of the study, calves weighed 248.4±5.9 kg on average, with an average daily gain of 1.5±0.1 kg/d. Roughage intake in the home pen during the weekend is shown in Table 1.

Double demand and cross points

The nine calves used in the present study were successfully trained to work economically on two panels delivering the same roughage reward (i.e. Lucerne hay), in that they consistently chose the panel with the lowest workload more often than the other panel (Table 2). Moreover, all calves were motivated to work for both hay and straw rewards throughout the study, despite high milk replacer and concentrate provision in the home pen (Table 3).

Calves showed a preference for long hay over chopped hay, indicated by an overall median cross point below the midpoint 21 and different from the midpoint (Table 2). The overall cross point for the comparison chopped straw versus long straw was not different from the midpoint (Table 2), which indicates that calves showed no preference for chopped or long straw. However, the confidence interval was wide, indicating large variation between individuals, and three calves seemed to have expressed a preference for chopped straw (calves no. 1 and 7) or long straw (calf no. 8) (Table 2). Calves showed a preference for chopped hay over chopped straw, indicated by an overall cross point higher than the midpoint, and different from the midpoint (Table 2). The cross point, i.e. 38.9, is higher than 35, which is the highest FR that was imposed in the present study, indicating that calves always achieved more hay rewards than straw rewards regardless of the costs. Median number of rewards consumed during one session was highest for the comparisons including hay rewards, and higher when the preferred resource was available at a low price for the comparison chopped versus long hay, and hay versus straw, i.e. comparisons where one resource was preferred over the other (Table 2).

Table 1. Roughage intake in the home pen (mean ± SEM g/d).

Period	Age (wk)	Chopped hay	Long hay	Chopped straw	Long straw	Lucerne hay
Start[1]	7–9	106±22	216±12	83±12	93±9	366±41
Training[2]	9–15	362±49	355±55	266±32	142±17	
Testing[3]	15–21	505±55	423±56	238±84	316±30	

[1]Roughage was provided ad libitum during the habituation period, one roughage type at a time.
[2]Roughage was provided ad libitum, one roughage at a time (2 days per week without training).
[3]Roughage was provided ad libitum, two roughage types at a time (1 day per week without testing). The two types of roughage provided were from the same source but had different particle lengths.

Post-mortem results

The calves in the present study had no overt health problems during the experiment. The results of the post-mortem gastrointestinal and respiratory health measurements showed no severe pneumonia, no rumen hyperkeratinisation, and relatively good rumen development (rumen development score [median] = 3.0). The median abomasal lesion score was 4.0 and was close to that found in European veal farms with large numbers of animals [34].

Discussion

The main aim of this study was to investigate the preferences of calves for different roughage particle lengths. Relative preference was quantified using a double demand operant conditioning paradigm. Double demand operant conditioning has previously been applied to rats [24,28,30], chickens [35], pigs [23,29], and adult cattle [36], but we could not find a study applying the double demand approach to calves. The methodology used to train the calves in the present study took 6 weeks in total, starting with 9 week-old calves (training started 2 weeks after the arrival of the calves, the first two weeks being used to familiarise calves to the roughages). The results showed that calves fed a high energy diet were willing to work for extra roughage rewards, including Lucerne hay, good quality hay and barley straw. The calves adjusted their efforts on the two panels according to their respective price such that when the two panels yielded the same roughage (Lucerne hay), they obtained more rewards from the panel with the lowest cost in all sessions. Calves expressed their preferences when two different rewards where available. It was possible to quantify the strength of preferences via the deviation of the cross point from the midpoint. This is clearly seen when comparing the deviations found for the preference of long hay over chopped hay (deviation of 6.5 from the midpoint) and the preference of hay over straw (deviation of 17.9 from the midpoint). This suggests that the preference of hay over straw is stronger than that of long hay over chopped hay in calves. Hay differs from straw in a number of ways apart from structure, as it contains more energy [22], has a different flavour [37] and is thought to have a beneficial influence on rumen function: due to increased fermentation, hay should lead to better papillae development [26]. However, the latter effect may be minimal in this study because of the high level of concentrate fed. The cross point for the comparison of hay versus straw was above 35, which is the highest cost imposed on resources in the present study. This indicates that for this comparison, the range of costs did not include a large enough difference in values. However, the results obtained do seem to confirm the hypothesis that hay is a preferred roughage compared to straw, even when energy is no limiting factor.

The statistical method used in this paper for cross point analysis of double demands differs from methods used in previous studies [23,24,28–30]. The presently applied method considers three

Table 2. Cross points of individual calves for each comparison, including training.

Calf	Training	Chopped vs. long hay	Chopped vs. long straw	Hay vs. straw
1	18.5	14.2	6.7	30.8
2	25.1	17.1	22.2	33.8
3	22.7	18.9	22.5	27.5
4	23.2	13.8	21.6	42.3
5	18.6	12.2	19.9	33.5
6	20.8	17.9	20.8	38.9
7	25.9	14.5	6.8	41.4
8	17.0	14.3	30.9	117.4
9	21.7	19.3	20.7	46.1
Median	21.7	14.5	20.8	38.9
Confidence interval	18.9–23.9	14.0–18.0	13.8–25.4[1]	32.3–42.0[2]
P-value	0.734	0.004	0.910	0.004

[1]Note that the confidence interval here includes 21 and is wide, indicating a large variation between individual calves and a difficulty in drawing conclusions on this particular comparison.
[2]Note that 42.0 is the largest value that the upper bound can take, since larger values would correspond to negative values for 42-x for the other resource.

Table 3. Total median number of rewards achieved (and total grams).

Comparison	FR	Median	Q1[3]	Q3[3]
Chopped vs. long hay[1]	7–35	57.0 (285)	45.0	86.5
	14–28	27.5 (138)	22.0	45.0
	21–21	26.0 (130)	15.0	42.0
	28–14	49.0 (245)	25.0	58.0
	35–7	81.5 (408)	45.0	100.8
Chopped vs. long straw[1]	7–35	22.0 (110)	11.0	43.0
	14–28	19.5 (98)	12.0	25.0
	21–21	17.0 (85)	10.0	24.0
	28–14	15.5 (78)	10.0	24.0
	35–7	31.0 (155)	17.0	52.0
Hay vs. straw[2]	7–35	79.7 (638)	60.6	105.8
	14–28	46.0 (368)	31.7	78.0
	21–21	28.1 (225)	19.0	36.0
	28–14	24.0 (192)	16.4	33.8
	35–7	18.7 (150)	14.6	31.6

[1]Reward size was 5 g.
[2]Reward size was 8 g.
[3]1st and 3rd quartile for the median.

aspects in the analysis of double demand functions. First, the dependence between data for the two resources offered simultaneously is included by using proportions as a response variable. Second, individual variation is expressed in an accessible and clear manner, and looking at individual cross points offers a clear picture of variation in preferences across animals [24,28]. Third, the analysis is robust, that is, not critically dependent upon complex model assumptions, and the use of Wilcoxon's signed rank test offers a conceptually and computationally straightforward statistical method.

Calves did not consistently prefer the roughage associated to the shortest ingestion and digestion time, i.e. chopped roughage; they did show a preference for long hay over chopped hay, but no preference was apparent for either long straw or chopped straw. Calves in this study were fed a high energy diet, consisting of milk replacer and concentrate, between testing sessions. It was, therefore, expected that these calves would not necessarily show a preference for the roughage permitting the best rate of energy gain. Furthermore, calves did not "abandon" the panel with the highest workload. This was the case when both panels provided the same reward, as well as when the "cheap" panel delivered the preferred reward. Contrafreeloading describes the concept where animals work for food when the same food is simultaneously freely available [2,38–40]. Although the food in the present study was never "free", it was sometimes very "cheap". Therefore, the animals displayed something very close to contrafreeloading, that we could term contracheaploading, and which most likely stems from the same motivations. Previous studies using double demand also observed this behaviour in their animals [29,30]. Contracheaploading in double demand operant setups most likely signals information gathering from various available resources, just like contrafreeloading [38] and could be an indication of animals' adaptation to a changing environment, e.g. the depletion of the highest quality food patch [2,38,39]. In nature, food patches used by animals will deplete over time, and gathering information about alternative patches may increase survival over the long term. In the present set-up the relative cost of the two resources were alternated between daily sessions and thus there was a high level of uncertainty, which is hypothesised to increase contrafreeloading [2]. In other contexts, contrafreeloading could be an indication of animals' need to express appetitive behaviour [10]. However, since calves had to work for all roughage resources, this is an unlikely explanation in the present set-up.

The preference for long hay found in the present study could be explained in two non-mutually-exclusive manners. First, calves may have preferred long hay because it required more chewing, and calves may have a high motivation for performing this behaviour [10]. The calves may have perceived the long hay portion as being larger than the chopped hay portion, through increased eating time [19], increased rumen fill [17], and slower clearance rate of the reticulorumen [20]. Long hay may also increase rumination as a post-ingestive consequence [13,41]. During the habituation period and in the home pen on days without training or testing, calves were fed each roughage type on separate occasions, which is assumed to have been sufficient for calves to learn post-ingestive consequences of all roughage types, including consequences for rumination [42].

Second, calves may have preferred long hay because it resulted in improved rumen function compared to the chopped hay, given that calves were indeed aware of post-ingestive consequences of each particle length. Longer particles of roughage take longer to chew and ruminate before the particle length is sufficiently reduced to move from the reticulorumen to the abomasum, and increased rumination increases salivation [7,19]. Saliva secretion increases the buffering capacity of rumen fluid [7,19], and prolonged presence of roughage particles in the rumen improves rumen motility and stimulates the removal of ingested hair and small feed particles from the rumen papillae [43]. This is especially important in calves fed large quantities of concentrate, and for which access to roughage is restricted. Therefore, longer roughage particles improve rumen muscularisation, papillae development, and rumen osmolality and pH [15,16], while preventing hairball and plaque development [13,21,43].

Interestingly, calves showed a preference for long over chopped roughage for hay but not for straw. Given the large variation between calves found in the comparison of chopped versus long straw (illustrated by the 95% confidence interval), it is difficult to conclude on this particular result. It is possible that with a larger sample of animals, a preference for one of the straws would have been observed. Straw is a coarse and low quality roughage with low energy and high fibre content, resulting in a low rate of energy gain [22]. Preference for shorter particles of straw was found to be stronger compared to preference for shorter particles of high quality roughage (such as hay) in sheep [22]. Therefore, ruminants can show preferences for different structures, even with low quality roughages. In our study, given the high energy feeding strategy provided outside of testing, calves were expected to show a preference for longer particles. Since this preference was not found for straw, we can only speculate that long straw was associated with some sort of cost that outweighed the benefits, and that this cost was not present, or present to a lesser extent in long hay. A possible cost could be worse abomasal damage [44]. Abomasal damage, i.e. lesions on the abomasal wall, could result from a combination of three factors: a) overfilling of the abomasum because of large milk meals causing local loss of blood supply of the abomasal wall (ischaemia), b) exacerbation of this damage from poorly digested feed particles coming from a poorly developed rumen, and c) exacerbation of this damage by coarse feed stuffs [13,45,46].

The post-mortem health measurements were carried out in the present study to check whether calves were healthy, and whether any underlying health problems could have explained any of the preferences. The feeding strategy combined with possibility to work for roughage in the operant pen aimed to permit a good growth, and this was successfully achieved. Looking at the numbers, rumen development seemed better than that found in European veal calves, but abomasal damage appeared comparable [34]. Similar abomasal damage could indicate that milk feeding was an important factor in causing abomasal damage [45], or that the improvement in rumen development was insufficient to minimise abomasal damage in the current study [46]. The infrequent feeding of large amounts of milk replacer in the present study may have caused the observed abomasal damage [45] (and could have further caused other physiological problems, such as for example insulin resistance [47,48], although this is not thought to have affected the results in any way). It is not known how abomasal damage may affect the preference for long or chopped particles of roughage. Despite these potential health issues, this feeding strategy was chosen to enable good control of milk intake (in terms of amount and time) before testing, in order to reduce inter- and intra-calf variation.

Conclusions

The present findings showed that 2–5 month old calves can learn a double demand operant setup and are motivated to work for roughage in addition to a high energy diet comprising of milk replacer and concentrate. Overall, calves preferred long particles of hay, but not straw, compared to chopped, and calves had a strong preference for chopped hay over chopped straw. These findings support the idea that ruminants are able to make choices based on rumen function and possibly also based on their motivation to chew and ruminate. These findings could be used to improve the welfare of calves in production systems: Farmed calves fed high energy diets alongside hay might benefit (e.g. in terms of rumen function) from being offered long hay instead of chopped hay.

Acknowledgments

The authors are grateful to MSc students Xun Li, Océane Schmitt, Lisanne Stadig, and Aurore Testemale, and to Fokje Steenstra (Wageningen University) for help with the collection of data. The authors thank Erik L. Decker (Aarhus University) for technical advice, the animal caretakers of Carus experimental facilities, Wageningen University, and the technical team of Tupola, Wageningen University. Finally, the authors thank Dennis Webb (University of Rennes 1) for language editing.

Author Contributions

Conceived and designed the experiments: LEW MBJ CGvR WJJG EAMB. Performed the experiments: LEW MBJ EAMB. Analyzed the data: LEW MBJ BE. Wrote the paper: LEW MBJ BE CGvR WJJG IJMdB EAMB.

References

1. Forkman B (1991) Some problems with current patch-choice theory: a study on the Mongolian gerbil. Behaviour 117: 243–254.
2. Inglis IR, Langton S, Forkman B, Lazarus J (2001) An information primacy model of exploratory and foraging behaviour. Anim Behav 62: 543–557.
3. Cooper SD, Kyriazakis I, Oldham JD (1996) The effects of physical form of feed, carbohydrate source, and inclusion of sodium bicarbonate on the diet selections of sheep. J Anim Sci 74: 1240–1251.
4. Cooper SDB, Kyriazakis I, Nolan JV (1995) Diet selection in sheep - The role of the rumen environment in the selection of a diet from 2 feeds that differ in their energy density. Brit J Nutr 74: 39–54.
5. Provenza FD (1995) Postingestive feedback as an elementary determinant of food preference and intake in ruminants. J Range Manage 48: 2–17.
6. Welch JG (1982) Rumination, particle size and passage from the rumen. J Anim Sci 54: 885–894.
7. González LA, Manteca X, Calsamiglia S, Schwartzkopf-Genswein KS, Ferret A (2012) Ruminal acidosis in feedlot cattle: Interplay between feed ingredients, rumen function and feeding behavior (a review). Anim Feed Sci Tech 172: 66–79.
8. Redbo I, Nordblad A (1997) Stereotypies in heifers are affected by feeding regime. Appl Anim Behav Sci 53: 193–202.
9. Redbo I (1990) Changes in duration and frequency of stereotypies and their adjoining behaviors in heifers, before, during and after the grazing period. Appl Anim Behav Sci 26: 57–67.
10. Hughes BO, Duncan IJH (1988) The notion of ethological need, models of motivation and animal-welfare. Anim Behav 36: 1696–1707.
11. Veissier I, Ramirez de la Fe AR, Pradel P (1998) Nonnutritive oral activities and stress responses of veal calves in relation to feeding and housing conditions. Appl Anim Behav Sci 57: 35–49.
12. Webb LE, Bokkers EAM, Engel B, Berends H, Gerrits WJJ, et al. (2012) Behaviour and welfare of veal calves fed different amounts of solid feed supplemented to a milk replacer ration adjusted for similar growth. Appl Anim Behav Sci 136: 108–116.
13. Webb LE, Bokkers EAM, Heutinck LFM, Engel B, Buist WG, et al. (2013) The effect of roughage type, amount and particle size on veal calf behavior and gastrointestinal health. J Dairy Sci 96: 7765–7776.
14. Mason GJ, Latham NR (2004) Can't stop, won't stop: is stereotypy a reliable animal welfare indicator? Anim Welfare 13: S57–S69.
15. Al-Saiady MY, Abouheif MA, Makkawi AA, Ibrahim HA, Al-Owaimer AN (2010) Impact of particle length of alfalfa hay in the diet of growing lambs on performance, digestion and carcass characteristics. Asian Australas J Anim Sci 23: 475–482.
16. Krause KM, Combs DK, Beauchemin KA (2002) Effects of forage particle size and grain fermentability in midlactation cows. II. Ruminal pH and chewing activity. J Dairy Sci 85: 1947–1957.
17. Kammes KL, Allen MS (2012) Nutrient demand interacts with grass particle length to affect digestion responses and chewing activity in dairy cows. J Dairy Sci 95: 807–823.
18. Kenney PA, Black JL, Colebrook WF (1984) Factors affecting diet selection by sheep. III Dry matter content and particle length of forage. Aust J Agric Res 35: 831–838.
19. de Boever JL, Andries JI, de Brabander DL, Cottyn BG, Buysse FX (1990) Chewing activity of ruminants as a measure of physical structure - A review of factors affecting it. Anim Feed Sci Tech 27: 281–291.
20. Wilson JR, Kennedy PM (1996) Plant and animal constraints to voluntary feed intake associated with fibre characteristics and particle breakdown and passage in ruminants. Aust J Agric Res 47: 199–225.
21. Suarez BJ, Van Reenen CG, Gerrits WJJ, Stockhofe N, van Vuuren AM, et al. (2006) Effects of supplementing concentrates differing in carbohydrate composition in veal calf diets: II. Rumen development. J Dairy Sci 89: 4376–4386.
22. Kenney PA, Black JL (1984) Factors affecting diet selection by sheep. I Potential intake rate and acceptability of feed. Aust J Agric Res 35: 551–563.
23. Jensen MB, Pedersen LJ (2007) The value assigned to six different rooting materials by growing pigs. Appl Anim Behav Sci 108: 31–44.
24. Sørensen DB, Ladewig J, Ersboll AK, Matthews L (2004) Using the cross point of demand functions to assess animal priorities. Anim Behav 68: 949–955.
25. Hursh SR (1993) Behavioral economics of drug self-administration - an introduction. Drug Alcohol Depend 33: 165–172.
26. Suarez BJ, Van Reenen CG, Stockhofe N, Dijkstra J, Gerrits WJJ (2007) Effect of roughage source and roughage to concentrate ratio on animal performance and rumen development in veal calves. J Dairy Sci 90: 2390–2403.
27. Welfare Quality (2009) Welfare Quality® Assessment Protocol for Cattle. Lelystad, The Netherlands: Welfare Quality Consortium.
28. Sørensen DB, Ladewig J, Matthews L, Ersboll AK, Lawson L (2001) Measuring motivation: using the cross point of two demand functions as an assessment of the substitutability of two reinforcers. Appl Anim Behav Sci 74: 281–291.
29. Pedersen LJ, Holm L, Jensen MB, Jorgensen E (2005) The strength of pigs' preferences for different rooting materials measured using concurrent schedules of reinforcement. Appl Anim Behav Sci 94: 31–48.
30. Holm L, Ritz C, Ladewig J (2007) Measuring animal preferences: Shape of double demand curves and the effect of procedure used for varying workloads on their cross-point. Appl Anim Behav Sci 107: 133–146.
31. McCullagh P, Nelder JA (1989) Generalized Linear Models. London: Chapman and Hall.
32. SAS Institute Inc. (2008) Statistical Analysis System Version 9.2. Cary, NC.
33. VSN-International (2012) GenStat for Windows 15th Edition. Hemel Hempstead, UK: VSN International.
34. Brscic M, Heutinck LFM, Wolthuis-Fillerup M, Stockhofe N, Engel B, et al. (2011) Prevalence of gastrointestinal disorders recorded at postmortem

inspection in white veal calves and associated risk factors. J Dairy Sci 94: 853–863.
35. McAdie TM, Foster TM, Temple W, Matthews LR (1993) A method for measuring the aversiveness of sounds to domestic hens. Appl Anim Behav Sci 37: 223–238.
36. Matthews LR, Temple W (1979) Concurrent schedule assessment of food preference in cows. J Exp Anal Behav 32: 245–254.
37. Provenza FD, Scott CB, Phy TS, Lynch JJ (1996) Preference of sheep for foods varying in flavors and nutrients. J Anim Sci 74: 2355–2361.
38. Inglis IR (2000) The central role of uncertainty reduction in determining behaviour. Behaviour 137: 1567–1599.
39. Inglis IR, Forkman B, Lazarus J (1997) Free food or earned food? A review and fuzzy model of contrafreeloading. Anim Behav 53: 1171–1191.
40. Osborne SR (1977) The free food (contrafreeloading) phenomenon: A review and analysis. Anim Learn Behav 5: 221–235.
41. Heinrichs J (2005) Rumen development in the dairy calf; Beauchemin K, editor. Edmonton: University Alberta Dept Agr, Food & Nutr Sci. 179–187 p.
42. Kyriazakis I, Anderson DH, Duncan AJ (1998) Conditioned flavour aversions in sheep: the relationship between the dose rate of a secondary plant compound and the acquisition and persistence of aversions. Brit J Nutr 79: 55–62.
43. Morisse JP, Cotte JP, Huonnic D, Martrenchar A (1999) Influence of dry feed supplements on different parameters of welfare in veal calves. Anim Welfare 8: 43–52.
44. Mattiello S, Canali E, Ferrante V, Caniatti M, Gottardo F, et al. (2002) The provision of solid feeds to veal calves: II. Behavior, physiology, and abomasal damage. J Anim Sci 80: 367–375.
45. Breukink HJ, Wensing T, Mouwen JMVM (1991) Abomasal ulcers in veal calves: pathogenesis and prevention. In: Metz JHM, Groenestein CM, editors. New Trends in Veal Calf Production: EAAP Publication no. 52, Pudoc, Wageningen, The Netherlands. pp. 118–122.
46. Berends H, van Reenen CG, Stockhofe-Zurwieden N, Gerrits WJJ (2012) Effects of early rumen development and solid feed composition on growth performance and abomasal health in veal calves. J Dairy Sci 95: 3190–3199.
47. Vicari T, van den Borne J, Gerrits WJJ, Zbinden Y, Blum JW (2008) Postprandial blood hormone and metabolite concentrations influenced by feeding frequency and feeding level in veal calves. Domest Anim Endocrinol 34: 74–88.
48. Bach A, Domingo L, Montoro C, Terre M (2013) Short communication: Insulin responsiveness is affected by the level of milk replacer offered to young calves. J Dairy Sci 96: 4634–4637.

9

Comparison of Seven Chemical Pretreatments of Corn Straw for Improving Methane Yield by Anaerobic Digestion

Zilin Song[1], GaiheYang[2], Xiaofeng Liu[1]*, Zhiying Yan[2]*, Yuexiang Yuan[1], Yinzhang Liao[1]

1 Chengdu Institute of Biology, Chinese Academy of Science, Chengdu, Sichuan, PR China, 2 Research Center of Recycle Agricultural Engineering Technology of Shaanxi Province, Northwest A&F University, Yangling, Shaanxi, PR China

Abstract

Agriculture straw is considered a renewable resource that has the potential to contribute greatly to bioenergy supplies. Chemical pretreatment prior to anaerobic digestion can increase the anaerobic digestibility of agriculture straw. The present study investigated the effects of seven chemical pretreatments on the composition and methane yield of corn straw to assess their effectiveness of digestibility. Four acid reagents (H_2SO_4, HCl, H_2O_2, and CH_3COOH) at concentrations of 1%, 2%, 3%, and 4% (w/w) and three alkaline reagents (NaOH, $Ca(OH)_2$, and $NH_3{\cdot}H_2O$) at concentrations of 4%, 6%, 8%, and 10% (w/w) were used for the pretreatments. All pretreatments were effective in the biodegradation of the lignocellulosic straw structure. The straw, pretreated with 3% H_2O_2 and 8% $Ca(OH)_2$, acquired the highest methane yield of 216.7 and 206.6 mL CH_4 g VS $^{-1}$ in the acid and alkaline pretreatments, which are 115.4% and 105.3% greater than the untreated straw. H_2O_2 and $Ca(OH)_2$ can be considered as the most favorable pretreatment methods for improving the methane yield of straw because of their effectiveness and low cost.

Editor: Dwayne Elias, Oak Ridge National Laboratory, United States of America

Funding: This study was supported by the National 973 of China (No. 2013CB733502), Applied Basic Research Program of Sichuan Province, China (No. 2013JY0050) and Deployment Project of Chinese Academy of Sciences (No. KGZD-EW-304-1). The funders had no role in study design, data collection and analysis, decision to publish, or preparation of the manuscript.

Competing Interests: The authors have declared that no competing interests exist.

* E-mail: lxf3636@163.com (XL); zhiyingyan2010@yeah.net (ZY)

Introduction

Biomass is considered as a valuable alternative energy source to fossil fuels worldwide because it can be converted into various available forms of energy, such as heat, electricity, steam, biogas, hydrogen, and liquid transportation biofuels [1,2]. As the largest agricultural country in the world, China has an abundance of biomass resources. Approximately 800 million tons of various crop residues are produced in China per year, of which corn and wheat straw account for 216 and 135 million tons, respectively [3]. Crop straws have not been widely used for bioenergy production because of the undeveloped conversion technology. Instead, many crop straws are burnt or directly dumped into the fields, causing serious environmental pollution and degraded soil conditions [4]. Therefore, the development of inexpensive and effective technologies for corn straw utilization is necessary.

Anaerobic digestion (AD) of agricultural straw for bioenergy production is widely used as a promising and alternative energy source to fossil fuels [5]. This technology has been considered as the main commercially viable option for the both treatment and recycling of biomass wastes, and thus is of great interest from an environmental and bioenergy source perspective [6]. However, the efficiency of this technology in treating agricultural straws is limited because the components of straw (lignin, cellulose, and hemicellulose) are difficult to degrade; thus, soluble compounds with low molecular weights are less available for anaerobic microorganisms [7]. Straw pretreatments prior to AD is a simple and effective method of improving the biodegradability of lignocellulosic materials because it can decompose cellulose and hemicellulose into relatively readily biodegradable components while breaking down the linkage between polysaccharide and lignin to make cellulose and hemicellulose more accessible to bacteria [8,9].

Pretreatment methods mainly include physical methods [2,10], chemical methods [11–14], biological methods [1,15], and a combination of the abovementioned methods [16,17]. Compared with physical and biological treatment methods, chemical pretreatment methods are predominantly used because they are inexpensive and are effective for enhancing the biodegradation of complex materials [18]. In chemical pretreatment methods, sulphuric acid (H_2SO_4), hydrochloric acid (HCl), hydrogen peroxide (H_2O_2), acetic acid (CH_3COOH), sodium hydroxide (NaOH), lime ($Ca(OH)_2$), and aqueous ammonia ($NH_3.H_2O$) are the common chemicals to improve AD performance of agricultural residues [19–26]. For instance, Fernández-Cegrí et al. [2] reported that the methane yield of sunflower oil cake with $Ca(OH)_2$ is 130 CH_4 g^{-1} COD, which is 25% higher that of the untreated sample. Zhu et al. [12] found that NaOH-pretreated corn stover yields 37.0% to 72.9% higher biogas productions than the untreated sample. Kang et al. [23] showed that the optimal conditions for the ethanol production of rapeseed straw is through immersion in aqueous ammonia containing 19.8% ammonia water at 69.0°C for 14.2 h. In addition, H_2SO_4, HCl, and

Table 1. Effect of acid pretreatment on the chemical composition of corn straw.

Pretreatment	Concentration	Cellulose %	Hemicellulose %	Lignin %	TC %	C/N
H_2SO_4	1%	47.1±2.5 a	26.9±1.2 a	7.5±0.5 a	37.3±2.0 b	45.5±2.0 b
	2%	41.3±1.8 b	22.5±1.8 b	7.3±0.4 a	30.6±1.9 c	38.7±1.6 c
	3%	38.0±1.6 bc	16.2±1.2 c	7.3±0.4 a	30.3±2.1 c	37.9±1.1 c
	4%	36.1±1.6 c	13.0±0.9 d	6.7±0.5 a	25.3±1.5 d	30.9±3.2 d
Untreated		49.3±1.8 a	28.8±1.4 a	7.5±0.4 a	42.3±2.8 a	51.4±3.6 a
HCl	1%	46.7±2.2 a	26.2±1.9 a	7.9±0.5 a	37.1±2.0 b	44.2±1.8 b
	2%	40.4±2.0 b	22.2±2.0 b	7.2±0.7 a	32.4±2.0 c	39.5±0.9 c
	3%	38.2±1.6 b	17.3±1.0 c	6.4±0.6 a	29.2±1.9 c	38.4±2.1 c
	4%	35.4±0.8 c	14.5±1.3 d	6.9±1.0 a	26.1±1.2 d	32.6±0.7 d
Untreated		49.3±1.8 a	28.8±1.4 a	7.5±0.4 a	42.3±2.8 a	51.4±3.6 a
CH_3COOH	1%	43.8±1.9 b	26.8±2.6 a	7.1±0.9 a	38.6±2.9 a	47.7±1.6 ab
	2%	37.4±2.4 c	21.7±1.1 b	6.7±0.5 a	34.8±0.9 b	46.4±1.9 b
	3%	34.2±0.9 d	18.1±1.4 c	6.8±0.5 a	29.5±0.9 c	36.9±2.6 c
	4%	30.4±1.5 e	15.1±0.5 d	6.7±0.7 a	26.4±1.6 d	32.2±0.7 d
Untreated		49.3±1.8 a	28.8±1.4 a	7.5±0.4 a	42.3±2.8 a	51.4±3.6 a
H_2O_2	1%	40.5±1.5 b	25.0±1.4 b	7.0±0.2 a	34.4±2.6 b	44.7±2.3 b
	2%	34.6±2.1 c	20.8±2.3 c	6.5±0.3 b	28.7±0.8 c	37.3±1.1 c
	3%	30.8±0.8 d	14.3±1.2 d	5.7±0.4 c	25.1±1.2 d	30.6±2.4 d
	4%	22.5±0.6 e	9.5±0.7 e	5.1±0.2 d	20.4±1.3 e	25.2±2.1 e
Untreated		49.3±1.8 a	28.8±1.4 a	7.5±0.4 a	42.3±2.8 a	51.4±3.6 a

Data are expressed as mean ± deviation of triplicate measurements. TC: Total carbon.
The ANOVA test was conducted to determine the differences between each pretreatment. Values with the same letters in each pretreatment indicate no significant difference at $P<0.05$.

CH_3COOH pretreatments have been used to improve the AD of lignocellulosic materials [24,25]. However, the most economically and effectively favorable treatments, among these, have yet to be identified. Additionally, the optimal concentration for the favorable pretreatment has been scarcely reported. Such information is important for the reasonable and efficient utilization of agricultural residues. The present study compared the effects of four acid and three alkaline pretreatments on the lignocellulosic compositions and methane yield of corn straws by AD. Our objective was to determine the most cost-effective pretreatment methods for enhancing the methane yield of straws.

Materials and Methods

Raw Material

Corn straw was obtained from a local villager near the Northwest A&F University (Yangling, Shaanxi, China). Prior to use, the straws were air dried, cut into lengths of 20 mm to 30 mm using a grinder, and then individually homogenized for further use. The full composition and main features of the corn straw were as follows (mean values of three determinations ± standard deviations): total solids (TS), 93.6%±2.8%; volatile solids (VS), 86.7%±1.9%; total carbon (TC), 42.3%±2.8%; total nitrogen (TN), 0.82%±0.05%; hemicellulose, 28.8%±1.4%; cellulose, 49.3%±1.8%; and lignin, 7.5%±0.4%.

Pretreatment Process

Seven pretreatment methods were used in this study, including four acid treatments (H_2SO_4, HCl, CH_3COOH, and H_2O_2) and three alkaline treatments (NaOH, $Ca(OH)_2$, and $NH_3{\cdot}H_2O$). The reagents were purchased from Sinophram Chemical Reagent Co. Ltd, Beijing, China. The chosen pretreatment conditions were based on previous studies [9,20] and carried out using different concentrations of reagents. Acid reagents (H_2SO_4, HCl, H_2O_2, and CH_3COOH) at concentrations of 1%, 2%, 3%, and 4% (w/w) and alkaline reagents (NaOH, $Ca(OH)_2$, and $NH_3{\cdot}H_2O$) at concentrations of 4%, 6%, 8%, and 10% (w/w) were used for the pretreatments. The corn straw not pretreated with any chemicals was used as the control. Each pretreatment was conducted in triplicate.

Dried corn straw (500 g) was soaked in the prepared 1.5 L solutions contained in beakers, yielding straw samples with 75% moisture. All prepared beakers were covered with plastic films, secured with a plastic ring, and then stored in a chamber at an ambient temperature of 25±2°C for 7 days. After the pretreatment, the straws were removed from the beakers, dried in an electronic oven (HengFeng SFG-02.600, Huangshi, China) at 80°C for 48 h, and then kept in a refrigerator for composition determination and AD experiments to investigate the effect of different chemical treatments on methane yield.

Anaerobic Digestion

The digestion experiment was conducted according to methods described by Song et al. [22] using laboratory-scale simulated anaerobic digesters in 1 L Erlenmeyer flasks. The batch reactors were used to determine the digestion levels of the straws with different pretreatments. Each pretreated straw was used as the digestion material, with the untreated straw as the control. The digestion inoculum was collected from an anaerobic digester in a model village powered by household biogas (Yangling, Shaanxi,

Table 2. Effect of alkaline pretreatment on the chemical composition of corn straw.

Pretreatment	Concentration	Cellulose %	Hemicellulose %	Lignin %	TC %	C/N
NaOH	4%	48.0±3.9 a	23.8±1.4 b	6.7±0.5 a	39.3±0.8 a	49.1±1.0 a
	6%	46.1±3.0 a	20.6±0.9 c	5.5±0.5 b	35.4±2.3 b	46.0±0.8 b
	8%	46.7±2.2 a	16.2±0.9 d	4.6±0.3 c	33.7±1.6 b	42.1±2.1 c
	10%	47.4±2.6 a	11.3±1.2 e	4.0±0.2 d	28.1±1.2 c	34.7±1.3 d
Untreated		49.3±1.8 a	28.8±1.4 a	7.5±0.4 a	42.3±2.8 a	51.4±3.6 a
$Ca(OH)_2$	4%	47.5±1.8 a	24.6±2.2 b	6.8±0.2 a	37.8±1.5 a	45.0±1.8 b
	6%	46.1±2.4 a	21.2±1.4 c	6.0±0.3 b	32.8±3.1 b	40.0±2.0 c
	8%	46.3±1.9 a	16.4±1.1 d	5.4±0.2 c	29.4±2.0 b	38.7±0.7 c
	10%	48.0±1.1 a	12.3±1.2 e	4.6±0.3 d	22.6±1.8 c	28.3±1.9 d
Untreated		49.3±1.8 a	28.8±1.4 a	7.5±0.4 a	42.3±2.8 a	51.4±3.6 a
$NH_3{\bullet}H_2O$	4%	48.1±1.2 a	25.7±1.9 a	7.0±0.6 ab	39.2±1.9 a	48.4±2.1 a
	6%	45.4±3.3 a	22.4±0.8 b	6.6±0.3 b	36.6±2.5 b	48.8±0.4 a
	8%	45.9±3.0 a	18.6±1.8 c	6.2±0.2 c	33.2±0.9 b	41.5±1.8 b
	10%	45.1±2.9 a	17.8±1.1 c	5.5±0.2 d	30.7±1.6 c	37.4±1.8 b
Untreated		49.3±1.8 a	28.8±1.4 a	7.5±0.4 a	42.3±2.8 a	51.4±3.6 a

Data are expressed as mean ± deviation of triplicate measurements. TC: Total carbon.
The ANOVA test was conducted to determine the differences between each pretreatment. Values with the same letters in each pretreatment indicate no significant difference at $P<0.05$.

China). This particular inoculum was selected because of its high methanogenic activity. The characteristics and features of the anaerobic inoculum used were as follows: pH, 7.6±0.1; TS, 86.6%; and VS, 47.5%. The digestion material (500 g) and inoculums (200 g) were added to each digester, followed by deionized water to obtain an 8% TS content. They were stirred and placed in a thermostatic water bath at the mesophilic condition of 37±1°C for 35 d of AD. All reactors were tightly sealed with rubber septa and screw caps. All reactors were gently mixed manually at approximately 1 min d^{-1} prior to biogas volume measurement to ensure mixing of the reactor contents. Moreover, 200 g of the inoculums was digested to serve as the blank in determining the normalized methane yield of the inoculum by itself. The digestion of each pretreatment was performed in triplicate.

Analysis and Calculations

The volume of biogas was measured by water displacement. The methane content in the produced biogas was analyzed with a fast methane analyzer (Model DLGA-1000, Infrared Analyzer,

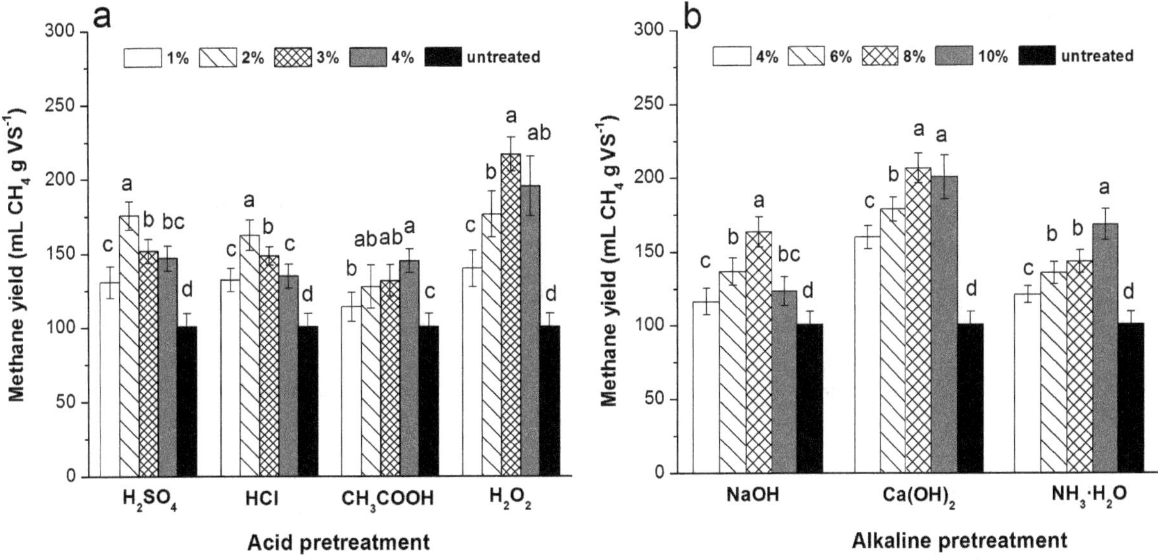

Figure 1. Effect of pretreatments on the methane yield of corn straw. (a) Acid pretreatment; (b) Alkaline pretreatment. Data was expressed at mean ± deviation of triplicate measurements. The ANOVA test was conducted to determine the differences between each pretreatment. Values with the same letters in each pretreatment indicate no significant difference at $P<0.05$.

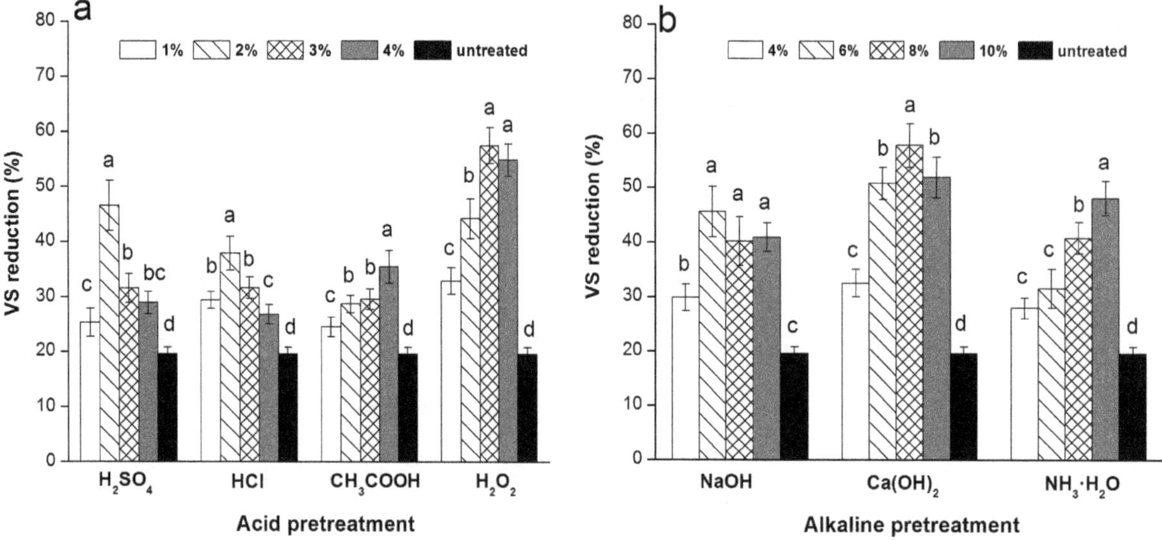

Figure 2. Effect of pretreatments on the VS consumption of corn straw. (a) Acid pretreatment; (b) Alkaline pretreatment. Data was expressed at mean ± deviation of triplicate measurements. The ANOVA test was conducted to determine the differences between each pretreatment. Values with the same letters in each pretreatment indicate no significant difference at $P<0.05$.

Dafang, Beijing, China). The TS, VS, TN, and pH of the materials were measured according to the *Standard Methods for the Examination of Water and Wastewater* of the American Public Health Association [27]. The pH was tested once every 5 d. TC content was analyzed using the method described by Cuetos et al. [28]. The C/N ratio was determined by dividing the total organic carbon content to the TN content. The volatile fatty acid (VFA) was analyzed using a colorimetric method [29], and the result was expressed in terms of acetic acid content. The cellulose, hemicellulose, and lignin contents were analyzed based on the methods previously described by Wang and Xu [30].

Data Analysis

Data is expressed as mean ± standard deviation (SD) of the triplicate measurements. Differences between mean values were examined by ANOVA. Comparisons among means were made using the Duncan multiple range test, and significance was set at $P<0.05$. All statistical analyses were performed using the software program SPSS 15.0. (SPSS Inc., Chicago, USA).

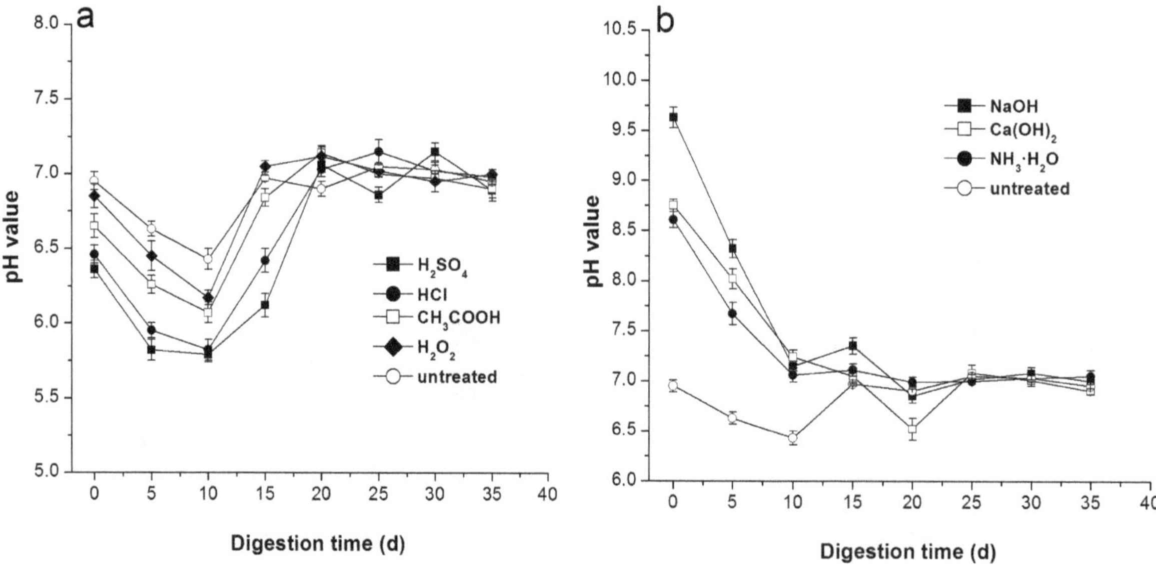

Figure 3. Change in the pH of pretreated corn straw during digestion. (a) Acid pretreatment; (b) Alkaline pretreatment. Data was expressed at mean ± deviation of triplicate measurements.

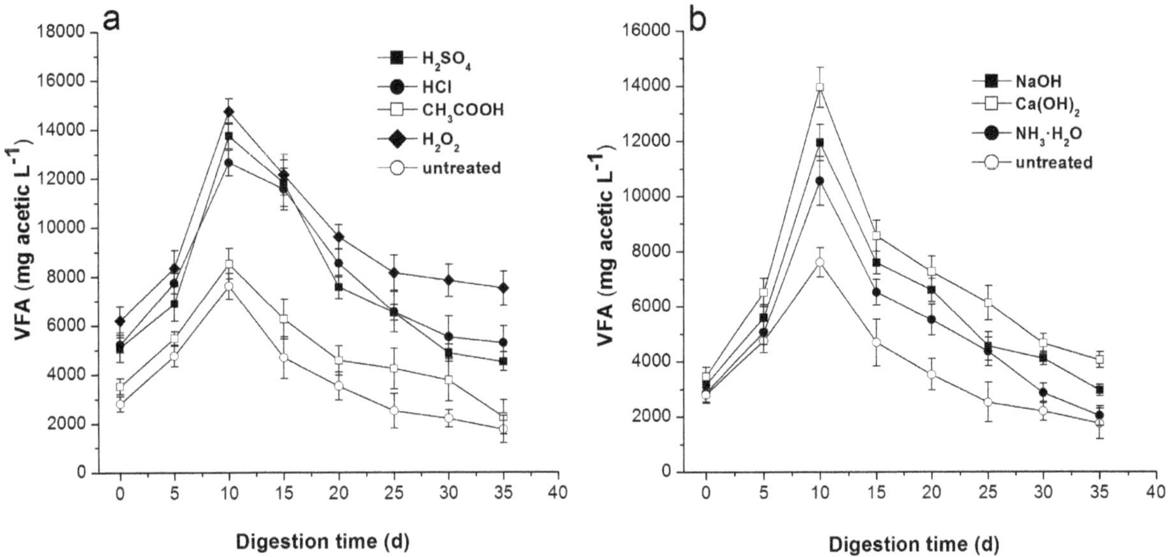

Figure 4. Change in the VFA of pretreated corn straw during digestion. (a) Acid pretreatment; (b) Alkaline pretreatment. Data was expressed at mean ± deviation of triplicate measurements.

Results and Discussion

Effects of Pretreatments on the Chemical Composition of Corn Straw

The aim of the pretreatments was to change the raw material properties, remove or dissolve lignin and hemicellulose, and reduce the crystallinity of cellulose [31]. In the present study, both acid and alkaline pretreatments changed the lignocellulosic composition of corn straw (Tables 1 and 2). Compared with the untreated straw, the hemicellulose and cellulose contents of the acid-treated straw significantly decreased by 6.6% to 66.0%, and 4.4% to 54.3% ($P<0.05$), and the hemicellulose and lignin contents of alkaline-treated corn straw decreased by 10.7% to 46.7%, and 10.8% to 60.7%. These results indicated that pretreatments are more effective in breaking down the lignocellulose matrix and in changing the chemical components of straw. Considerable amounts of lignocellulose appeared to be decomposed and converted into other soluble components that are available to anaerobic microorganisms [32].

Guo et al. [20] reported that corn stalk mainly lost its hemicellulose and cellulose fractions after the acid treatment and lost its lignin fraction after the alkaline treatment. Fernández-Cegrí et al. [2] observed that H_2SO_4 cannot dissolve the lignin of sunflower oil cake, maintaining the same proportion as that of the untreated case. They also found that alkali pretreatments give higher removal levels of lignin compared with other reagents regardless of the temperature effect. The present study revealed a similar phenomenon that acid and alkaline pretreatments had different effects on the lignocellulose composition. In the case of acid reagents, hemicellulose and cellulose contents significantly decreased while the lignin content remained constant in the treated and untreated samples, except when the H_2O_2 was used that the lignin content decreased by 6.7% to 32.0%. The alkaline treatment was mainly effective in removing the lignin fraction. The effectiveness of degrading the lignocellulosic structure usually depends on the type of pretreatment method used, because of the attack on the different parts of the substrate by different chemicals. Acid pretreatment results in disruption of covalent bonds, hydrogen bonds, and Van der Waals forces that hold together the biomass components, which consequently causes the solubilization of hemicellulose and the reduction of cellulose [33]. In contrast, alkali treatment breaks the links between lignin

Table 3. Economic performance of the different pretreatments.

	Chemicals	Concentration	Price [a](CNY)	Cost [b](CNY)	Methane yield (mL CH_4 gVS^{-1})
Acid	H_2SO_4	2%	21	2.57	175.6
	HCl	2%	15	4.92	163.4
	CH_3COOH	4%	12.5	9.34	145.1
	H_2O_2	3%	6	3.6	216.7
Alkaline	NaOH	8%	9	4.2	163.5
	$Ca(OH)_2$	8%	9.5	4.58	206.6
	NH_3•H_2O	10%	9	19.28	168.3

[a]The price was collected from the Sinophram Chemical Reagent Co. Ltd, Beijing China, and the unit of H_2SO_4, HCl, CH_3COOH, H_2O_2, and $NH_3.H_2O$ price was per 500 mL, NaOH and $Ca(OH)_2$ was per 500 g. CNY is the abbreviation for Chinese Yuan, and a dollar is equivalent to 6.12 CNY on Oct 1, 2012; Bank of China. [b] The cost was calculated based on the pretreatment of 1 kg corn straw.

monomers or between lignin and polysaccharides that makes the lignocelluloses swell through saponification reactions [34]. Among the pretreatments, H_2O_2 and NaOH showed the highest solubilization of hemicellulose cellulose, and lignin contents. This trend can be attributed to the strong oxidation ability of H_2O_2 [35] and the high alkalinity of NaOH that allow them to break down the lignocellulose matrix to change the chemical components of the straw. The increased degradation of lignocellulosic materials by H_2O_2 and NaOH suggests that these two chemicals are the most effective in degrading the lignocellulosic structure of corn straw.

The C/N ratio of anaerobic feedstock is significant for AD performance [36]. Analysis of the C/N ratio showed that the percentage of C in the pretreated straw significantly decreased with increasing chemical concentration ($P<0.05$, Tables 1 and 2). The decrease in TC content also affirmed this result. Although the C/N ratio in the pretreated straw was lower than that of the untreated sample, it was still higher than the optimum C/N ratio of feedstock materials (between 20 and 30) [36]. Therefore, the pretreated straw still represents a good co-digestion biomass because it provides a higher carbon fraction for digestion.

Effects of Pretreatments on the Methane Yield of Corn Straw

The methane yield, defined as CH_4 production per unit volatile solids (in mL CH_4 g VS^{-1}), was determined to compare the energy conversion efficiency and the improvement in biodegradability (Fig. 1). As shown in Fig. 1, the straws pretreated by acid and alkaline had significantly increased methane yields ($P<0.05$), i.e., an approximate 10.3% to 115.4% higher yield than for the untreated samples. These results are consistent with previous studies [11,17] which verified the effectiveness of chemical pretreatment in improving biodegradability and enhancing bioenergy production. This phenomenon can be explained by the fact that alkaline and acid pretreatments promote organic solubilization and increase the surface area available for enzymatic action [31]. Chemical pretreatments have different effects on the anaerobic digestibility of corn straw. The methane yield was not improved as the chemical concentration increased. The highest methane yield was achieved at different concentrations for the seven pretreatments. For instance, the highest methane yield was achieved by H_2SO_4 and HCl at 2% concentration, CH_3COOOH at 4%, H_2O_2 at 3%, $Ca(OH)_2$ and NaOH at 8%, and $NH_3{\cdot}H_2O$ at 10%. The reason may due to the fact that successful biogasification is not only affected by the sufficient soluble component available but also by anaerobic bacteria. More soluble components from the biodegradation of the lignocellulosic composition need more bacterial to assimilate them. In the present study, the same amount of inoculums (200g) was applied in each digestion experiment, thus, the relative shortage of inoculums could be responsible for the lower methane yield of the chemical pretreatment with high concentration. Among the acid and alkaline treatments, H_2O_2 and $Ca(OH)_2$ respectively produced the highest methane yield in the straw. This result suggests that H_2O_2 and $Ca(OH)_2$ are best for improving the methane yield of corn straws compared with the other pretreatments. The methane yield was significantly heightened as the H_2O_2 concentration increased from 1% to 3% and 4%. However, the methane yield did not increase with further dose increases, showing no significant difference between 3% and 4%. The same trend was also observed for the $Ca(OH)_2$ pretreatment at concentrations between 8% and 10%. The presence of excessive H^+ in 4% H_2O_2 and OH^- in the 10% $Ca(OH)_2$ pretreatment can cause toxicity to the methanogens thereby inhibiting their activity and interfering with their metabolism [37]. Therefore, 3% and 8% are the most suitable concentrations for the H_2O_2 and $Ca(OH)_2$ pretreatments of corn straw, respectively.

Effects of Pretreatments on VS Reduction of Corn Straw

Methane is generated from the conversion of substrates; thus, the methane yield can be determined by reductions in the amount of dry matter of the substrate, as represented by VS. The VS reductions in the straw are shown in Fig. 2. Consistent with previous studies [22], the chemically-treated corn straw obtained higher VS reductions than untreated samples and exhibited reduction of 57.3% to 70.0% for the acid pretreatment and 57.5% to 70.8% for the alkaline pretreatment. 3% H_2O_2 and 8% $Ca(OH)_2$ yielded the greatest reduction in the amount of dry matter of the substrate. The pretreatment triggers the conversion of VS into soluble compounds, including sugar, starch, pectin, tannin, cyclitol, and some inorganics, which become available to anaerobic microorganisms. Generally, this treatment contributes to a substantial improvement in the biodegradability of corn straw. High methane production requires more substrates for digestion; thus, increased VS reductions could explain why the methane yield of the treated straw was highly improved.

Effects of Pretreatments on pH during AD

To investigate the effect of pretreatment on the VFA and pH during the AD of corn straw, the optimal concentration of each pretreatment for methane production was selected as follows: 2% H_2SO_4, 2% HCl, 4% CH_3COOH, 3% H_2O_2, 8% NaOH, 8% $Ca(OH)_2$, and 10% $NH_3{\cdot}H_2O$.

Fermentative microorganisms can function in a wider pH range of between 4.0 and 8.5 [38]. In the present study over the first 10 d, the pH of the fermentation broth of the acid-pretreated corn straws was below 7.0 (Fig. 3), whereas that of the three alkaline-pretreated corn straws was over 7.0. The pH curves of all pretreatments were similar, showing a decreasing trend in the initial 10 d and an increasing trend thereafter, slight fluctuations between days 10 to 20. At the end of the fermentation, all pretreatments maintained a pH of approximately 7.0. This trend can be attributed to the variation in VFA concentration because the production of VFA during AD decreases pH. The highly concentrated substrate at the initial phase of AD supplies sufficient organic acid from the degradation of hemicellulose, cellulose, lignin, and VS for the methanogens [20], which decreases pH and accelerates methanogen growth. As digestion proceeded, the content of organic acid gradually decreased with the consumption by the methanogens, which increased the pH. The shortage in organic acid limited the activities of the methanogens but stimulated the acidogens, which increased the amount of organic acids and the dropped the pH. The activity of the methanogens increased again when the organic acid accumulated to an extent, which increased the pH. However, compared with the dramatic fluctuation in the initial phase of AD, the change in the pH in the middle–late phase was slightly heightened because the concentration of the organic acid in the substrate was not as high as the initial concentration. The lack of significant differences in the pH for all pretreatments at the end of AD indicates that these pretreatments can recover the pH. As shown in Fig. 3, the pH of the fermentation broth of the pretreated corn straw markedly declined compared with that of the untreated corn straw. This result can be ascribed to the various acids in the soluble substance of the pretreated straw being significantly higher than that of the untreated straw.

Effects of Pretreatments on VFA during AD

The VFA concentration of each pretreatment initially increased (Fig. 4) and then decreased, which is contrary to the trend of the pH curve. The VFA content of the fermentation broth from the pretreated straw increased more sharply than that of the untreated corn straw. This result can be attributed to the significantly higher soluble substance content of the pretreated corn straw compared with the untreated samples. Among the seven pretreatments, the average VFA concentrations (mg acetic L^{-1}) of the pretreatments during the AD were as follows: 7629 (H_2SO_4), 7879 (HCl), 4821 (CH_3COOH), 9321 (H_2O_2), 5810(NaOH), 6818 ($Ca(OH)_2$), and 4964 ($NH_3 \cdot H_2O$). The highest VFA values were observed for H_2O_2 in the acid treatment, whereas the lowest was observed for CH_3COOH. This result is consistent with the results of the hemicellulose, cellulose, and lignin decomposition and methane yield (Table 1), which further confirmed the effectiveness of H_2O_2 in biodegrading the lignocellulosic structure of straws. Large amounts of hemicellulose and cellulose are converted into simple sugars, lipids (fats) into fatty acids, amino acids, and short-chain organic acids (butyric acid, propionic acid, acetate, and acetic acid), all of which are utilized by methanogens for methane production [15]. In the alkaline pretreatments, the highest VFA content was observed after using $Ca(OH)_2$. This result was consistent with the observations from the methane yield experiments, but contradicted the lignocellulosic composition results where degradation of the lignin fraction was highest after NaOH pretreatments. This disparity can be explained by the fact that successful biogasification is not only affected by the sufficient soluble component available for the anaerobic bacteria but also by the balance between methanogens and acidogens [39]. The excessively high concentration of OH^- in NaOH likely inhibited acetogenesis and disturbed this balance. However, this hypothesis warrants further investigation.

Economic Performance of the Pretreatment Methods

The effectiveness of a pretreatment is not only based on the effectiveness of AD but also on the economic performance. Table 3 compares the economic performance of the pretreatments at the optimal concentrations for methane yield. H_2O_2 and H_2SO_4 showed the lowest costs among the acid pretreatments. However, H_2O_2 was more favorable because it produced higher methane yields than H_2SO_4. In the alkaline pretreatments, although no great difference in the expenses was observed between the $Ca(OH)_2$ and NaOH pretreatments, $Ca(OH)_2$ produced is slightly advantageous over NaOH as it generates a higher methane yield. Therefore, with respect to economic performance and effectiveness, H_2O_2 and $Ca(OH)_2$ can be considered as the most suitable pretreatments for corn straw.

Recently, some researchers combined chemical and physical treatments to improve the biodegradability of lignocellulose composition. High temperature (120–250°C) is often used in combination with dilute acids or base in a pressure cell for much shorter durations. For instance, Saha et al. [40] found the 74% higher saccharification yield wheat straw was subjected to 0.75% v/v of H_2SO_4 at 121°C for 1 h. Cara et al. [41] shown that olive tree biomass pretreated with 1.4% H_2SO_4 at 210°C resulted in 76.5% of hydrolysis yields. Rocha et al. [42] reported that ethanol yield as high as 0.47 g/g glucose was achieved in fermentation tests with cashew apple bagasse pretreated with diluted H_2SO_4 at 121°C for 15 min. These studies showed the advantage of combination treatment on solubilizing the lignocellulosic composition and shortening the pretreatment time. Nevertheless, depending on the process temperature, some sugar degradation compounds such as furfural and aromatic lignin degradation compounds are detected, and affect the microorganism metabolism in the fermentation step [40]. Furthermore, the pretreatment of high temperature combined with chemicals consumes a substantial amount of energy, and need high facility investment and high treatment cost.

In the present study, although pretreatment time (7 day) was longer than that of chemical treatment with the addition of heat and pressure, the contents of hemicelluloses, cellulose, and lignin fractions of corn straw was greatly reduced, which was contribute to the enhancement of methane production. Furthermore, using single chemicals have no excessive energy consumption and less operation cost. Since cost reduction and low energy consumption are required for an effective pretreatment, chemical pretreatment without the addition of heat and pressure would be desirable to optimize the effectiveness on the process. As for the longer incubation time of the chemical pretreatment, more efforts should be made to investigate the combination of chemicals and low temperature (Below 100°C) pretreatment to shorten the incubation time and improve the anaerobic digestion efficiency.

Conclusions

Four acid pretreatments (H_2SO_4, HCl, CH_3COOH, and H_2O_2) and three alkaline pretreatments (NaOH, $Ca(OH)_2$, and $NH_3 \cdot H_2O$) for improving the methane yield of corn straw were compared. All pretreatments were effective in the biodegradation of the lignocellulosic structure. Straw pretreated with 3% H_2O_2 and 8% $Ca(OH)_2$ elicited the highest methane yields of 216.7 and 206.6 mL CH_4 g VS^{-1}, which are 115.4% and 105.3% higher than that of the untreated straw, respectively. H_2O_2 and $Ca(OH)_2$ are economically and effectively superior to the other pretreatments. Therefore, H_2O_2 and $Ca(OH)_2$ are both recommended as the pretreatments for improving the methane yield of straw.

Acknowledgments

We thank Dr. Chao Zhang for assistance of improving the paper and advice on statistical treatments.

Author Contributions

Conceived and designed the experiments: ZLS GHY XFL. Performed the experiments: ZLS XFL. Analyzed the data: ZLS ZYY. Contributed reagents/materials/analysis tools: YXY YZL. Wrote the paper: ZLS.

References

1. Zhong WZ, Zhang ZZ, Luo YJ, Sun SS, Qiao W, et al. (2011) Effect of biological pretreatments in enhancing corn straw biogas production. Bioresour Technol 102: 11177–11182.
2. Fernández-Cegrí F, Raposo F, de la Rubia MA, Borja R (2013) Effects of chemical and thermochemical pretreatments on sunflower oil cake in biochemical methane potential assays. J Chem Technol Biotechnol 88: 924–929.
3. National Burean of Statistics of China (2010) China Statistical Yearbook of 2009. China Statistics Press, Beijing, China.
4. Pang YZ, Liu YP, Li XJ (2008) Improving biodegradability and biogas production of corn stover through sodium hydroxide solid state pretreatment. Energ Fuel 22: 2761–2766.
5. Murphy JD, Power NM (2009) An argument for using biomethane generated from grass as a biofuel in Ireland. Biomass Bioenerg 33: 504–512.
6. Amon T, Amon B, Kryvoruchko V, Machmüller A, Hopfner SK, et al. (2007) Methane production through anaerobic digestion of various energy crops grown in sustainable crop rotations. Bioresour Technol 98: 3204–3212.

7. Taherzadeh MJ, Karimi K (2008) Pretreatment of lignocellulosic waster to improve ethanol and biogas production: a review. Int J Mol Sci 9: 1621–1651.
8. Teghammar A, Yngvesson J, Lundin M, Taherzadeh MJ, Horváth IS (2010) Pretreatment of paper tube residuals for improved biogas production. Bioresour Technol 101: 1206–1212.
9. Ferreira LC, Donoso-Bravo A, Nilsen PJ, Fdz-Polanco F, Pérez-Elvira SI (2013) Influence of thermal pretreatment on the biochemical methane potential of wheat straw. Bioresour Technol 143: 251–257.
10. Sapci Z (2013) The effect of microwave pretreatment on biogas production from agricultural straws. Bioresour Technol 128: 487–494.
11. Zheng MX, Li XJ, Li LQ, Yang XJ, He YF (2009) Enhancing anaerobic biogasification of corn stover through wet state NaOH pretreatment. Bioresour Technol 100: 5140–5145.
12. Zhu J, Wan C, Li Y (2010) Enhanced solid-state anaerobic digestion of corn stover by alkaline pretreatment. Bioresour Technol 101: 7523–752.
13. Cao WX, Sun C, Liu RH, Yin RZ, Wu XW (2012) Comparison of the effects of five pretreatment methods on enhancing the enzymatic digestibility and ethanol production from sweet sorghum bagasse. Bioresour Technol 111: 215–221.
14. Michalska K, Miazek K, Krzystek L, Ledakowicz S (2012) Influence of pretreatment with Fenton's reagent on biogas production and methane yield from lignocellulosic biomass. Bioresour Technol 119: 72–78.
15. Gomez-Tovar F, Celis LB, Razo-Flores E, Alatriste-Mondragón F (2012) Chemical and enzymatic sequential pretreatment of oat straw for methane production. Bioresour Technol 116: 372–378.
16. Zhang QH, Tang L, Zhang JH, Mao ZG, Jiang L (2011) Optimization of thermal–dilute sulfuric acid pretreatment for enhancement of methane production from cassava residues. Bioresour Technol 102: 3958–3965.
17. Chandra R, Takeuchi H, Hasegawa T, Kumar R (2012) Improving biodegradability and biogas production of wheat straw substrates using sodium hydroxide and hydrothermal pretreatments. Energy 43: 273–282.
18. Zhou SX, Zhang YL, Dong YP (2012) Pretreatment for biogas production by anaerobic fermentation of mixed corn stover and cow dung. Energy 46: 644–648.
19. González G, Urrutia H, Roeckel M, Aspé E (2005) Protein hydrolysis under anaerobic, saline conditions in presence of acetic acid. J Chem Technol Biotechnol 80: 151–157.
20. Guo P, Mochidzuki K, Cheng W, Zhou M, Gao H, et al. (2011) Effects of different pretreatment strategies on corn stalk acidogenic fermentation using a microbial consortium. Bioresour Technol 102: 7526–7531.
21. Li Q, Gao Y, Wang HS, Li B, Liu C, et al. (2012) Comparison of different alkali-based pretreatments of corn stover for improving enzymatic saccharification. Bioresour Technol 125: 193–199.
22. Song ZL, Yang GH, Guo Y, Ren GX, Feng YZ (2012) Comparison of two chemical pretreatments of rice straw for biogas production by anaerobic digestion. BioResources 7: 3223–3236.
23. Kang KE, Jeong GT, Swoo CS, Park DH (2012) Pretreatment of rapeseed straw by soaking in aqueous ammonia. Bioprocess Biosyst Eng 35: 77–84.
24. Pakarinen OM, Kaparaju PLN, Rintala JA (2011) Hydrogen and methane yields of untreated, water-extracted and acid (HCl) treated maize in one-and two-stage batch assays. Int J Hydrogen Energy 36: 14401–14407.
25. Monlau F, Latrille E, Da Costa AC, Steyer JP, Carrère H (2013) Enhancement of methane production from sunflower oil cakes by dilute acid pretreatment. Appl Energ 102: 1105–1113.
26. Us E, Perendeci NA (2012) Improvement of methane production from greenhouse residues: optimization of thermal and H_2SO_4 pretreatment process by experimental design. Chem. Eng J 182: 120–131.
27. APHA (1998) Standard methods for the examination of water and wastewater, 20th ed, American Public Health Association, Washington, DC, USA.
28. Cuetos MJ, Fernandez C, Gomez X, Moran A (2011) Anaerobic co-digestion of swine manure with energy crop residues. Biotechnol Bioprocess Eng 16: 1044–1052.
29. Chengdu Institute of Biology, Chinese Academy of Sciences (1984) Routine Analysis of Biogas Fermentation, Science Technology Press, Beijing, China.
30. Wang YW, Xu WY (1987) The method for measurement the content of cellulose, hemicellulose and lignin in solid materials, Microbiology China 2: 81–85.
31. Silverstein RA, Chen Y, Sharma-Shivapp RR, Boyette MD, Osborne J (2007) A comparison of chemical pretreatment methods for improving saccharification of cotton stalks. Bioresour Technol 98: 3000–3011.
32. Lin YQ, Wang DH, Wu SQ, Wang CM (2009) Alkali pretreatment enhances biogas production in the anaerobic digestion of pulp and paper sludge. J Hazard Mater 170: 366–373.
33. Li C, Knierim B, Manisseri C, Arora R, Scheller HV, et al. (2010) Comparison of dilute acid and ionic liquid pretreatment of switchgrass: biomass recalcitrance, delignification and enzymatic saccharification. Bioresource Technol 101: 4900–4906.
34. Xiao B, Sun XF, Sun R (2001) Chemical, structural, and thermal characterizations of alkali-soluble lignins and hemicelluloses, and cellulose from maize stems, rye straw, and rice straw. Polymer Degrad Stability 74: 307–319.
35. Li ZL, Chen CH, Hegg EL, Hodge DB (2013) Rapid and effective oxidative pretreatment of woody biomass at mild reaction conditions and low oxidant loadings. Biotechnol Biofuels 6: 119. doi:10.1186/1754-6834-6-119.
36. Estevez MM, Linjordet R, Morken J (2012) Effects of steam explosion and codigestion in the methane production from Salix by mesophilic batch assays. Bioresour Technol 104: 749–756.
37. Chen Y, Cheng JJ, Creamer KS (2008). Inhibition of anaerobic digestion process: A review. Bioresour Technol 99: 4044–4064.
38. Hwang MH, Jang NJ, Hyum SH, Kim IS (2004) Anaerobic bio-hydrogen production from ethanol fermentation: the role of pH. J Biotechnol 111: 297–309.
39. Zhang T, Liu L, Song Z, Ren G, Feng Y (2013) Biogas production by co-digestion of goat manure with three crop residues. PLoS ONE 8(6): e66845.
40. Saha BC, Iten LB, Cotta MA, Wu YV (2005) Dilute acid pretreatment, enzymatic accharification and fermentation of wheat straw to ethanol. Process Biochem 40: 3693–3700.
41. Rocha MV, Rodrigues TH, de Macedo GR, Gonçalves LR (2009) Enzymatic hydrolysis and fermentation of pretreated cashew apple bagasse with alkali and diluted sulfuric acid for bioethanol production. Appl Biochem Biotechnol 155: 407–417.
42. Cara C, Ruiz E, Oliva JM, Sáez F, Castro E (2008) Conversion of olive tree biomass into fermentable sugars by dilute acid pretreatment and enzymatic saccharification. Bioresour Technol 99: 1869–1876.

10

Evaluation of Bacterial Expansin EXLX1 as a Cellulase Synergist for the Saccharification of Lignocellulosic Agro-Industrial Wastes

Hui Lin[1,3], Qi Shen[1], Ju-Mei Zhan[2], Qun Wang[1], Yu-Hua Zhao[1]*

1 Institute of Microbiology, College of Life Sciences, Zhejiang University, Hangzhou, China, **2** Institute of Plant Science, College of Life Sciences, Zhejiang University, Hangzhou, China, **3** Institute of Environment, Resource, Soil and Fertilizer, Zhejiang Academy of Agricultural Sciences, Hangzhou, China

Abstract

Various types of lignocellulosic wastes extensively used in biofuel production were provided to assess the potential of EXLX1 as a cellulase synergist. Enzymatic hydrolysis of natural wheat straw showed that all the treatments using mixtures of cellulase and an optimized amount of EXLX1, released greater quantities of sugars than those using cellulase alone, regardless of cellulase dosage and incubation time. EXLX1 exhibited different synergism and binding characteristics for different wastes, but this can be related to their lignocellulosic components. The cellulose proportion could be one of the important factors. However, when the cellulose proportion of different biomass samples exhibited no remarkable differences, a higher synergism of EXLX1 is prone to occur on these materials, with a high proportion of hemicellulose and a low proportion of lignin. The information could be favorable to assess whether EXLX1 is effective as a cellulase synergist for the hydrolysis of the used materials. Binding assay experiments further suggested that EXLX1 bound preferentially to alkali pretreated materials, as opposed to acid pretreated materials under the assay condition and the binding preference would be affected by incubation temperature.

Editor: Nikolas Nikolaidis, California State University Fullerton, United States of America

Funding: This study was supported by the Science and Technology Project of Zhejiang Province (2011C13016), National Natural Science Foundation of China (31070079; 41271335), the International Cooperation Project in Science and Technology of Zhejiang Province (2008C14038), High Technology Research and Development Program of China (863 Program)(2012AA06A203), and the National Key Technology Rand D Program (2012BAC17B04). The funders had no role in study design, data collection and analysis, decision to publish, or preparation of the manuscript.

Competing Interests: The authors have declared that no competing interests exist.

* E-mail: yhzhao225@zju.edu.cn

Introduction

Lignocellulosic waste is a promising resource for producing fuels and chemicals, both natural and man-made [1]. As the most abundant and renewable source on earth, lignocellulose consists of three major components: cellulose, hemicellulose and lignin [2]. Deconstruction of lignocellulose into fermentable sugars is a key process in its conversion to high-value chemicals and an array of glycoside hydrolases is required. There is no single enzyme that is able to hydrolyze lignocellulose biomass efficiently, due to its tightly-packed structure and complex components. Design of glycoside hydrolase mixtures that function synergistically to release sugars from biomass has been known to be an effective strategy [3,4]. Recently, combined utilization of proteins lacking glycoside hydrolase activity (non-GH) with glycoside hydrolases such as cellulase has been suggested as another effective option to facilitate the release of sugars from lignocellulosic biomass [5,6].

Expansins and expansin-like proteins are one kind of non-GH proteins that do not directly hydrolyze lignocellulose but can increase the hydrolysis efficiency of glycoside hydrolases in a synergistic manner [6,7,8]. The binding and loosening functions of expansin to cell wall components imply that it disrupts the hydrogen bonds in CPs and enhances the accessibility of cell wall degrading enzymes [9,10]. Many expansins or expansin-like proteins, such as LOOS1 [11], swollenin [12], AfSwo1 [13] and maize β-expansin [7], have been found that can enhance the activity of glycoside hydrolases in the saccharification of plant biomass. Quiroz-Castaneda et al. [11] showed that *Agave tequilana* fiber, extensively grown in some areas of Mexico, can be a susceptible substrate for a cocktail of commercial cellulases and xylanases in the presence of LOOS1. In the study of Chen et al. [13], cellulase used together with the expansin AfSwo1 from *Aspergillus fumigatus* to hydrolyze the avicel resulted in a 13.2% increase in the sugar yield in comparison with cellulase used alone, even though the cellulase loading in these trials was high.

EXLX1 is an expansin-like protein encoded by the *yoaJ* gene of *Bacillus subtilis* that can be produced by *Escherichia coli* [6,8,14]. EXLX1 has received increasing attentions as a result of emerging reports such as the binding characterization, the cell wall-creeping study and the structure-function analysis [6,14,15]. However, the synergistic characteristics of EXLX1 have not been well characterized, especially using natural plant biomass as substrates [16]. In most previous reports, synergistic effects of EXLX1 were evaluated using pure cellulose such as avicel and filter paper and investigated with a low loading of cellulase [6,16]. Actually, some reports have indicated that the weakening activity of expansin could be stronger in cellulose-xyloglucan composite materials than in cellulose only materials [15,17,18]. In this work, EXLX1 was characterized as a commercial cellulase synergist using various types of plant biomass wastes extensively used in biofuel production. The information provided would be useful for the evaluation of EXLX1 as a biochemical agent in cellulosic biomass conversion to reduce the cost of bioenergy production, although the mechanism of this process still needs further investigation.

Materials and Methods

Ethics Statement

All the materials used in this study such as Wheat straw collected from Fuyang city of Anhui province and Switchgrass obtained from Beijing City, were all agricultural waste widely distributed in the rural areas in China. As the materials collection activities had no conflicts of interest and environmental hazards, there were no specific permissions required. And the field studies did not involve endangered or protected species.

Cloning, expression and purification of EXLX1

The gene encoding EXLX1 protein (Primary accession No. in Uni-ProtKB: O34918) was amplified from genomic DNA of *Bacillus subtilis* subsp. *Subtilis* strain 168 M (ATCC27370) by PCR using 5′ CGGAATTCGCATATGACGACCTGCATG 3′ and 5′ CCGCTCGAGTTCAGGAAACTGAACATGGC 3′ as primers. The *EXLX1* gene was cloned into pET21a (+) vector (Novagen) between the EcoRI and XhoI sites, followed by transformation into *Escherichia coli* BL21 (DE3-pLys) for expression.

The original signal peptide was removed from the recombinant EXLX1 and a 6-histidine tag was added at the carboxyl terminus (Table S1). The recombinant cells harboring pET21a (+)-EXLX1 were cultivated in Luria-Bertani medium (pH 7.0) supplemented with 100 μg ml^{-1} ampicillin. Cultures were grown to $OD_{600}\approx 0.5$ at 37°C and then induced with 1 mM isopropyl β-D-1-thiogalactopyranoside (IPTG) for more than 5 h at 30°C. Cells were harvested by centrifugation at 12000 rpm for 10 min, washed twice with Buffer A (50 mM NaH_2PO_4, 500 mM NaCl, pH 8.0) and then disrupted by sonication. The cell debris was removed by centrifugation at 12000 rpm and 4°C for 30 min. The supernatant was loaded onto a column with Ni-NTA Agarose (Qiagen), which was pre-equilibrated with Buffer A. After binding, the column was subsequently washed with Buffer B (50 mM NaH_2PO_4, 500 mM NaCl, and 20 mM imidazole, pH 8.0) until no more protein was eluted. Finally, EXLX1 was eluted with Buffer C (50 mM NaH_2PO_4, 500 mM NaCl, and 250 mM imidazole, pH 8.0) and the eluted protein was stored in buffer D (50 mM NaH_2PO_4, 100 mM NaCl and 25% glycerol, pH 7.0) at −20°C after concentrated by ultrafiltration.

Untreated (UNT) and pretreated plant biomass

Except for Wheat straw and Switchgrass, other biomass samples such as Rice straw, Green Reed, Sugarcane bagasse and Chinese fir sawdust (Wood) were all collected from local markets in Hangzhou. Dilute acid pretreatment (0.7% H_2SO_4, 121°C, 1 h, ratio of straw to liquid 10%) and alkaline peroxide pretreatment (2.5% H_2O_2, pH 11.5, 37°C, 24 h, ratio of straw to liquid 10%) was applied to biomass samples [19], respectively. Both alkaline peroxide pretreated (ALKALI) materials and dilute acid pretreated (ACID) materials were washed with distilled water till neutral pH was achieved. All the materials were dried before being milled to lower than 20 meshes. Compositional analysis [20] was further conducted on the UNT, ACID and ALKALI biomass samples.

Enzymatic hydrolysis

Enzymatic hydrolysis was performed in a 1.5 ml microcentrifuge tube containing 5 mg of substrate (e.g. Whatman No. 1 filter paper, biomass samples) according to a previously published study with slight modifications [6]. Both the commercial cellulase (Celluclast 1.5L, Novozymes, Bagsvaerd, Denmark), EXLX1 and bovine serum albumin (BSA) were diluted to their corresponding concentrations in 750 μl final volume of citrate buffer (0.05 M, pH 4.8). The enzymatic hydrolysis was performed at 50°C before the reducing sugar (RS) was determined using dinitrosalicylic acid (DNS) reagent [21]. The synergetic activity of EXLX1 was calculated as described before [6]:

$$\text{Synergistic activity}(\%) = \left(\frac{\text{RS released byEXLX1 and cellulase}}{\text{RS released by cellulasealone}} - 1\right) \times 100$$

Thermal- and pH-stability of EXLX1

The temporary thermal-stability of EXLX1 was determined by assaying the residual synergistic activity of EXLX1 after incubating EXLX1 at temperatures ranging from 25°C to 70°C for 30 min. The long-term thermal-stability was determined from the residual synergistic activity after incubating EXLX1 at temperatures ranging from 37°C to 60°C for 48 h. In order to determine the effect of pH on the stability of EXLX1, the synergistic activity was assayed after incubating EXLX1 at 4°C for 48 h with the following buffers: 100 mM citrate buffer for pH 3.0–6.0, 200 mM sodium phosphate buffer for pH 6.0–8.0, and 100 mM glycine-NaOH for pH 8.6–10.0. Enzymatic hydrolysis was incubated for 36 h using Whatman No. 1 filter paper as the substrate. The concentration of the EXLX1 and the cellulase in the reaction system was 200 μg/g substrate (0.5 μg) and 0.06 FPU/g substrate, respectively. The residual activity (%) of the pretreated EXLX1 was calculated according to:

$$\text{Residual activity}(\%) = \frac{\text{Synergistic activity of the pretreated EXLX1}}{\text{Synergistic activity of the original EXLX1}} \times 100$$

Binding assay

All the UNT, ACID and ALKALI materials were provided as binding matrices. 30 μg of EXLX1 was mixed with 2.5 mg of binding matrix, respectively, suspended in 300 μl of citrate buffer (0.05 M, pH 4.8) and potassium phosphate buffer (0.05 M, pH 7.0). The mixture was incubated at 30°C for 4 h. After incubation, the supernatant was obtained by centrifuging at 12000 rpm for 5 min and subsequently quantified by the Bradford assay [22]. The amount of bound protein was determined by subtracting the amount of the unbound protein from the total amount of protein.

Statistical analysis

All the above-mentioned assays were repeated more than four times. Statistical analyses such as variance analysis (ANOVA) and correlation analysis were all completed with SPSS 16.0 (Chicago, IL).

Results and Discussion

Production of bioactive EXLX1

Recombinant EXLX1 was successfully expressed in *E. coli* and primarily exists in the soluble fraction (Fig.1a). As shown in Fig.1b, it could be found that the soluble fraction of IPTG induced cells exhibited remarkable disruptive activity against filter paper, which indicated the effective cell wall loosening of the recombinant EXLX1. Further, EXLX1 was purified using a 6-His tag located at the carboxyl-terminus of the protein (Fig.1a) and approximately 35 mg of purified EXLX1 was obtained from 1 L of culture broth. Determination of the synergistic activity of the purified EXLX1 with cellulase was performed with BSA used instead of EXLX1 in the negative control experiments. A synergistic activity (%) of 77.41±1.97 was obtained by incubating 5 mg filter paper with 0.5 μg of EXLX1 (200 μg/g substrate) and 0.06 FPU/g of cellulase for 36 h. The synergistic activity (%) was found to be 52.04±3.97 when the filter paper was incubated with 5 μg of EXLX1 (2000 μg/g substrate) and 0.6 FPU/g substrate of cellulase. Similar with the results published previously [6], BSA also exhibited a synergistic effect on cellulase activity but remarkably lower than EXLX1. The synergistic activity (%) was only 19.43±5.02 when 5 μg of BSA (2000 μg/g substrate) was incubated with 0.6 FPU/g substrate of cellulase for enzymatic hydrolysis. The above-mentioned results demonstrated that the recombinant EXLX1 in the present work exhibited similar synergism and loosening characteristics with those obtained in previous reports [6,8,14], although an additional polypeptide from the expression vector was involved at its amino terminus.

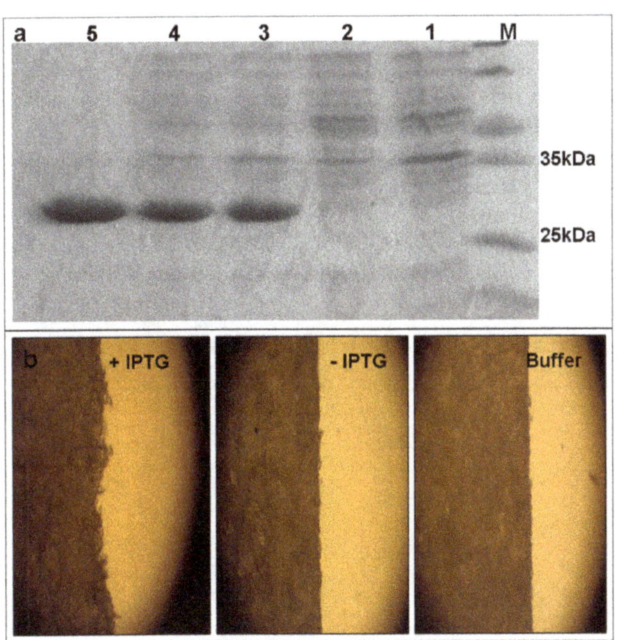

Figure 1. EXLX1 production, purification and filter paper disruption. (a) Sodium dodecyl sulfate polyacrylamide gel electrophoresis (SDS-PAGE) analysis: Lane M, protein molecular weight marker; Lane 1, the total cell lysate of the culture without IPTG; Lane 2, the soluble fraction of the culture without IPTG; Lane 3, the total cell lysate of the culture with IPTG; Lanes 4, the soluble fraction of the culture with IPTG; Lane 5, purified EXLX1 (about 26 kDa calculated based on amino acid sequences). (b) Light microscopy graphs of filter paper incubated with the soluble fraction of IPTG-induced culture, soluble fraction of the culture without IPTG and buffer A at 30°C for 12 h, respectively. Both the cell lysate and the soluble fraction were dissolved in buffer A (50 mM NaH_2PO_4, 500 mM NaCl, pH 8.0).

Thermal- and pH-stability of EXLX1

The sensitivity of EXLX1 to heat denaturation in aqueous solutions was assessed as the robustness of enzymes is an important factor for industrial applications. As shown in Fig.2a, EXLX1 was relatively stable when incubated at temperatures ranging from 25°C to 50°C for 30 min with more than 70% activities retained, while the residual synergistic activity decreased robustly when the temperature was higher than 50°C and completely disappeared at 70°C. The long-term thermal-stability of EXLX1 was also assessed due to the long reaction time required for lignocellulose hydrolysis of cellulase. The residual activity of EXLX1 declined significantly when the incubating time increased to 48 h and only about 50% of residual activity was retained after incubating at 50°C for 48 h (Fig.2a). Although temperature presented a significant effect on the activity of EXLX1, EXLX1 is still more thermal-stable than some previous reported expansins or expansin-like proteins [11,23]. The creep activity of EXLX1 has been found to be high in the reaction buffer with pH from 5.5–9.5 [18]. Similarly, the EXLX1 pretreated at various buffers with pH ranging from 4.8–9.5 for 48 h (Fig.2b) was found to be active at the reaction buffer in this study (pH 4.8).

Digestibility of UNT wheat straw with different cellulase and EXLX1 combinations

Enzymatic hydrolysis of UNT wheat straw was performed by using different cellulase and EXLX1 combinations with the corresponding RS yields shown in Fig.3. ANOVA analysis was subsequently provided to study the effects of cellulase dosage (0.3–5 FPU/g), EXLX1 amount (0–15 μg) and hydrolysis time (24–48 h) on the synergism of EXLX1 in the hydrolysis process. It could be found that all the factors and their interactions exhibited significant ($p<0.05$) effects on the synergism with the order Time>Cellulase>EXLX1>EXLX1×Time>EXLX1×Cellulase>EXLX1×Time×Cellulase. The results suggested that the effect of EXLX1 amount on the synergism would be interfered by both cellulase and incubation time so that the effective synergism of EXLX1 with cellulase occurred only when a specific dosage of cellulase was used combined with an appropriate amount of EXLX1. The optimal amount of EXLX1 for each treatment with specific cellulase dosage and incubation time can be observed in Fig.3. RS yields obtained from the treatments using the optimal cellulase and EXLX1 combinations were sorted out and fitted with logarithmic equation (Fig.4). As shown in Fig.4, the growth rate of RS yield at 24 h brought by the increase in cellulase dosage was found to be higher in the treatment using an optimal cellulase and EXLX1 mixture than that using cellulase alone, indicating that the hydrolysis efficiency of the treatment using the cellulase and EXLX1 mixture responds more sensitively to the changes in the cellulase dosage than that using pure cellulase. However, the growth rate of RS yield decreased with the proceeding of the incubation time in the treatments using optimal cellulase and EXLX1 mixture but increased in those using pure cellulase (Fig.4).

Generally, all the treatments using cellulase alone released lower quantities of RS than that using the optimal mixture of cellulase and EXLX1, regardless of cellulase amount and incubation time (Fig.4). The increasing percentage of RS yield brought by changing the catalyst from pure cellulase to cellulase and EXLX1 mixture was even higher than that brought by increasing the cellulase dosage in some cases (Fig.4). For example, using 1.2 FPU/g of cellulase in combination with 2.5 μg of EXLX1 gave 552.26 μg of RS after 24 h incubation, while only 480.91 μg of RS released using 5 FPU/g of pure cellulase under the same condition. The digestion time could also be shortened after the

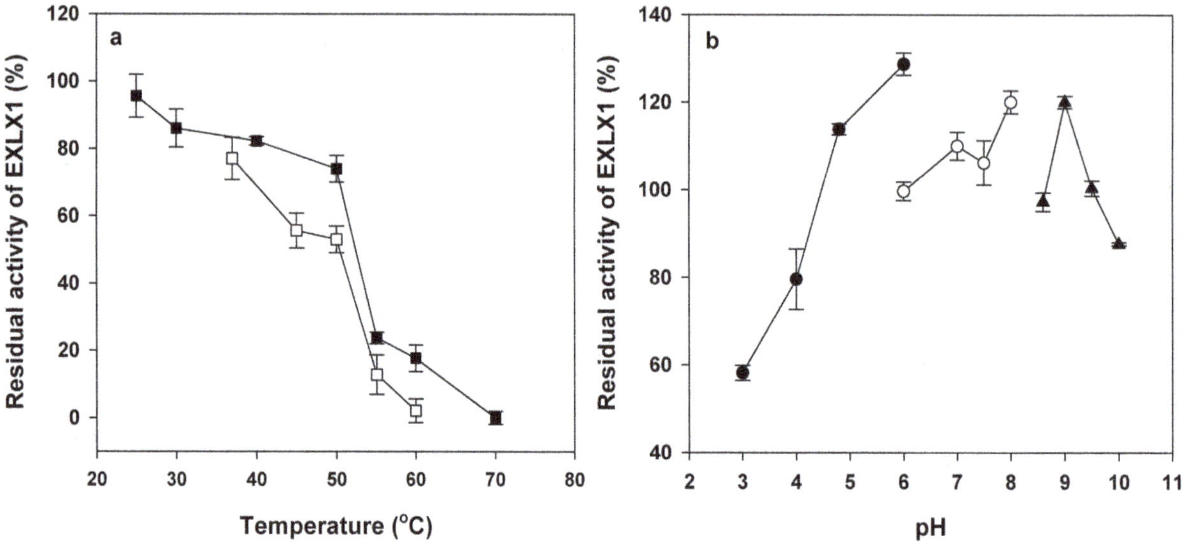

Figure 2. Thermal- and pH-stability of EXLX1. (a) Thermal-stability of EXLX1. EXLX1 was incubated at various temperatures for 30 min (black squares) and 48 h (open squares), respectively. The residual synergistic activity of the treated protein was then determined at 50°C after incubated for 36 h as described in Materials and Methods. (b) pH-stability of EXLX1. The protein was incubated in various buffers at 4°C for 48 h and the remaining activity was measured at 50°C as described in Materials and Methods. The used buffers were listed as follows: pH 3.0–6.0, 100 mM sodium citrate buffer (black circles); pH 6.0–8.0, 200 mM sodium phosphate buffer (open circles); pH 8.6–10, 100 mM glycine-NaOH buffer (black triangle). Data were shown as means ± standard deviations from data in triplicate.

addition of EXLX1. At the same cellulase loading, the RS yield obtained at 48 h using cellulase alone could be achieved at a shorter time owing to the synergism of EXLX1 (Fig.4). For example, 48 h of incubation was required for releasing 620.97 μg of RS when using 5 FPU/g of pure cellulase but only 24 h of incubation was required for releasing 699.77 μg of RS when using the mixture of 5 FPU/g cellulase and 5 μg EXLX1. All the results clearly demonstrated the enhancement effect of EXLX1 on cellulase activity using UNT wheat straw, which possesses more complexity than a pure cellulose fiber-like filter paper. According to the above-mentioned discussion, it could be suggested that EXLX1 might be an effective cellulase synergist to improve the RS released from lignocellulosic biomass wastes after appropriate optimization.

Characteristics of EXLX1 for various lignocellulosic wastes with different pretreatments

General composition of extractives free lignocellulosic wastes. Various types of lignocellulosic wastes such as wheat straw, rice straw, switchgrass, reed leaves, reed stalk, sugarcane bagasse and wood were used as substrates. Two representative pretreatment methods (i.e. acid pretreatment and alkali pretreat-

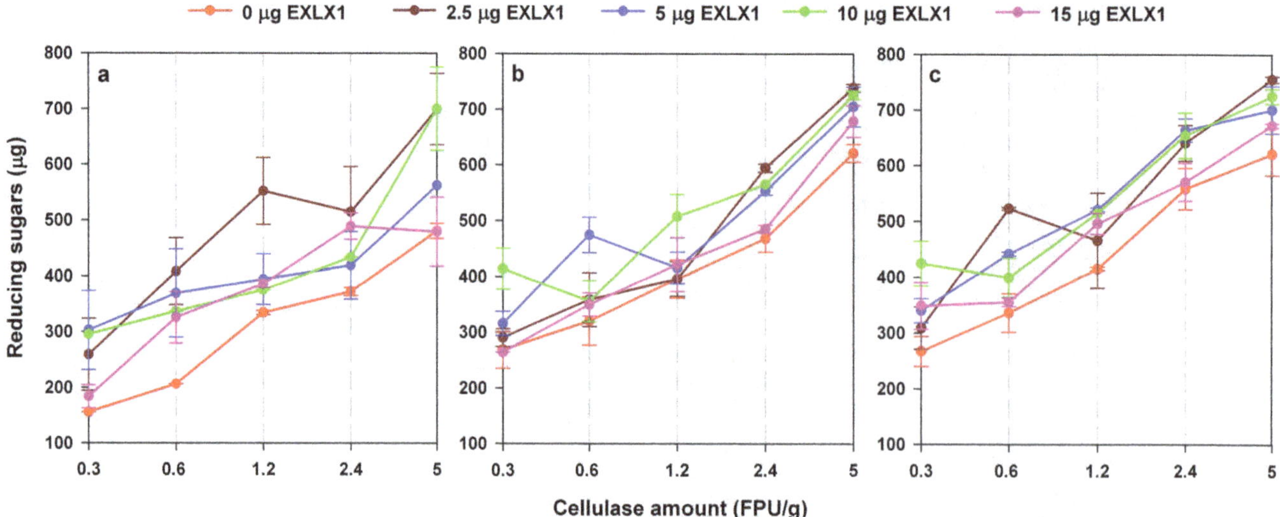

Figure 3. Reducing sugars (RS) released from UNT wheat straw using different combinations of cellulase and EXLX1 after incubation at 50°C for 24 h (a), 36 h (b) and 48 h (c). The dosage of cellulase for hydrolysis ranged from 0.3–5 FPU/gds, while the EXLX1 ranged from 0–15 μg. Four replicates were performed for each determination. Data were shown as mean ± standard error.

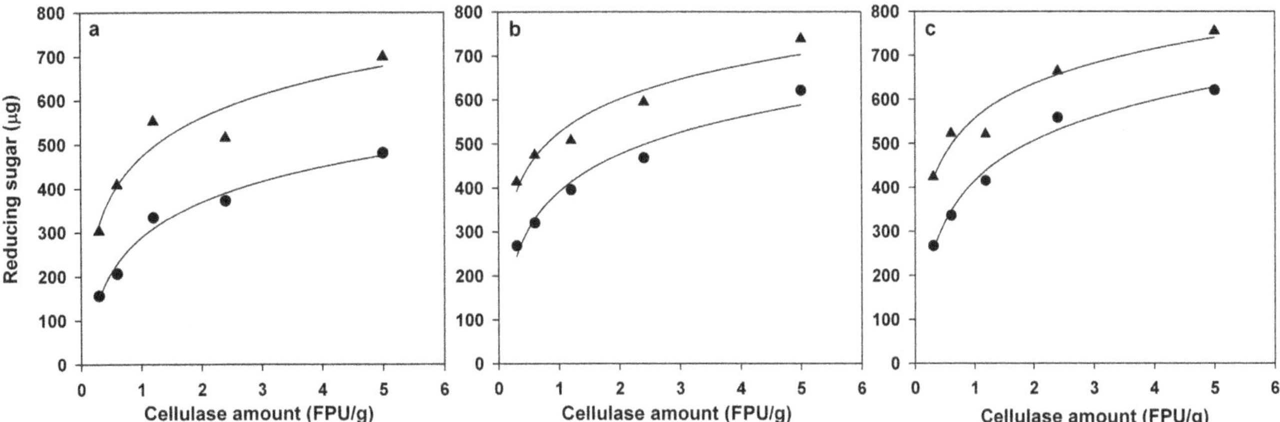

Figure 4. Comparison of reducing sugars (RS) yield of UNT wheat straw using optimized EXLX1 and cellulase combination (black triangles) and cellulase alone (lack circles), respectively. The hydrolysis was conducted at 50°C for 24 h (a), 36 h (b) and 48 h (c), respectively. Points and lines represented the mean values of four replicates and calculated values based on logarithmic equation ($Y = \alpha * \ln X + Y_0$), respectively. X is cellulase dosage (FPU/g). Y is RS yield. The estimated values of α, Y_0 and determination coefficient (R^2) under specific conditions were listed as follows: Cellulase alone for 24 h, $\alpha = 116.26$, $Y_0 = 287.92$, $R^2 = 0.9775$; Optimized cellulase and EXLX1 mixture for 24 h, $\alpha = 128.58$, $Y_0 = 471.14$, $R^2 = 0.9033$; Cellulase alone for 36 h, $\alpha = 122.02$, $Y_0 = 391.46$, $R^2 = 0.9612$; Optimized cellulase and EXLX1 mixture for 36 h, 110.10, $Y_0 = 524.92$, $R^2 = 0.9385$; Cellulase alone for 48 h, $\alpha = 132.54$, $Y_0 = 414.45$, $R^2 = 0.9813$; Optimized cellulase and EXLX1 mixture for 48 h: $\alpha = 114.64$, $Y_0 = 555.74$, $R^2 = 0.9405$.

ment) were used in the present work. Table 1 listed the average composition (dry basis) of the extractives free biomass with or without pretreatment. Biomass samples used in this study, except of wood and reed stalk, contained a large amount of extractives about over 25% of the day matter (data not shown). The content of crude fibers including both cellulose and hemicellulose were higher in ALKALI materials than in ACID materials, while the lignin content exhibited the opposite (Table 1). Almost all the hemicellulose was completely removed after the acid pretreatment.

Conversion efficiencies. As shown in Fig.5a, the RS yielded from UNT switchgrass and UNT sugarcane bagasse after the hydrolysis of pure cellulase were higher than that from other UNT materials under the same condition. It is interesting to found that hydrolysis of these two UNT materials using a cellulase and EXLX1 combination resulted in a higher RS yield in comparison with that using cellulase alone, although the cellulase dosage in pure cellulase treatments were 9-fold higher than that in EXLX1 and cellulase mixture treatments (Fig.5a). Take UNT switchgrass treatments for example, the RS yield increased by 30.0% by using the cellulase (0.06 FPU/g) and EXLX1 mixture instead of pure cellulase (0.06 FPU/g), while the RS yield increased by only 16.2% by increasing the dosage of the pure cellulase from 0.06 FPU/g to 0.6 FPU/g. The above-mentioned phenomenon was also found in UNT sugarcane bagasse treatments. However, when other UNT materials such as rice straw, wheat straw, reed stalk, reed leaves and wood were used as substrates, increasing the cellulase dosage in reaction systems could be more effective than adding EXLX1 into cellulase as a synergist (Fig.5a). Chemical-physical pretreatment is known to be an important process in the real-world lignocelluloses saccharification that can improve the conversion efficiency of lignocelluloses in some enzymatic hydrolysis processes. In this work, remarkable enhancement in RS yields brought by pretreatments was observed in treatments using a majority of plant samples (Fig.5b and Fig.5c). However, when the substrate was sugarcane bagasse or switchgrass, both ACID and ALKALI pretreatment exerted negative effects on the final RS yields. It is obvious that the combined utilization of EXLX1 and cellulase was more efficient for improving the conversion efficiency of sugarcane bagasse and switchgrass in comparison with chemically pretreating the plant samples. Indeed, we found that

Table 1. Average composition (dry basis) of the biomass samples with or without pretreatment.

	Cellulose (%)			Hemicellulose (%)			Lignin (%)		
Biomass	1	2	3	1	2	3	1	2	3
A	64.52±1.50	69.66±0.23	69.51±1.73	19.35±0.12	2.25±0.36	18.29±0.34	16.13±1.62	28.09±2.08	12.20±2.07
B	63.64±3.28	44.44±7.57	74.42±1.98	25.00±0.13	7.78±1.55	19.77±2.20	11.36±3.15	47.78±5.12	5.81±2.40
C	56.16±1.19	67.93±1.62	74.12±1.87	30.14±0.74	0.90±0.12	14.12±0.48	13.70±1.93	31.18±3.18	11.76±2.35
D	20.59±0.92	66.22±4.65	64.10±1.06	73.53±2.42	4.06±0.69	25.64±0.34	5.88±3.34	29.73±4.03	10.26±0.44
E	53.09±1.26	68.54±1.62	70.79±3.32	30.86±0.38	3.38±0.63	20.22±2.55	16.05±1.64	28.09±1.67	8.99±0.75
F	52.31±3.88	61.29±8.73	55.70±3.79	24.62±1.54	5.38±0.38	22.78±0.97	23.08±5.42	33.33±4.84	21.52±4.77
G	55.81±0.35	57.29±0.95	64.33±1.23	11.63±0.31	5.21±1.56	0.68±0.04	32.56±0.27	37.50±1.65	34.99±1.27

1: UNT materials; 2: ACID materials; 3: ALKALI materials.
A: Switchgrass; B: Sugarcane bagasse; C: Wheat straw; D: Reed leaves; E: Reed stalk; F: Rice straw; G: Wood.

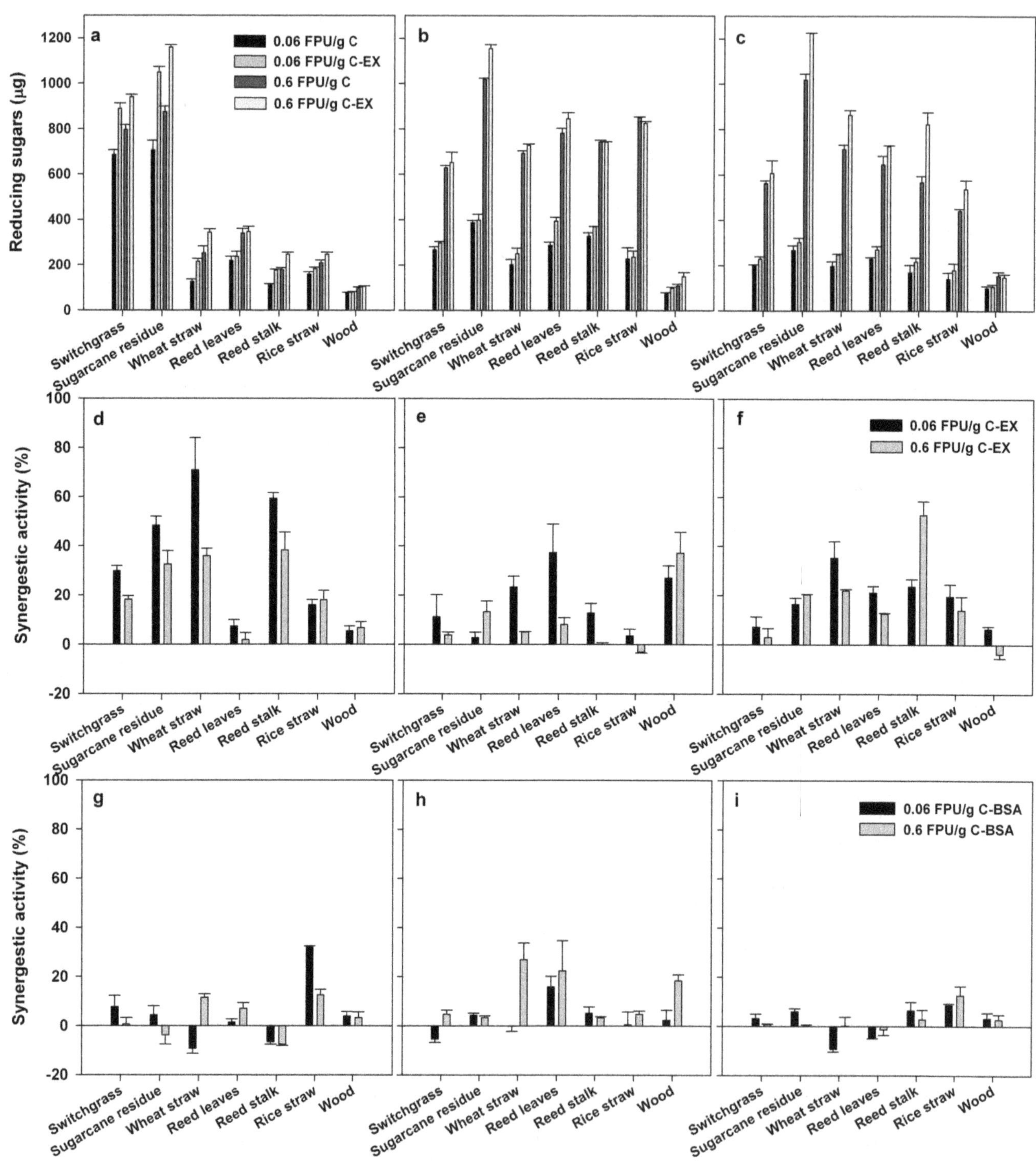

Figure 5. Hydrolysis activities of EXLX1 and cellulase combinations (a,b,c) and synergistic effects of EXLX1 (d,e,f) and BSA (g,h,i) on cellulase using various types of lignocellulosic biomass samples, including un-treated samples (a,d,g), dilute sulfuric acid pretreated samples (b,e,h) and alkali pretreated samples (c,f,i). The hydrolysis was conducted at 50°C for 36 h. 0.06 FPU/gds C: 0.06 FPU/gds of Cellulase; 0.6 FPU/gds C: 0.6 FPU/gds of Cellulase; 0.06 FPU/gds C-EX: 0.06 FPU/gds of Cellulase added with 0.5 µg EXLX1; 0.6 FPU/gds C-EX: 0.06 FPU/gds of Cellulase added with 5 µg EXLX1; 0.06 FPU/gds C-BSA: 0.06 FPU/gds of Cellulase added with 0.5 µg BSA; 0.6 FPU/gds C- BSA: 0.06 FPU/gds of Cellulase added with 5 µg BSA.

UNT switchgrass and UNT sugarcane bagasse contained extractives in a high proportion (Table1), which might contain free RS in a large content. During the pretreatment, free RS of extractives that affects the final RS yield was removed and finally resulted in the low RS yield after hydrolysis of the pretreated switchgrass and the pretreated sugarcane bagasse. In general, the enhancement induced by the synergistic effect of EXLX1 on the biomass conversion efficiency under some conditions would even higher

Table 2. Binding to various lignocellulosic biomass samples with different pretreatments.

	Relative amount of bound EXLX1 (%)		
Biomass	UNT	ACID	ALKALI
Switchgrass	28.36±5.40	25.69±3.19	64.29±5.85
Sugarcane bagasse	33.59±2.71	69.23±4.72	91.13±3.63
Wheat straw	100.43±0.29	58.69±2.47	83.47±2.46
Reed leaves	−8.07±0.35	24.69±2.40	46.42±3.98
Reed stalk	20.43±4.91	21.91±3.43	48.58±0.45
Rice straw	70.48±1.88	50.78±3.12	65.39±3.32
Wood	23.40±5.75	20.18±0.91	54.37±2.70

Incubation was performed in citric buffer (50 mM, pH 4.8) at 30°C for 4 h.

Table 3. The binding preference study of EXLX1 with ALKALI materials and ACID materials as substrate, respectively.

	ALKALI binding preference (%) [a]	
Biomass	30°C	4°C
Switchgrass	25.86±1.45	12.20±0.31
Sugarcane bagasse	12.20±6.33	−37.47±2.39
Wheat straw	20.00±2.91	4.41±1.07
Reed leaves	3.67±0.56	−2.65±1.32
Reed stalk	33.06±0.31	−21.90±0.52
Rice straw	20.57±2.70	11.17±1.03
Wood	35.41±2.81	26.30±0.52

Incubation was performed in potassium phosphate buffer (50 mM, pH 7) at 4°C and 30°C for 4 h, respectively.
[a]Binding preference (%) = The relative amount of the bound protein toward ALKALI materials (%)-The relative mount of the bound protein toward ACID materials (%).

than not only that induced by adding cellulase in a high amount but also that induced by chemically pretreating the plant samples before hydrolysis, although the conversion efficiency of the saccharification system exhibited a negative effect on the synergistic capacities of EXLX1 in most cases [6,16].

Synergistic characteristics of EXLX1. Synergistic activities of EXLX1 on cellulase for various biomass samples with different pretreatments were shown in Fig.5d, Fig.5e and Fig.5f. Negative control experiments using BSA instead of EXLX1 were also conducted with results shown in Fig.5g, Fig.5h and Fig.5i. Comparison of the synergism of EXLX1 and that of BSA showed that the effects of BSA on the cellulase activity in the hydrolysis process were completely different from that of EXLX1. For example, the higher synergism was more likely to appear in the BSA treatments with a higher amount of BSA but in the EXLX1 treatments with a lower amount of EXLX1. Besides, although BSA exhibited significantly positive effects on the cellulase activity in some cases, the average synergistic activity of EXLX1 was higher than that of BSA. All the above-mentioned results demonstrated that the synergistic effects observed in this work was specific to EXLX1. Indeed, the specific effect of EXLX1 or some other expansin-like proteins on glycoside hydrolase (e.g. cellulase, xylanase) has also been proposed by other researchers [6,13,24].

EXLX1 presented different synergisms for different lignocellulose biomass substrates (Fig.5d–f), but may be governed by the biomass composition. According to the results in Table 1, most of UNT materials contained more than 50% of cellulose without remarkable differences in the proportion except for UNT reed leaves. Insignificantly low correlations (0.06 FPU/g cellulase: Pearson Correlation $=0.055$, $p=0.918$; 0.6 FPU/g cellulase: Pearson Correlation $=-0.024$, $p=0.964$) were observed between the synergistic activity and the cellulose proportion of materials, when the synergistic activity data and the cellulose proportion data from all the treatments except for UNT reed leaves treatments were pooled for correlation analysis. Despite it, the cellulose proportion of materials may be an important factor affecting the synergistic activity of EXLX1, since EXLX1 exhibited a low synergism when UNT reed leaves that contain an extremely low proportion of cellulose were used as substrates. Correlation analysis of the synergistic activity of EXLX1 and the hemicellulose proportion was also performed using the pooled data from the treatments excluding UNT reed leaves treatments. A significantly positive relationship was found between the synergistic activity and the hemicellulose proportion (0.06 FPU/g cellulase: Pearson Correlation $=0.857$, $p<0.05$; 0.6 FPU/g cellulase: Pearson Correlation $=0.930$, $p<0.05$). Besides, correlation analysis results showed that the synergistic activity correlated negatively with the lignin content (0.06 FPU/g cellulase: Pearson Correlation $=-0.829$, $p<0.05$; 0.6 FPU/g cellulase: Pearson Correlation $=-0.845$, $p<0.05$). Indeed, the average synergistic activity (0.06 FPU/g cellulase: 18.52%; 0.6 FPU/g cellulase: 18.48%) of EXLX1 using ALKALI materials containing crude fibers in a high proportion and lignin in a low proportion was found to higher than that (0.06 FPU/g cellulase: 16.86%; 0.6 FPU/g cellulase: 9.24%) using ACID materials containing lignin in a high proportion. The results indicated that a higher synergism of EXLX1 is prone to appear in the treatment with the substrate that contains a high proportion of hemicellulose and a low proportion of lignin, when the cellulose proportion of different biomass materials exhibited no remarkable differences. The information gained in the present study would be favorable to assess whether EXLX1 is effective as a cellulase synergist for the hydrolysis of the determined lignocellulosic biomass. Although the mechanism for the relationship between the lignocellulose composition and the synergism effect of EXLX1 is still unknown, we supposed that it might be linked to the unique feature of expansin in breaking hydrogen bonds between hemicellulose and cellulose, which has been suspected of facilitating the cellulase targeting of the wall polysaccharide network [25,26].

Binding characteristics of EXLX1. The binding capacity of EXLX1 is known to affect its wall extension activity [18]. As shown in Table 2, the binding affinity of EXLX1 to different biomass samples was significantly different. EXLX1 displayed much higher binding to UNT wheat straw and UNT rice straw than to other UNT materials, while no adsorption of EXLX1 to UNT reed leaves was found. The binding affinity of EXLX1 to different biomass samples might be related to the lignocellulosic components as the UNT reed leaves contained an extremely lower proportion of cellulose than other UNT materials (Table 1). EXLX1 bound to all the ACID and ALKALI biomass samples and exhibited much higher binding to ALKALI materials than to ACID materials at 30°C in the acidic reaction buffer with pH 4.8, regardless of biomass types (Table 2). Similar results were also found at 30°C in the neutral reaction buffer with pH 7.0 (Table 3). It has been suggested that lignin is nonspecifically bind with

EXLX1 [15], while cellulose specifically binds with this protein and is controlled by the D2 domain of EXLX1 with hydrophobic amino acid residues responsible for binding cellulose [18,25]. In this work, EXLX1 exhibited more preferential binding for ALKALI materials at 30°C than ACID materials, indicating that the binding activity of EXLX1 to polysaccharide components might be more active than that to lignin under the assay conditions. Kim et al [15], however, proposed that EXLX1 has preferential binding for *M. xgiganteu* after alkali pretreatment than that after acid pretreatment at 4°C. The variability may be due to that many factors would affect the binding behavior of EXLX1 [14,27]. For example, the plant biomass wastes generally contained a high content of chemicals involved in the soluble extractives such as divalent cations and NaCl, which would affect the binding affinity of EXLX1 [27]. Besides, it was further found that the quantities of EXLX1 bound to ACID reed stalk and ACID sugarcane bagasse exceeded the amounts bound to the corresponding ALKALI samples when the temperature was set at 4°C (Table 3). The result suggested that the binding affinity of EXLX1 might also be affected by incubation temperature.

Conclusions

As a kind of expansin can be actively produced in a large amount and purified effectively, EXLX1 has the promising potentiality as a cellulase synergist to improve the RS yielded from lignocellulosic wastes after optimization. The potential of EXLX1 as a cellulase synergist was related to lignocellulosic substrate type and would be governed by their composition. A higher synergism of EXLX1 with cellulase is prone to appear in the treatment using a material that contains a high proportion of hemicellulose and a low proportion of lignin when the cellulose proportion of different biomass materials exhibited no remarkable differences.

Author Contributions

Conceived and designed the experiments: HL. Performed the experiments: HL QS JZ. Analyzed the data: HL. Contributed reagents/materials/analysis tools: QW. Wrote the paper: HL YZ.

References

1. Detroy RW, Stjulian G (1983) Biomass conversion - fermentation chemicals and fuels. Crc Critical Reviews in Microbiology 10: 203–228.
2. Dashtban M, Schraft H, Qin WS (2009) Fungal Bioconversion of Lignocellulosic Residues; Opportunities & Perspectives. International Journal of Biological Sciences 5: 578–595.
3. Gusakov AV, Salanovich TN, Antonov AI, Ustinov BB, Okunev ON, et al. (2007) Design of highly efficient cellulase mixtures for enzymatic hydrolysis of cellulose. Biotechnol Bioeng 97: 1028–1038.
4. Lin H, Wang B, Zhuang RY, Zhou QF, Zhao YH (2011) Artificial construction and characterization of a fungal consortium that produces cellulolytic enzyme system with strong wheat straw saccharification. Bioresource Technology 102: 10569–10576.
5. Carrard G, Koivula A, Soderlund H, Beguin P (2000) Cellulose-binding domains promote hydrolysis of different sites on crystalline cellulose. Proc Natl Acad Sci U S A 97: 10342–10347.
6. Kim ES, Lee HJ, Bang WG, Choi IG, Kim KH (2009) Functional characterization of a bacterial expansin from *Bacillus subtilis* for enhanced enzymatic hydrolysis of cellulose. Biotechnology and Bioengineering 102: 1342–1353.
7. Baker JO, King MR, Adney WS, Decker SR, Vinzant TB, et al. (2000) Investigation of the cell-wall loosening protein expansin as a possible additive in the enzymatic saccharification of lignocellulosic biomass. Appl Biochem Biotechnol 84–86: 217–223.
8. Suwannarangsee S, Bunterngsook B, Arnthong J, Paemanee A, Thamchaipenet A, et al. (2012) Optimisation of synergistic biomass-degrading enzyme systems for efficient rice straw hydrolysis using an experimental mixture design. Bioresour Technol 119: 252–261.
9. Sampedro J, Cosgrove DJ (2005) The expansin superfamily. Genome Biology 6: 242.241–242.246.
10. McQueen-Mason SJ, Cosgrove DJ (1995) Expansin mode of action on cell walls. Analysis of wall hydrolysis, stress relaxation, and binding. Plant Physiol 107: 87–100.
11. Quiroz-Castaneda RE, Martinez-Anaya C, Cuervo-Soto LI, Segovia L, Folch-Mallol JL (2011) Loosenin, a novel protein with cellulose-disrupting activity from *Bjerkandera adusta*. Microb Cell Fact 10: 8.
12. Wang Y, Tang R, Tao J, Gao G, Wang X, et al. (2011) Quantitative investigation of non-hydrolytic disruptive activity on crystalline cellulose and application to recombinant swollenin. Appl Microbiol Biotechnol 91: 1353–1363.
13. Chen XA, Ishida N, Todaka N, Nakamura R, Maruyama J, et al. (2010) Promotion of efficient Saccharification of crystalline cellulose by *Aspergillus fumigatus* Swo1. Appl Environ Microbiol 76: 2556–2561.
14. Kerff F, Amoroso A, Herman R, Sauvage E, Petrella S, et al. (2008) Crystal structure and activity of *Bacillus subtilis* YoaJ (EXLX1), a bacterial expansin that promotes root colonization. Proc Natl Acad Sci U S A 105: 16876–16881.
15. Kim IJ, Ko HJ, Kim TW, Nam KH, Choi IG, et al. (2013) Binding characteristics of a bacterial expansin (BsEXLX1) for various types of pretreated lignocellulose. Appl Microbiol Biotechnol 97: 5381–5388.
16. Lee HJ, Lee S, Ko HJ, Kim KH, Choi IG (2010) An expansin-like protein from *Hahella chejuensis* binds cellulose and enhances cellulase activity. Mol Cells 29: 379–385.
17. Whitney SEC, Gidley MJ, McQueen-Mason SJ (2000) Probing expansin action using cellulose/hemicellulose composites. Plant Journal 22: 327–334.
18. Georgelis N, Tabuchi A, Nikolaidis N, Cosgrove DJ (2011) Structure-function analysis of the bacterial expansin EXLX1. J Biol Chem 286: 16814–16823.
19. Saha BC, Cotta MA (2010) Comparison of pretreatment strategies for enzymatic saccharification and fermentation of barley straw to ethanol. N Biotechnol 27: 10–16.
20. Van Soest PJ, Robertson JB, Lewis BA (1991) Methods for dietary fiber, neutral detergent fiber, and nonstarch polysaccharides in relation to animal nutrition. Journal of Dairy Science 74: 3583–3597.
21. Adney B, Baker J (1996) Measurement of cellulase activities. Golden: CO: National Renewable Energy Laboratory.
22. Bradford MM (1976) A rapid and sensitive method for the quantitation of microgram quantities of protein utilizing the principle of protein-dye binding. Anal Biochem 72: 248–254.
23. Li LC, Bedinger PA, Volk C, Jones AD, Cosgrove DJ (2003) Purification and characterization of four beta-expansins (Zea m 1 isoforms) from maize pollen. Plant Physiol 132: 2073–2085.
24. Gourlay K, Hu J, Arantes V, Andberg M, Saloheimo M, et al. (2013) Swollenin aids in the amorphogenesis step during the enzymatic hydrolysis of pretreated biomass. Bioresour Technol 142: 498–503.
25. Georgelis N, Yennawar NH, Cosgrove DJ (2012) Structural basis for entropy-driven cellulose binding by a type-A cellulose-binding module (CBM) and bacterial expansin. Proc Natl Acad Sci U S A 109: 14830–14835.
26. Cosgrove DJ (2000) Loosening of plant cell walls by expansins. Nature 407: 321–326.
27. Kim IJ, Ko HJ, Kim TW, Choi IG, Kim KH (2013) Characteristics of the binding of a bacterial expansin (BsEXLX1) to microcrystalline cellulose. Biotechnol Bioeng 110: 401–407.

11

Effects of Biochar on Soil Microbial Biomass after Four Years of Consecutive Application in the North China Plain

Qing-zhong Zhang[1]*, Feike A. Dijkstra[2], Xing-ren Liu[1], Yi-ding Wang[1], Jian Huang[1], Ning Lu[1]

1 Key Laboratory of Agricultural Environment, Ministry of Agriculture, Sino-Australian Joint Laboratory For Sustainable Agro-Ecosystems, Institute of Environment and Sustainable Development in Agriculture, Chinese Academy of Agricultural Sciences, Beijing, China, **2** Centre for Carbon, Water and Food, Department of Environmental Sciences, The University of Sydney, Camden, New South Wales, Australia

Abstract

The long term effect of biochar application on soil microbial biomass is not well understood. We measured soil microbial biomass carbon (MBC) and nitrogen (MBN) in a field experiment during a winter wheat growing season after four consecutive years of no (CK), 4.5 (B4.5) and 9.0 t biochar ha^{-1} yr^{-1} (B9.0) applied. For comparison, a treatment with wheat straw residue incorporation (SR) was also included. Results showed that biochar application increased soil MBC significantly compared to the CK treatment, and that the effect size increased with biochar application rate. The B9.0 treatment showed the same effect on MBC as the SR treatment. Treatments effects on soil MBN were less strong than for MBC. The microbial biomass C:N ratio was significantly increased by biochar. Biochar might decrease the fraction of biomass N mineralized (K_N), which would make the soil MBN for biochar treatments underestimated, and microbial biomass C:N ratios overestimated. Seasonal fluctuation in MBC was less for biochar amended soils than for CK and SR treatments, suggesting that biochar induced a less extreme environment for microorganisms throughout the season. There was a significant positive correlation between MBC and soil water content (SWC), but there was no significant correlation between MBC and soil temperature. Biochar amendments may therefore reduce temporal variability in environmental conditions for microbial growth in this system thereby reducing temporal fluctuations in C and N dynamics.

Editor: Andrew C. Singer, NERC Centre for Ecology & Hydrology, United Kingdom

Funding: This work was supported by the S & T Innovation Program of Chinese Academy of Agricultural Sciences. The funders had no role in study design, data collection and analysis, decision to publish, or preparation of the manuscript.

Competing Interests: The authors have declared that no competing interests exist.

* Email: ecologyouth@126.com

Introduction

Biochar is the product of the thermal degradation of organic materials in the absence of air (pyrolysis) and is distinguished from charcoal by its use as a soil amendment [1]. Application of biochar has been proposed as a novel approach to improve soil fertility, increase soil carbon sequestration and mitigate the greenhouse effect [2]. Nevertheless, the turnover of soil organic matter and biochar and their interaction in soils remain poorly understood. Positive and negative priming effects on native organic carbon mineralization in biochar-amended soils have been reported [3–6], and the priming direction is thought to be controlled by the biochar type (determined by production conditions and sources used) and soil pH. Moreover, the presence of soil organic matter can stimulate the mineralization of the more labile components of biochar over the short term, but over the long term, the biochar-soil interaction may enhance soil C storage via processes of organic matter sorption to biochar and physical protection [3,5,6]. There is strong evidence that priming effect on soil organic matter decomposition relies on microbial biomass [7]. Research on soil microbial biomass dynamics with biochar application will help the understanding of priming effect on soil organic matter and biochar decomposition. Biochar addition can increase soil microbial biomass, and may also affect the soil biological community composition, which in turn will affect nutrient cycling, plant growth, and greenhouse gas emission, as well as soil organic carbon mineralization mentioned above [1].

However, the present knowledge on soil microbial biomass dynamics due to biochar application is mainly based on the comparison between biochar application and no biochar treatment without considering crop residue return practice, and lacks long-term field experimental data [8–12]. Our field experiment with consecutive biochar amendments and crop residue return lasted nearly 4 years in a winter wheat-maize relay cropping system until October 2010, with total inputs of biochar ranging from 18.0 to 36.0 t ha^{-1}. The aim of this paper is to examine effects of biochar and wheat straw residue incorporation on the temporal and spatial variation in microbial biomass carbon (MBC) and microbial biomass nitrogen (MBN) measured within a wheat growing season under different soil depths after four years of consecutive application.

Materials and Methods

Site description

The field experiment was conducted at an experimental station (36°58′N, 117°59′E, elevation 17 m) for ecological and sustainability research in Huantai County, Shandong Province, China, and was begun in 2007 [13]. This site has a warm, temperate, continental monsoon climate with a mean annual temperature of 12.4°C. The mean annual precipitation was 600 mm, with most rainfall occurring in June, July, and August. The soil is classified to Fluvic Cambisol according to the USDA system. The soil was a sandy loam, and the proportion of sand, silt, and clay particles in the top 20 cm of soil was 70.8, 26.9, and 2.3%, respectively. The soil bulk density of the top 20 cm of soil before the biochar amendment was 1.52 g cm^{-3}, and the soil organic matter (SOM) content was 15 g kg^{-1} of soil. The soil pH was 8.1 before experiment, and was not significantly changed due to biochar amendment [13].

Biochar

Milled biochar was purchased from a local company (Jinfu Biochar Company, Liaoning Province), with a density of 0.297 g cm^{-3} and a pH of 8.2 [14]. The biochar was made by incomplete self-combustion of crushed corncob in an open-top concrete tank with an igniting apparatus at the bottom for about 24 hours at about 360°C. The incomplete self-combustion process did not need an external energy source. The biochar contained 65.7% carbon and 0.909% N (analyzed with an elemental analysis apparatus, Flash EA 2000, Thermo Electron Corporation, Italy). Available phosphorus content was 0.08% (extracted with 0.5 M $NaHCO_3$ at a pH of 8.5, and analyzed with a colorimetric method), and available potassium content was 1.60% (extracted with 2.0 M HNO_3, and analyzed with a flare photometer, FP640, Cany, China). Ash content of the biochar was 72.0% (determined by dry combustion in a muffle furnace at 550°C for 2 h).

Experimental design

The experimental crops were winter wheat (*Triticum aestivum* L.) and maize (*Zea mays* L.) in relay cropping. Generally, the maize was sown in early June and matured in late September. The winter wheat was then sown in early October and harvested in early June of the next year. The field experiment was a randomized complete block design and each of the experimental plots was 6 m×6 m = 36 m^2, four treatments (CK, B4.5, B9.0, SR) with three replications. The experiment included a control treatment with no biochar addition (CK), two biochar treatments with 4.5 and 9.0 t ha^{-1} yr^{-1} biochar applied (B4.5 and B9.0, respectively), and a treatment where all wheat and maize straw residue produced in the plot was returned to the field and incorporated into the soil after harvest by a straw returning machine (SR), which allowed us to compare biochar vs. fresh litter effects on microbial biomass. Based on the estimated total amount of aboveground crop residues of about 15 t ha^{-1} yr^{-1} for local normal croplands and the empirical value of about 30% in weight of crop residues left as biochar from the local biochar company, biochar amount can be obtained from the field crop residues is equivalent to that used in the B4.5 treatment.

The biochar was distributed equally to each crop (half applied before sowing of wheat and the other half before sowing of maize). Inorganic basal N fertilizers as urea of 200 kg N ha^{-1} yr^{-1} (165 kg N ha^{-1} yr^{-1} was used before 2009), P as superphosphate of 52.5 kg P_2O_5 ha^{-1} yr^{-1}, and K as potassium sulfate of 37.5 kg K_2O ha^{-1} yr^{-1} were applied in all treatments. Before 2009, all fertilizers were used as base fertilizer. From 2009, half of nitrogen fertilizer was applied as base fertilizer, and the other half was applied as topdressing. Biochar and the base fertilizers were broadcast on the soil surface and incorporated into the soil by rotary tillage to depth of 15 cm before seeding.

Soil sampling

Soil samples were collected during the winter wheat growing period (from October 2010 to June 2011) divided into five functional stages: (1) wheat post-seeding stage; (2) before freezing stage, (3) reviving stage; (4) shooting stage, and (5) harvest stage. Soil samples were collected at depths of 0–5, 5–10, 10–20, and 20–30 cm in each experimental plot. The soil samples were collected from five points randomly, and mixed into one sample, each mixed soil sample was divided into two parts. One part of the soil sample was determined for soil water content (SWC) and the other part was prepared for microbial analysis. All samples were immediately stored in sealed plastic bags in a cooler and transported to laboratory and stored in refrigerator at 4°C. All microbiological determinations were performed within one week of sampling.

Measurement and monitoring

Microbial biomass carbon (MBC) and nitrogen (MBN) in soil were determined by fumigation extraction method [15,16], and the value of 0.45 was used for both the fraction of biomass C mineralized (K_C) and the fraction of biomass N mineralized (K_N). Soil samples were thoroughly mixed and ground to pass through a 2-mm sieve, and then the soil moisture was adjusted to about 40% water holding capacity. We fumigated 20.0 g (dry weight equivalent) of soil with ethanol-free chloroform for 24 h. Both fumigated and non-fumigated soils were extracted with 80 ml of 0.5 M K_2SO_4 by shaking for 30 min on a reciprocating shaker at 40 cycles per minute and then filtered (soil:water = 1:4). The TOC analyzer (Multi N/C 2100, Jena, Germany) was used to determine the C and N in the extracts.

SWC was determined gravimetrically by oven-drying at 105°C for 48 h. Soil temperature at 5-cm depth in each replication was monitored hourly by a temperature probe (tolerance: ±0.2°C over the 0–70°C range; temperature measurement range: −50 to +70°C; model 109, Campbell Scientific, Logan, UT, USA) connected to a datalogger (model U12-006, Jimuduoli, Beijing, China) from 22 October 2010, except during harvesting and sowing (June 10–July 2) of each year.

Data analyses

We used repeated measures ANOVA to test for main effects of treatment (between-subjects factor), soil depth (within-subjects factor) and date (within-subjects factor), and their interactive effects on MBC, MBN and microbial biomass C:N ratio. When necessary, data were log-transformed to reduce heteroscedasticity and improve assumptions of normality. We used Pearson's test to determine whether there was a significant correlation between microbial and environmental soil properties (moisture and temperature) and Fisher's least significant difference (LSD) test to determine the significant difference in coefficient of seasonal variability (CV) of MBC and MBN between different treatments. All statistical analyses were performed with JMP (version 4.0.4; SAS Institute, Cary, NC, USA).

Results

Microbial biomass C (MBC)

Variability of MBC under different rates of biochar and straw residue addition was large (Fig. 1). Treatment effects on MBC were significant (Table 1), and on average, largest in the B9.0 treatment, followed by the SR treatment, and lowest in CK. There were significant treatment*date and treatment*date*depth interactive effects ($P = 0.02$ and $P < 0.0001$ respectively, Table 1). The greatest MBC was found at depth of 10–20 cm treated with SR on 20 October of 2010 (post-seeding stage), followed by the same treatment in 5 November of 2010 (before freezing stage) among all treatments and both soil layers during the whole experimental period. At depth of 0–5 and 5–10 cm, the MBC content in the treatments of B9.0 and SR in general showed the largest increases compared to CK. The MBC in the B9.0 treatment increased by 118% to 763%, while the MBC in the SR treatment was 2% lower to 722% higher compared to CK, depending on the time of year (Fig. 1a, b). The MBC at 0–5 and 5–10 cm in the B4.5 treatment also increased by 6% to 246%, depending on the time of year.

The MBC in the B9.0 treatment increased by 45% to 294% at soil depths of 10–20 and 20–30 cm compared to CK (Fig. 1c, d). The MBC in the SR treatment decreased by 62% during the shooting stage (25 April) at 20–30 cm soil depth, but showed some of the largest increases (between 50% and 408%) at other times at this depth. The MBC in the B4.5 treatment at 20–30 cm soil depth increased in most of the stages, but decreased at 10–20 cm soil depth by 19%, 40%, 6%, and 10% in the winter wheat post-seeding stage (20 October 2010), before freezing stage (5 November 2010), reviving stage (26 March 2011), and harvest stage (5 June 2011), respectively.

Table 2 presents the coefficients of seasonal variation of the MBC under different treatments. The CV of CK varied between 20% and 38%, and the overall CV was 25%. The overall CV of B4.5 and B9.0 treatments was 11% and 16%, respectively. The CVs were significantly smaller than that of CK. But the CV of SR treatment varied from 40% to 63%, which was significantly higher than that of CK. The CV of the surface soil tended to be higher than that of the subsoil in the same treatment, but not for the SR treatment.

Microbial biomass N (MBN)

As with MBC, there was large variability in MBN among the different biochar and SR treatments, dates, and depths (Fig. 2, Table 1). However, unlike the MBC results, across date and depth the largest MBN was observed in the SR treatment (on average 35, 106, and 31% higher than the CK, B4.5 and B9.0 treatment respectively). The MBN in the B4.5 treatment decreased at most dates and soil depths, and across all dates compared to the CK treatment. Across date, the MBN in the B4.5 treatment decreased by 40, 47, 3, and 30% at 0–5, 5–10, 10–20, and 20–30 cm soil depth. The MBN in the B9.0 treatment often increased at 0–5, 5–10, and 20–30 cm soil depth during the stages of wheat post-seeding stage, before freezing stage, and reviving stage compared to the CK treatment. Large MBN pools were particularly observed in the SR treatment at 10–20 and 20–30 cm soil depth. At the

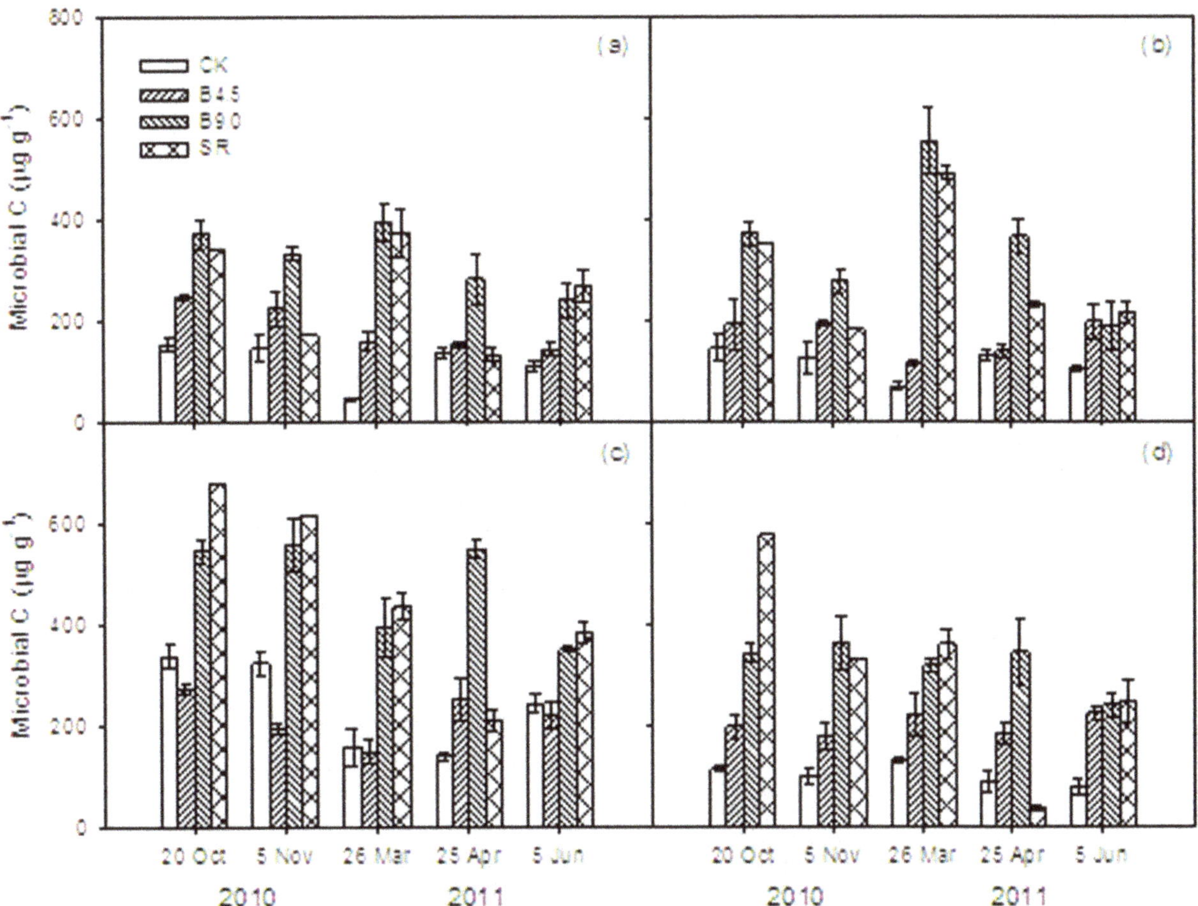

Figure 1. Microbial biomass carbon (MBC) under different treatments (CK: control, B4.5: 4.5 t ha^{-1} yr^{-1} biochar addition, B9.0: 9.0 t ha^{-1} yr^{-1} biochar addition, SR: incorporation of wheat straw) and different soil depths (a: 0–5, b: 5–10, c: 10–20, and d: 20–30 cm) during winter wheat season. Vertical bars represent the standard errors for means of each treatment ($n = 3$).

Table 1. Repeated Measures ANOVA P values.

Variable	Microbial C	Microbial N	Microbial C:N
Treatm	0.0003	<0.0001	<0.0001
Date	<0.0001	<0.0001	0.002
Depth	<0.0001	0.0001	<0.0001
Treatm*date	0.02	<0.0001	0.0005
Treatm*depth	0.12	<0.0001	0.0004
Date*depth	<0.0001	0.0001	0.005
Treatm*date*depth	<0.0001	0.0003	0.0007

shooting stage, the MBN of the four treatments were all relatively low.

The CV of MBN in the CK treatment varied from 24% to 44%, and the overall CV was 32% (Table 3). The overall CV of B4.5 was 19%, which was significantly lower than that of other treatments. The overall CV of B9.0 and SR treatments were 49% and 65% respectively, which were significantly larger than that of CK. From this, it can be seen that addition of 4.5 t ha^{-1} of biochar decreased the temporal and spatial fluctuation of MBN, but that a higher biochar addition increased the fluctuation of MBN.

Microbial biomass C:N ratio

Variability in microbial biomass C:N ratio was large among the different treatments. We observed significant treatment effects on microbial biomass C:N ratio (Table 1, Fig. 3), with largest ratios in the B9.0 treatment (on average 181% greater than CK) and B4.5 treatment (on average 93% higher than CK), suggesting that the increase in MBC with biochar addition was larger compared to the increase in MBN. The greatest microbial biomass C:N ratio ratios were found at 10–20 cm and 20–30 cm soil depth in the B9.0 treatment.

Discussion

Soil microbial biomass

The change in MBC reflects the process of microbial growth, death and organic matter degradation. Our results showed that MBC increased with biochar amendment compared to CK, which suggested that microbial growth could be accelerated by biochar addition. Reported biochar effects on soil MBC are quite inconsistent. Several studies found that there was no significant effect of biochar amendment on soil MBC [8,11]. Dempster et al. [12] found that MBC significantly decreased with biochar addition while MBN was unaltered in a coarse textured soil, and others observed the same positive effects of biochar addition on microbial biomass as ours [1,17–19]. Moreover, a positive linear relationship between microbial biomass and biochar concentration was also observed in a highly weathered soil [20]. Biochar type was thought as the driving parameter for any effects on soil microbial biomass, community, and activity [9,21].

Compared to wheat straw residue return, after 4 years of annual application, biochar increased MBC in 0–30 cm soil when applied at a high rate (9.0 t ha^{-1} yr^{-1}, $p<0.05$), but decreased MBC when applied at a low rate (4.5 t ha^{-1} yr^{-1}, $p<0.01$). The amount of C added in the B4.5 treatment most likely did not induce a similar increase in MBC as in the SR treatment, while the amount of C added in the B9.0 treatment did.

The MBN at 0–30 cm soil depth significantly decreased in the B4.5 treatment, but increased in the B9.0 treatment compared to CK ($p<0.01$), whereas the MBN at 0–30 cm soil depth in both biochar treatments were lower compared to SR treatment ($p<0.05$). Zavalloni et al. [8] found that biochar amendment at a rate of 5% had no significant influence on soil MBN in an incubation experiment. The increase in MBC and the decrease in MBN for biochar treatments indicate that biochar in soil acted as a carbon source rather than a nitrogen source for soil microbes. In return, this could have consequences for N cycling (e.g., increased microbial N immobilization with biochar addition). Our previous study in the field showed that biochar addition decreased soil available N by 7–10% compared to CK ($p>0.05$), but increased total N by 14–21% ($p<0.05$) in the 0–15 cm soil [22].

In the relatively unfertile and coarse-textured soil of our study, microbial biomass was probably limited by both a suitable growth environment and by C availability [17]. The labile fraction of biochar has been shown to stimulate microbial activity and abundance in some cases. Sorption of comparatively polar organic matter and nutrients could support energy for microorganisms, while macro-and micropores of biochar, which hold air and water, likely support microorganisms' livable habitat [1].Our results

Table 2. Coefficients of seasonal variation of MBC under different treatments (%).

Treatment	0–5 cm	5–10 cm	10–20 cm	20–30 cm	0–30 cm
CK	37.0ab	25.4ab	38.1ab	20.3b	25.0b
B4.5	25.7bc	22.9b	22.6bc	10.4c	11.2c
B9.0	19.4c	34.2ab	20.5c	15.3b	15.6c
SR	40.6ab	43.1a	40.4a	63.3a	41.6a

Note: Different small letters indicate significant differences among treatments (CK: control, B4.5: 4.5 t ha^{-1} yr^{-1} biochar addition, B9.0: 9.0 t ha^{-1} yr^{-1} biochar addition, SR: incorporation of wheat straw) at the $P<0.05$ level (L.S.D.).

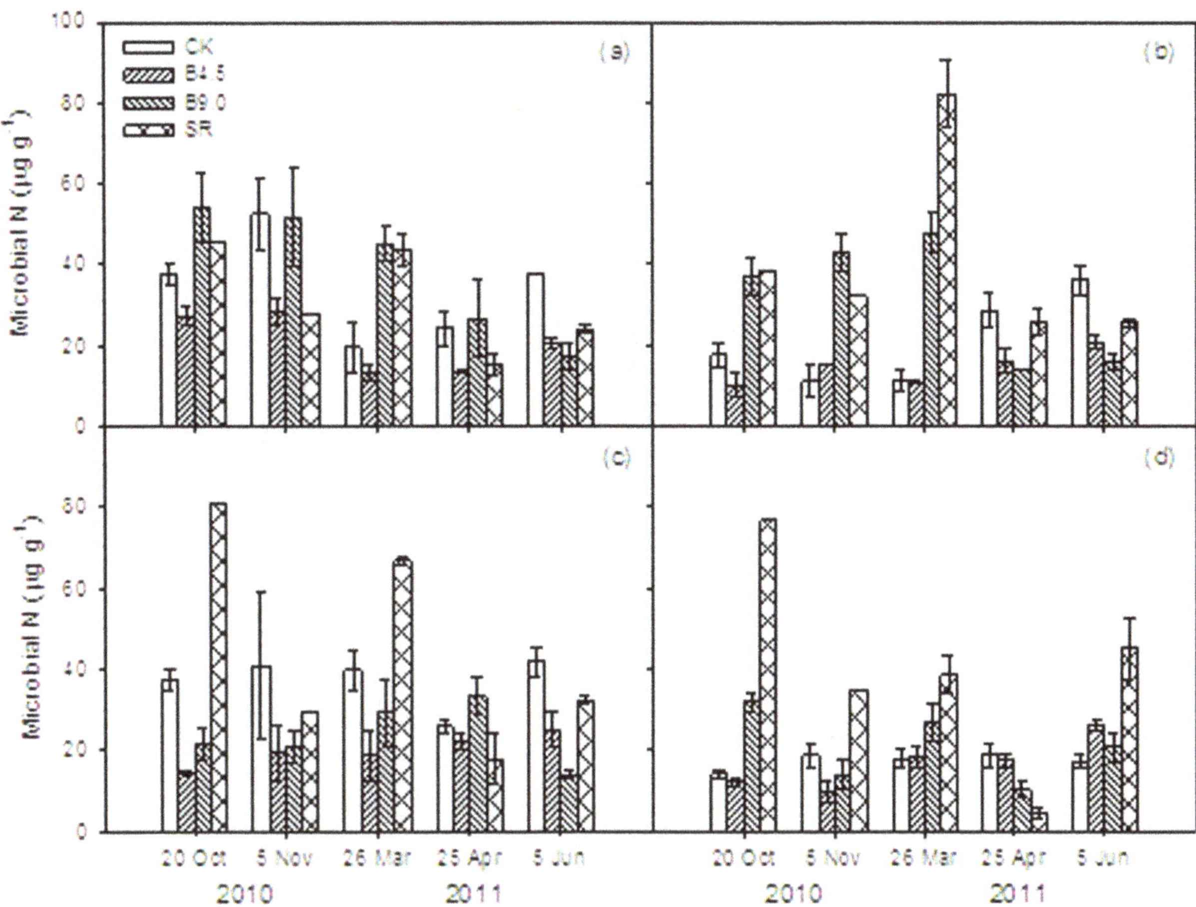

Figure 2. Microbial biomass nitrogen (MBN) under different treatments (CK: control, B4.5: 4.5 t ha^{-1} yr^{-1} biochar addition, B9.0: 9.0 t ha^{-1} yr^{-1} biochar addition, SR: incorporation of wheat straw) and different soil depths (a: 0–5, b: 5–10, c: 10–20, and d: 20–30 cm) during winter wheat season. Vertical bars represent the standard errors for means of each treatment ($n = 3$).

suggest that biochar supplied livable habitat for microorganisms stimulating microorganisms which can use the carbon source from the labile fraction of biochar.

Soil microbial biomass C:N

Changes in the microbial biomass C:N ratio can reflect changes in the relative availability of C and N to microbes, but could also reflect changes in the microbial community structure. In our study, the microbial biomass C:N ratio significantly increased with increasing biochar application and the microbial biomass C:N ratio more than doubled in the B9.0 treatment compared to the control (Fig. 3). A pot experiment showed that there was no significant effect on the microbial biomass C:N ratio of biochar addition at a rate of 25 t ha^{-1} [12]. Nicolardot et al. [23] determined that there was a significantly positive correlation between soil microbial biomass C:N ratio and the C:N ratio of the added organic matter. In contrast, Kushwaha et al. [24] reported a decrease in microbial biomass C:N ratio after straw return, while Kallenbach and Grandy [25] found no effect on microbial biomass C:N ratio after application of organic carbon, suggesting that changes in microbial biomass C:N ratio cannot solely be explained by organic amendments with relatively high C content. Several studies observed a change in soil microbial structure and abundance with biochar application [1]. In our study the

Table 3. Coefficients of seasonal variation of MBN under different treatments (%).

Treatment	0–5 cm	5–10 cm	10–20 cm	20–30 cm	0–30 cm
CK	38.1ab	43.6ab	23.7b	31.4a	31.7b
B4.5	34.9b	23.9b	6.9c	34.9c	18.6c
B9.0	49.5a	52.5a	43.7ab	58.8b	49.1ab
SR	42.0ab	55.3a	66.9a	93.6a	65.4a

Note: Different small letters indicate significant differences among treatments (CK: control, B4.5: 4.5 t ha^{-1} yr^{-1} biochar addition, B9.0: 9.0 t ha^{-1} yr^{-1} biochar addition, SR: incorporation of wheat straw) at the $P<0.05$ level (L.S.D.).

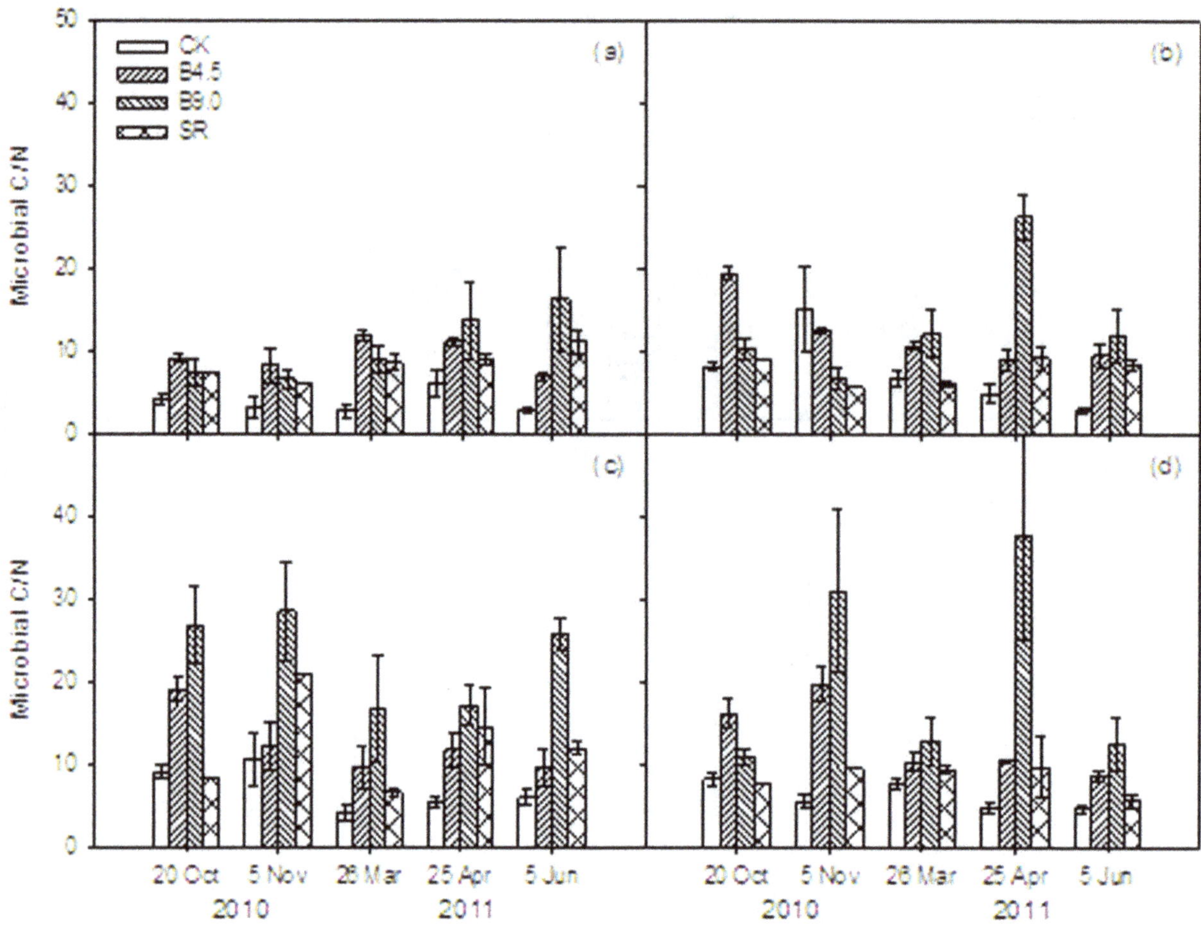

Figure 3. The ratio of soil microbial biomass carbon (MBC) to soil microbial biomass nitrogen (MBC/MBN) under different treatments (CK: control, B4.5: 4.5 t ha^{-1} yr^{-1} biochar addition, B9.0: 9.0 t ha^{-1} yr^{-1} biochar addition, SR: incorporation of wheat straw) and different soil depths (a: 0–5, b: 5–10, c: 10–20, and d: 20–30 cm) during winter wheat season. Vertical bars represent the standard errors for means of each treatment ($n = 3$).

increased microbial C:N ratio with biochar suggests increased microbial N limitation, while it is unclear if the increased microbial C:N ratio was also due to changes in microbial community structure. Moreover, biochar application may alter the release of MBN or the fraction of biomass N mineralized (K_N). The value of K_N had a wide range from 0.28–0.81 [26–28], and it was difficult to be measured precisely. The high values of microbial C:N ratio for biochar treatments especially at 10–20 and 20–30 cm soil depths might be due to the same value of K_N used for all treatments herein. It is worthy of further study on the effect of biochar on the fraction of biomass N mineralized in the fumigation-extraction method.

Relationship between MBC and environmental factors

Data analysis showed that there was a significant positive correlation between SWC and MBC (**Fig. 4**, $R^2 = 0.172$, $P < 0.001$). The positive relationship between SWC and MBC suggests that SWC may have limited microbial activity at the experimental site. Biochar addition can increase soil water content [29,30]. Therefore, biochar application could influence soil microbial biomass via variation in soil water content.

However, the increases in MBC with biochar addition did not always coincide with an increase in SWC. For instance, MBC increased sharply in the reviving stage (March 2011) in the B9.0 treatment while SWC did not. Moreover, our study showed that biochar application showed a limited effect on soil water content [13,14], which also did not coincide with the higher MBC in the biochar treatments than in the CK treatment.

There was no significant correlation between MBC and soil temperature, and our result was supported by other studies [31,32]. For instance, Contin et al. [32] found that MBC did not change significantly at different incubation temperatures in arable and grassland soils. But we should point out that the spike in MBC at 5–10 cm depth of soil on March 26 may be related to the soil temperature increase that occurred during the same time.

The response of microbial biomass to seasonal changes is important in regulating microbial turnover, which may in turn influence nutrient availability and ultimately plant nutrition [33]. Seasonal fluctuations in MBC in the biochar treatments were smaller than in the CK and SR treatments (Table 2), whereas, seasonal fluctuations in MBN showed no obvious differences among the CK, B9.0 and SR treatments, and only the seasonal fluctuations in the B4.5 treatment was lower than in the other treatments. Biochar may have reduced extreme environmental conditions for microbial growth, which could have important implications for C and N dynamics, and crop yield [34,35].

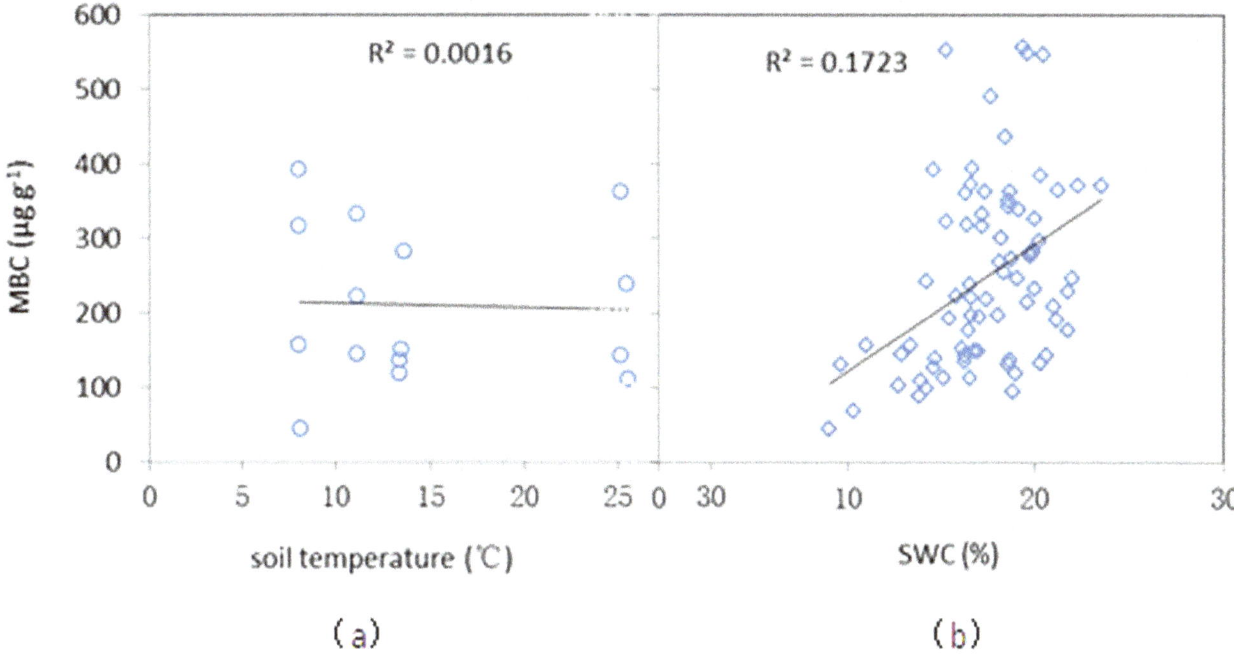

Figure 4. The correlations between soil microbial biomass carbon (MBC) and soil temperature (a), and between MBC and soil water content (SWC, b).

Conclusion

After 4 consecutive years of application, biochar increased soil MBC significantly compared to CK. The biochar application at a rate of 9.0 t ha^{-1} yr^{-1} reached and even exceeded the effect of wheat residue return, whereas smaller differences in soil MBN were found among treatments. Soil MBC for biochar treatments showed less seasonal fluctuation compared to CK and SR treatments, suggesting that biochar provided a more suitable habitat for soil microorganisms. Biochar treatment showed the highest value of soil microbial biomass C:N ratio, and one possible reason might be that biochar could decease the fraction of biomass N mineralized.

Acknowledgments

This work was supported by the S&T Innovation Program of Chinese Academy of Agricultural Sciences.

Author Contributions

Conceived and designed the experiments: QZZ. Performed the experiments: JH NL QZZ. Analyzed the data: QZZ FAD XRL YDW. Contributed reagents/materials/analysis tools: QZZ FAD YDW XRL. Contributed to the writing of the manuscript: QZZ FAD XRL.

References

1. Lehmann J, Rillig MC, Thies J, Masiello CA, Hockaday WC, et al. (2011) Biochar effects on soil biota – A review. Soil Biology and Biochemistry 43: 1812–1836.
2. Lehmann J (2007) A handful of carbon. Nature 447: 143–144.
3. Zimmerman AR, Gao B, Ahn M-Y (2011) Positive and negative carbon mineralization priming effects among a variety of biochar-amended soils. Soil Biology and Biochemistry 43: 1169–1179.
4. Luo Y, Durenkamp M, De Nobili M, Lin Q, Brookes PC (2011) Short term soil priming effects and the mineralisation of biochar following its incorporation to soils of different pH. Soil Biology and Biochemistry 43: 2304–2314.
5. Cross A, Sohi SP (2011) The priming potential of biochar products in relation to labile carbon contents and soil organic matter status. Soil Biology and Biochemistry 43: 2127–2134.
6. Keith A, Singh B, Singh BP (2011) Interactive priming of biochar and labile organic matter mineralization in a smectite-rich soil. Environmental Science & Technology 45: 9611–9618.
7. Thiessen S, Gleixner G, Wutzler T, Reichstein M (2013) Both priming and temperature sensitivity of soil organic matter decomposition depend on microbial biomass – An incubation study. Soil Biology & Biochemistry 57: 739–748.
8. Zavalloni C, Alberti G, Biasiol S, Vedove GD, Fornasier F, et al. (2011) Microbial mineralization of biochar and wheat straw mixture in soil: A short-term study. Applied Soil Ecology 50: 45–51.
9. Steinbeiss S, Gleixner G, Antonietti M (2009) Effect of biochar amendment on soil carbon balance and soil microbial activity. Soil Biology and Biochemistry 41: 1301–1310.
10. Luo Y, Durenkamp M, De Nobili M, Lin Q, Devonshire BJ, et al. (2013) Microbial biomass growth, following incorporation of biochars produced at 350°C or 700°C, in a silty-clay loam soil of high and low pH. Soil Biology and Biochemistry 57: 513–523.
11. Castaldi S, Riondino M, Baronti S, Esposito FR, Marzaioli R, et al. (2011) Impact of biochar application to a Mediterranean wheat crop on soil microbial activity and greenhouse gas fluxes. Chemosphere 85: 1464–1471.
12. Dempster DN, Gleeson DB, Solaiman ZM, Jones DL, Murphy DV (2012) Decreased soil microbial biomass and nitrogen mineralisation with Eucalyptus biochar addition to a coarse textured soil. Plant and Soil 354: 311–324.
13. Zhang QZ, Wang XH, Du ZL, Liu XR, Wang YD (2013) Impact of biochar on nitrate accumulation in an alkaline soil. Soil Research 51: 521–528.
14. Zhang QZ, Wang YD, Wu YF, Wang XH, Du ZL, et al. (2013) Effects of biochar amendment on soil thermal conductivity, reflectance, and temperature. soil Science Society of America Journal 77: 1478–1487.
15. Brookes PC, Landman A, Pruden G, Jenkinson DS (1985) Chloroform fumigation and the release of soil nitrogen: A rapid direct extraction method to measure microbial biomass nitrogen in soil. Soil Biology and Biochemistry 17: 837–842.
16. Wu J, Joergensen RG, Pommerening B, Chaussod R, Brookes PC (1990) Measurement of soil microbial biomass C by fumigation-extraction—an automated procedure. Soil Biology and Biochemistry 22: 1167–1169.
17. Kolb SE, Fermanich KJ, Dornbush ME (2008) Effect of charcoal quantity on microbial biomass and activity in temperate soils. Soil Science Society of America Journal 73: 1173–1181.

18. O' Neill B, Grossman J, Tsai MT, Gomes JE, Lehmann J, et al. (2009) Bacterial community composition in Brazilian Anthrosols and adjacent soils characterized using culturing and molecular identification. Microbial Ecology 58: 23–35.
19. Liang B, Lehmann J, Sohi SP, Thies JE, O'Neill B, et al. (2010) Black carbon affects the cycling of non-black carbon in soil. Organic Geochemistry 41: 206–213.
20. Steiner C, Das KC, Garcia M, Förster B, Zech W (2008) Charcoal and smoke extract stimulate the soil microbial community in a highly weathered xanthic Ferralsol. Pedobiologia 51: 359–366.
21. Lehmann J, Joseph S (2009) Biochar for Environmental Management: An Introduction. In: Lehmann J, Joseph S, editors. Biochar for Environmental Management: Science and Technology. London: Earthscan. pp. 1–12.
22. Guo W, Chen HX, Zhang QZ, Wang YD (2011) Effects of biochar application on total nitrogen and alkali-hydrolyzable nitrogen content in the topsoil of the high-yield cropland in north China Plain. Ecology and Environmental Sciences 20: 425–428.
23. Nicolardot B, Recous S, Mary B (2001) Simulation of C and N mineralisation during crop residue decomposition: a simple dynamic model based on the C: N ratio of the residues. Plant and Soil 228: 83–103.
24. Kushwaha CP, Tripathi SK, Singh KP (2000) Variations in soil microbial biomass and N availability due to residue and tillage management in a dryland rice agroecosystem. Soil and Tillage Research 56: 153–166.
25. Kallenbach C, Grandy AS (2011) Controls over soil microbial biomass responses to carbon amendments in agricultural systems: A meta-analysis. Agriculture, Ecosystems and Environment 144: 241–252.
26. Joergensen RG (1996) The fumigation-extraction method to estimate soil microbial biomass: Calibration of the kEC value. Soil Biology and Biochemistry 28: 25–31.
27. Greenfield LG (1995) Release of microbial cell N during chloroform fumigation. Soil Biology and Biochemistry 27: 1235–1236.
28. Sparling G, Zhu C (1993) Evaluation and calibration of biochemical methods to measure microbial biomass C and N in soils from western australia. Soil Biology and Biochemistry 25: 1793–1801.
29. Chen Y, Shinogi Y, Taira M (2010) Influence of biochar use on sugarcane growth, soil parameters, and groundwater quality. Australian Journal of Soil Research 48: 526–530.
30. Piccolo A, Pietramellara G, Mbagwu JSC (1996) Effects of coal derived humic substances on water retention and structural stability of Mediterranean soils. Soil Use and Management 12: 209–213.
31. Joergensen RG, Brookes PC, Jenkinson DS (1990) Survival of the soil microbial biomass at elevated temperatures. Soil Biology and Biochemistry 22: 1129–1136.
32. Contin M, Corcimaru S, De Nobili M, Brookes PC (2000) Temperature changes and the ATP concentration of the soil microbial biomass. Soil Biology and Biochemistry 32: 1219–1225.
33. Wardle DA (1992) A comparative assessment of factors which influence microbial biomass carbon and nitrogen levels in soil. Biological Reviews of the Cambridge Philosophical Society 67: 321–358.
34. Kumar K, Goh KM (1999) Crop residues and management practices: effects of soil quality, soil nitrogen dynamics, crop yield, and nitrogen recovery. Advances In Agronomy 68: 197–319.
35. Lou Y, Li Z, Zhang T, Liang Y (2004) CO_2 emissions from subtropical arable soils of China. Soil Biology & Biochemistry 36: 1835–1842.

12

Zinc, Iron, Manganese and Copper Uptake Requirement in Response to Nitrogen Supply and the Increased Grain Yield of Summer Maize

Yanfang Xue[1], Shanchao Yue[1], Wei Zhang[1], Dunyi Liu[1], Zhenling Cui[1], Xinping Chen[1], Youliang Ye[2], Chunqin Zou[1]*

1 Center for Resources, Environment and Food Security, China Agricultural University, Beijing, China, 2 College of Resources and Environmental Sciences, Henan Agricultural University, Zhengzhou, China

Abstract

The relationships between grain yields and whole-plant accumulation of micronutrients such as zinc (Zn), iron (Fe), manganese (Mn) and copper (Cu) in maize (*Zea mays* L.) were investigated by studying their reciprocal internal efficiencies (RIEs, g of micronutrient requirement in plant dry matter per Mg of grain). Field experiments were conducted from 2008 to 2011 in North China to evaluate RIEs and shoot micronutrient accumulation dynamics during different growth stages under different yield and nitrogen (N) levels. Fe, Mn and Cu RIEs (average 64.4, 18.1and 5.3 g, respectively) were less affected by the yield and N levels. ZnRIE increased by 15% with an increased N supply but decreased from 36.3 to 18.0 g with increasing yield. The effect of cultivars on ZnRIE was similar to that of yield ranges. The substantial decrease in ZnRIE may be attributed to an increased Zn harvest index (from 41% to 60%) and decreased Zn concentrations in straw (a 56% decrease) and grain (decreased from 16.9 to 12.2 mg kg^{-1}) rather than greater shoot Zn accumulation. Shoot Fe, Mn and Cu accumulation at maturity tended to increase but the proportions of pre-silking shoot Fe, Cu and Zn accumulation consistently decreased (from 95% to 59%, 90% to 71% and 91% to 66%, respectively). The decrease indicated the high reproductive-stage demands for Fe, Zn and Cu with the increasing yields. Optimized N supply achieved the highest yield and tended to increase grain concentrations of micronutrients compared to no or lower N supply. Excessive N supply did not result in any increases in yield or micronutrient nutrition for shoot or grain. These results indicate that optimized N management may be an economical method of improving micronutrient concentrations in maize grain with higher grain yield.

Editor: Paul A. Cobine, Auburn University, United States of America

Funding: This research was supported by the National Natural Science Foundation of China (31272252), China Agriculture Research System (CARS-02), the 973 project (No. 2009CB18606) and the Innovative Group Grant of NSFC (31121062). The funders had no role in study design, data collection and analysis, decision to publish, or preparation of the manuscript.

Competing Interests: The authors have declared that no competing interests exist.

* E-mail: zcq0206@cau.edu.cn

Introduction

Maize (*Zea mays* L.), as one of the world's leading cereal grains along with rice and wheat, is very popular due to its diverse functionality as a food source for both humans and animals [1]. It is estimated that maize together with rice and wheat provide at least 30% of the food calories to more than 4.5 billion people in 94 developing countries [2]. Increased maize production is also required to meet the demands for animal feed and biofuel [3].

In China, maize accounts for more than one-third of Chinese cereal production and China is responsible for 20.9% of global maize output from 2009 to 2011 [4]. Pursuing high grain yields in China has been the top priority in policy. However, a 71% increase in total annual grain production from 1977 to 2005 was accompanied by a 271% increase in nitrogen (N) fertilizer input, which resulted in serious environmental problems such as eutrophication, greenhouse gas emissions and soil acidification in China [5–8]. Intensive cultivation of high-yielding cultivars with over-applications of N, phosphorous (P) and potassium (K) fertilizers also leads to micronutrient, especially zinc (Zn) and iron (Fe) deficiencies in many countries [9]. For example, in China, approximately 40% of the soils are Zn and Fe deficient, and about 30% are manganese (Mn) and copper (Cu) deficient [10]. Therefore, improving N management and developing related policies are major issues in crop production and environmental protection in China. Several studies have reported that increasing N fertilization has little or no effect on grain micronutrient concentrations, such as Zn, Fe and Cu in maize [11–13]. Therefore, it is necessary to investigate whether optimal N management by decreasing excessive N supply, would result in negative impacts on micronutrient nutrition of maize plants and their allocation into grains under field conditions.

Maize grain yield has steadily increased over the last century through both conventional breeding and agronomic practices [14]. Increased grain yield and biomass production may be associated with greater uptake of total plant nutrients, such as N, P and K [15,16]. A double peak and a higher N and P accumulation rates in maize with higher yield have been found compared to lower yielding maize [15]. Little information is available about the accumulation dynamics of micronutrients such as Zn, Fe, Mn and Cu in response to an increased grain yield or increased biomass production. However, the patterns of micronutrient accumulation during different growth stages have been quantified [17,18]. It has been suggested that further investigation is needed to understand the timing of micronutrient uptake in response to yields, which varies between crop species, environments and management practices [19]. Knowledge of the dynamics of micronutrient accumulation associated with yield–trait relationships in crops would provide an efficient tool to synchronize micronutrient demand and supply, thus improving nutrient management efficiencies (e.g. Zn fertilizer) and benefiting sustainable production without harming the environment.

The study of the nutrient reciprocal internal efficiencies (RIEs, g of nutrient requirement in plant dry matter per Mg grain yield) could be used to determine the relationships between grain yield and whole-plant nutrient accumulation [19]. Some studies have reported large variations in the relationship between maize grain yield and nutrient accumulation in response to nutrient supplies, grain yields, genotypes, and environments [20–23]. For example, N reciprocal internal efficiency (N requirement per Mg of grain yield) decreased from 19.8 to 16.9 kg with increasing grain yield while it was higher with N application than no N application [22]. These studies mainly focused on N, P and K. Little information is available on micronutrient accumulation (such as Zn, Fe, Mn and Cu) in response to different grain yields, N fertilization rates or different cultivars. Understanding the relationships between micronutrient accumulation, especially Zn, and grain yield with different N management practices should provide valuable information leading to improvements in micro-fertilizer management practices. It is reported that crop recovery efficiency of micronutrients ranged from only 5% to 10% [24].

Recent studies have shown that increased maize grain yield decreases grain N concentration [14,22]. A negative correlation between grain yield and concentrations of Zn and Fe in maize grain has also been reported [25]. Therefore, it is important to investigate the concentrations of Zn and Fe, together with Mn and Cu, in maize grain in response to increased grain yield.

The objectives of this study were to (i) quantify RIEs (ZnRIE, FeRIE, MnRIE and CuRIE) in relation to increased grain yield, different cultivars and N management practices, (ii) estimate shoot Zn, Fe, Mn and Cu accumulation dynamics in response to increased grain yields and different N management practices, and (iii) investigate the effects of improved grain yields, different cultivars and N management practices on the concentrations of Zn, Fe, Mn and Cu in maize grain.

Materials and Methods

Experimental design

Field experiment I was conducted in four consecutive years (2008–2011) in Quzhou county (36°53′60″ N, 115°0′ E) in Hebei province, on a calcareous alluvial soil typical of the North Plain of China (NCP), pH 8.3 (1:2.5 w/v in water) and 1.4% organic matter [26]. One cultivar (Zhengdan958) of maize was planted in a randomized complete block design with four replicates (300 m^2 $plot^{-1}$) in 2008 and 2009 while three maize cultivars (Zhengdan958, Xianyu335 and Yedan13, recorded as ZD, XY, YD, respectively) were planted in 2010 and 2011 in split-plots (N fertilizer as the main plots and cultivars as sub-plots) with three or four replicates. Nitrogen rates were applied as urea as follows: no N application (recorded as N-0), 40–70% of optimized N treatment (recorded as N-low), optimized N treatment (240, 150, 105 and 193 kg N ha^{-1} in four consecutive years, respectively, recorded as N-opt) based on the in-season root-zone N-management approach, as previously reported [3,27], and 130 or 150% of optimized N treatment and farmer's nitrogen practice (250 kg N ha^{-1}, recorded as N-over). Nitrogen fertilizer was applied in split doses as described in Table S1. Before sowing, 45 kg ha^{-1} of P_2O_5, 45 kg ha^{-1} of K_2O, and 30 kg ha^{-1} of $ZnSO_4 \cdot 7H_2O$ (without application of $ZnSO_4 \cdot 7H_2O$ in 2011) were broadcast and incorporated into the upper 0–15 cm of the soil by rotary tillage. Another 45 kg ha^{-1} of K_2O was applied by hand as a top dressing at the V10 (ten-leaf) growth stage. The climate conditions for the experiment have been recently reported by Zhang et al. [28].

Field experiment II was conducted in 2009 in Wenxian (34°52′42″ N, 112°57′29″ E), Henan province in the middle of the NCP. One maize cultivar (Fengyu 4, recorded as FY) was planted in a randomized complete block design with three replicates (20 m^2 $plot^{-1}$). Nitrogen rates were applied as urea as follows: no N application (recorded as N-0), 50% and 75% of optimized N application (recorded as N-low), optimized N application (240 kg N ha^{-1}, recorded as N-opt) and 150% of optimized N application (recorded as N-over). Nitrogen fertilizer was applied in split doses as described in Table S1. In addition, 90 kg ha^{-1} of P_2O_5 and 90 kg ha^{-1} of K_2O were applied by hand at the V5 (five-leaf) growth stage.

The two experiments were conducted in a winter wheat-summer maize rotation system. Nitrogen fertilizer, P_2O_5 and K_2O were applied as urea, superphosphate and potassium chloride, respectively. Weeds were well controlled, and no obvious water or pest stress was observed during the maize growing season. No specific permissions were required for these locations. The field studies did not involve endangered or protected species.

Sampling and nutrients analysis

6-plant (experiment I) and 2-plant (experiment II) samples were collected at V6 (six-leaf), V12 (twelve-leaf), R1 (silk emerging), R3 (milk stage) and R6 (physiological maturity, divided into straw and grain) growth stages. All plant samples were collected at V6, V12, R1, R3 and R6 stages in experiment I in 2008 and 2009. In 2010, plant samples were only collected at R1 and R3 stages with three N treatments (no N application, optimized N application and farmer's nitrogen practice) and at R6 with five N treatments. In 2011, plant samples were only collected at V6 with one cultivar (ZD) and N treatments excluding 130% of optimized N treatment, at R1 with three cultivars (ZD, XY and YD) and three N treatments (no N application, optimizer N application and farmer's nitrogen practice) and R6 stages with three cultivars (ZD, XY and YD) and five N treatments. All plant samples were collected at V6, V12, R1, R3 and R6 stages in experiment II. The shoots were rapidly washed with deionized water and then oven-dried at 70°C to determine dry weight. Plant samples were ground with a stainless steel grinder (RT-02B, Taiwan, China) and digested with HNO_3-H_2O_2 in a microwave accelerated reaction system (CEM, Matthews, NC, USA). The concentrations of Zn, Fe, Mn and Cu in the digested solutions were determined by inductively coupled plasma atomic emission spectroscopy (ICP-AES, OPTIMA 3300 DV, Perkin-Elmer, USA). IPE556 grain and IPE883 straw (Wageningen University, The Netherlands) were used as reference materials.

To estimate grain yields, ears in the central part of 60 or 180 m^2 (experiment I) and 10 m^2 (experiment II) areas were harvested at maturity. At harvest, sub-samples of 6 plants in the two experiments were collected and divided into grain and straw parts to determine above-ground biomass and the harvest index (HI) after shoots were oven-dried at 70°C.

Data analysis

According to grain yield, all data collected from 2008 to 2011 were divided into four yield ranges: <7.5, 7.5–9, 9–10.5 and >10.5 Mg ha^{-1}. According to these yield ranges, maize biomass and shoot micronutrient (Zn, Fe, Mn and Cu) accumulation as well as micronutrient RIEs (ZnRIE, FeRIE, MnRIE and CuRIE) were analysed. For each of the four different cultivars (YD, XY, ZD and FY), RIEs were analysed. According to N management practices, all data were also divided into four groups: N-0 (no N application), N-low (30–75% of optimized N treatment), N-opt (optimized N treatment) and N-over (130% or 150% of optimized N treatment as well as farmer's nitrogen practice). The number of observations for shoot samples at different growth stages with different yield ranges and N levels was shown in Table S2. All the above data analysis referred to others [22,29,30]. The following parameters were calculated: Micronutrient harvest indices (ZnHI, FeHI, MnHI and CuHI) = grain micronutrient accumulation/ shoot micronutrient accumulation (1); Grain micronutrient accumulation = grain micronutrient concentration x grain dry weight (2); Shoot micronutrient accumulation = shoot micronutrient concentration x biomass (3); Micronutrient RIE = shoot micronutrient accumulation/grain yield = grain micronutrient accumulation/(grain yield x micronutrient harvest index) [16,31] (4).

The Pearson correlation procedure was used to evaluate the correlations among the measured parameters of all data using SAS software (SAS 8.0, USA). Means of different yield ranges and N levels were compared using one-way ANOVA at a 0.05 level of probability followed by Duncan test of SPSS 13.0 for Windows. All results were expressed on a dry weight basis with an exception of grain yields. Grain yields were reported at standard moisture of 15.5%.

Results

Grain yields, yield components, biomass production and micronutrient (Zn, Fe, Mn and Cu) accumulation in shoot and grain

Overall, maize grain yield (n = 149) averaged 8.0 Mg ha^{-1} with a range from 3.9 to 12.8 Mg ha^{-1} (Table 1). The total biomass production at physiological maturity averaged 13.2 Mg ha^{-1}, with a harvest index (HI) of 51% (Table 2). The Zn, Fe, Mn and Cu RIEs averaged 31.0, 64.4, 18.1, and 5.3 g, respectively. The concentrations of grain Zn, Fe, Mn and Cu averaged 16.3, 15.8, 3.2 and 1.4 mg kg^{-1}, respectively. The harvest indices of Zn, Fe, Mn and Cu averaged 48%, 23%, 16% and 24%, respectively (Table 2). Yield was significantly negatively associated with grain Zn concentration ($r = -0.36^{***}$) but positively associated with grain Mn concentration ($r = 0.22^{**}$). However, yield was not correlated with grain concentrations of Fe and Cu (Table 3).

Overall, seasonal biomass production and shoot Zn accumulation continued to increase throughout the growing season. Seasonal accumulation of shoot Fe and Cu showed a slight decrease from R3 to R6 while shoot Mn accumulation showed a large decrease from R3 to R6 (Figure 1 and Figure S1). At silking stage (R1), more than three quarters of shoot Zn, Fe and Cu accumulation, and more than 100% of shoot Mn accumulation, had occurred compared to only half of the biomass accumulation (Figure 1 and Figure S1). Generally, throughout the growing season, the biomass and shoot micronutrient accumulation, as a percentage of the total (at maturity) decreased in the order Mn>Cu>Fe≥Zn>biomass (Figure S1).

During the growth stages, the highest accumulation rates of biomass and shoot Fe and Zn consistently occurred between V12 and R1. During this period, biomass accumulated 3.5 Mg ha^{-1} with an average growth rate of 251 kg ha^{-1} d^{-1}; shoot Zn accumulated 98 g ha^{-1} with an average accumulation rate of 7.0 g ha^{-1} d^{-1}; shoot Fe accumulated 215 g ha^{-1} with an average accumulation rate of 15.3 g ha^{-1} d^{-1}. In contrast, the highest accumulation rates of shoot Mn (83 g ha^{-1} in total, with an average accumulation rate of 6.0 g ha^{-1} d^{-1}) and Cu (15 g ha^{-1} in total, with an average accumulation rate of 1.1 g ha^{-1} d^{-1}) occurred between V6 and V12 (Figure 1).

Relationships between yield and shoot micronutrient accumulation in response to different yield ranges

To understand the relationship between yield and shoot micronutrient accumulation in relation to grain yield, all the data in this study was grouped into 4 yield ranges as follows: <7.5 Mg ha^{-1} (the number of observations, n = 58, mean yield 5.9 Mg ha^{-1}, recorded as GY1), 7.5–9.0 Mg ha^{-1} (n = 45, mean yield 8.2 Mg ha^{-1}, recorded as GY2), 9.0–10.5 Mg ha^{-1} (n = 26, mean yield 9.7 Mg ha^{-1}, recorded as GY3) and >10.5 Mg ha^{-1} (n = 20, mean yield 11.4 Mg ha^{-1}, recorded as GY4) (Table 1). As shown in Table 2, the increase in grain yield from GY1 to GY2 was mainly attributed to thousand grain weight (an increase of 12%) and grains per ear (an increase of 11%). The increase in grain yield from GY2 to GY3 was mainly attributed to grains per ear (an increase of 14%) and HI (which increased from 50% to 53%). The further increase in grain yield from GY3 to GY4 was mainly attributed to HI (which increased from 53% to 55%).

With the increase of grain yield from GY1 to GY4, ZnRIE decreased gradually from 36.3, 32.4, 26.6 to 18.0 g. However, FeRIE (ranging from 57.4 to 68.6 g), MnRIE (ranging from 17.5 to 19.0 g) and CuRIE (ranging from 4.9 to 5.6 g) were not significantly affected by yields (Table 2). With increasing yield from GY1 to GY3, grain Zn concentration (average 16.9 mg kg^{-1}) was not affected, but it was significantly decreased to 12.2 mg kg^{-1} when grain yield increased from GY3 to GY4. Similar results were found for grain Fe concentration. Straw Zn concentration showed a decreasing trend from 24.4 to 10.7 mg kg^{-1} while straw Mn concentration showed an increasing trend from 17.3 to 23.9 mg kg^{-1} with the increasing yield from GY1 to GY4. In contrast, straw Fe and Cu concentrations were less affected by yields. Grain Mn concentration was significantly higher in GY2 and GY3 than GY1 and GY4. Similarly, grain Cu concentration was the highest in GY3 than the other three yield ranges (Table 2). With an increasing grain yield from GY1 to GY4, ZnHI increased gradually from 41%, 48%, 54% to 60% while FeHI (ranging from 21% to 27%), MnHI (ranging from 14% to 18%) and CuHI (ranging from 22% to 29%) were the highest in GY3 than the other three yield ranges (Table 2).

The response of ZnRIE to increasing grain yield could be classified into two response stages. When grain yield was increased from GY1 to GY3, ZnRIE decreased from 36.3 g to 26.6 g (a decrease of 27%) because of increasing ZnHI (which increased from 41% to 54%) and significantly declining straw Zn concentration (a decrease of 30%) coupled with the relatively constant grain Zn concentration (average 16.9 mg kg^{-1}) (Table 2). With the increasing yield from GY3 to GY4, ZnRIE further declined from 26.6 to 18.0 g (a decrease of 32%) mainly because of a further increasing ZnHI from 54% to 60% and the decreasing Zn concentrations in both straw (which decreased by 38%) and grain (which decreased by 26%) (Table 2). Furthermore, across all the grain yield ranges, ZnRIE was significantly positively correlated with Zn concentrations of grain ($r = 0.54^{***}$) and especially straw ($r = 0.90^{***}$) but negatively associated with grain yield ($r = -0.58^{***}$) and especially ZnHI ($r = -0.85^{***}$) (Table 3).

Table 1. Descriptive statistics of yield for total samples, four yield ranges (15.5% moisture), four different cultivars and four nitrogen (N) levels.

	n[a]	Mean	SD[b]	Minimum	25% Q[c]	Median	75% Q	Maximum
		---Mg ha^{-1}---						
Total	149	8.0	2.1	3.9	6.4	8.1	9.5	12.8
Yield ranges (Mg ha^{-1})								
<7.5	58	5.9	1.0	3.9	5.2	5.9	6.7	7.4
7.5–9	45	8.2	0.4	7.5	7.8	8.3	8.5	9.0
9–10.5	26	9.7	0.4	9.0	9.2	9.6	10.0	10.4
>10.5	20	11.4	0.6	10.5	11.0	11.3	11.8	12.8
Cultivars								
YD	30	5.6	1.1	3.9	4.7	5.4	6.5	8.1
XY	30	8.0	1.5	3.9	7.3	8.4	9.1	10.8
ZD	75	8.5	1.9	5.0	7.2	8.3	9.8	12.8
FY	14	10.2	1.2	8.1	9.5	10.4	11.0	12.0
N levels								
N-0	30	6.1	1.5	3.9	5.1	5.8	6.7	9.5
N-low	32	8.3	2.1	4.3	7.3	8.2	9.8	11.5
N-opt	30	8.9	2.1	4.4	7.4	9.2	10.4	12.3
N-over	57	8.3	1.8	4.5	7.2	8.3	9.3	12.8

n[a]: number of observations.
SD[b]: standard deviation.
Q[c]: quartile.

Similarly, FeRIE, MnRIE and CuRIE were consistently positively associated with their respective straw concentrations and straw yields (for MnRIE and CuRIE) but negatively correlated with their respective harvest index and grain yield (for FeRIE and CuRIE) (Table 3).

With the increase in grain yield from GY1 to GY4, the biomass production showed increasing trends, especially during the reproductive development (e.g. R3 and R6) (Figure 2A). Similar results were also found for shoot Mn and Cu accumulation due to the similar or lower shoot Mn and Cu concentrations in GY1 than the other three yield ranges, where shoot Mn and Cu concentrations were similar during the reproductive development (Figure 2D, E and Figure 3C, D). Shoot Fe accumulation among the four yield ranges were similar at V12 and R3 and tended to increase at V6 and especially at R6 with the increasing grain yield (Figure 2C). However, shoot Zn accumulation in GY4 was slightly (e.g. V6, R3 and R6 stages) or even significantly lower than the other two or three (e.g. V12 and R1 stages) yield ranges (Figure 2B).

In agreement with the overall trends shown in Figure S1, seasonal biomass production and shoot Zn accumulation continued to accumulate throughout the growing season irrespective of yield ranges (Figure 2B and Figure S2). Shoot Fe accumulation showed a decreasing trend from R3 to R6 for GY1, GY2 and GY3 but continued to accumulate throughout the growing stages for GY4 (Figure 2C and Figure S2). Shoot Mn accumulation decreased to a great extent from R3 to R6 irrespective of yields (Figure 2D and Figure S2). Similarly, shoot Cu accumulation showed a decreasing trend from R3 to R6 except GY1 (Figure 2E and Figure S2).

According to the increased grain yield from GY1 to GY4, nearly 75%, 86%, 91% and 66% of pre-silking shoot Zn accumulation, 81%, 95%, 76% and 59% of pre-silking shoot Fe accumulation, 62%, 90%, 81% and 71% of pre-silking shoot Cu accumulation and more than 100% of pre-silking shoot Mn accumulation occurred compared to 52%, 61%, 54% and 42% of pre-silking biomass accumulation (Figure 2 and Figure S2).

RIEs in response to different cultivars

To understand micronutrient RIEs in relation to different cultivars, all data were divided into four groups: YD (the number of observations, n = 30, mean yield 5.6 Mg ha^{-1}), XY (the number of observations, n = 30, mean yield 8.0 Mg ha^{-1}), ZD (the number of observations, n = 75, mean yield 8.5 Mg ha^{-1}) and FY (the number of observations, n = 14, mean yield 10.2 Mg ha^{-1}) (Table 1). Grain yield was in the order YD<XY<ZD<FY.

The effects of cultivars on ZnRIE were similar to those of yield ranges (described above). ZnRIE decreased gradually from 41.0, 36.1, 26.6 to 22.0 g for YD, XY, ZD and FY, respectively. The response of ZnRIE to cultivars could also be classified into two response stages. ZnRIE decreased from 41.0 g for YD to 26.6 g for ZD (a decrease of 35%) because of significantly increasing ZnHI (from 37% to 55%) and declining straw Zn concentration (a decrease of 45%) coupled with the relatively constant grain Zn concentration (average 16.7 mg kg^{-1}) (Table 2). ZnRIE further decreased from 26.6 g for ZD to 22.0 g for FY (a decrease of 17%) mainly because of decreasing Zn concentrations in both grain (which decreased by 17%) and straw (which decreased by 14%). FeRIE (ranging from 54.4 to 97.4 g) and MnRIE (ranging from 16.5 to 25.9 g) were in the order XY≤ZD≤YD<FY. The

Table 2. Biomass production, yield components (including ears number, grains per ear and thousand grain weight (TGW)), harvest index (HI), micronutrient reciprocal internal efficiencies (ZnRIE, FeRIE, MnRIE and CuRIE), grain micronutrient concentrations (GZnC, GFeC, GMnC and GCuC), straw micronutrient concentrations (SZnC, SFeC, SMnC and SCuC) and micronutrient harvest index (ZnHI, FeHI, MnHI and CuHI) of summer maize as affected by different yield ranges, cultivars and N levels.

Parameters	Total	Yield ranges (Mg ha^{-1})				Cultivars				N levels			
		<7.5	7.5–9	9–10.5	>10.5	YD	XY	ZD	FY	N-0	N-low	N-opt	N-over
	n[a] 149	58	45	26	20	30	30	75	14	30	32	30	57
Biomass (Mg ha^{-1})	13.2	10.2d	14.0c	14.9b	17.5a	9.7c	14.2b	13.5b	16.4a	10.3b	13.9a	14.5a	13.6a
Ears number (10^4 ha^{-1})	7.4	7.3a	7.5a	7.5a	7.5a	6.8b	7.8a	7.9a	5.4c	7.5a	7.3a	7.3a	7.6a
Grains per ear	425	379c	420b	477a	500a	429b	352c	416b	616a	352b	444a	450a	439a
TGW (g)	275	257b	288a	283a	292a	233c	314a	277b	277b	268a	276a	277a	278a
HI (%)	51	50c	50c	53b	55a	49b	48b	53a	53a	50a	51a	52a	51a
ZnRIE (g)	31.0	36.3a	32.4a	26.6b	18.0c	41.0a	36.1b	26.6c	22.0d	28.9a	28.4a	31.2a	33.4a
GZnC (mg kg^{-1})	16.3	16.7a	17.4a	16.6a	12.2b	17.6a	16.0a	16.4a	13.6b	15.2b	15.2b	16.6ab	17.3a
SZnC (mg kg^{-1})	20.0	24.4a	20.2b	17.1b	10.7c	29.0a	23.7b	16.1c	13.8c	18.2ab	17.6b	20.9ab	21.8a
ZnHI (%)	48	41d	48c	54b	60a	37b	40b	55a	53a	48a	49a	49a	48a
FeRIE (g)	64.4	68.6a	64.6a	59.7a	57.4a	66.8b	54.4c	61.2bc	97.4a	65.8a	69.0a	59.9a	63.4a
GFeC (mg kg^{-1})	15.8	15.9a	16.2a	16.6a	13.8b	17.6a	13.9b	15.9a	16.0a	13.4c	15.5b	16.0ab	17.2a
SFeC (mg kg^{-1})	62.6	63a	62a	61a	66a	58.6b	46.6c	61.9b	109.4a	63.3a	69.6a	58.3a	60.6a
FeHI (%)	23	21b	23b	27a	22b	23a	22a	24a	14b	19b	21ab	25a	25a
MnRIE (g)	18.1	17.8a	18.3a	17.5a	19.0a	17.8b	16.5b	17.3b	25.9a	16.8a	18.3a	18.0a	18.7a
GMnC (mg kg^{-1})	3.2	2.9b	3.5a	3.4a	3.0b	2.7c	3.7a	3.1b	3.3b	2.95c	3.03bc	3.35a	3.28ab
SMnC (mg kg^{-1})	19.0	17.3b	18.4b	19.9b	23.9a	17.4b	14.4c	19.3b	30.2a	16.8b	19.4ab	19.1ab	19.8a
MnHI (%)	16.0	15bc	17ab	18a	14c	14c	20a	16b	11d	16a	15a	17a	16a
CuRIE (g)	5.3	5.6a	5.4a	5.1a	4.9a	6.7a	5.5b	4.6c	6.0ab	4.5b	5.4a	5.4a	5.7a
GCuC (mg kg^{-1})	1.4	1.39b	1.43b	1.68a	1.26b	1.61a	1.38b	1.40ab	1.34b	1.24b	1.27b	1.57a	1.55a
SCuC (mg kg^{-1})	5.0	4.9a	5.0a	4.9a	5.5a	6.0a	4.7b	4.5b	6.4a	4.0b	5.2a	5.1a	5.4a
CuHI (%)	24	23b	23b	29a	22b	21b	22b	27a	19b	26a	21b	26a	24ab

n[a]: number of observations. Means in a row followed by different lowercase letters are significantly different at different yield ranges, cultivars and N levels ($P<0.05$).

significantly lower FeRIE in XY than FY is due to a lower grain Fe concentration (13.9 and 16.0 mg kg^{-1} for XY and FY, respectively) and a higher FeHI (22% and 14% for XY and FY, respectively). CuRIE was in the order ZD<XY≤FY≤YD (Table 2).

Relationship between yield and shoot micronutrient accumulation in response to different N rates

To understand the relationship between yield and shoot micronutrient accumulation in relation to N management practices, all data were divided into four groups: N-0 (the number of observations, n = 30, mean yield 6.1 Mg ha^{-1}), N-low (the

Table 3. Correlative coefficients (r) among the measured parameters at maturity (n = 149).

	ZnRIE		FeRIE		MnRIE		CuRIE		GY
GY	−0.58***	GY	−0.20*	GY	ns	GY	−0.18*	GZnC	−0.36***
SY	ns	SY	ns	SY	0.41***	SY	0.30***	GFeC	ns
GZnC	0.54***	GFeC	ns	GMnC	ns	GCuC	0.29***	GMnC	0.22**
SZnC	0.90***	SFeC	0.89***	SMnC	0.78***	SCuC	0.76***	GCuC	ns
ZnHI	−0.85***	FeHI	−0.77***	MnHI	−0.79***	CuHI	−0.58***	—	—

GY: grain yield based on 15.5% of moisture (Mg ha^{-1}); SY: straw yield based on dry weight (Mg ha^{-1}); ZnRIE, FeRIE, MnRIE and CuRIE: reciprocal internal efficiencies (g micronutrient requirement per Mg grain yield); GZnC, GFeC, GMnC and GCuC: grain micronutrient concentrations (mg kg^{-1}); SZnC, SFeC, SMnC and SCuC: straw micronutrient concentrations (mg kg^{-1}); ZnHI, FeHI, MnHI and CuHI: micronutrient harvest indices (%).
*Significant at $P<0.05$.
**Significant at $P<0.01$.
***Significance at $P<0.001$.
ns: not significant at $P<0.05$.

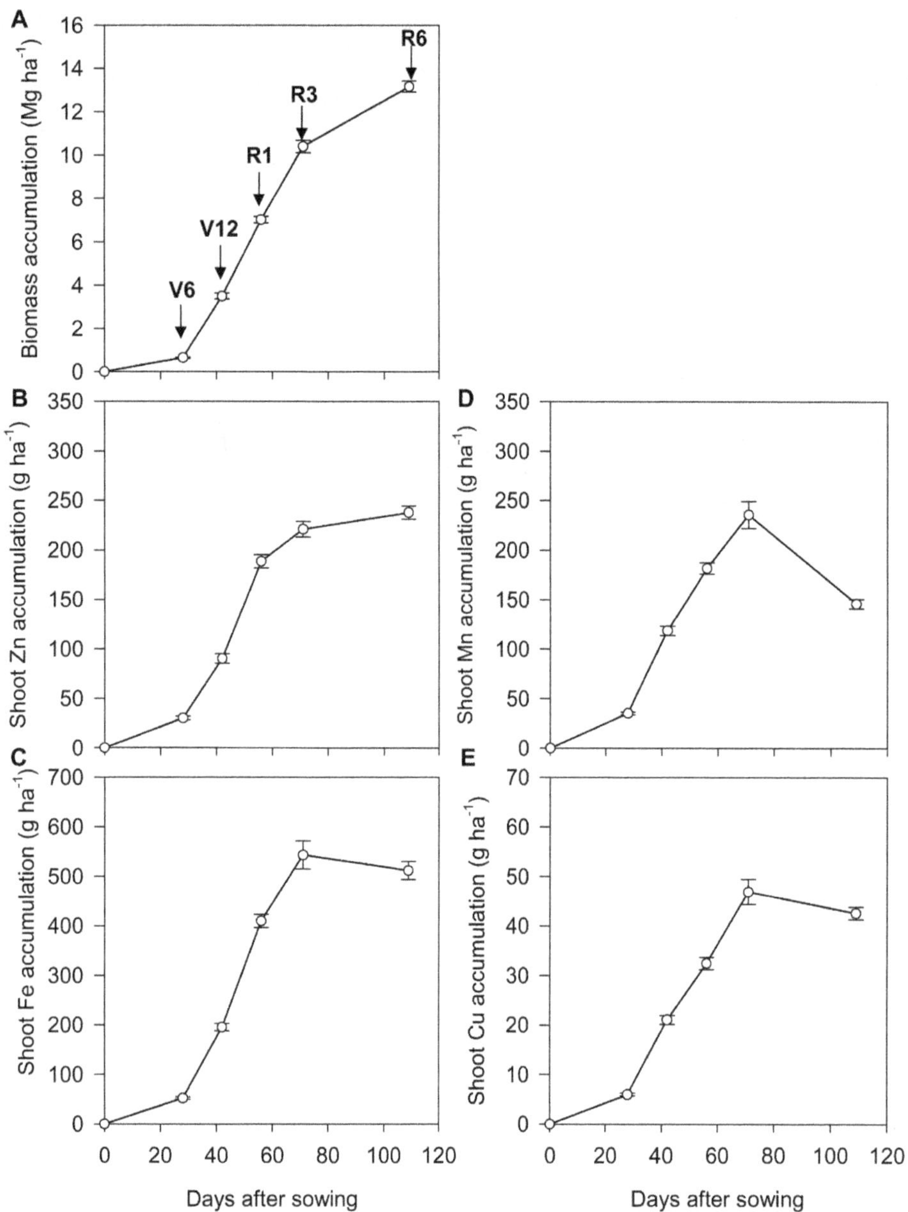

Figure 1. Dynamics of biomass (A), shoot Zn accumulation (B), shoot Fe accumulation (C), shoot Mn accumulation (D) and shoot Cu accumulation (E) of summer maize at V6 (six-leaf stage, n = 70), V12 (12-leaf stage, n = 54), R1 (silk emerging, n = 115), R3 (milk stage, n = 81) and R6 (physiological maturity, n = 149) stages, respectively. The bars represent the standard error of the mean.

number of observations, n = 32, mean yield 8.3 Mg ha^{-1}), N-opt (the number of observations, n = 30, mean yield 8.9 Mg ha^{-1}) and N-over (the number of observations, n = 57, mean yield 8.3 Mg ha^{-1}) (Table 1 and Table S1). The N-0 application rate resulted in significantly lower grain yield than the other three N application rates, while the N-opt treatment resulted in the highest grain yield. Compared to N-0 treatment, the higher grain yields with N application were mainly attributed to more biomass production and grains per ear. However, average ear numbers, thousand grain weight and HI were less affected by N rates (Table 2). Similarly, ZnRIE, FeRIE and MnRIE were also less affected by N rates. CuRIE was significantly lower in N-0 treatment compared to the other three N treatments. Grain concentrations of Zn (increased from 15.2 to 17.3 mg kg^{-1}), Fe (increased from 13.4 to 17.2 mg kg^{-1}), Mn (increased from 2.95 to 3.35 mg kg^{-1}) and Cu (increased from 1.24 to 1.57 mg kg^{-1}) showed increasing trends with the increase of N applied, although there were no significant differences between the N-opt and N-over treatments (Table 2). ZnHI and MnHI were less affected by N rates. FeHI increased from 19% to 25% with the increase of N levels from N-0 to N-opt while a further increase of N from N-opt to N-over treatments did not affect it. CuHI ranged from 21% to 26% and was inconsistently affected by N treatments (Table 2).

Before silking (R1) stage, with the increase of N levels from N-0 to N-opt treatment, the biomass production showed progressive enhancements. However, a further increasing N from N-opt to N-over treatment did not make an extra contribution to growth (Figure 4A). During the reproductive growth stages (e.g. R3 and R6), biomass production in the N-0 treatment was significantly lower than the other three N rates where there were no significant

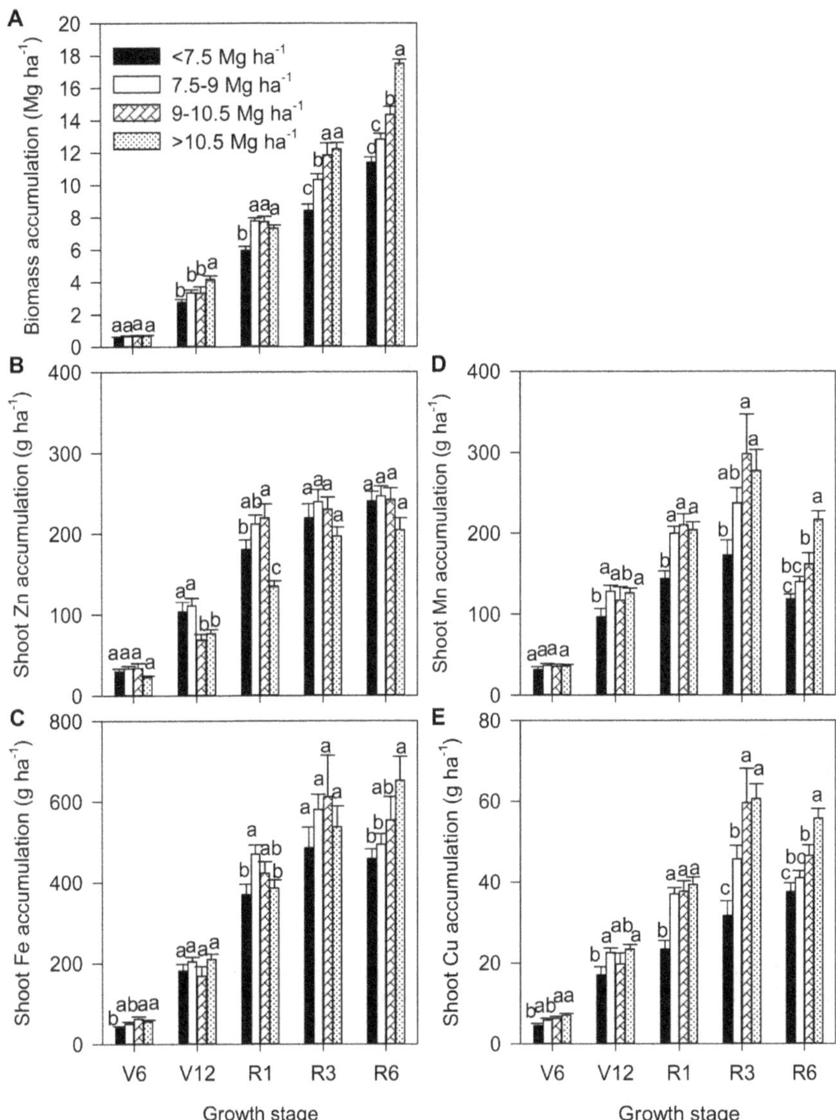

Figure 2. Dynamics of biomass (A), shoot Zn accumulation (B), shoot Fe accumulation (C), shoot Mn accumulation (D) and shoot Cu accumulation (E) of summer maize at V6 (six-leaf stage), V12 (12-leaf stage), R1 (silk emerging), R3 (milk stage) and R6 (physiological maturity) stages, respectively, with different yield ranges. The number of observations at each stage was shown in Table S2. The bars represent the standard error of the mean. Bars with different lowercase letters are significantly different at different yield ranges ($P<0.05$).

differences in biomass production (Figure 4A). Similar results were found in shoot Fe accumulation, as there were similar shoot Fe concentrations for each N rate at most of the growth stages (Figure 4C and Figure 5B). Similar results were also found for shoot Mn and Cu accumulation (Figure 4D, E). Shoot Mn and Cu concentrations were significantly lower in the N-0 treatment than the other three N treatments where shoot Mn and Cu concentrations were similar during the reproductive growth stages (Figure 5C, D). Shoot Zn accumulation tended to increase gradually with the increase of N from N-0 to N-opt throughout the growth stages, but there was a lower shoot Zn concentration in the N-low treatment compared to N-0 and N-opt treatments (Figure 4B and Figure 5A). A further increase in N from the N-opt to the N-over treatment did not affect shoot Zn accumulation as there was a similar biomass and shoot Zn concentration in both treatments (Figure 4B and Figure 5A).

With the increase of N rates from the N-0 to the N-over treatment, nearly 74%, 67%, 86% and 84% of pre-silking shoot Zn accumulation, 74%, 77%, 84% and 87% of pre-silking shoot Fe accumulation, 59%, 83%, 78% and 84% of pre-silking shoot Cu accumulation and more than 100% of pre-silking shoot Mn accumulation occurred compared to 48%, 54%, 54% and 57% of pre-silking biomass accumulation (Figure 4).

Discussion

Overall, maize grain yield ($n = 149$) averaged 8.0 Mg ha^{-1} with a range from 3.9 to 12.8 ha^{-1} (Table 1), which was 49% higher than the average yields in China (5.4 Mg ha^{-1}) and 61% higher than the world (5.0 Mg ha^{-1}) during 2006 to 2009 [4,32].

The relationships between grain yield and whole-plant micronutrient accumulation can be explained by studying the RIEs [19]. In this study, the average Zn, Fe, Mn and Cu RIEs were lower

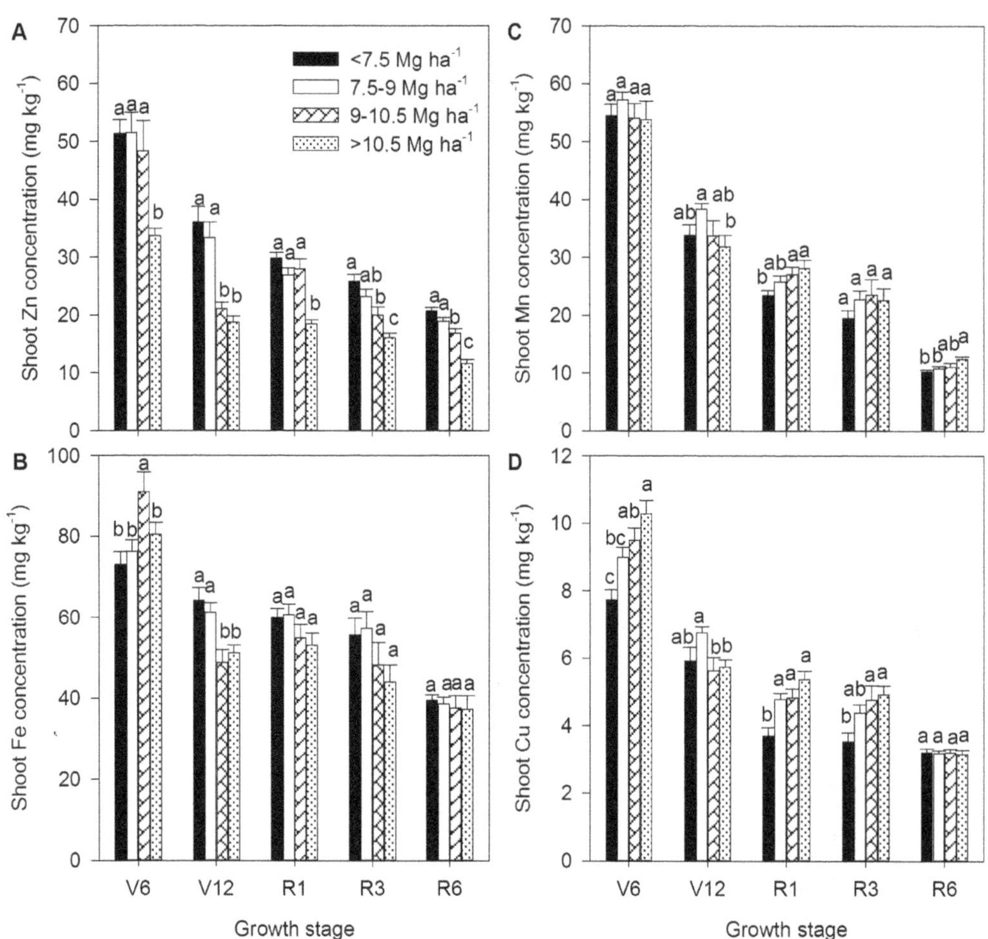

Figure 3. Dynamics of shoot Zn concentration (A) shoot Fe concentration (B), shoot Mn concentration (C) and shoot Cu concentration (D) of summer maize at V6 (six-leaf stage), V12 (12-leaf stage), R1 (silk emerging), R3 (milk stage) and R6 (physiological maturity) stages, respectively, with different grain yield ranges. The number of observations was shown in Table S2.The bars represent the standard error of the mean. Bars with different lowercase letters are significantly different at different yield ranges ($P<0.05$).

than corresponding values [17,19], but ZnRIE was comparable to that from Potarzycki [33] and Bender et al. [18], indicating that RIEs varied among different environments, cultivars, management regimes (such as N management) and yields [20,22,23]. With the increased N rate from N-0 to N-over treatments, there was 15% (non significant) increase in ZnRIE (Table 2). Potarzycki [33] also found a slight increase in ZnRIE with an increased supply of N from 115 to 175 kg N ha^{-1}. However, there was a significant 50% decrease in ZnRIE with the increase of grain yield from GY1 to GY4, although the data for GY4 excluded N-0 treatment (Tables 1 and 2). Similarly, with an increasing grain yield in the order YD<XY<ZD<FY, there was a significant 46% decrease in ZnRIE (Table 2). Furthermore, a significant negative relationship ($r = -0.58$***) between grain yield and ZnRIE was found (Table 3). Similarly, N reciprocal internal efficiency has also been found to decrease with increasing grain yields in spring maize and in winter wheat [22,29]. These results indicate yields and cultivars, indeed, exert a greater influence on ZnRIE than N management practices. The substantial decrease in the ZnRIE with the increase in grain yield was predominantly associated with an increase in ZnHI and a decrease of Zn concentrations in grain and especially in straw. Cultivars showed a similar effect on ZnRIE (Table 2). An increased ZnHI with increased grain yield was also reported [19]. Behera et al. [34] reported that ZnHI was only around 30% with 6.5 Mg ha^{-1} grain yield. These results suggest grains provide a large sink for Zn translocation.

Increased biomass production may be a driving force for the uptake and assimilation of mineral nutrients such as N, P and K [15]. Similarly, our results showed, with the increases of yield and biomass production, there was a concomitant increase in the accumulation of shoot Fe, Mn and Cu at physiological maturity (Figure 2C, D, E). Furthermore, with the increase in yield from GY2 to GY4, the declining proportions of pre-silking Fe (which decreased from 95% to 59%), Cu (which decreased from 90% to 71%) and biomass production (which decreased from 61% to 42%) were consistently found (Figure 2A, C, E and Figure S2). These results indicate the significance of increasing proportions of post-silking shoot Fe and Cu accumulation, possibly because of the significantly enhanced shoot biomass production by photosynthesis during reproductive development (which increased from 39% to 58%) with increasing yields. Due to the stable or decreasing shoot Fe and Cu accumulation between R3 and R6 (Figure 2C, E and Figure S2), more supplies of Fe and Cu during the vegetative stages (V12 to R1) and reproductive stages (R1 to R3) are necessary to maximize yields. Although shoot Mn accumulation at physiological maturity increased with the increase of yield, at least 94% of pre-silking shoot Mn accumulation (Figure 2D and Figure S2) indicates that more Mn should be applied before silking,

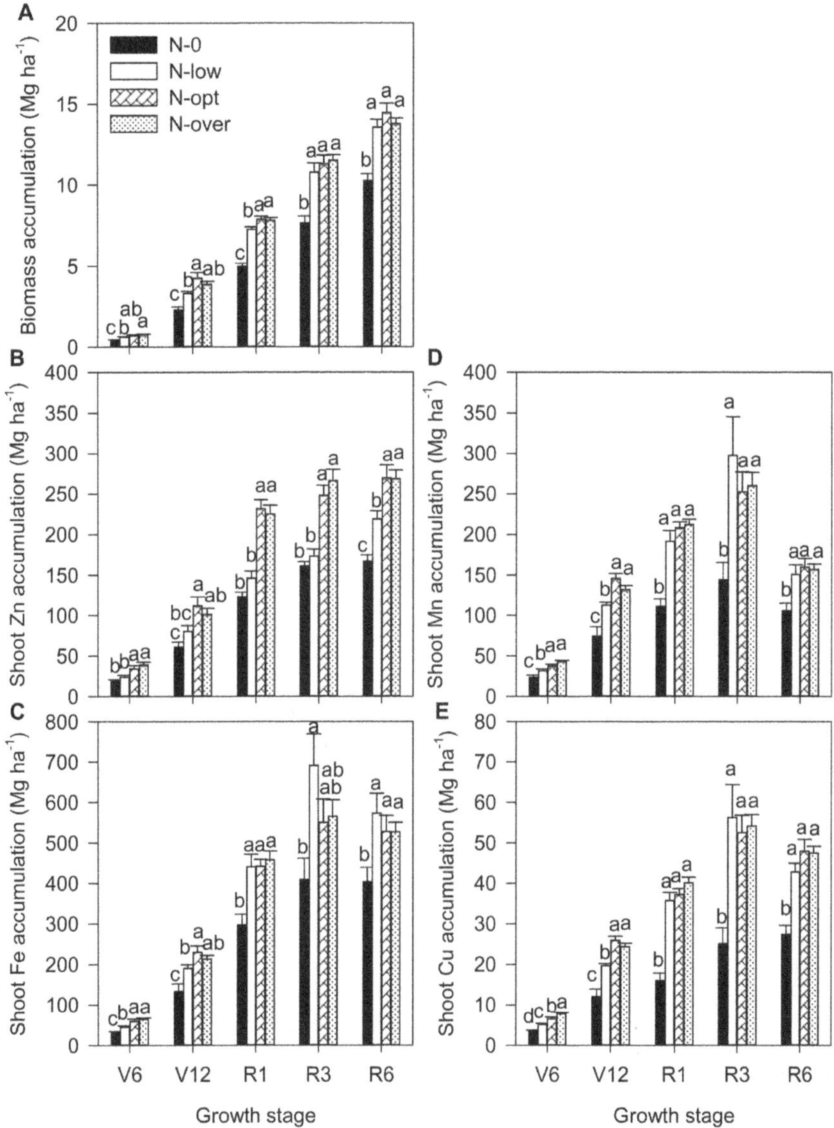

Figure 4. Dynamics of biomass (A), shoot Zn accumulation (B), shoot Fe accumulation (C), shoot Mn accumulation (D) and shoot Cu accumulation (E) of summer maize at V6 (six-leaf stage), V12 (12-leaf stage), R1 (silk emerging), R3 (milk stage) and R6 (physiological maturity) stages, respectively, with different N levels. The number of observations was shown in Table S2.The bars represent the standard error of the mean. Bars with different lowercase letters are significantly different at different N levels ($P<0.05$).

especially during V6 to V12 when the highest accumulation rate of shoot Mn generally occurred. Similarly, it has been reported that maximum Mn and Cu accumulation was achieved at stage R3, possibly due to losses during leaf senescence, leaf leaching and the low uptake rates resulting from root senescence during late-reproductive stages [19].

The similar or lower shoot Zn accumulation in GY4 compared to the other three yield ranges during all growth stages was at least in part due to the significantly lower shoot Zn concentration in GY4 compared to the other three yield ranges at each growth stage (Figure 2B and Figure 3A), probably due to a dilution effect as a result of the increased biomass production. Additionally, the shoot Zn concentration in GY4 was below the critical Zn-deficient range of 15–20 mg kg^{-1} [35], at almost all of the growing stages (Figure 3A), which indicates a potential Zn deficiency. Therefore, enhanced grain yield and biomass production may be involved in increasing Zn internal utilization efficiency (expressed by decreased ZnRIE) and Zn remobilization efficiency (expressed by increased ZnHI) at the cost of decreased Zn concentrations in straw and grain under potentially Zn deficient conditions. The higher plant Zn internal utilization efficiency in maize compared to legumes and rice was previously reported [36]. The higher Zn internal efficiency of maize with a greater yield and biomass production may be related to higher activity of carbonic anhydrase (CA) and Cu/Zn superoxide dismutase. Rengel [37] suggested that Zn-efficient wheat genotypes had an ability to maintain greater CA activity under Zn deficiency that could help maintain higher photosynthesis rates and dry matter production.

Although shoot Zn accumulation at physiological maturity was not significantly affected by the increased yield, the decreasing proportions of pre-silking shoot Zn accumulation (which decreased from 91% to 66%) together with the continued accumulation of shoot Zn throughout the growth stages with the increase of grain yield (Figure 2B and Figure S2) indicate the high reproductive-

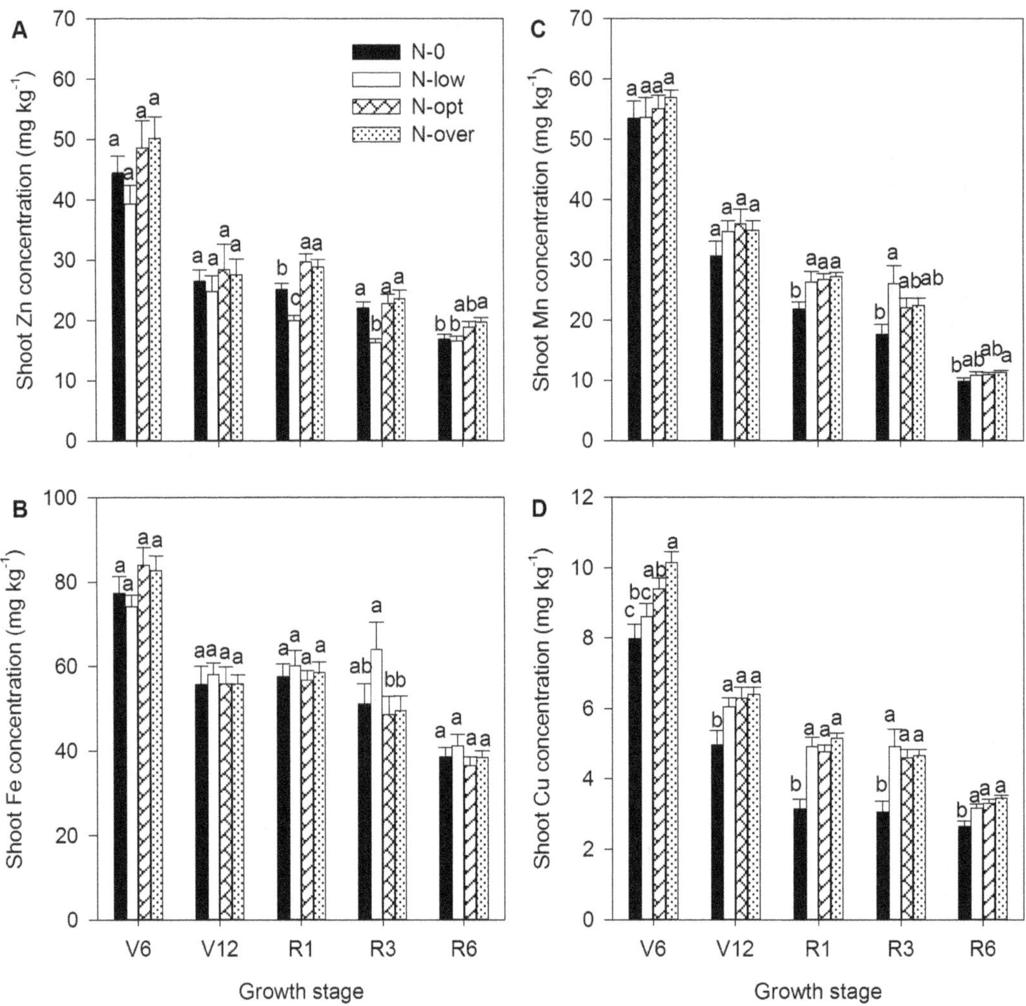

Figure 5. Dynamics of shoot Zn concentration (A) shoot Fe concentration (B), shoot Mn concentration (C) and shoot Cu concentration (D) of summer maize at V6 (six-leaf stage), V12 (12-leaf stage), R1 (silk emerging), R3 (milk stage) and R6 (physiological maturity) stages, respectively, with different N levels. The number of observations was shown in Table S2.The bars represent the standard error of the mean. Bars with different lowercase letters are significantly different at different N levels ($P<0.05$).

stage Zn demand. Further decreasing pre-silking accumulation of shoot Zn (48%) and Cu (45%) was found with 14.2 Mg ha^{-1} (15.5 of moisture) grain yield [18]. Grzebisz [38] also reported that vegetative tissues were only minor Zn sources while post-silking shoot Zn uptake from the soil during grain filling was the major source for growing kernels. Therefore, in order to gain higher grain yield and avoid a potential Zn deficiency, more Zn and season-long supply of Zn is necessary for maximum yields, especially during the V12 to R1 growth stages when the highest Zn accumulation rate occurs. The higher Zn and Fe accumulation rates during V12 to R1 may be because ear size and number of ovules are being established while the stalk and leaves are accumulating more photosynthates [15].

Excessive N fertilization in intensive agricultural areas of China has resulted in serious environmental problems. Currently, improving N management and developing related policies are major issues in crop production and environmental protection in China. In this study, increasing N supply from the N-0 to N-opt treatment tended to increase not only grain Zn, Fe, Mn and Cu concentrations, but also grain yields while further increased N from N-opt to N-over treatment did not result in an increase in grain concentrations but resulted in a decrease in yield (Tables 1 and 2). Similarly, there were also not any differences between N-opt to N-over treatments in shoot concentrations and the accumulation of micronutrients during any of the growth stages (Figures 4 and 5). These results indicate that optimized N management by reasonably decreasing N input was able to maintain plant nutrition of micronutrients for maximum yields as well as grain concentrations. In agreement with our results, Oktem et al. [39] previously reported the positive effects of N on Zn, Fe and Cu, but not Mn, concentrations in maize grain. Very recently, the positive effects of N on grain micronutrient concentrations (less pronounced for Zn) were also found [19]. It has been reported that application of Zn-enriched urea results in higher productivity and grain Zn concentrations in a rice–wheat cropping system than the same rate of $ZnSO_4$ and urea applied separately [40,41]. It is also convenient to apply a split application of Zn later to maintain season-long Zn supply for maize by overcoming the rapid fix on a calcareous soil because urea is often split applied. Therefore, an optimized Zn-enriched urea management strategy may be an economical method of improving Zn concentration in maize and maintaining sufficient Zn nutrition for higher grain yield.

Conclusions and Remarks

With increases in yield and biomass production, maize shoots contained more Fe, Mn and Cu at maturity but the pre-silking proportions of shoot Fe and Cu decreased. These results indicate that with increasing yield, more Fe and Cu would be needed, not only during the vegetative stages, but also reproductive stages (e.g. R1 to R3) for maximum yields. In contrast, more Mn should be applied before silking, especially during V6 to V12 when the highest accumulation rate of shoot Mn generally occurred. Shoot Zn accumulation was non-significant among the yield ranges possibly due to a dilution effect or a potential Zn deficiency, or because of the improvements in both Zn utilization efficiency (e.g. a decrease in ZnRIE) and Zn remobilization efficiency (e.g. an increase in ZnHI). Furthermore, the substantial decrease in ZnRIE with the increase of yield was largely associated with an increase in ZnHI, at the cost of the decreased Zn concentrations in grain and especially in straw. Cultivars had the similar effects on ZnRIE. Increasing N supply generally increased micronutrient concentrations of maize grain [19,39]. However, optimizing N management by reasonably decreasing N input did not result in any negative effects on plant and grain nutrition of micronutrients, while achieving the highest yield. These results showed that optimized N management is an applicable strategy to improve micronutrient nutrition for maximum yield.

Supporting Information

Figure S1 Changes in biomass and micronutrient (Zn, Fe, Mn and Cu) accumulation expressed as biomass and micronutrient accumulation at each stage divided by their corresponding values at maturity. V6: six-leaf stage; V12: 12-leaf stage; R1: silk emerging; R3: milk stage; R6: physiological; the number of observations was 70, 54, 115, 81 and 149 at V6, V12, R1, R3 and R6, respectively, as shown in Table S2.

Figure S2 Changes in biomass and micronutrient (Zn, Fe, Mn and Cu) accumulation expressed as biomass and micronutrient accumulation at each stage divided by their corresponding values at maturity for (A) yield <7.5 Mg ha^{-1}, (B) yield between 7.5 to 9 Mg ha^{-1}, (C) yield between 9 to 10.5 Mg ha^{-1}, (D) yield >10.5 Mg ha^{-1}. V6: six-leaf stage; V12: 12-leaf stage; R1: silk emerging; R3: milk stage; R6: physiological; the number of observations was 70, 54, 115, 81 and 149 at V6, V12, R1, R3 and R6, respectively, as shown in Table S2.

Table S1 N split supply as urea during the vegetative period of summer maize from 2008 to 2011based on determination of soil mineral N (N_{min}) of 0–90 cm at sowing, V5, V6, V10 and V12 stages in the field.

Table S2 Number of observations at different growth stages with different yield ranges and N levels.

Acknowledgments

The authors are grateful to Meng-Long Qiu and Zhong-Xiang Li from China agricultural university for their assistance with this work. The authors are also grateful to Tom Sizmur of Rothamsted Research, UK for his improvements to the English in the manuscript.

Author Contributions

Conceived and designed the experiments: CZ XC ZC. Performed the experiments: YX SY WZ DL YY. Analyzed the data: YX CZ. Contributed reagents/materials/analysis tools: CZ XC ZC YY. Wrote the paper: YX CZ.

References

1. Nuss ET, Tanumihardjo SA (2010) Maize: a paramount staple crop in the context of global nutrition. Compr Rev Food Sci F 9: 417–436.
2. Shiferaw B, Prasanna B, Hellin J, Bänziger M (2011) Crops that feed the world 6. Past successes and future challenges to the role played by maize in global food security. Food Sec 3: 307–327.
3. Chen XP, Cui ZL, Vitousek PM, Cassman KG, Matson PA, et al. (2011) Integrated soil-crop system management for food security. Proc Natl Acad Sci USA 108: 6399–6404.
4. FAO FAOSTAT–Agriculture Database. Available: http://faostat3.fao.org/faostat-gateway/go/to/download/Q/*/E. Accessed 2013 Sep 13.
5. Ju XT, Xing GX, Chen XP, Zhang SL, Zhang LJ, et al. (2009) Reducing environmental risk by improving N management in intensive Chinese agricultural systems. Proc Natl Acad Sci USA 106: 3041–3046.
6. Guo JH, Liu XJ, Zhang Y, Shen JL, Han WX, et al. (2010) Significant acidification in major Chinese croplands. Science 327: 1008–1010.
7. Le C, Zha Y, Li Y, Sun D, Lu H, et al. (2010) Eutrophication of lake waters in China: cost, causes, and control. Environ Manage 45: 662–668.
8. Zheng XH, Han SH, Huang Y, Wang YS, Wang MX (2004) Re-quantifying the emission factors based on field measurements and estimating the direct N_2O emission from Chinese croplands. Global Biogeochemical Cycles 18.
9. Cakmak I (2002) Plant nutrition research: priorities to meet human needs for food in sustainable ways. Plant Soil 247: 3–24.
10. Yang XE, Chen WR, Feng Y (2007) Improving human micronutrient nutrition through biofortification in the soil-plant system: China as a case study. Environ Geochem Health 29: 413–428.
11. Losak T, Hlusek J, Martinec J, Jandak J, Szostkova M, et al. (2011) Nitrogen fertilization does not affect micronutrient uptake in grain maize (*Zea mays* L.). Acta Agric Scand B Soil Plant Sci 61: 543–550.
12. Yu WT, Zhou H, Zhu XJ, Xu YG, Ma Q (2011) Field balances and recycling rates of micronutrients with various fertilization treatments in Northeast China. Nutr Cycl Agroecosyst 90: 75–86.
13. Feil B, Moser SB, Jampatong S, Stamp P (2005) Mineral composition of the grains of tropical maize varieties as affected by pre-anthesis drought and rate of nitrogen fertilization. Crop Sci 45: 516–523.
14. Ciampitti IA, Vyn TJ (2012) Physiological perspectives of changes over time in maize yield dependency on nitrogen uptake and associated nitrogen efficiencies: A review. Field Crops Res 133: 48–67.
15. Karlen DL, Flannery RL, Sadler EJ (1987) Nutrient and dry matter accumulation rates for high yielding maize. J Plant Nutr 10: 1409–1417.
16. Ciampitti IA, Camberato JJ, Murrell ST, Vyn TJ (2013) Maize nutrient accumulation and partitioning in response to plant density and nitrogen rate: I. Macronutrients. Agron J 105: 783–795.
17. Karlen DL, Flannery RL, Sadler EJ (1988) Aerial accumulation and partitioning of nutrients by corn. Agron J 80: 232–242.
18. Bender RR, Haegele JW, Ruffo ML, Below FE (2013) Nutrient uptake, partitioning, and remobilization in modern, transgenic insect-protected maize hybrids. Agron J 105: 161–170.
19. Ciampitti IA, Vyn TJ (2013) Maize nutrient accumulation and partitioning in response to plant density and nitrogen rate: II. Calcium, magnesium, & micronutrients. Agron J. 105:1645–1657.
20. Setiyono TD, Walters DT, Cassman KG, Witt C, Dobermann A (2010) Estimating maize nutrient uptake requirements. Field Crops Res 118: 158–168.
21. Hirel B, Le Gouis J, Ney B, Gallais A (2007) The challenge of improving nitrogen use efficiency in crop plants: towards a more central role for genetic variability and quantitative genetics within integrated approaches. J Exp Bot 58: 2369–2387.
22. Hou P, Gao Q, Xie RZ, Li SK, Meng QF, et al. (2012) Grain yields in relation to N requirement: optimizing nitrogen management for spring maize grown in China. Field Crops Res 129: 1–6.
23. Zhang Y, Hou P, Gao Q, Chen XP, Zhang FS, et al. (2012) On-farm estimation of nutrient requirements for spring corn in North China. Agron J 104: 1436–1442.
24. Mortvedt JJ (1994) Needs for controlled-availability micronutrient fertilizers. Fertilizer Res 38: 213–221.
25. Banziger M, Long J (2000) The potential for increasing the iron and zinc density of maize through plant-breeding. Food Nutr Bull 21: 397–400.

26. Xue YF, Yue SC, Zhang YQ, Cui ZL, Chen XP, et al. (2012) Grain and shoot zinc accumulation in winter wheat affected by nitrogen management. Plant Soil 361: 153–163.
27. Cui ZL, Zhang FS, Chen XP, Miao YX, Li JL, et al. (2008) On-farm evaluation of an in-season nitrogen management strategy based on soil Nmin test. Field Crops Res 105: 48–55.
28. Zhang SS, Yue SC, Yan P, Qiu ML, Chen XP, et al. (2013) Testing the suitability of the end-of-season stalk nitrate test for summer corn (*Zea mays* L.) production in China. Field Crops Res 154:153–157.
29. Yue SC, Meng QF, Zhao RF, Ye YL, Zhang FS, et al. (2012) Change in nitrogen requirement with increasing grain yield for winter wheat. Agron J 104: 1687–1693.
30. Meng QF, Yue SC, Chen XP, Cui ZL, Ye YL, et al. (2013) Understanding dry matter and nitrogen accumulation with time-course for high-yielding wheat production in China. PLoS One 8: e68783.
31. Sadras VO (2006) The N:P stoichiometry of cereal, grain legume and oilseed crops. Field Crops Res 95: 13–29.
32. Wang TY, Ma XL, Li Y, Bai DP, Liu C, et al. (2011) Changes in yield and yield components of single-cross maize hybrids released in China between 1964 and 2001. Crop Sci 51: 512–525.
33. Potarzycki J (2010) The impact of fertilization systems on zinc management by grain maize. Fertilizers Fertilization. 39: 78–89.
34. Behera SK, Shukla AK, Wanjari RH, Singh MV (2011) Influence of different sources of zinc fertilizer on yield and zinc nutrition of maize (*Zea mays* L.). 3rd International Zinc Symposium: improving crop production and human health; October 2011; India. Available: http://www.Zinccrops2011.Org/presentations/2011-zinccrops2011-behera-1-abstract.pdf.
35. Marschner (2011) Mineral nutrition of higher plants, 3rd edn. Academic, London.
36. Fageria NK, Barbosa MP, Santos AB (2008) Growth and zinc uptake and use efficiency in food crops. Commun Soil Sci Plant Anal 39: 2258–2269.
37. Rengel Z (1995) Carbonic anhydrase activity in leaves of wheat genotypes differing in Zn efficiency. J Plant Physiol 147: 251–256.
38. Grzebisz W (2008) Effect of Zinc foliar application at an early stage of maize growth on patterns of nutrients and dry matter accumulation by the canopy. Part II. Nitrogen uptake and dry mattern accumulation patterns. J Elementol 13: 17–39.
39. Oktem A, Oktem AG, Emeklier HY (2010) Effect of nitrogen on yield and some quality parameters of sweet corn. Commun Soil Sci Plant Anal 41: 832–847.
40. Shivay Y, Kumar D, Prasad R (2008) Effect of zinc-enriched urea on productivity, zinc uptake and efficiency of an aromatic rice–wheat cropping system. Nutr Cycl Agroecosyst 81: 229–243.
41. Shivay Y, Kumar D, Prasad R, Ahlawat IPS (2008) Relative yield and zinc uptake by rice from zinc sulphate and zinc oxide coatings onto urea. Nutr Cycl Agroecosyst 80: 181–188.

Comparison of Soil Respiration in Typical Conventional and New Alternative Cereal Cropping Systems on the North China Plain

Bing Gao[1], Xiaotang Ju[1]*, Fang Su[1], Fengbin Gao[1], Qingsen Cao[1], Oene Oenema[2], Peter Christie[1], Xinping Chen[1], Fusuo Zhang[1]

1 College of Resources and Environmental Sciences, China Agricultural University, Beijing, China, **2** Wageningen University and Research Center, Alterra, Wageningen, The Netherlands

Abstract

We monitored soil respiration (Rs), soil temperature (T) and volumetric water content (VWC%) over four years in one typical conventional and four alternative cropping systems to understand Rs in different cropping systems with their respective management practices and environmental conditions. The control was conventional double-cropping system (winter wheat and summer maize in one year - Con.W/M). Four alternative cropping systems were designed with optimum water and N management, i.e. optimized winter wheat and summer maize (Opt.W/M), three harvests every two years (first year, winter wheat and summer maize or soybean; second year, fallow then spring maize - W/M-M and W/S-M), and single spring maize per year (M). Our results show that Rs responded mainly to the seasonal variation in T but was also greatly affected by straw return, root growth and soil moisture changes under different cropping systems. The mean seasonal CO_2 emissions in Con.W/M were 16.8 and 15.1 Mg CO_2 ha^{-1} for summer maize and winter wheat, respectively, without straw return. They increased significantly by 26 and 35% in Opt.W/M, respectively, with straw return. Under the new alternative cropping systems with straw return, W/M-M showed similar Rs to Opt.W/M, but total CO_2 emissions of W/S-M decreased sharply relative to Opt.W/M when soybean was planted to replace summer maize. Total CO_2 emissions expressed as the complete rotation cycles of W/S-M, Con.W/M and M treatments were not significantly different. Seasonal CO_2 emissions were significantly correlated with the sum of carbon inputs of straw return from the previous season and the aboveground biomass in the current season, which explained 60% of seasonal CO_2 emissions. T and VWC% explained up to 65% of Rs using the exponential-power and double exponential models, and the impacts of tillage and straw return must therefore be considered for accurate modeling of Rs in this geographical region.

Editor: Xiujun Wang, University of Maryland, United States of America

Funding: This work was funded by the National Natural Science Foundation of China (41230856, 31172033), the Special Fund for the Agricultural Profession (201103039), the '973' Project (2009CB118606) and the Innovation Group Grant of the National Natural Science Foundation of China (31121062). The funders had no role in study design, data collection and analysis, decision to publish, or preparation of the manuscript.] into [This work was funded by the '973' Project (2012CB417105, 2009CB118606), the National Natural Science Foundation of China (41230856, 31172033), the Special Fund for the Agricultural Profession (201103039), China Postdoctoral Science Foundation (2013M 530778), and the Innovation Group Grant of the National Natural Science Foundation of China (31121062). The funders had no role in study design, data collection and analysis, decision to publish, or preparation of the manuscript.

Competing Interests: The authors have declared that no competing interests exist.

* E-mail: juxt@cau.edu.cn

Introduction

Soils provide a very large sink of carbon (C) in terrestrial ecosystems with C reserves of about 1500 Pg C (1 Pg = 10^{15} g) and make a major contribution to the global carbon equilibrium [1]. Slight changes in soil C might therefore lead to significant changes in the concentration of CO_2 in the atmosphere. Soil respiration is the main terrestrial source of C return to the atmosphere with a flux reaching 98±12 Pg C in 2008 and increasing at a rate of 0.1 Pg C y^{-1} from 1989 [2]. Agricultural soils play a very important role in the global C cycle [3,4] and account for 11% of global anthropogenic CO_2 emissions [5]. It is therefore important to minimize soil respiration and retain more C sequestered in agricultural soils.

Soil respiration comprises mainly autotrophic respiration by plant roots and heterotrophic respiration of plant residues, root litter and exudates, and soil organic matter by soil microorganisms [6,7]. Its magnitude is affected mainly by soil and climatic conditions [8] such as soil temperature and moisture [1,9,10], vegetation characteristics and management practices [11–14]. Soil respiration therefore shows high spatial and temporal variation [1]. Understanding this variation in different cropping systems in specific region will make a large contribution to the efficient management of C flow in agricultural ecosystems.

Soil respiration in cropland is greatly affected by tillage practices and straw management, with the greatest increase occurring immediately after tillage operations, and cumulative soil CO_2 emissions can be lowered significantly by reducing the intensity of tillage [15,16]. Daily CO_2 fluxes can differ significantly at some sampling dates between conventional moldboard plow tillage and no tillage in continuous corn [17]. Soil CO_2 emission can be enhanced in the short term after crop residues are returned to the field [15,18] but this practice may build up the soil organic carbon (SOC) pools in the long term and may therefore be regarded as a more sustainable way of managing SOC compared to straw burning or other uses for straw [19]. Differences in fertilizer N rates had no significant effect on the CO_2 exchange rates in the same crop rotation and CO_2 fluxes did not differ with crop rotation under no till practices [16]. In addition, crop species and/or other management practices affect soil CO_2 emission as a result of their influence on soil biological and biochemical properties [14,20].

Soil temperature and moisture are two of the most important environmental factors controlling soil respiration [1,21,22]. Soil temperature is significantly positively correlated with soil respiration using linear [7], exponential [1,7], improved Arrhenius [8], power and quadratic [9] and Q_{10} [10] models in different regions. Soil moisture is also a key factor controlling soil respiration, especially in arid or semiarid regions where it can be more important than temperature and become the dominant factor [11]. This shows that when one factor linking soil temperature and moisture is in a higher or lower range, the other might become a major factor controlling soil respiration [13,23,24]. The respiration rate will be limited when soil volumetric water content (VWC%) drops below a threshold of 15% [20]. Soil CO_2 emission increased significantly with increasing temperature up to 40°C, with emissions reduced at the lowest and highest soil moisture contents [20]. Therefore, the single-factor models cannot describe soil CO_2 emission well because they neglect the impacts of interactions between factors. The multiple polynomial models considering both soil temperature and moisture result in a much better description of CO_2 ($r^2 = 0.70$–0.78, $P<0.0001$) emissions than using temperature ($r^2 = 0.27$–0.54, $P<0.01$) or moisture ($r^2 = 0.29$–0.45, $P<0.01$) alone [20].

China has broad climatic regimes and the different ecosystems depend on regional climatic conditions [25]. The North China Plain (NCP) is a major agricultural region. The soil type is Fluvo-aquic soil and the climate is sub-humid temperate monsoon with abundant solar radiation but with cold and dry conditions in winter and spring and warm and wet weather in summer. Evapotranspiration is intense and the spring drought is an important feature. Winter wheat-summer maize is the typical double cropping system and current farming practice involves application of 300 kg N ha^{-1} yr^{-1} for winter wheat and 250 kg N ha^{-1} yr^{-1} for summer maize with a ratio of basal to topdressing applications of 1:1 and 1:1.5, respectively [26,27]. The soil is rotary tilled to 20 cm depth after maize straw removal for sowing winter wheat, and maize is sown directly after removing the wheat straw. Generally, wheat is irrigated three to four times and maize once or twice depending on precipitation. The amount of irrigation water ranges from 60 to 100 mm on each occasion [26]. About 30–60% of N input could be saved without sacrificing yields while significantly reducing environmental risk by adopting optimum N management in the winter wheat-summer maize system as shown by our earlier study [28]. However, over-exploitation of groundwater has become the main factor restricting sustainable agricultural development [29]. There is therefore concern to explore new alternative cropping systems for sustainable use of groundwater and optimum N fertilization to reduce pollution. Winter wheat–summer maize–spring maize with three harvests over two years and a single spring maize system have shown great potential to reduce water use and N use and can achieve balanced use of groundwater [26], and this cropping system may serve as a new alternative system for efficient resource use and sustainable development. However, it is still unclear how these changes will affect soil respiration in the study region.

Low frequency of measurement, lack of data at some growth stages, and failure to consider the interactive effects of soil moisture and temperature on soil respiration may lead to failure to describe the characteristics of soil respiration in this region [18,30,31]. There are indications that the correlation between soil respiration and soil temperature to 5 cm depth is 0.51 but the study that produced this result involved measurement only 21 times over one year [30]. Meng et al. [31] found that soil respiration had a higher correlation with soil temperature to 5 cm depth using the exponential model through weekly measurements of soil respiration under the typical double-cropping system over a whole year. Soil temperature at 5 cm depth explained 63–74% of soil respiration using the exponential model except during the winter, and the application of crop residues had significant positive impacts on soil respiration [18]. The management of N and water, crop residues and tillage practices will change significantly after conversion to new alternative cropping systems [26], an effect closely related to soil respiration. However, no quantitative information is yet available regarding soil respiration in new alternative cropping systems in this region.

In the present study we have compared soil respiration characteristics in different cropping systems with their respective management practices and environmental variables and we explore the factors affecting these differences. We have also analyzed the effects of straw return on variation in seasonal CO_2 emissions on the North China Plain.

Materials and Methods

Site description

A long-term field experiment was set up in October 2007 at Quzhou experimental station (36.87°N, 115.02°E) of China Agricultural University in Hebei province. The site is a sub-humid temperate monsoon area at an altitude of 40 m. The annual mean temperature is 13.2°C. Annual mean precipitation was 494 mm from 1980 to 2010 with a range of 213–840 mm, and 68% of precipitation falls from June to September [26]. The typical double-cropping system is a winter wheat and summer maize rotation which accounts for >80% of agricultural fields in Quzhou county. The soil type is Fluvo-aquic soil and the bulk density of the top 30 cm of the soil profile is 1.37 g cm^{-3}, soil pH is 7.72 (1:2.5, soil:water), SOC content 7.31 g kg^{-1}, total N 0.7 g kg^{-1}, Olsen-P 4.8 mg kg^{-1} and available K 72.7 mg kg^{-1}. Fig. 1 shows the daily mean air temperatures and precipitation during the measurement period (also see Table S1).

Field experiment treatments and management

A completely randomized design was employed with five treatments and four replicates. Each plot is 1800 m^2 (30×60 m). The control is conventional winter wheat and summer maize based on local farming practice (Con.W/M). Four new alternative cropping systems were designed with high-yielding varieties (using optimum planting density and crop management) and optimum water and N fertilizer management compared with conventional practice. They are: optimized two harvests in one year (winter wheat and summer maize - Opt.W/M), three harvests within two years (first year, winter wheat and summer maize or winter wheat and summer soybean; second year fallow then spring maize - W/M-M and W/S-M) and single spring maize per year (M).

Nitrogen input and irrigation for Con.W/M were described in the Introduction above. The basal fertilizer for wheat was surface broadcast before rotary tillage to 20 cm depth after removal of maize straw from the soil and topdressing was broadcast at shooting for wheat followed by irrigation, with both fertilizer

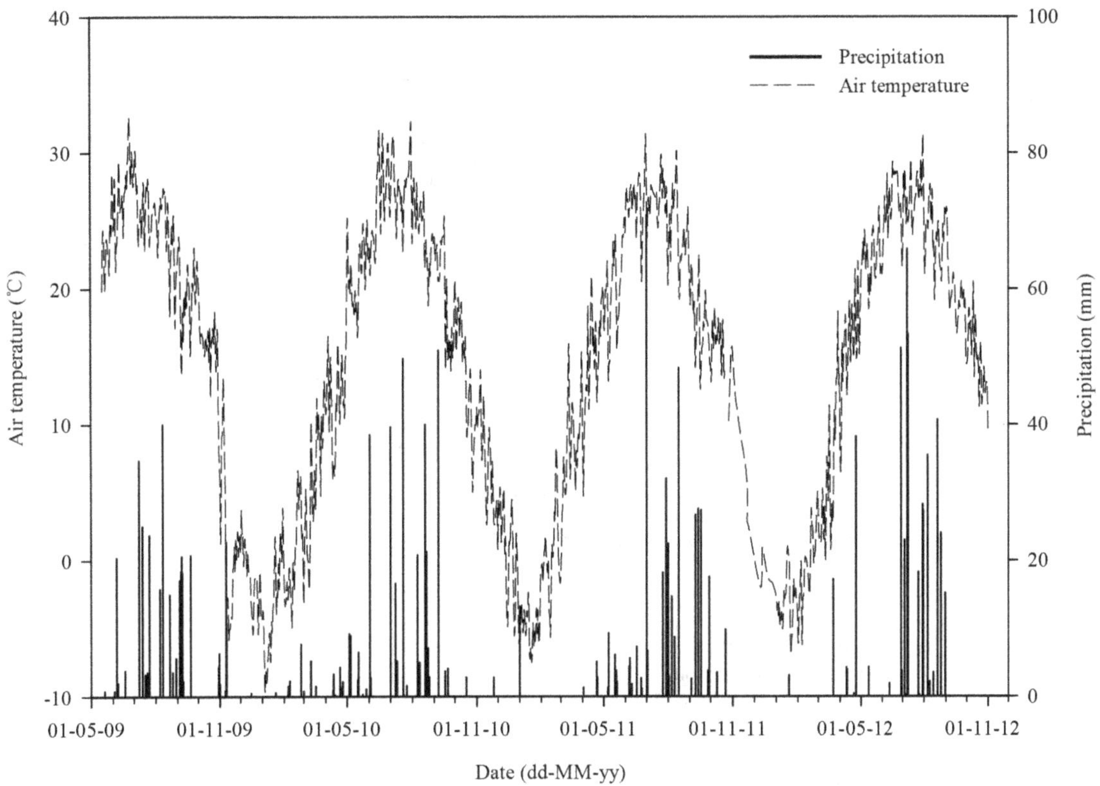

Figure 1. Daily mean air temperature (°C) and precipitation (mm) during the field experiment.

applications at 150 kg N ha^{-1} in the form of urea. The basal application for summer maize comprised 45 kg N ha^{-1} applied to the soil as 15-15-15 compound fertilizer with a seed drill after removing the wheat straw from the soil, and 55 kg N ha^{-1} surface broadcast as urea followed by irrigation and the topdressing 150 kg N ha^{-1} was applied at the ten-leaf stage of summer maize in the form of urea. In the other systems optimized N management was devised according to the N target values minus the soil nitrate-N content in the root zone before side-dressing as described by Cui et al. [27]. For summer maize 45 kg N ha^{-1} was applied as a basal dressing in the same way as for Con.W/M and 80 and 60 kg N ha^{-1} were side-dressed using a soil cover of 0–5 cm after band application at the six- and ten-leaf stages of summer maize, respectively. No other N fertilizer except 45 kg N ha^{-1} was applied as a basal application for soybean as for Con.W/M. Irrigation times and rates were determined by testing the soil water content before the critical growing seasons as described by Meng et al. [26]. The details of nitrogen input and irrigation rate over the whole study are shown in Table 1. Wheat straw was mulched after chopping into 5–10 cm pieces and summer maize or soybean was sown directly. Summer and spring maize and soybean residues were also chopped into 5–10 cm pieces and mechanically ploughed into the top 30 cm of the soil after maize and soybean were harvested and then winter wheat was sown if there was no fallow the following season. The soil was rotavated to 20 cm depth before sowing spring maize.

We measured soil respiration in each plot of the experiment from May 2009 to October 2012. Crops present in the different treatments during gas measurement are shown in Fig. 2.

Soil respiration measurement

Soil respiration was representatively determined in every plot using an automatic soil CO_2 flux system (LI-COR LI-8100, Lincoln, NE). Measurements were carried out daily for 10 days after fertilization events and 3–5 days after irrigation or precipitation events (>10 mm) depending on the size of gas fluxes; for the remaining periods emissions were measured twice per week and once a week when the soil was frozen. Two bases were used in each plot, one on a row and the other in the middle of the row during the maize and soybean seasons. Each base was a PVC tube with an inner diameter of 20 cm and a height of 13 cm inserted 9 cm into the soil for measurement and was removed only before sowing. Soil respiration was measured directly by LI-8100 in units of μmol CO_2 m^{-2} s^{-1} in the field between 08:30 and 11:00 am. Soil respiration is presented as the mean values of four replicated measurements on four different plots. The seasonal amounts of CO_2 emissions were sequentially linearly determined from the emissions between every two adjacent intervals of the measurements.

Auxiliary measurements

Soil temperature to 5 cm depth was measured directly by Li-8100 through a temperature sensing probe during the measurement time. Soil moisture at 0–5 cm is expressed as volumetric water content (VWC%) and was measured directly by Li-8100 through an ECH_2O type of EC-10 soil water sensing probe (Decagon Devices, Inc, Pullman, WA). We also measured the top 20 cm depth SOC content in each plot of this field experiment after summer maize harvest in 2011 using the method described by Huang et al. [19]. The daily mean air temperatures and precipitation data during the field experiment were obtained from

Table 1. Nitrogen fertilizer rates and irrigation rates throughout the study period.

Year	N application rate (kg N ha^{-1})					Irrigation rate (mm)				
	Con.W/M[1]	Opt.W/M	W/M-M	W/S-M	M	Con.W/M	Opt.W/M	W/M-M	W/S-M	M
2009	W[2] 300	W 263	F -[3]	F -	F -	W 250	W 215	F -	F -	F -
	M_1 250	M_1 185	M_2 135	M_2 210	M_2 95	M_1 60	M_1 60	M_2 125	M_2 135	M_2 125
2010	W 300	W 100	W 140	W 140	F -	W 180	W 120	W 120	W 120	F -
	M_1 250	M_1 185	M_1 185	S 45	M_2 150	M_1 60	M_1 60	M_1 60	S 60	M_2 110
2011	W 300	W 139	F -	F -	F -	W 240	W 275	F -	F -	F -
	M_1 250	M_1 185	M_2 162	M_2 178	M_2 150	M_1 70	M_1 70	M_2 60	M_2 60	M_2 60
2012	W 300	W 140	W 162	W 158	F -	W 180	W 160	W 160	W 170	F -
	M_1 250	M_1 185	M_1 185	S 45	M_2 266	M_1 90	M_1 90	M_1 90	S 90	M_2 120
Total	2200	1382	969	776	661	1130	1050	615	635	415

[1]Con.W/M, Opt.W/M, W/M-M, W/S-M and M represent conventional and optimized winter wheat–summer maize, winter wheat–summer maize–spring maize, winter wheat–summer soybean–spring maize and spring maize treatment, respectively.
[2]W, M_1, M_2, S and F represent winter wheat, summer maize, spring maize, summer soybean and fallow.
[3]Denotes no data in the fallow season.

an automatic weather station located 50 m from our experimental site as shown in Fig. 1. Soil respiration and environmental variable data from the present study are presented in the Supplementary Data (Table S1).

Correlations between soil respiration and soil temperature and moisture

The compound factor models of soil respiration with soil temperature and moisture (equations 1–4) were employed as follows:

$$Rs = a + bTs + cWs \quad (1)$$

$$Rs = aTs^{b}Ws^{c} \quad (2)$$

$$Rs = ae^{bTs}Ws^{c} \quad (3)$$

$$Rs = ae^{bTs+cWs} \quad (4)$$

We established the four compound factor models above among soil respiration (Rs), soil temperature (Ts) and VWC(%) (Ws) using the measured fluxes from May 2009 to May 2012, and compared MAE (the mean absolute error), ME (model efficiency, the ratio of difference in measured and predicted flux in total variation in measured flux, expressed as significant correlation coefficient from −1 to 1), d (the percentage of mean square error and potential error, expressed as significant correlation from 0 to 1) [32,33], RMSE (root mean square error, reflecting the degree of dispersion of one variable), MSE_s (systematic error) and MSE_u (random error) [34] among the models. We comprehensively evaluated the model performances by the sizes of MAE, ME, d, RMSE, MSE_s and MSE_u, and the value of $MSE_s/(MSE_u+MSE_u)$. In general, MSE_u is close to RMSE in a well fitting model.

Statistical analysis

The primary data were processed using Microsoft Excel 2003 spreadsheets. Total CO_2 emissions in the different treatments were tested by analysis of variance and mean values were compared using SAS statistical software (Version 9.2; SAS Institute, Inc., Cary, NC) to calculate least significant difference (LSD) at the 5% level. Compound factor regression analysis among soil respiration, T and VWC% were performed using Sigmaplot 12.0 (Systat Software Inc., Erkrath, Germany).

Results

Characteristics of soil respiration in the different cropping systems

Over a complete rotation cycle soil respiration gradually increased from March, reached a maximum in July and gradually decreased from August to November, and then remained at the lowest values during winter, in a pattern similar to soil temperature (Figs. 2 and 3A). The mean soil respiration values were 3.35, 4.55, 4.03, 3.35 and 3.25 μmol CO_2 m^{-2} s^{-1} for Con.W/M, Opt.W/M, W/M-M, W/S-M and M throughout the study period, with ranges of 0.02–12.4, 0.26–14.9, 0.31–12.1, 0.34–11.3 and 0.30–11.2 μmol CO_2 m^{-2} s^{-1}, respectively. Three peaks per year occurred in the typical double-cropping system, at the shooting stage of winter wheat, six-leaf of summer maize and the period after winter wheat sowing, the first two peaks caused by rapid crop growth and the last by the return of summer maize straw combined with soil tillage. Soil respiration of Opt.W/M was higher than of Con.W/M at the six-leaf stage of summer maize in the middle of July and the period after winter wheat sowing. The maximum peaks of soil respiration in Con.W/M were 8.2, 7.7, 7.8, 12.4 and 4.9, 2.8, 3.6, 2.9 μmol CO_2 m^{-2} s^{-1} during these two periods for four growing seasons, respectively, and they increased to 10.8, 9.6, 10.1, 14.9 and 7.6, 10.3, 6.5, 11.2 μmol CO_2 m^{-2} s^{-1} in Opt.W/M during the corresponding periods. Under the new alternative cropping systems one peak disappeared in the fallow season (season with no winter wheat planted). The highest value of soil respiration was around 7.0 μmol CO_2 m^{-2} s^{-1} in the spring maize season under the new alternative cropping systems, but it increased to more than 10.0 μmol CO_2 m^{-2} s^{-1} for summer maize in Opt.W/M at the corresponding time (Fig. 2).

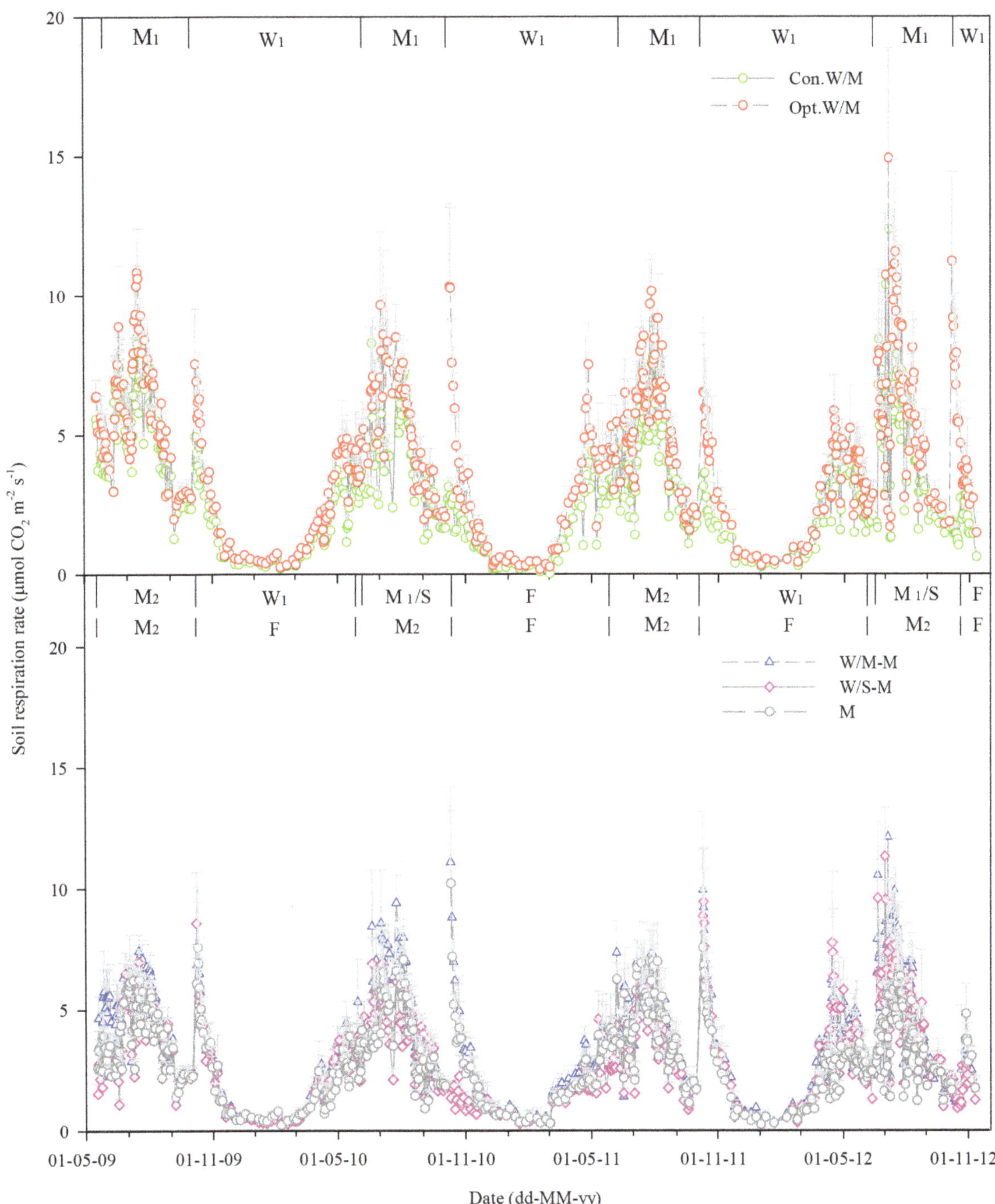

Figure 2. CO_2 emissions of different cropping systems. Con.W/M, Opt.W/M, W/M-M, W/S-M and M represent conventional winter wheat–summer maize in one year, optimized winter wheat–summer maize in one year, winter wheat–summer maize (or summer soybean) –spring maize three harvests in two years and single spring maize system in one year; W, M_1, M_2, S and F represent winter wheat, summer maize, spring maize, soybean and fallow.

Soil respiration was very low even after summer soybean stover return to the field in W/S-M in mid-November 2010 when the soil temperature in the top 5 cm ranged from −2.3 to +4.7°C within a month of soil tillage. A similar phenomenon occurred at the end of October 2012 due to the late spring maize and summer soybean harvests and the soil was tilled when soil temperature to 5 cm depth was around 10°C, and the peaks of W/M-M, W/S-M and M were only one third of the values of those at the corresponding times in other years. In addition, soil respiration showed large between-year change, so that peaks of soil respiration occurred after irrigation at shooting of winter wheat in other years, but not in winter wheat in 2010.

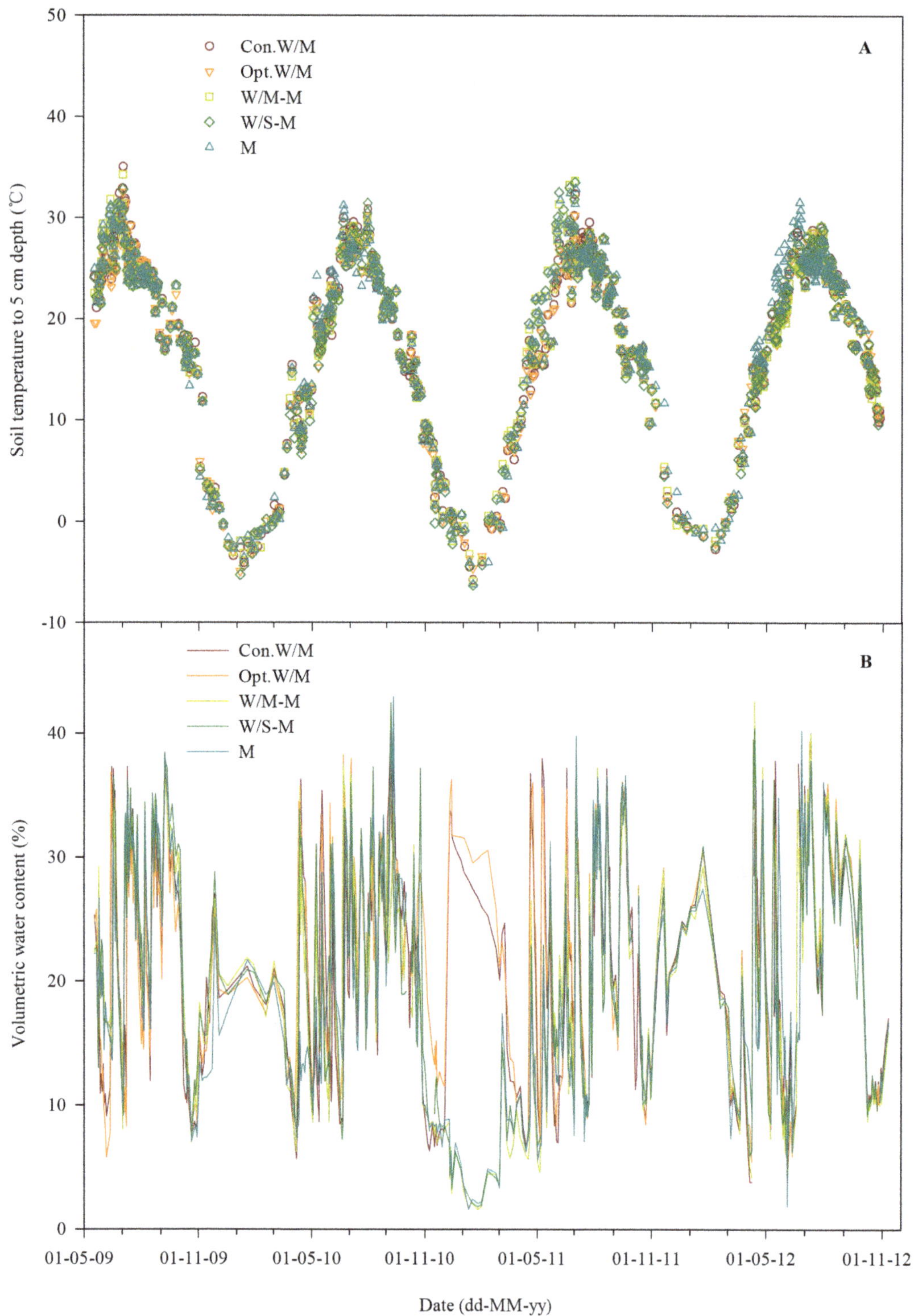

Figure 3. Dynamics of (A) soil temperature and (B) soil VWC% to 5 cm depth.

Total CO_2 emissions in each cropping season and each rotation cycle

Total CO_2 emissions in each cropping season and each rotation cycle were system dependent (Table 2). The mean seasonal total CO_2 emissions of Con.W/M were 16.8 and 15.1 Mg CO_2 ha^{-1} for summer maize and winter wheat, respectively. They increased significantly by 26 and 35% in Opt.W/M in the corresponding season. Under the new alternative cropping systems W/M-M showed similar results to Opt.W/M, and the seasonal total CO_2 emission of W/M-M was significantly higher than the corresponding season of Con.W/M except spring maize in 2009. However, W/S-M showed no significant difference from Con.W/M in each cropping season and the total CO_2 emissions in the fallow season and spring maize of W/S-M were clearly affected by summer soybean planting. Total CO_2 emission of M in each cropping season also showed no clear difference from Con.W/M except spring maize in 2011. In order to compare the impacts of cropping systems on CO_2 emissions of each rotation cycle we calculated the total CO_2 emissions during the period 2011–2012, which included two rotation cycles of Con.W/M, Opt.W/M, and M and a completely rotation cycle of W/M-M and W/S-M. Total CO_2 emission of Con.W/M was 61.9 Mg CO_2 ha^{-1}, and increased significantly by 37 and 29% in Opt.W/M and W/M-M treatment, respectively. The total CO_2 emission of W/M-M was not significantly different from Opt.W/M when there was only one season of winter wheat in two years, but total CO_2 emission of W/S-M decreased sharply in contrast to Opt.W/M when summer soybean was planted to replace summer maize of W/M-M because soil respiration was reduced significantly in the following fallow and spring maize seasons after the low biomass of soybean straw was returned to the soil. Total CO_2 emissions expressed as one complete rotation cycle of W/S-M, Con.W/M and M treatments were not significantly different (Table 2).

Soil respiration as affected by C input in each growth season

The measured soil respiration rates in this study consisted mainly of autotrophic respiration by crop roots in the current season, heterotrophic respiration of root litter and exudates in the current season, and heterotrophic respiration of crop straw return to the soil from the previous season and soil organic matter. As Fig. 2 and Table 2 show, the characteristics and total seasonal cumulative CO_2 emissions were greatly affected by straw return and crop growth status. To further explain soil respiration driven by C input in each growing season, we analyzed the correlation of seasonal cumulative CO_2 emissions with: (1) current-season aboveground biomass only; (2) the sum of C input of straw return from the previous season and the aboveground biomass in the current season. The relationship is improved significantly by inclusion of straw inputs (Fig. 4, equation A) compared to current-season aboveground biomass only (Fig. 4, equation B). Carbon input of straw return from the previous season and the aboveground biomass in the current season explains up to 60% of seasonal cumulative CO_2 emissions, much higher than that of 27% with current-season aboveground biomass only, which demonstrates that straw C inputs from the previous season can significantly affect soil respiration.

Correlation between soil respiration and soil temperature and VWC% to 5 cm depth

Large changes in soil respiration followed the variation in temperature over a complete year (Figs. 2 and 3A). Soil temperature explained 45% of soil respiration using the quadratic model (Fig. 5, equation A). In addition, soil moisture exerted some impacts on soil respiration under our climatic conditions such as inhibition within a short period after irrigation at shooting and grain filling stages of winter wheat and then a sharp increase which was derived from the effects of drying and wetting cycles. Soil respiration showed significant correlations with soil VWC% using the linear ($R_s = 2.7535+0.0447V$, $R^2 = 0.04$, $n = 2282$) and power ($R_s = 1.7708V^{0.2488}$, $R^2 = 0.06$, $n = 2282$) models at $P<0.001$. However, soil VWC% explained only 4–6% of soil respiration. We further examined the combined effects of soil temperature and VWC% using four compound models, namely the linear, power, exponential-power and double exponential models (Table 3). The results indicate that the R^2 values combining temperature and VWC% are significantly higher than using the quadratic model only considering soil temperature. Soil temperature and VWC% explained up to 65% of soil respiration using the exponential-power and double exponential models.

The exponential-power and double exponential models (Table 3) gave significant improvements compared to the linear and power models. We again compared MAE, ME and d, RMSE, MSE_s and MSE_u among the four models and comprehensively evaluated the model performances by the sizes of these indicators and the value of $MSE_s/(MSE_u+MSE_u)$. The exponential-power model was much better for description of soil respiration in response to soil temperature and VWC% (both in the top 5 cm) in our study because it had lower MAE, RMSE, MSE_u, and higher

Table 2. Total CO_2 emissions in each cropping season and each rotation cycle (Mg CO_2 ha^{-1}).

Year	Con.W/M		Opt.W/M	W/M-M		W/S-M		M	
	Crop	CO_2	CO_2	Crop	CO_2	Crop	CO_2	Crop	CO_2
2009	M_1	19.1±0.8bc[1]	22.0±1.5a	M_2	21.4±1.9ab	M_2	17.5±1.3c	M_2	17.4±1.4c
2010	W	15.5±1.2bc	17.2±0.6ab	W	17.8±1.2a	W	17.0±0.9abc	F	14.9±0.5c
	M_1	16.0±0.9b	21.4±1.7a	M_1	20.6±1.8a	S	15.7±0.9b	M_2	18.6±2.5ab
2011	W	13.4±0.7bc	22.4±0.4a	F	19.6±0.4a	F	11.1±1.6c	F	15.9±1.6b
	M_1	15.5±1.4b	20.4±1.9a	M_2	19.6±2.1a	M_2	16.0±1.1b	M_2	18.7±1.2a
2012	W	16.5±0.9bc	21.7±3.5a	W	21.3±2.4a	W	19.3±1.6ab	F	15.0±1.8c
	M_1	16.4±0.6c	20.4±1.9a	M_1	19.4±1.4ab	S	17.0±2.0bc	M_2	18.0±0.7abc
2009-2012 Mean	M_1	16.8	21.1	M_1	20.0	S	16.4	-	
	W	15.1	20.4	W	19.6	W	18.2	-	
	-	-	-	M_2	20.5	M_2	16.8	M_2	18.2
	-	-	-	F	19.6	F	11.1	F	15.3
2011-2012	2 W-M_1[2]	61.9±1.3b	84.9±8.2a	F-M_2-W-M_1	79.8±5.6a	F-M_2-W-S	64.1±2.7b	2 F-M_2	67.1±2.0b

[1]The same letter in the same line denotes no significant difference in different cropping systems by LSD at $P<0.05$.

[2]2 W-M_1, F-M_2-W-M_1 (or S) and 2 F-M_2 represent two winter wheat-summer maize rotation cycles, fallow-spring maize-winter wheat-summer maize (or summer soybean) rotation cycle and two fallow-spring maize rotation cycles.

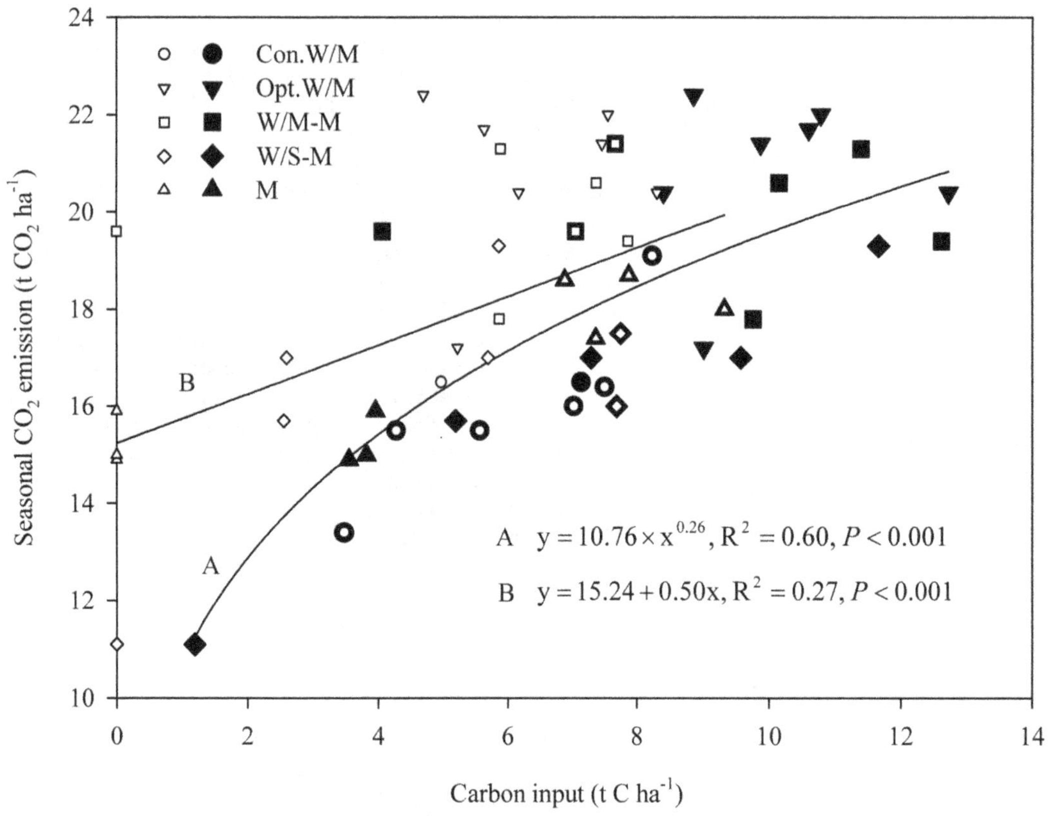

Figure 4. Correlation between seasonal CO_2 emission and carbon input. Carbon input was calculated from current-season aboveground biomass only (A); and calculated from straw return of the previous season and the aboveground biomass in the current season (B); the abbreviations of the treatment are shown in the footnotes in Fig. 2.

ME than the double exponential model and the values of MSE_s/(MSE_s+MSE_u) were similar using both models.

Discussion

Soil respiration in croplands is affected mainly by soil properties, cropping system (which is related to crop species), tillage and straw management, water and nutrient management, and environmental variables (soil temperature, moisture etc.) [1,6,20,35]. There is temporal variation within the same cropping system and spatial variation among different cropping systems [16,17,31]. Changes in soil respiration in our sub-humid temperate monsoon region are largely affected by the seasonal variation in temperature, which is in line with most previous reports [30,31,36]. However, soil respiration responded little to soil temperature as shown in Fig. 5, equation A using the quadratic model because some data points did not fit the model with the impacts of soil tillage before the wheat crop was sown. The R^2 value improved by 18%, and up to 53% when the data within one month after tilling were excluded (Fig. 5, equation B). Moreover, we found that soil respiration tended to follow the variation in temperature from August to the following March when the data after tillage were excluded (Figs. 2 and 3A). Soil temperature explained 74% of soil respiration when only the data from August to March were included (Fig. 6, equation A). Therefore, the impacts of tillage must be considered for modeling soil respiration on the NCP.

The short decline in soil respiration after irrigation might be attributable to blocked diffusion of CO_2 with high moisture and limited oxygen concentrations in the soil matrix [37], and the flushes afterwards may be due to the stimulation of decomposition of plant residues [21], root litter and exudates or autotrophic respiration of rapid root growth, which taken together induced the effects of drying and wetting cycles. Soil respiration would be limited when soil moisture was too high or too low and the maximum range is usually close to field water holding capacity [38]. The disappearance of respiration flushes was due to the low soil temperatures within a week after irrigation at the shooting stage of wheat in 2010 relative to other years (6–10°C in 2010 vs 12–21°C in 2011 and 12–19°C in 2012) (Fig. 3A). Soil moisture was not the key driving factor over the whole study period but did affect soil respiration slightly at particular stages and therefore only explained a very small proportion of the variation in soil respiration in our study area.

Numerous studies have reported that soil respiration is significantly affected by tillage practices combined with straw management [14–16]. Total soil respiration was significantly higher in Opt.W/M than Con.W/M as the latter soil was rotary tilled to 20 cm depth after maize straw removal and Opt.W/M was ploughed into the top 30 cm of the soil after maize straw return to the soil, soil respiration increased sharply after soil disturbance by tillage operations possibly because increased soil aeration accelerated the decomposition rate of crop residues which was associated with higher microbial activity [14,15,39]. However, the impacts of maize straw return and tillage were lowered by delaying tillage until the soil temperature to a depth of 5 cm reached 10°C or lower.

Although seasonal cumulative CO_2 emission in Opt.W/M and W/M-M increased significantly relative to Con.W/M as a result of

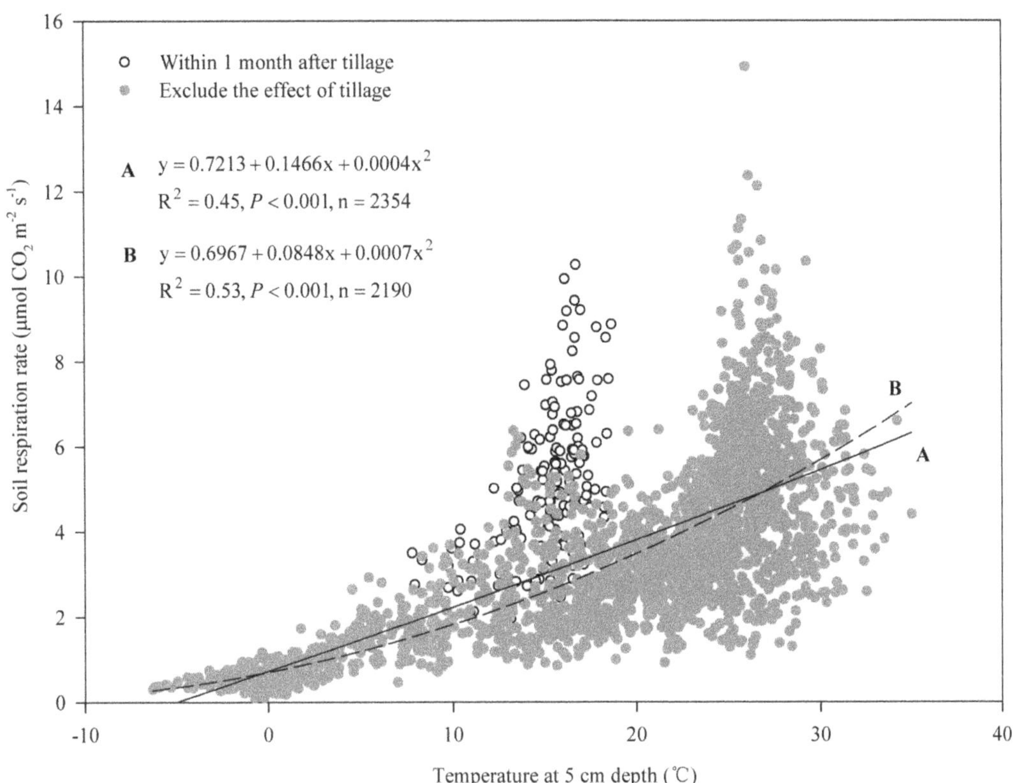

Figure 5. Impacts of soil tillage combined with straw return on soil respiration. Correlation between soil respiration and soil temperature at 5 cm depth over the whole year (equation A) and the correlation between soil respiration and soil temperature at 5 cm depth excluding the data within one month of tillage (equation B).

straw return [14,15], this practice also increases the SOC content over the long term [16,17,40]. The SOC content in the top 20 cm of the soil profile in straw return treatments increased by 3.9–16.5% relative to the straw removal treatments in winter wheat–summer maize double-cropping systems on the NCP, with a mean increase in rate of 0.04 to 1.44 t C ha^{-1} y^{-1} over a six-year period as shown by Huang et al. [19]. We also measured SOC to a depth of 20 cm in the present field experiment after summer maize harvest in 2011 and all values increased to 8.07, 8.71, 7.93 and 7.52 g kg^{-1} in Con.W/M, Opt.W/M, W/M-M and W/S-M, respectively, with the sole exception of a slight decrease to 7.18 g kg^{-1} in M (from 7.31 g kg^{-1} at the start of the field experiment in 2007). Although there was no crop straw return, Con.W/M also showed a clear increment relative to the initial value in line with Huang et al. [19], and this may have been due to the large amounts of crop roots and rhizo-deposited carbon. Con.W/M showed a greater increase in SOC than W/M-M, W/S-M and M, possibly due to the lower intensity of tillage in Con.W/M than in W/M-M, W/S-M and M. Our results show that soil respiration responded mainly to the seasonal variation in soil temperature but was also greatly affected by straw return, root growth and soil moisture changes under the different cropping systems.

Supporting Information

Table S1 Measured soil respiration fluxes, soil temperature and soil volumetric water content to 5 cm depth, daily mean air temperature and precipitation in the field experiments from 18th May 2009 to 11th November 2012.

Table 3. Correlation between soil respiration and soil temperature and VWC(%) to 5 cm depth.

Model	Fitting equation	n	R^2	MAE	ME	RMSE	MSE_s	MSE_u	d
Linear	R_s[1] = 0.7712+0.1581T − 0.0030V	1905	0.47*[2]	1.11	0.48	1.48	1.28	1.03	0.80
Power (T>0)	$R_s = 0.6291T^{0.5929}V^{-0.0096}$	1811	0.56*	1.17	0.36	1.57	2.13	0.59	0.70
Exponential-power	$R_s = 0.9347e^{0.069T}V^{-0.0464}$	1905	0.65*	1.14	0.38	1.61	1.41	1.67	0.80
Double exponential	$R_s = 0.8924e^{0.0693T-0.0045V}$	1905	0.65*	1.20	0.31	1.69	1.36	1.81	0.78

[1] R_s, T and V represent soil respiration, soil temperature and VWC% to 5 cm depth, respectively.
[2] * represents highly significant correlation at $P<0.001$.

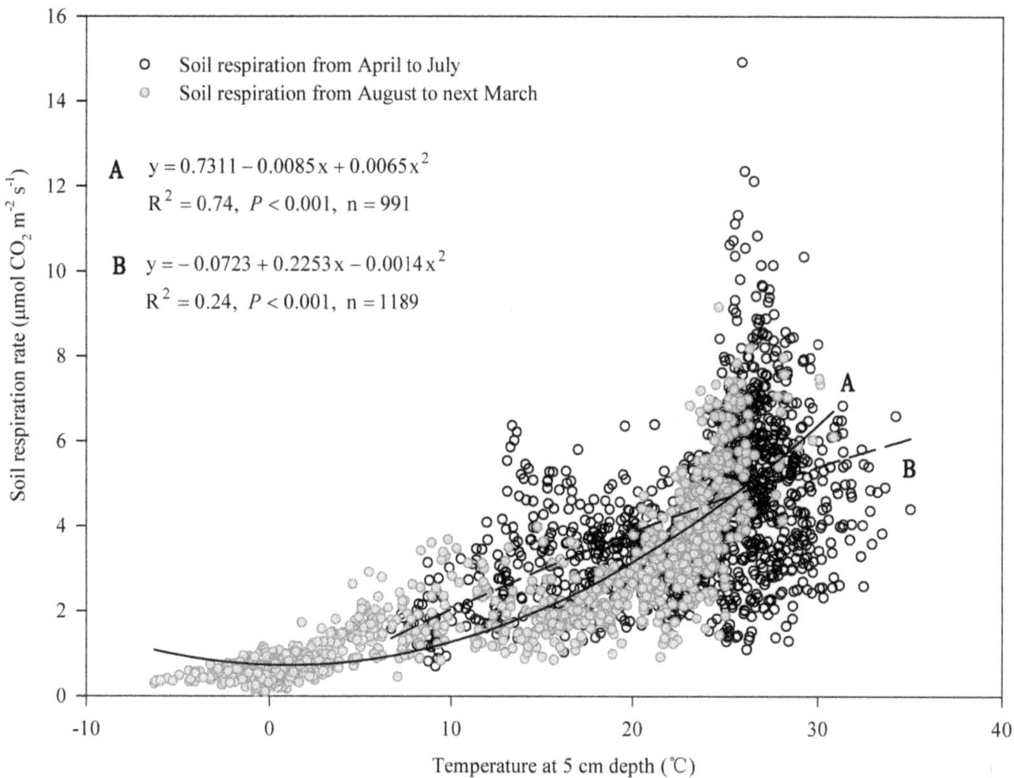

Figure 6. Correlation between soil respiration and soil temperature at 5 cm depth. Equations A and B represent the correlations between soil respiration and soil temperature at 5 cm depth from August to March after removing the impacts of tillage and from April to July, respectively.

Author Contributions

Conceived and designed the experiments: XJ XC FZ. Performed the experiments: BG FS FG QC. Analyzed the data: BG XJ. Contributed reagents/materials/analysis tools: XJ. Wrote the paper: BG XJ. Given some suggestion and modified the language for the manuscript OO PC.

References

1. Davidson EA, Belk E, Boone RD (1998) Soil water content and temperature as independent or confounded factors controlling soil respiration in temperate mixed hardwood forest. Global Change Biol 4: 217–227.
2. Bond-Lamberty B, Thomson A (2010) Temperature-associated increases in the global soil respiration record. Nature 464: 579–582.
3. Robertson GP, Paul EA, Harwood RR (2000) Greenhouse gases in intensive agriculture: contributions of individual gases to the radiative forcing of the atmosphere. Science 289: 1922–1926.
4. Mahecha MD, Reichstein M, Carvalhais N, Lasslop G, Lange H, et al. (2010) Global convergence in the temperature sensitivity of respiration at ecosystem level. Science 329: 838–840.
5. Grace J, Rayment M (2000) Respiration in the balance. Nature 404: 819–820.
6. Raich JW, Schlesinger WH (1992) The global carbon dioxide flux in soil respiration and its relationship to vegetation and climate. Tellus B 44: 81–99.
7. Melling L, Hatano R, Goh KJ (2005) Soil CO_2 flux from three ecosystems in tropical peatland of Sarawak, Malaysia. Tellus B 57: 1–11.
8. Lloyd J, Taylor JA (1994) On the temperature dependence of soil respiration. Funct Ecol 8: 315–323.
9. Fang C, Moncrieff JB (2001) The dependence of soil efflux on temperature. Soil Biol Biochem 33: 155–165.
10. Li HJ (2008) Studies on soil respiration and its relations to environmental factors in different ecosystems. Shanxi: PhD thesis, Shanxi University (in Chinese).
11. Wang WJ, Dalal RC, Moody PW, Smith CJ (2003) Relationships of soil respiration to microbial biomass, substrate availability and clay content. Soil Biol Biochem 35: 273–284.
12. Carlisle EA, Steenwerth KL, Smart DR (2006) Effects of land use on soil respiration: Conversion of Oak woodlands to vineyards. J Environ Qual 35: 1396–1404.
13. Rey A, Pegoraro E, Tedeschi V, De Parri I, Jarvis PG, et al. (2002) Annual variation in soil respiration and its components in a coppice oak forest in central Italy. Global Change Biol 8: 851–866.
14. Bavin TK, Griffis TJ, Baker JM, Venterea RT (2009) Impact of reduced tillage and cover cropping on the greenhouse gas budget of a maize/soybean rotation ecosystem. Agric Ecosyst Environ 134: 234–242.
15. Al-Kaisi MM, Yin XH (2005) Tillage and crop residue effects on soil carbon and carbon dioxide emission in corn-soybean rotations. J Environ Qual 34: 437–445.
16. Mosier AR, Halvorson AD, Reule CA, Liu XJ (2006) Net global warming potential and greenhouse gas intensity in irrigated cropping systems in Northeastern Colorado. J Environ Qual 35: 1584–1598.
17. Alluvione F, Halvorson AD, Del Grosso SJ (2009) Nitrogen, tillage, and crop rotation effects on carbon dioxide and methane fluxes from irrigated cropping systems. J Environ Qual 38: 2023–2033.
18. Zhang QZ, Wu WL, Wang MX, Zhou ZR, Chen SF (2005) The effects of crop residue amendment and N rate on soil respiration. Acta Ecologica Sinica 25: 2883–2887 (in Chinese with English abstract).
19. Huang T, Gao B, Christie P, Ju XT (2013) Net global warming potential and greenhouse gas intensity in a double cropping cereal rotation as affected by nitrogen and straw management. Biogeosciences Discuss 10: 13191–13229.
20. Iqbal J, Hu RG, Lin S, Ahamadou B, Feng ML (2009) Carbon dioxide emissions from Ultisol under different land uses in mid-subtropical China. Geoderma 152: 63–73.
21. Xu LK, Baldocchi DD, Tang JW (2004) How soil moisture, rain pulses, and growth alter the response of ecosystem respiration to temperature. Global Biogeochem Cycl 18: GB4002.
22. Conant RT, Dalla-Bett P, Klopatek CC, Klopatek JM (2004) Controls on soil respiration in semiarid soils. Soil Biol Biochem 36: 945–951.
23. Howard DM, Howard PJA (1993) Relationships between CO_2 evolution, moisture content, and temperature for a range of soil types. Soil Biol Biochem 25: 1537–1546.
24. Lellei-Kovács E, Kovács-Láng E, Botta-Dukát Z, Kalapos T, Emmett B, et al. (2011) Thresholds and interactive effects of soil moisture on the temperature response of soil respiration. Eur J Soil Biol 47: 247–255.
25. Fang JY, Chen AP, Peng CH, Zhao SQ, Ci LJ (2001) Changes in forest biomass carbon storage in China between 1949 and 1998. Science 292: 2320–2322.

26. Meng QF, Sun QP, Chen XP, Cui ZL, Yue SC, et al. (2012) Alternative cropping systems for sustainable water and nitrogen use in the North China Plain. Agric Ecosyst Environ 146: 93–102.
27. Cui ZL, Chen XP, Zhang FS (2010) Current Nitrogen management status and measures to improve the intensive wheat-maize system in China. Ambio 39: 376–384.
28. Ju XT, Xing GX, Chen XP, Zhang SL, Zhang LJ, et al. (2009) Reducing environmental risk by improving N management in intensive Chinese agricultural systems. Proc Natl Acad Sci USA 106: 3041–3046.
29. Hu C, Delgado JA, Zhang X, Ma L (2005) Assessment of groundwater use by wheat (*Triticum aestivum* L.) in the Luancheng Xian Region and potential implications for water conservation in the Northwestern North China Plain. J Soil Water Conserv 60: 80–88.
30. Niu LA, Hao JM, Zhang BZ, Niu XS, Lu ZY (2009) Soil respiration and carbon balance in farmland ecosystems on North China Plains. Ecol Environ Sci 18: 1054–1060 (in Chinese with English abstract).
31. Meng FQ, Guan GH, Zhang QZ, Si YJ, Qu B, et al. (2006) Seasonal variation in soil respiration under different long-term cultivation practices on high yield farm land in the North China Plain. Acta Scientiae Circumstantiae 26: 992–999 (in Chinese with English abstract).
32. Guo RY, Nendel C, Rahn C, Jiang CG, Chen Q (2010) Tracking nitrogen losses in a greenhouse crop rotation experiment in North China using the EU-Rotate_N simulation model. Environ Pollut 158: 2218–2229.
33. Willmott CJ (1982) Some comments on the evaluation of model performance. Am Meteorological Soc 63: 1309–1313.
34. Xiao M (2006) Effect of long-term fertilization on soil carbon and nitrogen storages and dynamics of nitrogen mineralization. Beijing: China Agricultural University (in Chinese).
35. Lohila A, Aurela M, Regina K, Laurila T (2003) Soil and total ecosystem respiration in agricultural fields: effect of soil and crop type. Plant Soil 251: 303–317.
36. Han GX, Zhou GS, Xu ZZ, Yang Y, Liu JL, et al. (2007) Soil temperature and biotic factors drive the seasonal variation of soil respiration in a maize (*Zea mays* L.) agricultural ecosystem. Plant Soil 291: 15–26.
37. Gaumont-Guay D, Black TA, Griffis TJ, Barr AG, Jassal RS, et al. (2006) Interpreting the dependence of soil respiration on soil temperature and water content in a boreal aspen stand. Agric For Mete 140: 220–235.
38. Davidson EA, Verchot LV, Cattanio JH, Ackerman IL, Carvalho JEM (2000) Effects of soil water on soil respiration in forests and cattle pastures of eastern Amazonia. Biogeochemistry 48: 53–69.
39. Jackson LE, Calderon FJ, Steenwerth KL, Scow KM, Rolston DE (2003) Responses of soil microbial processes and community structure to tillage events and implications for soil quality. Geoderma 114: 305–317.
40. Huang Y, Sun WJ (2006) Changes in topsoil organic carbon of croplands in mainland China over the last two decades, Chinese Sci Bull 51: 1785–1803.

14

Functional Groups Determine Biochar Properties (pH and EC) as Studied by Two-Dimensional ^{13}C NMR Correlation Spectroscopy

Xiaoming Li[1,2], Qirong Shen[1,2], Dongqing Zhang[3], Xinlan Mei[1,2], Wei Ran[1,2], Yangchun Xu[1,2], Guanghui Yu[1,2]*

1 Agricultural Ministry Key Lab of Plant Nutrition and Fertilization in Low-Middle Reaches of the Yangtze River, Nanjing, PR China, **2** College of Resources and Environmental Sciences, Nanjing Agricultural University, Nanjing, PR China, **3** Baoshan Environmental Protection Bureau, Shanghai, PR China

Abstract

While the properties of biochar are closely related to its functional groups, it is unclear under what conditions biochar develops its properties. In this study, two-dimensional (2D) ^{13}C nuclear magnetic resonance (NMR) correlation spectroscopy was for the first time applied to investigate the development of functional groups and establish their relationship with biochar properties. The results showed that the agricultural biomass carbonized to biochars was a dehydroxylation/dehydrogenation and aromatization process, mainly involving the cleavage of O-alkylated carbons and anomeric O-C-O carbons in addition to the production of fused-ring aromatic structures and aromatic C-O groups. With increasing charring temperature, the mass cleavage of O-alkylated groups and anomeric O-C-O carbons occurred prior to the production of fused-ring aromatic structures. The regression analysis between functional groups and biochar properties (pH and electrical conductivity) further demonstrated that the pH and electrical conductivity of rice straw derived biochars were mainly determined by fused-ring aromatic structures and anomeric O-C-O carbons, but the pH of rice bran derived biochars was determined by both fused-ring aromatic structures and aliphatic O-alkylated (HCOH) carbons. In summary, this work suggests a novel tool for characterising the development of functional groups in biochars.

Editor: Andrea Motta, National Research Council of Italy, Italy

Funding: The work was funded by the National Basic Research Program of China (2011CB100503), the National Natural Science Foundation of China (21007027), the 111 Project (B12009), the Qing Lan Project of Jiangsu Province, and the Priority Academic Program Development of Jiangsu Higher Education Institutions (PAPD). The funders had no role in study design, data collection and analysis, decision to publish, or preparation of the manuscript.

Competing Interests: The authors have declared that no competing interests exist.

* E-mail: yuguanghui@njau.edu.cn

Introduction

Agricultural biomass is one of the most abundant renewable resources on Earth. The thermal transformation of agricultural biomass into biochar in an oxygen-depleted atmosphere has been extensively studied, because of concerns over enhancement of soil fertility and mitigation of climate change by long-term carbon sequestration [1,2]. The main functional groups of biochar are aromatic and heterocyclic carbons, which are believed to be stable in soil due to their chemical recalcitrance [3,4]. Therefore, the characterization of functional groups in biochar is critical to understanding the reaction mechanisms of the charring processes. Furthermore, acquiring clear knowledge of biochar functional groups is important for the beneficial use of biochar products.

The properties of biochar greatly depend on the production conditions (i.e., charring temperature) and the biomass type used to produce biochar [5,6,7]. In general, biochar produced below 400°C had a low pH, low electrical conductivity (EC), and small surface area [5]. The properties of biochar were closely related to its functional groups in the charring temperatures range of 100~600°C [7]. However, it is unclear under what conditions biochar develops its properties [5,8].

Solid-state ^{13}C nuclear magnetic resonance (NMR) spectroscopy has been frequently employed to characterize the functional groups of biochar [9,10,11,12]. However, the individual spectral features of ^{13}C NMR often overlap because of the extreme heterogeneity of biochar. Therefore, it is difficult to establish the relationship between biochar properties and functional groups. Recently, we have applied two-dimensional (2D) correlation spectroscopy to environmental sciences for the purpose of unrevealing the overlapped peaks [13,14,15]. The main advantages of 2D correlation spectroscopy are as follows: 1) providing a better resolution of significant peaks; and 2) the ability to allow for probe the specific sequencing of spectral intensity changes through asynchronous analysis [13,14,15]. By applying 2D correlation spectroscopy, it is expected that the best possible biochar may be "designed" for a given application, based on assessments of the dynamic functional groups of biochar and their relationship with the corresponding properties.

The objectives of the present study were to 1) investigate the development of functional groups in biomass-derived biochar by applying 2D correlation spectroscopy and 2) establish the relationship between functional groups and biochar properties (i.e., pH and EC). For these purposes, two types of agricultural biomass, i.e. rice straw and rice bran, were selected to produce biochars.

Materials and Methods

Biochar Preparation

Rice straw and rice bran were collected from Yixing city, Jiangsu Province, China. No specific field permits were required for this study. The land accessed is not privately owned or protected. No protected species were sampled. Biochars were produced via pyrolyzing biomass in a furnace (SXL-1200, Shanghai Daheng Optics and Fine Mechanics Co., Ltd., Shanghai, China) at various temperatures under oxygen-limited conditions. The furnace could provide a constant temperature environment of 100–800°C at an averagely 10°C min^{-1} rate of temperature increase. All biochar samples were then examined for their physical and chemical characteristics.

Solid-state ^{13}C NMR Spectroscopy

Solid-state ^{13}C NMR spectroscopy was conducted on a Bruker AV-400, equipped with a 4-mm wide-bore MAS probe [13]. NMR spectra were obtained by applying the following parameters: rotor spin rate of 13000 Hz, 1 s recycle time, 1 ms contact time, 20 ms acquisition time, and 4000 scans. Samples were packed in 4-mm zirconia rotors with Kel-F caps. The pulse sequence was applied with a ^{1}H ramp to account for non-homogeneity of the Hartmann-Hahn condition at high spin rotor rates. Chemical shifts were calibrated with adamantine. It should be noted that solid-state ^{13}C NMR spectroscopy is only a quanlitative rather than quantitative method [16].

Analysis of 2D Correlation Spectroscopy

The 2D correlation spectra were produced according to the method of Noda and Ozaki [17]. In the present study, the charring temperature was applied as an external perturbation, and a set of temperature-dependent NMR spectra was obtained. Let us consider the analytical spectrum $I(x, t)$. The variable x is the index variable representing the NMR spectra induced by the perturbation variable t. We intentionally use x instead of the general notation used in conventional 2D correlation equations based on the spectral index v. The analytical spectrum $I(x, t)$ at m with evenly spaced points in t (between T_{min} and T_{max}) can be represented as

$$I_j(x)=I(x,t_j), j=1,2,\cdots,m. \quad (1)$$

A set of dynamic spectra is given by:

$$\tilde{I}(x,t)=I(x,t_j)-\bar{I}(x) \quad (2)$$

where $\bar{I}(x)$denotes the reference spectrum, which is typically the average spectrum and is expressed as $\bar{I}(x)=\frac{1}{m}\sum_{j=1}^{m} I(x,t_j)$. The synchronous correlation intensity can be directly calculated from the following dynamic spectra:

$$\varnothing(x_1,x_2)=\frac{1}{m-1}\sum_{j=1}^{m}\tilde{I}_j(x_1)\tilde{I}_j(x_2) \quad (3)$$

Asynchronous correlation can be obtained by:

$$\varphi(x_1,x_2)=\frac{1}{m-1}\sum_{j=1}^{m}\tilde{I}_j(x_1)\sum_{k=1}^{m}N_{jk}\tilde{I}_j(x_2) \quad (4)$$

The term N_{jk} corresponds to the jth column and the kth raw element of the discrete Hilbert-Noda transformation matrix, which is defined as

$$N_{jk}=\begin{cases}0 & if\ j=k\\ \frac{1}{\pi(k-j)} & otherwise\end{cases} \quad (5)$$

The intensity of a synchronous correlation spectrum ($\varnothing(x_1,x_2)$) represents simultaneous changes in two spectral intensities measured at x_1 and x_2 during the interval between T_{min} and T_{max}. In contrast, an asynchronous correlation spectrum ($\varphi(x_1,x_2)$) includes out-of-phase or sequential changes in spectral intensities measured at x_1 and x_2.

Prior to 2D analysis, the NMR spectra were normalized by summing the absorbance from 200-0 ppm and multiplying by 1000. Subsequently, 2D correlation spectroscopy was conducted using 2Dshige software (Kwansei Gakuin University, Japan).

Chemical Characterization Assay

Biochar (<0.25 mm) was soaked with deionized water at a 1:5 solid/water ratio for 24 h with agitation. The slurry was then measured for pH using a pH electrode and for electrical conductivity (EC) using a conductivity meter (LF91, German). Elemental composition (CHNSO) was determined by dry combustion using a Perkin–Elmer2400 CHNS elemental analyzer. The O content was calculated by O% = 100-C%-H%-N%-S%.

Statistical Analysis

Statistical analysis was performed using the software SPSS version 16.0 for Windows (SPSS, Chicago, IL). Pearson's correlation coefficient (R) was used to evaluate the linear correlation between biochar properties (pH and EC) and functional groups. Pearson's coefficient is always between −1 and +1, where −1 denotes a perfect negative correlation, +1 denotes a perfect positive correlation, and 0 denotes the absence of a relationship. The correlations were considered to be statistically significant at a 95% confidence interval ($p<0.05$).

Results and Discussion

Properties of Biochars at Various Charring Temperatures

With increasing charring temperature, yields declined rapidly at 300°C and remained relatively stable above 600°C for both biochars (Table 1). Final yields of biochars were approximately 10–14% for both biochars. The C content increased, whereas the H, N, S and O contents decreased with increasing charring temperature. All of the atomic ratios [i.e., H/C, O/C and (O+N)/C] sharply decreased, suggesting that with increasing charring temperature, the relative degree of aromaticity (H/C ratio) and polarity [O/C and (O+N)/C ratios] markedly decreased, which could be attributable to the development of functional groups and will be discussed in the next sections.

The typical van Krevelen plot showed that the progressive decrease in the H/C and O/C atomic ratios with temperature followed the trajectory associated with dehydration reactions (Figure 1). The most dramatic loss of H and O occurred in the range of 300–400°C for both biochars. Moreover, the biochars produced at 400 and 500°C were expected to have characteristics of chars, which is consistent with biochar yields (Table 1). Above 600°C, the biochars may be considered to be soot [18]. These trends in elemental composition and atomic ratios with the various

Table 1. Yields, elemental composition, atomic ratios, pH and EC of rice straw and rice bran derived biochars prepared at various charring temperatures.[a]

material	charring temperature (°C)	yield/%	elemental composition (%)					atomic ratios			pH	EC (mS/cm)
			C	H	N	S	O	H/C	O/C	(O+N)/C		
Rice straw	100	97.2±0.0	39.13	4.58	3.72	3.41	49.16	1.40	0.94	1.02	5.25±0.04	4.09±0.10
	200	77.1±0.9	42.02	3.34	3.04	2.15	49.45	0.95	0.88	0.94	5.93±0.01	5.64±0.02
	300	30.1±1.8	49.68	2.17	2.07	1.68	44.40	0.52	0.67	0.70	7.16±0.05	5.57±0.02
	400	26.0±1.7	61.24	1.51	1.07	1.02	35.16	0.30	0.43	0.44	8.41±0.11	6.22±0.12
	500	19.7±2.3	69.78	0.99	0.98	0.89	27.36	0.17	0.29	0.31	10.1+0.19	6.59±0.01
	600	13.5±2.2	77.24	0.87	0.81	0.72	20.36	0.14	0.20	0.21	10.6±0.05	7.31±0.05
	700	12.8±0.6	86.99	0.85	0.88	0.75	10.53	0.12	0.09	0.10	10.3±0.03	7.51±0.07
	800	10.7±0.8	88.12	0.83	0.84	0.81	9.40	0.11	0.08	0.09	10.5±0.02	7.72±0.02
	100	97.4±0.2	42.39	4.21	3.24	2.17	47.99	1.19	0.85	0.91	5.48±0.01	2.84±0.02
Rice bran	200	83.9±0.0	46.53	2.47	2.15	1.84	47.01	0.64	0.76	0.79	6.16±0.01	2.56±0.02
	300	31.1±1.9	50.68	2.14	1.58	1.51	44.09	0.51	0.65	0.68	6.89±0.02	1.12±0.01
	400	24.5±1.1	64.05	1.16	1.34	0.92	32.53	0.22	0.38	0.40	7.43±0.02	2.20±0.03
	500	18.5±2.4	71.72	1.01	1.03	0.84	25.40	0.17	0.27	0.28	8.95±0.02	3.24±0.071
	600	18.0±1.9	77.58	0.79	0.72	0.79	20.12	0.12	0.19	0.20	10.04±0.09	3.67±0.03
	700	15.4±2.1	82.03	0.84	0.74	0.81	15.52	0.12	0.14	0.15	10.28±0.08	3.74±0.03
	800	14.4±3.1	83.14	0.84	0.77	0.80	14.45	0.12	0.13	0.14	10.92±0.06	3.44±0.03

[a]**All chemical analyses were carried out in triplicate.**

charring temperatures were consistent with the previous investigations [19,20,21].

The charring of rice straw and rice bran increased their pH (Table 1). The pH of charred rice straw and sawdust increased from acidic (approximately 5) to alkaline (approximately 10) for both biochars. However, the charring of rice straw increased its EC, but the charring of rice bran decreased its EC before 300°C and then increased its EC (except for 800°C). The EC of charred rice straw gradually increased from 4.09±0.10 at 100°C to 7.72±0.02 at 800°C. However, the development of EC for charred rice bran fluctuated. Specifically, the EC decreased from 2.84±0.02 at 100°C to 1.12±0.01 at 300°C and then increased to 3.44±0.03 at 800°C. The differences between charred rice straw and rice bran should be attributed to the nature of plant biomass. These results were similar to those of Lehmann [5], supporting that the properties of biochar were greatly dependent on the production procedure and raw material.

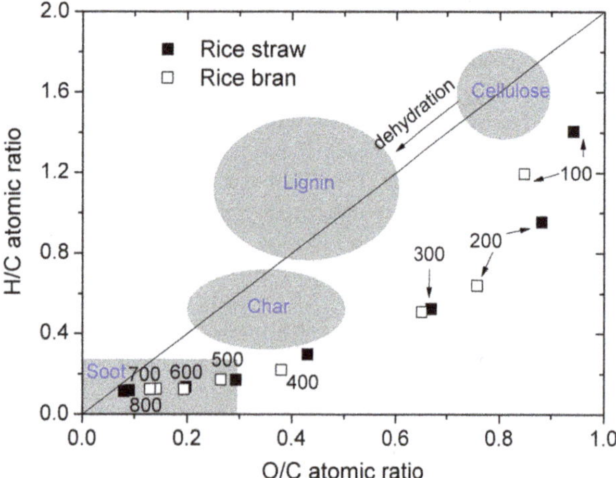

Figure 1. Van Krevelen plot of elemental ratios for rice straw (■) and sawdust (□) derived biochars prepared at various charring temperatures. Thick line represents the direction for dehydration reactions and grey shadings highlight approximate elemental ratios of unaltered biomacromolecules (cellulose and lignin) and black carbon materials (char and soot) following Hammes et al. [16].

Development of Functional Groups during the Charring Process

One-dimensional NMR spectra of rice straw and rice bran derived biochars (Figures 2-a and 2-b) showed a two-phase characteristic. Note that no NMR signal was detected for biochars at 700 and 800°C, which is most likely attributable to the high EC that affected the alignment of aromatic sheets [22], an increase in conductivity of the sample, or the occurrence of a larger number of delocalized π electrons around the ^{13}C nuclei arising from the growth of the aromatic planes of the chars at higher temperature [23]. The transition charring temperature occurred at 300°C. In the first phase, aliphatic O-alkylated (HCOH) (73 ppm) carbons were predominant in biochars. However, in the second phase, O-alkylated (HCOH) carbons were gradually eliminated and fused-ring aromatic structures (128 ppm) became the predominant component. Meanwhile, the NMR bands strongly overlapped in the region of 50–100 ppm for biochars less than 300°C and in the region of 100–150 ppm for biochars more than 300°C. Similar results were observed by Zhang et al. [24], who investigated charred corn straw in the range of 100–600°C by ^{13}C NMR spectroscopy and found that with increasing charring temperature, the trends of total aliphatic C (0–93 ppm) contents decreased, while the total aromatic C (93–165 ppm) contents increased.

Rutherford et al. also demonstrated that the initial loss of material was attributable to aliphatic components, which were either lost or converted to aromatic carbon in the charring process [25]. Recently, Cao et al. also observed that the similar temperature-dependent ^{13}C NMR spectroscopy of wood chars, suggesting that heat treatment at 300°C resulted in a material that was composed primarily of residues of biopolymers such as lignin and cellulose; carbohydrates were completely lost for char prepared at 350°C; at 400°C and above, the char lost ligno-cellulosic features and consisted predominantly of aromatic structures [23].

Because the functional groups in one-dimensional ^{13}C NMR spectra strongly overlapped, no further analysis is possible from cursory observations of the original spectra. Therefore, in this study, 2D correlation analysis was applied to further investigate the development of functional groups with charring temperature.

Analysis of 2D ^{13}C NMR correlation spectroscopy produces synchronous and asynchronous maps. The correlation peaks in the synchronous map are composed of autopeaks and crosspeaks, which appear at both diagonal and off-diagonal positions, respectively. Autopeaks represent the overall susceptibility of the corresponding spectral region to change in spectral intensity as an external perturbation is applied to the system, while crosspeaks represent simultaneous or coincidental changes of spectral intensities observed at two different spectral variables. An asynchronous spectrum has no autopeaks and consists exclusively of crosspeaks located at off-diagonal positions. This spectrum provides useful information on the sequential order of events observed by the spectroscopic technique along the external variable.

The 2D ^{13}C NMR correlation spectroscopy showed that for both biochars, the synchronous and asynchronous maps were similar (Figure 3). The synchronous maps displayed two major positive autopeaks at 128 ppm and 73 ppm, indicating that of all the functional groups, both fused-ring aromatic structures (128 ppm) and aliphatic O-alkylated (HCOH) (73 ppm) carbons intensity changed with increasing charring temperature. Therefore, the production of biochars is a process of dealkylation, dehydroxylation/dehydrogenation and aromatization of plant biopolymers, which was consistent with the results of the elemental analysis (Figure 1). It was noted from the synchronous maps that the highest change in intensity was in the band at 128 ppm, followed by 73 ppm, suggesting that with increasing charring temperature, fused-ring aromatic structures changed much more than aliphatic O-alkylated (HCOH) carbons. Moreover, one positive crosspeak at $\Phi(105, 73)$ and two negative crosspeaks at $\Phi(128, 73)$ and $\Phi(128, 105)$ were identified, reflecting that aliphatic O-alkylated (HCOH) carbons and anomeric O-C-O carbons at approximately 105 ppm co-vary with increasing charring temperature, whereas aliphatic O-alkylated (HCOH) carbons had an opposite variation trend with fused-ring aromatic structures. This result was consistent with the observation from the one-dimensional spectrum that aliphatic O-alkylated (HCOH) (i.e., 73 ppm) carbons decreased but fused-ring aromatic structures (i.e., 128 ppm) increased (Figure 2).

The asynchronous maps exhibited two major positive crosspeaks at $\Psi(128, 73)$ and $\Psi(128, 105)$ above the diagonal line for both biochars. Based on the Noda's rules [17], the sequence of the change of bands with increasing charring temperature followed the order 73 ppm (105 ppm)→128 ppm. Therefore, with increasing charring temperature, the mass cleavage of aliphatic O-alkylated (HCOH) groups and anomeric O-C-O carbons occurred prior to the production of fused-ring aromatic structures, which was consistent with the observation of one-dimensional ^{13}C NMR spectra (Figure 2).

Although the asynchronous maps derived from the NMR region of 65–140 ppm were similar for both biochars (Figure 3), subtle differences could be found in the asynchronous map in the small NMR regions (Figure 4). Of these changes, none of the bands was common to both plant biomasses, reflecting that all of the changes in the functional groups were specific to the nature of the plant

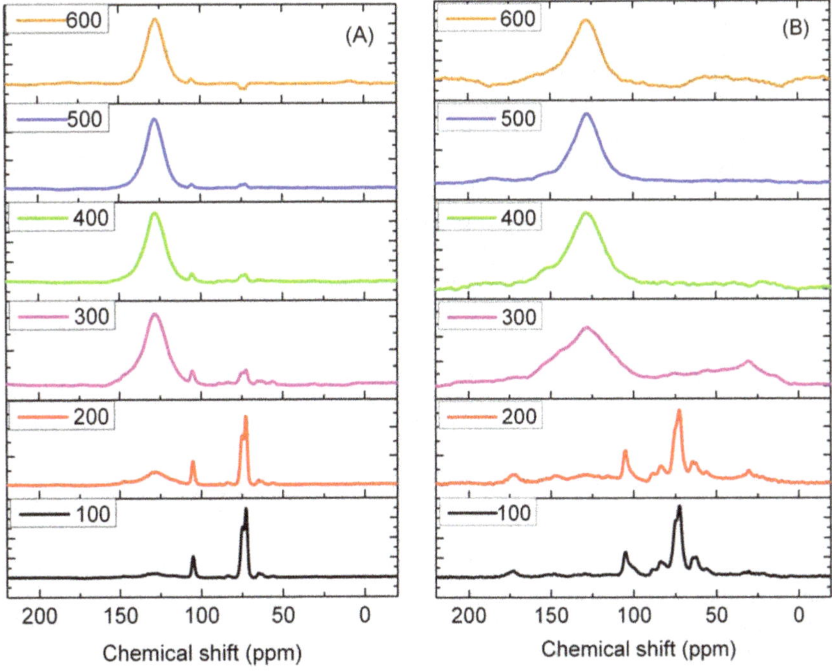

Figure 2. Solid-state ^{13}C NMR spectra of rice straw (A) and rice bran (B) derived biochars prepared at various charring temperatures.

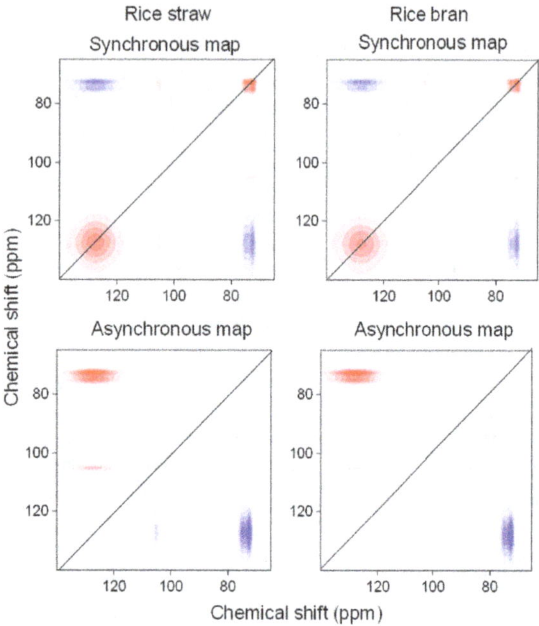

Figure 3. Synchronous and asynchronous 2D ^{13}C NMR maps of rice straw and rice bran derived biochars over charring temperatures (100~600°C). Red represents positive correlation, and blue represents negative correlation; a higher color intensity indicates a stronger positive or negative correlation.

materials. The band at 127 ppm (Figure S1) was overlapped by 140 and 127 ppm for rice straw derived biochars, but by 140, 127 and 117 ppm for rice bran derived biochars. Only one major negative crosspeak located at Ψ(105, 104) was found in charred rice straw; however, both negative and positive crosspeaks located at Ψ(106.5, 105) and Ψ(105, 104) were observed in charred rice bran, showing that the band at 105 ppm presented in the synchronous map (Figure S1) was overlapped by 2 bands (i.e., 105 and 104 ppm) and 3 bands (i.e., 106.5, 105 and 104 ppm) for rice straw and rice bran derived biochars, respectively. Similarly, in the region of 78–70 ppm, 2 [i.e., Ψ(76.4, 73) and Ψ(73, 72)] and 6 [Ψ(76.4, 75.3), Ψ(76.4, 73), Ψ(75.3, 74), Ψ(75.3, 71.5), Ψ(74, 73) and Ψ(73, 71.5)] major crosspeaks were found in rice straw and rice bran derived biochars, respectively, demonstrating the band at 73 ppm (Figure S1) overlapping by 3 bands (76.4, 73 and 72 ppm) and 5 bands 76.4, 75.3, 74, 73, 72 and 71.5 ppm) for rice straw and rice bran derived biochars, respectively.

The sequencing of the change of overlapping bands with increasing charring temperature could also be identified based on Noda's rules [17]. Specifically, the production of aromatic C-O groups (140 ppm) occurred prior to the removal of fused-ring aromatic structures (127 ppm) for rice straw derived biochars; however, the removal of fused-ring aromatic structures (117 and 127 ppm) occurred prior to the production of aromatic C-O groups (140 ppm) for rice bran derived biochars. For anomeric O-C-O carbons the mass cleavage of the band at 104 ppm occurred prior to that of the band at 105 ppm for rice straw derived biochars; however, it follows the order: 105 ppm>104 ppm or 106.5 ppm for rice bran derived biochars. The mass cleavage of aliphatic O-alkylated (HCOH) groups follows the order: 73 ppm>72 ppm>76.4 ppm for rice straw derived biochars and 73 ppm>75.3 ppm>72 ppm>74 ppm>76.4 ppm for rice bran derived biochars. In summary, the asynchronous maps could provide the information about the overlapping bands and be used to identify the plant materials.

The regression analysis between functional groups and charring temperatures demonstrated that functional groups were linear with charring temperatures (Table S1). The slope of the regression equation suggested that the development rate of functional groups followed the order: fused-ring aromatic structures (127 and 128 ppm, slope≈0.03)>aromatic C-O groups (140 ppm, slope≈0.02)>aliphatic O-alkylated (HCOH) carbons (72, 73, 75.3 and 76.4 ppm, slope≈0.002)≈anomeric O-C-O carbons (104, 105 and 106.5 ppm, slope≈0.002). This result reveals that with increasing charring temperatures, the aromatization process is faster than the dealkylation and dehydroxylation/dehydrogenation processes. Meanwhile, one could design the "desired" biochars by choosing a suitable charring temperature.

In summary, 2D ^{13}C NMR correlation spectroscopy demonstrated that the agricultural biomass carbonized to biochars consisted of dealkylation, dehydroxylation/dehydrogenation and aromatization processes, mainly involving the cleavage of aliphatic O-alkylated (HCOH) carbons and anomeric O-C-O carbons, as well as the production of fused-ring aromatic structures and aromatic C-O groups. When compared to one-dimensional ^{13}C NMR spectroscopy, 2D ^{13}C NMR correlation spectroscopy solves the problem of overlapping bands and enables researchers to probe the specific sequencing of spectral intensity changes of biochars.

Establishment of the Relationship between Biochar Properties (pH and EC) and Functional Groups

The development of functional groups determined the pH and EC of biochars. Pearson correlation analysis was used to establish the relationship between biochar properties (pH and EC) and functional groups (Table 2). For rice straw derived biochars, the pH was negatively correlated ($R>-0.82$, $p<0.05$) with aliphatic O-alkylated (HCOH) carbons (72, 73 and 76.4 ppm) and anomeric O-C-O carbons (104 and 105 ppm), but positively correlated with fused-ring aromatic structures (120 and 128 ppm) ($R=0.99$, $p<0.05$) and aromatic C-O groups (140 ppm) ($R=0.94$, $p<0.01$) (Table 2). The EC of charring rice straw was also negatively correlated ($R>-0.86$, $p<0.05$) with aliphatic O-alkylated (HCOH) carbons (72, 73, and 76.4 ppm), but positively correlated with fused-ring aromatic structures (127 and 128 ppm) ($R=0.92$, $p<0.01$) and aromatic C-O groups (140 ppm) ($R=0.89$, $p<0.05$). For rice bran derived biochars, the pH was negatively correlated ($R>-0.82$, $p<0.05$) with aliphatic O-alkylated (HCOH) carbons (72, 73, 75.3 and 76.4 ppm) and anomeric O-C-O carbons (104, 105 and 106.5 ppm), but positively correlated with fused-ring aromatic structures (117, 127 and 128 ppm) ($R>0.93$, $p<0.01$) and aromatic C-O groups (140 ppm) ($R=0.95$, $p<0.01$) (Table 2). However, the EC of charring rice bran had no correlation with any of the functional groups.

The regression analysis between functional groups and biochar properties (pH and EC) (Table S2) further demonstrated that the pH and EC of rice straw derived biochars were mainly determined by fused-ring aromatic structures (127 and 128 ppm) and anomeric O-C-O carbons (104 and 105 ppm), which had a high slope when compared to other functional groups. However, the pH of rice bran derived biochars was determined by both fused-ring aromatic structures (127 and 128 ppm) and aliphatic O-alkylated (HCOH) carbons (72, 73 and 75.3 ppm) with a slope of 0.7.

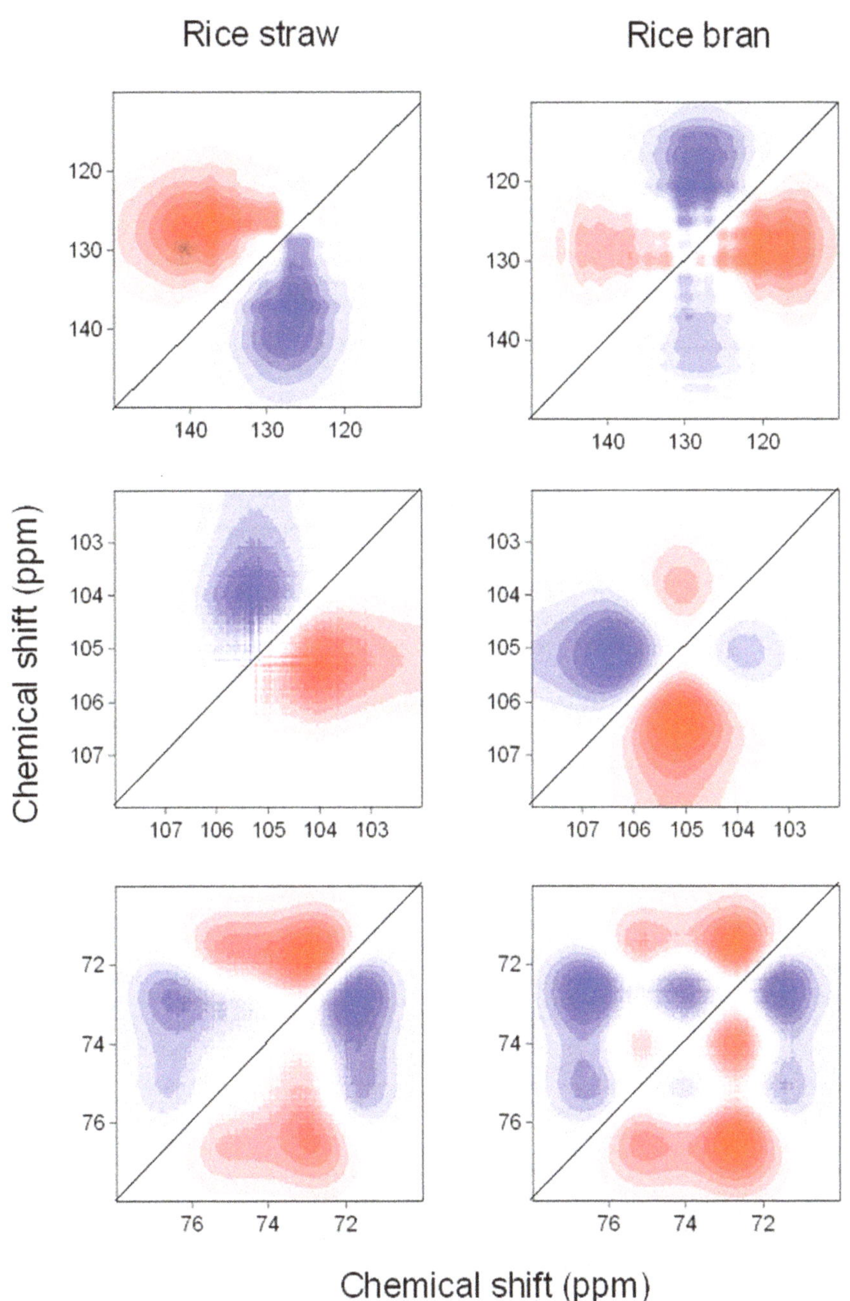

Figure 4. Asynchronous 2D ^{13}C NMR maps of rice straw and rice bran derived biochars over charring temperatures (100~600°C) in a shorten range of bands. Red represents positive correlation, and blue represents negative correlation; a higher color intensity indicates a stronger positive or negative correlation.

Significance of This Work

The application of biochar to soil will enhance soil fertility and carbon sequestration [26]. However, understanding the development of functional groups and the methods that are most appropriate to determining them has lagged [8]. This type of information is needed to optimize the properties of biochar for specific purposes, such as pH amelioration, nutrient retention, or sequestration of soil organic matter. In this study, two-dimensional correlation spectroscopy was employed to investigate the development of functional groups with increasing charring temperature. Two-dimensional correlation spectroscopy could solve the peak overlapping problem of biochars, which is helpful for establishing the relationship between biochar properties and functional groups. It is possible to "design" the desired biochars for a given application by applying 2D correlation spectroscopy. For example, one could produce biochars with a high pH and EC to apply them to acidic soils for pH amelioration, by controlling the high percentage of aromatic carbons (fused-ring aromatic structures and aromatic C-O groups). Moreover, 2D correlation spectroscopy could also give the sequencing of functional groups changes. One could produce the desired biochars with special functional groups based on the sequencing of functional groups changes. In summary, two-dimensional correlation spectroscopy provides

Table 2. Pearson correlation coefficiency (*R*) between properties (pH, EC) and functional groups derived from solid state ^{13}C NMR spectroscopy.[a]

		NMR band (ppm)										
		72	73	75.3	76.4	104	105	106.5	117	127	128	140
pH	Rice straw	−0.87*	−0.87**	NA	−0.88*	−0.82*	−0.82*	NA	NA	0.99**	0.99**	0.94**
	Rice bran	−0.83*	−0.82*	−0.84*	−0.86*	−0.89*	−0.86*	−0.95**	0.93**	0.99**	0.99**	0.95**
EC	Rice straw	−0.87*	−0.86*	NA	−0.88*	−0.79	−0.79	NA	NA	0.92**	0.92**	0.89*
	Rice bran	−0.01	0.01	−0.02	−0.07	−0.14	−0.07	−0.30	0.24	0.52	0.52	0.33

[a]Note:
*$p<0.05$;
**$p<0.01$; EC, electrical conductivity; NA, not available.

novel insights into the development of functional groups in biochars.

Supporting Information

Table S1 Regression Analysis between Functional Groups (I/I_0) with Charring Temperatures (T).

Table S2 Regression Analysis between Biochar Properties (pH and EC) and Functional Groups (I/I_0) Derived from Solid State ^{13}C NMR Spectroscopy.

Figure S1 Synchronous 2D ^{13}C NMR maps of rice straw and sawdust derived biochars over charring temperatures (100~600°C). Red represents positive correlation, and blue represents negative correlation; a higher color intensity indicates a stronger positive or negative correlation.

Author Contributions

Conceived and designed the experiments: GY QS. Performed the experiments: XL XM. Analyzed the data: GY QS WR YX DZ. Contributed reagents/materials/analysis tools: GY. Wrote the paper: GY XL.

References

1. Lehmann J, Gaunt J, Rondon M (2006) Biochar sequestration in terrestrial ecosystems-a review. Mitigation and Adaptation Strategies for Global Change 11: 403–427.
2. Cao XY, Ro KS, Chappell M, Li Y, Mao JD (2011) Chemical structures of swine-manure chars produced under different carbonization conditions investigated by advanced solid-state ^{13}C nuclear magnetic resonance (NMR) spectroscopy. Energy Fuels 25: 388–397.
3. Lehmann J, Joseph S (2009) Biochar for environmental management: an introduction. In Biochar for Environmental Management: Science and Technology; Lehmann, J., Joseph, S., Eds.; Earthscan: London, pp 1–10.
4. Mao JD, Johnson RL, Lehmann J, Olk DC, Neves EG, et al. (2012) Abundant and stable char residues in soils: implications for soil fertility and carbon sequestration. Environ Sci Technol. 46: 9571–9576.
5. Lehmann J (2007) A handful of carbon. Nature 447: 143–144.
6. Meyer S, Glaser B, Quicker P (2011) Technical, economical, and climate-related aspects of biochar production technologies: a literature review. Environ Sci Technol. 45: 9473–9483.
7. Harvey OR, Herbert BE, Rhue RD, Kuo LJ (2011) Metal interactions at the biochar-water interface: energetics and structure-sorption relationships elucidated by flow adsorption microcalorimetry. Environ Sci Technol. 45: 5550–5556.
8. Manya JJ (2012) Pyrolysis for biochar purposes: a review to establish current knowledge gaps and research needs. Environ Sci Technol. 46: 7939–7954.
9. Knicker H, Hilscher A, Gonzalez-Vila FJ, Almendros G (2008) A new conceptual model for the structural properties of char produced during vegetation fires. Org Geochem. 39: 935–939.
10. Cao X, Ma L, Liang Y, Gao B, Harris W (2011) Simultaneous immobilization of lead and atrazine in contaminated soils using dairy manure biochar. Environ Sci Technol. 45: 4884–4889.
11. Kinney TJ, Masiello CA, Dugan B, Hockaday WC, Dean MR, et al. (2012) Hydrologic properties of biochars produced at different temperatures. Biomass and Bioenergy 41: 34–43.
12. Cheng HN, Wartelle LH, Klasson KT, Edwards JC (2010) Solid-state NMR and ESR studies of activated carbons produced from pecan shells. Carbon 48: 2455–2469.
13. Yu GH, Tang Z, Xu YC, Shen QR (2011) Multiple fluorescence labeling and two dimensional FTIR–^{13}C NMR heterospectral correlation spectroscopy to characterize extracellular polymeric substances in biofilms produced during composting. Environ Sci Technol. 45: 9224–9231.
14. Wang LP, Shen QR, Yu GH, Ran W, Xu YC (2012) Fate of biopolymers during rapeseed meal and wheat bran composting as studied by two-dimensional correlation spectroscopy in combination with multiple fluorescence labeling techniques. Bioresour Technol. 105: 88–94.
15. Yu GH, Wu MJ, Wei GR, Luo YH, Ran W, et al. (2012) Binding of organic ligands with Al(III) in dissolved organic matter from soil: implications for soil organic carbon storage. Environ Sci Technol. 46: 6102–6109.
16. Mao JD, Hu WG, Ding GW, Schmidt-Rohr K, Davies G, et al. (2002). Suitability of different ^{13}C solid-state NMR techniques in the characterization of humic acids. Intern J Environ Anal Chem. 82: 183–196.
17. Noda I, Ozaki Y, Eds. (2004) Two-Dimensional Correlation Spectroscopy-Applications in Vibrational and Optical Spectroscopy; John Wiley & Sons: England.
18. Hammes K, Smernik RJ, Skjemstad JO, Schmidt MWI (2008) Characterisation and evaluation of reference materials for black carbon analysis using elemental composition, colour, BET surface area and ^{13}C NMR spectroscopy. Appl Geochem. 23: 2113–2122.
19. Chen BL, Zhou DD, Zhu LZ (2008) Transitional adsorption and partition of nonpolar and polar aromatic contaminants by biochars of pine needles with different pyrolytic temperatures. Environ Sci Technol. 42: 5137–5143.
20. Keiluweit M, Kleber M, Sparrow MA, Simoneit BRT, Prahl FG (2012) Solvent-extractable polycyclic aromatic hydrocarbons in biochar: influence of pyrolysis temperature and feedstock. Environ Sci Technol. 46: 9333–9341.
21. Cantrell KB, Hunt PG, Uchimiya M, Novak JM, Ro KS (2012) Impact of pyrolysis temperature and manure source on physiochemical characteristics of biochar. Bioresour Technol. 107: 419–428.
22. Freitas JCC, Bonagamba TJ, Emmerich FG (1999) ^{13}C high-resolution solid-state NMR study of peat carbonization. Energy Fuels 13: 53–59.
23. Cao XY, Pignatello JJ, Li Y, Lattao C, Chappell MA, et al. (2012) Characterization of wood chars produced at different temperatures using advanced ^{13}C solid-state NMR spectroscopic techniques. Energy & Fuels 26: 5983–5991.
24. Zhang GC, Zhang Q, Sun K, Liu XT, Zheng WJ, et al. (2011) Sorption of simazine to corn straw biochars prepared at different pyrolytic temperatures. Enviorn Pollut. 159: 2594–2601.
25. Rutherford DW, Wershaw RL, Rostad CE, Kelly CN (2012) Effect of formation conditions on biochars: compositional and structural properties of cellulose, lignin, and pine biochars. Biomass and Bioenergy 46: 693–701.
26. Mukherjee A, Zimmerman AR, Harris W (2011) Surface chemistry variations among a series of laboratory-produced biochars. Geoderma 163: 247–255.

15

Potential of Cometabolic Transformation of Polysaccharides and Lignin in Lignocellulose by Soil *Actinobacteria*

Tomáš Větrovský[1], Kari Timo Steffen[2], Petr Baldrian[1]*

1 Laboratory of Environmental Microbiology, Institute of Microbiology of the ASCR, v.v.i., Praha, Czech Republic, 2 Department of Applied Chemistry and Microbiology, University of Helsinki, Helsinki, Finland

Abstract

While it is known that several *Actinobacteria* produce enzymes that decompose polysaccharides or phenolic compounds in dead plant biomass, the occurrence of these traits in the environment remains largely unclear. The aim of this work was to screen isolated actinobacterial strains to explore their ability to produce extracellular enzymes that participate in the degradation of polysaccharides and their ability to cometabolically transform phenolic compounds of various complexities. Actinobacterial strains were isolated from meadow and forest soils and screened for their ability to grow on lignocellulose. The potential to transform ^{14}C-labelled phenolic substrates (dehydrogenation polymer (DHP), lignin and catechol) and to produce a range of extracellular, hydrolytic enzymes was investigated in three strains of *Streptomyces* spp. that possessed high lignocellulose degrading activity. Isolated strains showed high variation in their ability to produce cellulose- and hemicellulose-degrading enzymes and were able to mineralise up to 1.1% and to solubilise up to 4% of poplar lignin and to mineralise up to 11.4% and to solubilise up to 64% of catechol, while only minimal mineralisation of DHP was observed. The results confirm the potential importance of *Actinobacteria* in lignocellulose degradation, although it is likely that the decomposition of biopolymers is limited to strains that represent only a minor portion of the entire community, while the range of simple, carbon-containing compounds that serve as sources for actinobacterial growth is relatively wide.

Editor: Michael Freitag, Oregon State University, United States of America

Funding: This work was supported by the Academy of Sciences of the Czech Republic (IAA603020901) and by the research concept of the Institute of Microbiology of the ASCR, v.v.i. (RVO61388971). The funders had no role in study design, data collection and analysis, decision to publish, or preparation of the manuscript.

Competing Interests: The authors have declared that no competing interests exist.

* E-mail: baldrian@biomed.cas.cz

Introduction

Lignocellulose represents the dominant portion of plant biomass and is thus a key pool of carbon in terrestrial ecosystems. The decomposition of lignocellulose in soil environments, where it originates as aboveground or belowground litter, is thus an essential process of the carbon cycle. Microorganisms represent the key decomposers of lignocellulose in soils and especially fungi are often regarded as major lignocellulose decomposers [1], most likely because their larger, multicellular and often filamentous bodies are better suited for the exploitation of bulky substrates [2]. This potential has led to the evolution of efficient enzymatic systems responsible for the decomposition of biopolymers in several fungi [1,3–5].

The process of lignocellulose decomposition is mediated by extracellular enzymes that target its main components: the polysaccharides cellulose and hemicelluloses and polyphenolic lignin [6]. A wide array of enzymes is necessary for the complete decomposition of lignocellulose. The system for cellulose decomposition typically consists of endocellulases, cellobiohydrolases (exocellulases) and β-glucosidases. The hemicellulolytic system is composed of multiple glycosyl hydrolases that are specific for xylose-, mannose-, arabinose- and galactose-containing polysaccharides; and lignin degradation is mediated by oxidative enzymes, such as oxidases (laccases), peroxidases and auxiliary enzymes, that produce hydrogen peroxide [1,7,8].

Although fungi vary largely in their production of extracellular enzymes, several groups, including saprotrophic wood decomposers and cord-forming fungi, that inhabit litter and soil were shown to produce complete arrays of extracellular enzymes that decompose all of the components of lignocellulose [4,9]. Current advances in genome sequencing indicate that the theoretical potential of bacteria to degrade certain components of lignocellulose, e.g., cellulose, is relatively widespread [10]; and, for certain taxa, enzymes involved in decomposition were characterised [11]. Moreover, recent reports also show that bacteria may play a significant role in cellulose decomposition in soil environments [12]. However, the composition of bacterial enzymatic systems has not been systematically addressed, and it is difficult to estimate their potential to transform individual lignocellulose components.

Actinobacteria seem to be good candidates for efficient lignocellulose decomposition, and their filamentous growth may help them access and utilise polymeric substrates [13]. Therefore, the involvement of certain *Actinobacteria* in the degradation of polysaccharides or phenolic compounds in dead plant biomass is generally accepted [14,15]. This is based on previous reports that suggest the presence of decomposer traits in several actinobacterial taxa. In the case of cellulose, the production of endocellulase by the genera *Streptomyces*, *Cellulomonas* and *Acidothermus* was reported [16–18], while efficient exocellulases often combined with xylanase activity were found in *Thermobifida*, *Cellulomonas* and *Cellulosimicrobium* [19–21]. β-Glucosidases have been characterised in the above genera as well as in *Clavibacter*, *Terrabacter*, *Micrococcus*, *Microbacterium* and *Bifidobacterium* [22–26]. Recently, many putative cellulose-degrading enzymes were found in the sequenced genomes of several *Actinobacteria* [27], and this phylum showed the highest percentage of genomes that harboured putative cellulolytic enzymes. Approximately 1/3 of the 514 characterised genomes harboured at least one putative cellulase [10]. Although information on hemicellulose-degrading enzymes is scarce, individual enzymes were reported in multiple genera, including *Streptomyces*, *Cellulomonas*, *Cellulosimicrobium* and *Kocuria* [28–31].

Although the major lignin degraders are white-rot fungi, there are also many reports about bacterial strains that are able to degrade lignin. In addition to the *Proteobacteria* and *Firmicutes* [32–34], these reports also mention actinobacterial taxa. Studies on the decomposition of natural and synthetic lignins, for example, isolated lignin, prepared ^{14}C-synthetic lignins or model compounds, indicated that the genera *Arthrobacter*, *Nocardia* and *Streptomyces* were capable of lignin utilisation, although their efficiencies varied widely and did not reach the level that was observed in ligninolytic fungi [35–38].

Despite the relative abundance of reports on the decomposer abilities of *Actinobacteria*, the current information remains rather fragmented. Different strains have been studied for the production of individual enzymes, and the abilities to use plant lignocellulose as a growth substrate were not studied in much detail. The aim of this study was to explore the ability of soil *Actinobacteria* to act as decomposers of dead plant biomass. To achieve this goal, a set of natural isolates was screened for their production of cellulolytic enzymes, and active strains were tested for their ability to use lignocellulose in the form of wheat straw as a growth source. The potential decomposers were further screened for the production of multiple lignocellulose-degrading enzymes and for their ability to cometabolically transform ^{14}C-labelled phenolic substances (lignin, DHP and catechol) during their growth on wheat straw. The *Actinobacteria* in this study were isolated from soil with mixed-metal pollution. Due to their heavy metal resistance, *Actinobacteria* are frequently found in such soils, and this group may replace the more sensitive fungi as the main decomposers in such soils [39,40]. We thus expected that the decomposition of lignocellulose would be a common trait of strains isolated from this environment.

Materials and Methods

Isolation of Bacterial Strains

The study was carried out on private land. No specific permission was required for the activities covered by this study. However, for more intensive research activities, individual private owners have to grant the permit to conduct such research. The field studies did not involve endangered or protected species. Bacterial strains were isolated from the organic horizons of the grassland and forest soils near Příbram, Czech Republic (49°42′22.207″N, 13°58′27.296″E). The soil is a cambisol with pH of approximately 5.5 and has a clay/silt/sand ratio of approx. 40:30:30% and an elevated heavy metal content (namely, Cu, Zn, Cd, Zn, Pb and As) due to its location near a polymetallic smelter [41]. The study was performed on private land; no specific permit was required for the activities performed as part of this study.

Physical and chemical treatments were used for the selective isolation of soil *Actinobacteria*. Soil samples were pre-treated by dry heating (120°C) and phenol treatment (1.5%), and water extracts

Table 1. Screening of actinobacterial strains for their ability to produce cellobiohydrolase and 1,4-β-glucosidase.

Strain	Cellobiohydrolase	1,4-β-Glucosidase
pl18	++	+++
pl21	–	+
pl23	+	++
pl28	+	++
pl36	+++	+++
pl41	+++	+++
pl67	++	–
pl70	–	++
pl73	–	+
pl75	–	+
pl77	–	+++
pl80	–	+
pl81	–	+
pl84	+	++
pl86	+	–
pl88	+++	+++
pl95	+++	+++
pl98	–	+
pl100	–	+
pl107	–	++
pl112	–	+
pl116	+++	+++
pl118	+	+++
pl123	+++	+++
pl124	++	+
pl129	+	+
pl131	–	++
pl134	–	+++
pl136	–	++
pl138	–	+++
pl149	+	++
pl150	–	+++
pl153	+++	+++
pl154	++	+++
pr10	+	–
pr22	–	+++
pr24	+	–
pr3	+	+
pr30	+	+++
pr4	–	+
pr40	++	–
pr41	+++	++
pr45	–	+++
pr48	–	++
pr49	+	+
pr52	++	–
pr55	+++	+
pr57	+	++

Table 1. Cont.

Strain	Cellobiohydrolase	1,4-β-Glucosidase
pr6	+++	++
pr7	+	–
pr9	++	–

Twenty-five isolates produced neither of the enzymes. Legend: "+++" high production of enzyme (activity >100 mU/mL), "++" average production (100> activity >10), small production (activity <10) and "–" no enzyme production.

were used to inoculate plates containing selective media, either humic acid-vitamin agar (1 g L^{-1} humic acid, 0.5 g L^{-1} Na_2HPO_4, 7.7 g L^{-1} KCl, 0.05 g L^{-1} $MgSO_4.7H_2O$, 0.01 g L^{-1} $FeSO_4.7H_2O$, 0.02 g L^{-1} $CaCO_3$, B-vitamins: 0.5 mg L^{-1} thiamine-HCl, riboflavin, niacin, pyridoxine, inositol, Ca-pantothenate, *p*-aminobenzoic acid and 0.25 mg L^{-1} biotin, 18 g L^{-1} agar, pH 7.2) or lignin-soy bean flour-vitamin agar containing soil extract (1 g L^{-1} lignin, 0.2 g L^{-1} soy bean flour, 0.5 g L^{-1} Na_2HPO_4, 7.7 g L^{-1} KCl, 0.05 g L^{-1}, $MgSO_4.7H_2O$, 0.01 g L^{-1} $FeSO_4.7H_2O$, 0.02 g L^{-1} $CaCO_3$, B-vitamins (see above), 100 mL L^{-1} soil extract, 18 g L^{-1} agar, pH 7.5) that was supplemented with the antibiotics kanamycin (20 mg L^{-1}) and nalidic acid (10 mg L^{-1}) [42,43]. Pure cultures of bacteria were obtained from agar plates, and those strains that were identified as *Actinobacteria* were retained. Strains were stored in a sterile, 50% glycerol solution in 25 mM Tris at −20°C and subcultured on GYM agar (4 g L^{-1} glucose, 4 g L^{-1} yeast extract, 10 g L^{-1} malt extract, 2 g L^{-1} $CaCO_3$, 12 g L^{-1} agar, pH 7.2) at 25°C.

Screening for Lignocellulose-degrading Strains

Efficient decomposition of cellulose, which is the major component of dead plant biomass, depends on the production of cellobiohydrolase and β-glucosidase. To screen for the production of these two enzymes, isolated actinobacterial strains were cultivated in liquid GYM medium for 14 days at 25°C without agitation (three replicates). The cultivation liquid was collected, and the activities of the extracellular enzymes were measured spectrophotometrically as described previously [44]. Cellobiohydrolase (exocellulase, EC 3.2.1.91) and 1,4-β-glucosidase (EC 3.2.1.21) were assayed using 4-methylumbelliferyl-β-D-cellobioside and 4-methylumbelliferyl-β-D-glucopyranoside, respectively, in 50 mM sodium acetate buffer (pH 5.0). The reaction mixtures were incubated at 40°C for 120 min and terminated by sodium carbonate addition [44].

The activity of each strain was ranked on a scale of negative, low, medium or high. Cellobiohydrolase: negative = 0−0.5 μmol min^{-1} mL^{-1}, low = from >1 to 15 μmol min^{-1} mL^{-1}, medium from >15 to 50 μmol min^{-1} mL^{-1}, and high >50 μmol min^{-1} mL^{-1}. 1,4-β-glucosidase: negative = 0–1 μmol min^{-1} mL^{-1}, low = from >1 to 50, medium from >50 to 200 μmol min^{-1} mL^{-1}, and high >200 μmol min^{-1} mL^{-1}. Strains that produced cellobiohydrolase and 1,4-β-glucosidase and exhibited high activity of at least one of those enzymes were identified, and their ability to grow on lignocellulose as a carbon source was examined. One gram of air-dried, milled wheat straw was added into 100-mL, thick-walled flasks to form a uniform layer. Each flask was supplemented with 5 mL of distilled water and sterilised by autoclaving (2×30 min at 121°C with cooling to room temperature between the two cycles). The flasks were inoculated with 1 mL of cell suspension that had been pre-grown for three days on liquid GYM media. Triplicate flasks for each strain were incubated for 21 days at 25°C. After incubation, the enzymes were extracted in 15 mL of distilled water, and the extracts were filtered and used for enzyme activity measurements.

The activities of 1,4-β-glucosidase, cellobiohydrolase and 1,4-β-xylosidase in the extracts were assessed using 4-methylumbelliferyl-β-D-glucopyranoside, MUF-β-D-cellobioside and MUF-β-D-xylopyranoside, respectively, in 50 mM sodium acetate buffer, pH 5.0, as previously described [44]. Substrates (100 μL in DMSO) at a final concentration of 500 μM were combined with the three technical replicates of the 100-μL extracts in a 96-well multiwell plate. For the background fluorescence measurement, 100 μL of sodium acetate buffer was combined with 100 μL of the 4-methylumbelliferol standards to correct for fluorescence quenching. The multiwell plates were incubated at 40°C, and fluorescence was recorded from 5 min to 125 min using the Infinite microplate reader (TECAN, Austria) at an excitation wavelength of 355 nm and an emission wavelength of 460 nm. The quantitative enzymatic activities after blank subtraction were calculated based on standard curves of 4-methylumbelliferone, and enzyme activity was expressed per g of straw dry mass.

Of the 14 strains, only six exhibited visual growth and produced extracellular enzymes in straw. Among these, three strains, pl88, pr6 and pr55, that highly produced cellobiohydrolase were selected for the detailed characterisation of glycosyl hydrolase production and the decomposition of phenolic compounds.

To analyse the spectra of the extracellular enzymes that were produced by the bacterial strains, pl88, pr6 and pr55 were cultivated on diluted GYM media with either cellulose as a specific inducer or with finely milled wheat straw (mesh size 0.2 mm) as a complex inducer. In 50-mL flasks, 10 mL of 10× diluted GYM media was combined with 50 mg of cellulose or wheat straw and sterilised by autoclaving. The flasks were inoculated with 100 μL of cell suspension that had been pre-grown for three days in liquid GYM media. Triplicate flasks for each strain were incubated for 21 days at 25°C. After incubation, the 1,4-β-glucosidase, cellobiohydrolase and 1,4-β-xylosidase activities were measured, as described above. The activities of 1,4-α-glucosidase, 1,4-α-arabinosidase, 1,4-β-galactosidase, 1,4-β-mannosidase and 1,4-β-glucuronidase in the extracts were assessed using 4-methylumbelliferyl-α-D-glucopyranoside, 4-methylumbelliferyl-α-L-arabinopyranoside, 4-methylumbelliferyl-β-D-galactopyranoside, 4-methylumbelliferyl-β-D-mannopyranoside and 4-methylumbelliferyl-β-D-glucuronide, respectively, and the same method. The activities of endo-1,4-β-glucanase (endocellulase) and endo-1,4-β-xylanase (endoxylanase) were assayed using azo-dyed carboxymethyl cellulose and birchwood xylan, respectively, according to the manufacturer's instructions (Megazyme, Ireland). Reaction mixture containing 0.2 mL of a 2% dyed substrate in 200 mM sodium acetate buffer, pH 5.0, and 0.2 mL of sample was incubated at 40°C for 60 min, and the reaction was ended by adding 1 mL of ethanol followed by 10 s of vortexing and 10 min of centrifugation (10,000×g) [45]. The amount of released dye was measured at 595 nm, and the enzyme activity was calculated according to standard curves that correlated dye release with the release of reducing sugars.

Transformation of ^{14}C-labelled Phenolic Compounds

The transformation of phenolic compounds of various complexities was studied using ^{14}C-labelled compounds. $^{14}C_{\beta}$-labelled dehydrogenation polymer (^{14}C-DHP) was synthesised according to Brunow [46] and dissolved in a N,N-dimethylformamide-water suspension (1:20 v/v) [47]. $^{14}C_{\beta}$-labelled lignin was extracted from labelled poplar trees that were prepared according to Odier [48] and used as a solid material. $^{14}C_{\beta}$-labelled catechol in an ethanol

Table 2. Identification of selected actinobacterial strains and the accession numbers of their partial 16S rRNA gene sequences.

Strain	Accession No	Closest hit	Accession No	Similarity (%)	Coverage (%)
pl18	KC789721	*Micromonospora saelicesensis* strain L6	JN862845	99.6	99.3
pl36	KC789723	*Streptomyces ciscaucasicus* strain HBUM83169	EU841585	99.9	99.8
pl41	KC789724	*Curtobacterium flaccumfaciens* strain LMG 3645	NR025467	100.0	99.3
pl88	KC789730	*Streptomyces atratus* strain HBUM173340	FJ486302	100.0	94.2
pl95	KC789731	*Streptomyces aureus* strain HBUM174596	EU841581	99.4	100.0
pl116	KC789734	*Kribbella antibiotica* strain YIM 31530	NR029048	98.8	99.8
pl118	KC789719	*Streptomyces sanglieri* strain IHB B 6004	KF475877	99.6	99.9
pl123	KC789735	*Nocardia exalbida* W9709	GQ376167	99.2	100.0
pl153	KC789738	*Streptomyces hygroscopicus* subsp. *geldanus* strain NBRC14620	AB184606	99.8	99.9
pl154	KC789739	*Microbispora rosea* subsp. *rosea* strain A011	AB369120	99.8	95.6
pr6	KC789740	*Streptomyces mauvecolor* strain 7534	JN180187	99.2	99.3
pr30	KC789741	*Amycolatopsis saalfeldensis* strain HKI 0474	DQ792502	99.2	100.0
pr41	KC789742	*Streptomyces setonensis* strain 17-1	EU367980	99.4	99.1
pr55	KC789744	*Streptomyces sannanensis* strain 126195	JN180213	98.9	99.4

solution (Sigma) that was mixed with water (1:18.75 v/v) was used directly.

The cometabolic transformation of phenolic compounds was studied in 100-mL, thick-walled flasks containing 1 g of air-dried and milled wheat straw, which was added to form a uniform layer. Each flask was supplemented with 5 mL of distilled water and sterilised by autoclaving (2×30 min at 121°C with cooling to room temperature between the two cycles). The flasks were inoculated with 1 mL of cell suspension that had been pre-grown for three days in liquid GYM media (five flasks per strain). Control flasks were left uninoculated. The following day, ^{14}C-labelled DHP, catechol and poplar lignin were added. In the DHP flasks, 750 µL of ^{14}C-DHP in a N,N-dimethylformamide-water suspension was added drop-wise onto the surface of the straw layer, which resulted in a final radioactivity of 127,500 dpm per flask. In the catechol flasks, 750 µL of ^{14}C-catechol in an ethanol-water solution was added, which resulted in a final radioactivity of 550,500 dpm per flask. In the lignin flasks, 10 mg of fine, ^{14}C-poplar lignin powder was added onto the surface of the straw layer, which resulted in a final radioactivity of 230,000 dpm per flask, and then, 750 µL of sterile water was added. The flasks were sealed with rubber septa and aluminium caps.

Incubation proceeded for 76 days at 24°C in the dark. Volatile compounds were flushed out of the flasks every week using sterile air, and CO_2 was trapped by bubbling the released air through two sequential flasks containing Opti-Fluor and Carbosorb/Opti-Fluor (Packard Instruments) every week. A liquid scintillation counter (Wallac 1411, WallacOy, Finland) was used to quantify the trapped $^{14}CO_2$ during the experiment. After incubation, the flasks supplemented with the ^{14}C-labelled substances were stored at −18°C until mass-balance extraction.

The flasks containing the residual ^{14}C material were extracted twice with 6 mL of distilled water. After the addition of water, the incubation flasks were shaken for 2 h (280 rpm) at room temperature on a table rotary shaker. The suspension was then poured into a 60-mL syringe containing a pre-weighed cotton plug, and the aqueous extract was pushed through the syringe. For measurements, 1 mL of water extract was diluted with 7 mL of distilled water, and the dilutions were mixed with 10 mL of LUMAGEL. The radioactivity of each extract was measured using a liquid scintillation counter (Model 1411, WallacOy, Finland). The cotton plugs that were used in the filtrations and the residual solid material were air dried and then burned in a combustion chamber (Junitek, Finland). The radioactivity was then counted using a liquid scintillation counter as reported previously [49]. The efficiency of combustion was verified using ^{14}C-labelled standards.

A one-way analysis of variance with the Fisher's least significant difference *post hoc* test was used to analyse the statistical significance of differences among treatments. Differences with a $P<0.05$ were regarded as statistically significant.

Identification of Actinobacterial Strains

DNA was isolated from the actinobacterial biomass that was obtained by cultivation in liquid GYM medium using the modified Miller-SK method [50]. Isolated genomic DNA was used as a template in PCR reactions using universal primers for the bacterial

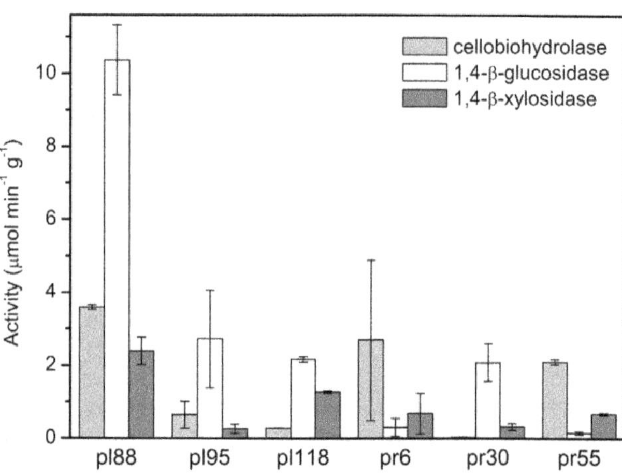

Figure 1. Production of cellobiohydrolase, β-glucosidase and β-xylosidase by *Actinobacteria*. Activity of cellobiohydrolase, 1, 4-β-glucosidase and 1,4-β-xylosidase after a 21-day cultivation of the selected actinobacterial strains on wheat straw. The data represent the means and standard errors.

and archaeal 16 S rRNA gene, pH-T7 (5'-TAATACGACTCACTATAGAGAGTTTGATCCTGGCTCAG-3') and pA (5'-AAGGAGGTGATCCAGCCGCA-3'). Each 50-µl reaction mixture contained 5 µl of 10× buffer for DyNAzyme DNA Polymerase (Finnzymes), 3 µl of purified BSA (10 mg mL^{-1}), 2 µl of each primer (0.01 mM), 1 µl of PCR Nucleotide Mix (10 mM each), 1.5 µl of DyNAZyme II DNA Polymerase (2 U μl^{-1}, Finnzymes) and 1 µl of isolated genomic DNA. The cycling conditions were as follows: 1× 94°C 5 min, 35× (94°C 1 min, 57°C for 45 s min and 72°C for 90 s) followed by 72°C for 10 min. The PCR products were directly sequenced by Macrogen (Seoul, Korea), and the sequences were manually edited using the BioEdit program (http://www.mbio.ncsu.edu/bioedit/bioedit.html) and corrected prior to a BLASTn search against the nucleotide database at the NCBI (http://www.ncbi.nlm.nih.gov/blast).

Results and Discussion

Seventy-six strains of soil *Actinobacteria* were isolated from the soils of the study area and screened for their ability to produce enzymes involved in cellulose decomposition, including cellobiohydrolase and 1,4-β-glucosidase. Of these strains, 32% did not produce any of the tested enzymes, while 31% produced both of them. The production of 1,4-β-glucosidase was more common (57% of strains) than that of cellobiohydrolase (41% of strains; Table 1). The percentage of strains that produced cellobiohydrolase roughly corresponded to the percentage of actinobacterial strains that harboured a gene for endocellulase or cellobiohydrolase (i.e., the glycosyl hydrolase family GH5, 6, 8, 9, 12, 44, 45 or 48), which was one-third of all of the sequenced actinobacterial genomes that were analysed in a recent study [10]. The percentage of strains that did not produce detectable amounts of any enzyme (32%) was higher than what was inferred from the analysis of the genomes, which was less than 20% of the genomes [10]. It is thus possible that some of the strains that harbour 1,4-β-glucosidase do not express the gene or show only low expression levels.

Fourteen strains that produced both enzymes, and highly produced at least one, were selected for further studies and were identified by 16S rRNA sequencing. Of these, eight strains showed highest similarity with members of the genus *Streptomyces*, while the best hits for the others were from the genera *Amycolatopsis*, *Curtobacterium*, *Kribbella*, *Microbispora*, *Micromonospora* and *Nocardia* (Table 2). The activity of cellobiohydrolase and 1,4-β-glucosidase in the genera *Amycolatopsis*, *Kribella*, *Micromonospora*, *Nocardia* and *Streptomyces* corresponded well with the presence of the corresponding genes in their genomes [10]. The currently analysed genomes of *Nocardia* did not contain a cellobiohydrolase gene, and the genomes of *Curtobacterium* and *Microbispora* have not been analysed. Despite their high cellulolytic activity, only six of the fourteen analysed isolates (pl88, pl95, pl118, pr6, pr30 and pr55) showed visually detectable growth on milled wheat straw after 21 days of culturing. All of these strains produced extracellular glycosyl hydrolases: cellobiohydrolase, 1,4-β-glucosidase and 1,4-

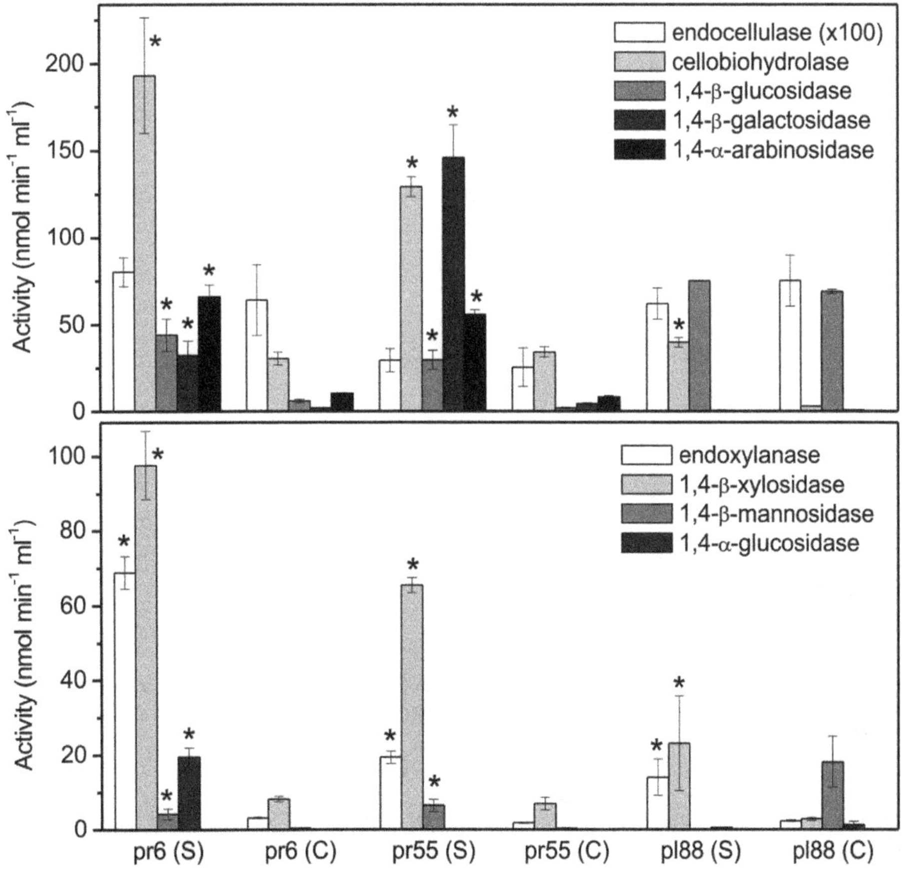

Figure 2. Production of hydrolytic enzymes by selected *Actinobacteria*. Activity of glycosyl hydrolases after a 21-day cultivation of the selected actinobacterial strains on wheat straw (S) and cellulose (C). The data represent the means and standard errors. The activity of endocellulase was multiplied 100× to fit the same scale. Asterisks indicate significant difference ($P<0.05$) in enzyme activity among treatments.

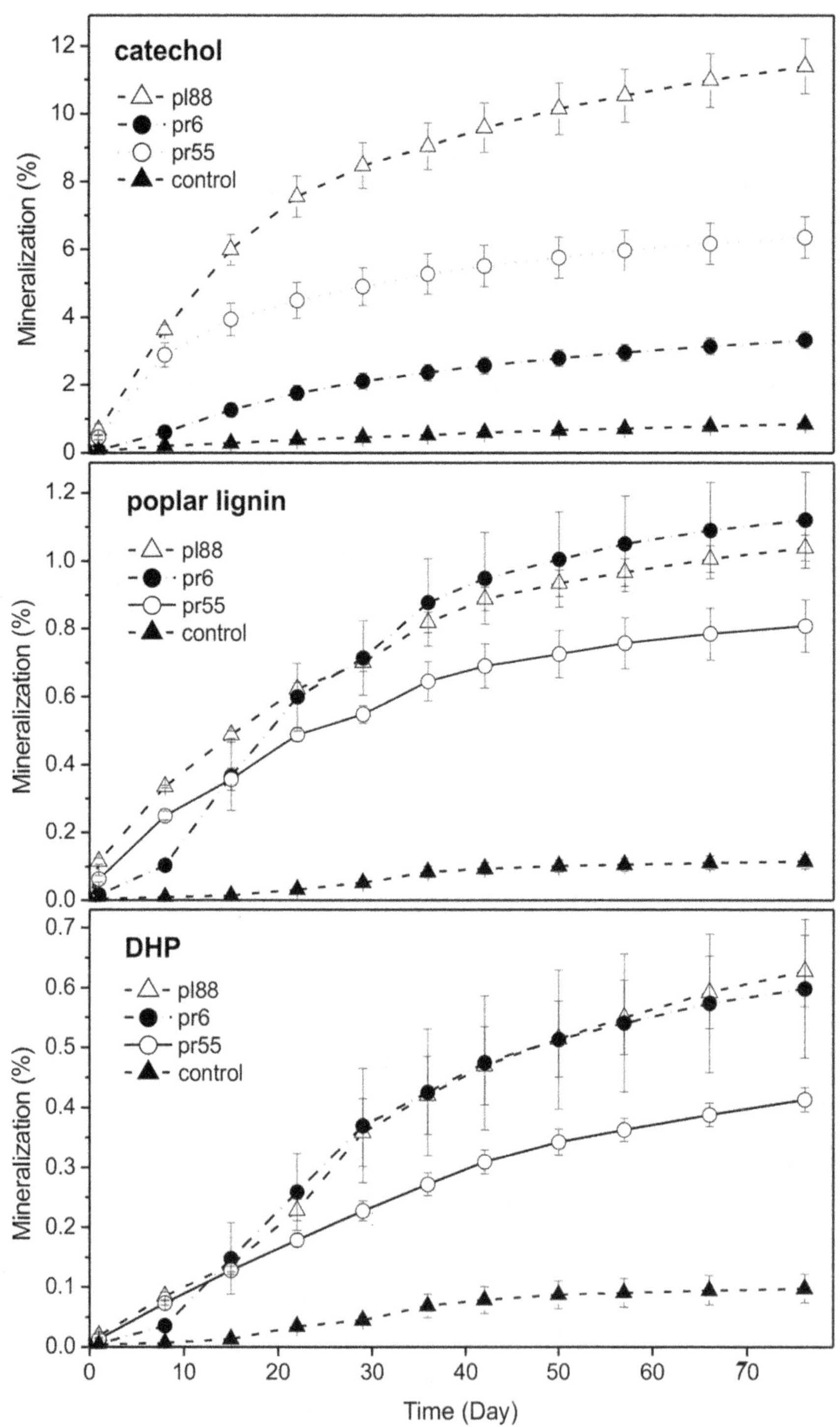

Figure 3. Mineralization of phenolic compounds by selected *Actinobacteria*. Time course of $^{14}CO_2$ production during the transformation of ^{14}C-catechol, ^{14}C-poplar lignin and ^{14}C-DHP in wheat straw microcosms by the selected actinobacterial strains. Control treatments contained sterile straw. The data represent the means and standard errors.

β-xylosidase, although their activities differed (Figure 1). The strains that highly produced cellobiohydrolase (>2 μmol min^{-1} g^{-1} straw dry mass) were selected for further experiments. Cellobiohydrolase represents the rate-limiting enzyme in the decomposition of cellulose, the most abundant and rapidly decomposable polysaccharide in plant litter [12,51].

All studied strains produced a complete set of cellulolytic enzymes: endocellulase, cellobiohydrolase and 1,4-β-glucosidase, and all strains produced the xylanolytic enzymes endoxylanase and 1,4-β-xylosidase as well as the hemicellulases 1,4-β-galactosidase and 1,4-β-mannosidase. 1,4-α-Arabinosidase was only produced by pr6 and pr55, while amylase was produced by pr6

Table 3. Mass balance of ^{14}C-catechol, ^{14}C-poplar lignin and ^{14}C-DHP lignin after a 76-day incubation in wheat straw microcosms with the selected actinobacterial strains.

Substrate	Strain	Respired		Soluble		Bound	
catechol	pl88	11.64±0.85	a	65.2±0.7	c	23.2±0.6	c
	pr6	3.33±0.29	b	70.6±1.1	b	26.1±0.8	b
	pr55	6.36±0.60	c	74.4±0.5	a	19.2±1.1	d
	control	0.94±0.02	d	52.2±0.1	d	46.8±0.1	a
poplar lignin	pl88	1.04±0.03	a	2.7±0.1	c	96.2±0.1	b
	pr6	1.12±0.13	a	4.0±0.2	a	94.9±0.2	c
	pr55	0.81±0.07	b	3.4±0.2	b	95.8±0.2	b
	control	0.09±0.04	c	2.0±0.2	d	97.9±0.3	a
DHP	pl88	0.67±0.06	a	21.7±0.4	a	77.7±0.4	b
	pr6	0.64±0.11	a	22.4±0.5	a	76.9±0.6	b
	pr55	0.45±0.02	b	23.1±0.5	a	76.4±0.6	b
	control	0.11±0.03	c	18.9±1.1	b	81.0±1.1	a

The data (% of the total) represent the means and standard errors. Different letters indicate statistically significant differences at $P<0.05$.

and pl88. The bacteria produced the same enzymes, regardless of whether cellulose or milled wheat straw was used as the carbon source (Figure 2). However, wheat straw, which is a complex substrate that contains various polysaccharides, increased the production of most hemicellulases and cellobiohydrolase and, in the case of pr6 and pr55, the production of 1,4-β-glucosidase. The titres of exocellulase were similar in both treatments. Remarkably, the production of endocellulase was 10–100× lower than that of endoxylanase, and the activities of 1,4-β-xylosidase on straw were higher than that of 1,4-β-glucosidase. Unfortunately, the production of xylanase was not studied in *Actinobacteria* in sufficient detail; however, studies on environmental isolates indicated that this ability might be relatively common [52]. When the genes of the most common endoxylanase, GH10, were screened across a range of different soils, actinobacterial genes were recovered with a relatively high frequency [53]. The activity of 1,4-β-arabinosidase was higher than that of 1,4-β-glucosidase in the pr6 and pr55 strains. It is thus possible that hemicellulose, and particularly xylan, represents a preferred source for lignocellulose-degrading *Actinobacteria*. Although cellulose decomposition was found to be more rapid in decomposing plant litter than that of hemicelluloses, the utilisation of xylose and arabinose-containing hemicelluloses was relatively fast [51]. Although the production of 1,4-β-mannosidase, 1,4-β-galactosidase and 1,4-α-arabinosidase by individual taxa of *Actinobacteria* was reported previously [54–56], we show that they belong to a set of enzymes that are simultaneously produced by saprotrophic taxa during their growth on lignocellulose. The enzyme 1,4-β-glucuronidase, which is involved in the degradation of pectins, was not produced by the tested strains.

To assess whether bacteria that utilise polysaccharides are also able to cometabolise, that is, transform or even mineralise the phenolic compounds within lignocellulose, their transformation of ^{14}C-labelled phenolic compounds was studied. Transformation was studied using compounds of various complexities that included the monomeric catechol (the precursor of lignin), the nonspecifically labelled poplar lignin that was isolated from plant tissues and $^{14}C_{\beta}$-labelled DHP, which is a specifically labelled lignin model compound. After 76 days of cultivation of the bacteria on wheat straw supplemented with the ^{14}C-labelled compounds, considerable mineralisation was only found for catechol: 11.4% for strain pl88, 6.4% for pr55 and 3.33% for pr6 (Figure 3). Mineralisation of poplar lignin was higher in pr6 and pl88 (>1%) than in pr55 (0.81%), and the same result was found for DHP, where mineralisation by pr55 was also lower than that of the other two strains. In all cases, inoculation with *Actinobacteria* resulted in significantly higher mineralisation than that of control flasks. The mass balance extraction of wheat straw showed that, in addition to the mineralisation of phenolic compounds, *Actinobacteria* increased the relative share of water-soluble phenolics, and this was most apparent for catechol, where insoluble ^{14}C compounds represented 47% of the total in the control, while they were 19–26% of the control after incubation with bacteria. The amounts of soluble ^{14}C in the DHP- and lignin-containing flasks also increased after bacterial treatment, but only slightly, by 2–4% and 0.7–2.0%, respectively (Table 3).

The mineralisation of lignin by *Streptomycetes* was first reported by Crawford, who found that mineralisation rates of ^{14}C-lignin-labelled fir varied from 1.5 to 3% after nearly 42 days [57]. Similar rates were also reported by Pasti et al. [58]. Occasionally, higher lignin mineralisation rates were reported, such as 2.9% after 10 days by *Arthrobacter* sp. [59] or up to 5% after 15 days by *Nocardia* sp. [38]. In comparison, the mineralisation rates that were observed in our isolates seemed to be relatively low, but care must be taken in these comparisons because the modes of lignin labelling and preparation greatly affect mineralisation rates. The mineralisation rates of the recalcitrant lignin model compound ^{14}C-DHP that were reported in this study indicates that the ability of the studied *Actinobacteria* to catalyse the complete decomposition of lignin is low and that cometabolic lignin degradation during the growth of active decomposers on lignocellulose is rather negligible.

Although solubilisation is the first important step in lignin decomposition because it makes the polar lignin residues available to other microorganisms, only a few studies have reported the amount of solubilised lignin. Our results indicate only a limited solubilisation of lignin, and these data are comparable to those published for several *Streptomycetes* (6–10%) [58].

Catechol is the first intermediate product of phenol degradation, and its cleavage is a critical step in the aerobic degradation of aromatic compounds in microorganisms [60]. Several bacterial strains that are capable of catechol degradation belong to various phyla, such as *Actinobacteria*, *Alphaproteobacteria*, *Betaproteobacteria* or *Gammaproteobacteria*. Catechol degradation was investigated mainly in studies that focused on the treatment of phenolic wastes, and efficient degraders were found in the genera *Pseudomonas*, *Acinetobacter* and *Klebsiella* [61]. All of the isolates in our study were able to efficiently degrade or solubilise catechol. Although the importance of phenolics as a growth substrate for bacteria is not clear, low molecular mass phenolic compounds are relatively common in litter [62] and their transformation might be of importance.

The results confirm the potential importance of *Actinobacteria* in lignocellulose degradation, although it is likely that the decomposition of biopolymers is limited to strains that represent only a minor portion of the entire community. Our results indicate that strains that are capable of growth on a complex lignocellulose substrate exist; these taxa are able to decompose cellulose and hemicelluloses, and at least some of them may prefer the latter. The importance of cometabolic degradation of lignin seems to be limited when compared with the abilities of saprotrophic fungi, but *Actinobacteria* may still contribute to the solubilisation of phenolics, especially low-molecular-mass compounds. Future studies using genomic, transcriptomic or proteomic approaches are needed to explore the link between the genetic potential of *Actinobacteria* and their actual activities as decomposers of organic matter.

Acknowledgments

Markéta Marečková and Jan Kopecký are acknowledged for their help with the isolation of *Actinobacteria*.

Author Contributions

Conceived and designed the experiments: TV PB. Performed the experiments: TV KS. Analyzed the data: TV KS PB. Wrote the paper: TV PB.

References

1. Baldrian P, Šnajdr J (2011) Lignocellulose-Degrading Enzymes in Soils. In: Shukla G, Varma A, editors. Soil Enzymology. Berlin: Springer-Verlag Berlin. p.167–186.
2. de Boer W, Folman LB, Summerbell RC, Boddy L (2005) Living in a fungal world: impact of fungi on soil bacterial niche development. FEMS Microbiol Rev 29: 795–811.
3. Baldrian P (2008) Enzymes of Saprotrophic Basidiomycetes. In: Boddy L, Frankland JC, VanWest P, editors. Ecology of Saprotrophic Basidiomycetes. San Diego: Elsevier Academic Press Inc. p.19–41.
4. Eastwood DC, Floudas D, Binder M, Majcherczyk A, Schneider P, et al. (2011) The Plant Cell Wall–Decomposing Machinery Underlies the Functional Diversity of Forest Fungi. Science 333: 762–765.
5. Floudas D, Binder M, Riley R, Barry K, Blanchette RA, et al. (2012) The paleozoic origin of enzymatic lignin decomposition reconstructed from 31 fungal genomes. Science 336: 1715–1719.
6. Sjöström E (1993) Wood chemistry: Fundamentals and applications. 2nd edition ed: Academic Press (San Diego).
7. Martinez AT, Speranza M, Ruiz-Duenas FJ, Ferreira P, Camarero S, et al. (2005) Biodegradation of lignocellulosics: microbial chemical, and enzymatic aspects of the fungal attack of lignin. Int Microbiol 8: 195–204.
8. Theuerl S, Buscot F (2010) Laccases: toward disentangling their diversity and functions in relation to soil organic matter cycling. Biol Fertil Soils 46: 215–225.
9. Baldrian P, Voříšková J, Dobiášová P, Merhautová V, Lisá L, et al. (2011) Production of extracellular enzymes and degradation of biopolymers by saprotrophic microfungi from the upper layers of forest soil. Plant Soil 338: 111–125.
10. Berlemont R, Martiny AC (2013) Phylogenetic Distribution of Potential Cellulases in Bacteria. Appl Environ Microbiol 79: 1545–1554.
11. Lynd LR, Weimer PJ, van Zyl WH, Pretorius IS (2002) Microbial cellulose utilization: Fundamentals and biotechnology. Microbiol Mol Biol Rev 66: 506–577.
12. Štursová M, Žifčáková L, Leigh MB, Burgess R, Baldrian P (2012) Cellulose utilization in forest litter and soil: identification of bacterial and fungal decomposers. FEMS Microbiol Ecol 80: 735–746.
13. Chater KF, Biro S, Lee KJ, Palmer T, Schrempf H (2010) The complex extracellular biology of *Streptomyces*. FEMS Microbiol Rev 34: 171–198.
14. Warren RAJ (1996) Microbial hydrolysis of polysaccharides. Annu Rev Microbiol 50: 183–212.
15. McCarthy AJ (1987) Lignocellulose-degrading actinomycetes. FEMS Microbiol Rev 46: 145–163.
16. Enkhbaatar B, Temuujin U, Lim JH, Chi WJ, Chang YK, et al. (2012) Identification and Characterization of a Xyloglucan-Specific Family 74 Glycosyl Hydrolase from *Streptomyces coelicolor* A3(2). Appl Environ Microbiol 78: 607–611.
17. Sun Y, Cheng JJ, Himmel ME, Skory CD, Adney WS, et al. (2007) Expression and characterization of *Acidothermus cellulolyticus* E1 endoglucanase in transgenic duckweed *Lemna minor* 8627. Biores Technol 98: 2866–2872.
18. Yin LJ, Huang PS, Lin HH (2010) Isolation of Cellulase-Producing Bacteria and Characterization of the Cellulase from the Isolated Bacterium *Cellulomonas* Sp YJ5. J Agric Food Chem 58: 9833–9837.
19. Song JM, Wei DZ (2010) Production and characterization of cellulases and xylanases of *Cellulosimicrobium cellulans* grown in pretreated and extracted bagasse and minimal nutrient medium M9. Biomass Bioenergy 34: 1930–1934.
20. Yoon MH, Choi WY (2007) Characterization and action patterns of two beta-1,4-glucanases purified from *Cellulomonas uda* CS1-1. J Microbiol Biotechnol 17: 1291–1299.
21. Zhang F, Hu SN, Chen JJ, Lin LB, Wei YL, et al. (2012) Purification and partial characterisation of a thermostable xylanase from salt-tolerant *Thermobifida halotolerans* YIM 90462(T). Process Biochem 47: 225–228.
22. An DS, Cui CH, Lee HG, Wang L, Kim SC, et al. (2010) Identification and Characterization of a Novel *Terrabacter ginsenosidimutans* sp nov beta-Glucosidase That Transforms Ginsenoside Rb1 into the Rare Gypenosides XVII and LXXV. Appl Environ Microbiol 76: 5827–5836.
23. Fan HX, Miao LL, Liu Y, Liu HC, Liu ZP (2011) Gene cloning and characterization of a cold-adapted beta-glucosidase belonging to glycosyl hydrolase family 1 from a psychrotolerant bacterium *Micrococcus antarcticus*. Enzyme Microb Technol 49: 94–99.
24. Nakano H, Okamoto K, Yatake T, Kiso T, Kitahata S (1998) Purification and characterization of a novel beta-glucosidase from *Clavibacter michiganense* that hydrolyzes glucosyl ester linkage in steviol glycosides. J Ferment Bioeng 85: 162–168.
25. Nunoura N, Ohdan K, Yano T, Yamamoto K, Kumagai H (1996) Purification and characterization of beta-D-glucosidase (beta-D-fucosidase) from *Bifidobacterium breve* clb acclimated to cellobiose. Biosci Biotechnol Biochem 60: 188–193.
26. Quan LH, Min JW, Jin Y, Wang C, Kim YJ, et al. (2012) Enzymatic Biotransformation of Ginsenoside Rb1 to Compound K by Recombinant beta-Glucosidase from *Microbacterium esteraromaticum*. J Agric Food Chem 60: 3776–3781.
27. Anderson I, Abt B, Lykidis A, Klenk HP, Kyrpides N, et al. (2012) Genomics of Aerobic Cellulose Utilization Systems in *Actinobacteria*. PLOS One 7: e39331.
28. Khanna S, Gauri (1993) Regulation, purification, and properties of xylanase from *Cellulomonas fimi*. Enzyme Microb Technol 15: 990–995.
29. Li CJ, Hong YZ, Shao ZZ, Lin L, Huang XL, et al. (2009) Novel Alkali-Stable, Cellulase-Free Xylanase from Deep-Sea *Kocuria* sp Mn22. J Microbiol Biotechnol 19: 873–880.
30. Oh HW, Heo SY, Kim DY, Park DS, Bae KS, et al. (2008) Biochemical characterization and sequence analysis of a xylanase produced by an exo-symbiotic bacterium of *Gryllotalpa orientalis*, *Cellulosimicrobium* sp HY-12. Ant Leeuw Int J Gen Mol Microbiol 93: 437–442.
31. Petrosyan P, Luz-Madrigal A, Huitrón C, Flores M (2002) Characterization of a xylanolytic complex from *Streptomyces* sp. Biotechnol Lett 24: 1473–1476.
32. Robinson LE, Crawford RL (1978) Degradation of ^{14}C-labelled lignins by *Bacillus megaterium*. FEMS Microbiol Lett 4: 301–302.
33. Vicuna R (1988) Bacterial degradation of lignin. Enzyme Microb Technol 10: 646–655.
34. Zimmermann W (1990) Degradation of lignin by bacteria. J Biotechnol 13: 119–130.
35. Cartwrig N, Holdom KS (1973) Enzymic lignin, its release and utilization by bacteria. Microbios 8: 7–14.
36. Crawford DL, Barder MJ, Pometto AL, Crawford RL (1982) Chemistry of softwood lignin degradation by *Streptomyces viridosporus*. Arch Microbiol 131: 140–145.
37. Sutherland JB, Blanchette RA, Crawford DL, Pometto AL (1979) Breakdown of Douglas fir phloem by a lignocellulose-degrading *Streptomyces*. Curr Microbiol 2: 123–126.
38. Trojanowski J, Haider K, Sundman V (1977) Decomposition of ^{14}C-labelled lignin and phenols by a *Nocardia* sp. Arch Microbiol 114: 149–153.
39. Caliz J, Montserrat G, Marti E, Sierra J, Chung AP, et al. (2013) Emerging resistant microbiota from an acidic soil exposed to toxicity of Cr, Cd and Pb is mainly influenced by the bioavailability of these metals. J Soils Sediments 13: 413–428.
40. Harichova J, Karelova E, Pangallo D, Ferianc P (2012) Structure analysis of bacterial community and their heavy-metal resistance determinants in the heavy-metal-contaminated soil sample. Biologia 67: 1038–1048.
41. Mühlbachová G (2011) Soil microbial activities and heavy metal mobility in long-term contaminated soils after addition of EDTA and EDDS. Ecol Eng 37: 1064–1071.
42. Hayakawa M, Nonomura H (1987) Humic acid-vitamin agar, a new medium for the selective isolation of soil actinomycetes. J Ferment Technol 65: 501–509.
43. Hayakawa M, Momose Y, Yamazaki T, Nonomura H (1996) A method for the selective isolation of *Microtetraspora glauca* and related four-spored actinomycetes from soil. J Appl Bacteriol 80: 375–386.
44. Baldrian P (2009) Microbial enzyme-catalyzed processes in soils and their analysis. Plant Soil Environ 55: 370–378.
45. Valášková V, Šnajdr J, Bittner B, Cajthaml T, Merhautová V, et al. (2007) Production of lignocellulose-degrading enzymes and degradation of leaf litter by saprotrophic basidiomycetes isolated from a *Quercus petraea* forest. Soil Biol Biochem 39: 2651–2660.
46. Brunow G, Ammalahti E, Niemi T, Sipila J, Simola LK, et al. (1998) Labelling of a lignin from suspension cultures of *Picea abies*. Phytochem 47: 1495–1500.
47. Tuomela M, Oivanen P, Hatakka A (2002) Degradation of synthetic ^{14}C-lignin by various white-rot fungi in soil. Soil Biol Biochem 34: 1613–1620.
48. Odier E, Janin G, Monties B (1981) Poplar Lignin Decomposition by Gram-Negative Aerobic Bacteria. Appl Environ Microbiol 41: 337–341.
49. Šnajdr J, Steffen KT, Hofrichter M, Baldrian P (2010) Transformation of ^{14}C-labelled lignin and humic substances in forest soil by the saprobic basidiomycetes *Gymnopus erythropus* and *Hypholoma fasciculare*. Soil Biol Biochem 42: 1541–1548.
50. Sagova-Mareckova M, Cermak L, Novotna J, Plhackova K, Forstova J, et al. (2008) Innovative methods for soil DNA purification tested in soils with widely differing characteristics. Appl Environ Microbiol 74: 2902–2907.
51. Šnajdr J, Cajthaml T, Valášková V, Merhautová V, Petránková M, et al. (2011) Transformation of *Quercus petraea* litter: successive changes in litter chemistry are

reflected in differential enzyme activity and changes in the microbial community composition. FEMS Microbiol Ecol 75: 291–303.
52. She YL, Li XT, Sun BG, Lv YG, Song HX (2012) Screening of Actinomycetes with High Producing Xylanase. Adv Mat Res 365: 332–337.
53. Wang G, Meng K, Luo H, Wang Y, Huang H, et al. (2012) Phylogenetic Diversity and Environment-Specific Distributions of Glycosyl Hydrolase Family 10 Xylanases in Geographically Distant Soils. PLOS One 7: e43480.
54. Post DA, Luebke VE (2005) Purification, cloning, and properties of beta-galactosidase from *Saccharopolyspora erythraea* and its use as a reporter system. Appl Microbiol Biotechnol 67: 91–96.
55. Shi P, Yao G, Cao Y, Yang P, Yuan T, et al. (2011) Cloning and characterization of a new β-mannosidase from *Streptomyces* sp. S27. Enzyme Microb Technol 49: 277–283.
56. Tajana E, Fiechter A, Zimmermann W (1992) Purification and characerization of two alpha-L-arabinofuranosidases from *Streptomyces diastaticus*. Appl Environ Microbiol 58: 1447–1450.
57. Crawford DL (1978) Lignocellulose decomposition by selected streptomyces strains. Appl Environ Microbiol 35: 1041–1045.
58. Pasti MB, Pometto AL, Nuti MP, Crawford DL. Lignin-solubilizing ability of actinomycetes isolated from termite (*Termitidae*) gut. Appl Environ Microbiol 56: 2213–2218.
59. Kerr TJ, Kerr RD, Benner R (1983) Isolation of a Bacterium Capable of Degrading Peanut Hull Lignin. Appl Environ Microbiol 46: 1201–1206.
60. Krastanov A, Alexieva Z, Yemendzhiev H (2013) Microbial degradation of phenol and phenolic derivatives. Eng Life Sci 13: 76–87.
61. El Azhari N, Devers-Lamrani M, Chatagnier G, Rouard N, Martin-Laurent F (2010) Molecular analysis of the catechol-degrading bacterial community in a coal wasteland heavily contaminated with PAHs. J Hazard Mater 177: 593–601.
62. Osono T, Takeda H (2005) Decomposition of organic chemical components in relation to nitrogen dynamics in leaf litter of 14 tree species in a cool temperate forest. Ecol Res 20: 41–49.

16

Biogas Production by Co-Digestion of Goat Manure with Three Crop Residues

Tong Zhang[1], Linlin Liu[2], Zilin Song[1], Guangxin Ren[2], Yongzhong Feng[2], Xinhui Han[2], Gaihe Yang[2]*

1 College of Forestry and the Research Center of Recycle Agricultural Engineering and Technology of Shaanxi Province, Northwest A&F University, Yangling, Shaanxi, People's Republic of China, **2** College of Agronomy and the Research Center of Recycle Agricultural Engineering and Technology of Shaanxi Province, Northwest A&F University, Yangling, Shaanxi, People's Republic of China

Abstract

Goat manure (GM) is an excellent raw material for anaerobic digestion because of its high total nitrogen content and fermentation stability. Several comparative assays were conducted on the anaerobic co-digestion of GM with three crop residues (CRs), namely, wheat straw (WS), corn stalks (CS) and rice straw (RS), under different mixing ratios. All digesters were implemented simultaneously under mesophilic temperature at 35±1 °C with a total solid concentration of 8%. Result showed that the combination of GM with CS or RS significantly improved biogas production at all carbon-to-nitrogen (C/N) ratios. GM/CS (30:70), GM/CS (70:30), GM/RS (30:70) and GM/RS (50:50) produced the highest biogas yields from different co-substrates (14840, 16023, 15608 and 15698 mL, respectively) after 55 d of fermentation. Biogas yields of GM/WS 30:70 (C/N 35.61), GM/CS 70:30 (C/N 21.19) and GM/RS 50:50 (C/N 26.23) were 1.62, 2.11 and 1.83 times higher than that of CRs, respectively. These values were determined to be the optimal C/N ratios for co-digestion. However, compared with treatments of GM/CS and GM/RS treatments, biogas generated from GM/WS was only slightly higher than the single digestion of GM or WS. This result was caused by the high total carbon content (35.83%) and lignin content (24.34%) in WS, which inhibited biodegradation.

Editor: Chenyu Du, University of Nottingham, United Kingdom

Funding: This work was supported by science and technology support projects "the biological technology integration and demonstration of high yield biogas digestion from the mix ingredients" (2011BAD15B03) from Ministry of Science and Technology Department of the People's Republic of China, and the Fundamental Research Funds for the Central Universities (QM2012002) from Ministry of Education of the People's Republic of China. The funders had no role in study design, data collection and analysis, decision to publish, or preparation of the manuscript.

Competing Interests: The authors have declared that no competing interests exist.

* E-mail: ygh@nwsuaf.edu.cn

Introduction

China is one of the largest agricultural countries in the world. The production of net available crop residues (CRs) in China is estimated to be over 800 million t/yr [1], which ranks first in the world. The use of agricultural waste as a major component of renewable energy is suitable for improving energy security and decreasing environmental disruption caused by carbon emissions [2,3]. Wheat straw (WS), rice straw (RS) and corn stalks (CS) are the top three agricultural wastes in China and account for 80.5% of the total output (15.7%, 24.2% and 40.6%, respectively) [1]. Thus, studying the energy generation potential of these three wastes is important.

Anaerobic digestion (AD) is a biological process that produces biogas from bio-degradable wastes by bacteria under poor or no oxygen conditions. In the past two decades, AD has been applied as an effective technology for solving the energy shortage and environmental pollution problems of biotechnology industries and residential activities caused by heating and electricity generation [4,5,6].

CRs and animal manure have recently been used together to produce biogas by AD. Compared with the single digestion of feedstock, the co-digestion of CRs and animal manures increases the rate of biogas production because of the greater balance between carbon and nitrogen [7] and improves AD efficiency [8]. Annual goat manure (GM) yield in China is approximately 3.21×10^8 t followed by dairy manure, swine manure and chicken manure [9]. The total nitrogen (TN) contents of fresh GM (1.01%) and chicken manure (1.03%) are significantly higher than those of dairy manure (0.35%) and swine manure (0.24%) [10]. High TN content is beneficial to co-digestion with CRs because it decreases the carbon-to-nitrogen (C/N) ratios of single CRs substrates. GM is also insensitive to acidification during anaerobic fermentation [11,12]. Hence, GM is an excellent raw material for AD. Although various raw materials, such as agricultural waste, animal manures, sewage sludge and food waste have been reported as potentially feasible for co-digestion [7,13,14,15,16,17,18,19,20], the suitable mixing ratios of multi-component substrates between GM and various CRs are largely unknown.

We investigated the biogas-producing efficiency of anaerobic co-digestion influenced by different GM and CR mixing ratios. The best ratio between these substrates was obtained by comparing the results. Furthermore, an optimum co-digestion condition for biogas production was proposed.

Materials and Methods

Feedstocks and inocula

GM was obtained from a local livestock farm near Northwest Agriculture and Forestry University (NWAFU), Yangling Shaanxi,

China. WS, CS and RS were collected from the experimental field of NWAFU. All of these straws were cut into sections at lengths of 2 cm to 3 cm by using a grinder. Inoculum was the anaerobic sludge of dairy manure, which was obtained from an anaerobic digester in a local village.

Experimental digester and design

The experiment was conducted according to Song et al. (2012) by using lab-scale anaerobic digesters fabricated from 1 L Erlenmeyer flasks. Batch reactors were used to determine the co-digestions of GM mixed with three CRs. The working volume of each digester was 700 mL, including 140 g inocula and an appropriate amount of digesting material. Deionized water was added to digesters to maintain a total solid (TS) content of 8% [5]. All reactors were gently mixed manually for approximately 1 min/d prior to biogas volume measurement.

To obtain the best mixing ratio of the co-digestion of GM supplemented with three CRs as external carbon sources, five different mixing mass ratios at 90:10, 70:30, 50:50, 30:70 and 10:90 were tested under mesophilic condition (35±1°C) for 55 d. Unmixed GM (100:0) and CR (0:100) were anaerobically digested as controls. Each treatment was performed thrice with a control to investigate the effect of different mixed ratios on biogas production.

Analysis and statistics

The TS, volatile solids (VS), pH, volatile fatty acid (VFA), and TN content of the materials were determined in accordance with the *Standard Methods for the Examination of Water and Wastewater* of the American Public Health Association [21]. Total carbon (TC) and lignin contents were analyzed by using the method described by Cuetos et al. and Song et al. [5,22]. The amount of biogas produced from each digester was recorded every day by using the water displacement method during the digestion period. Each batch experiment was deemed complete when a clear downward trend in daily biogas volume produced was observed for 10 d.

ANOVA was performed to determine the significant differences among each treatment by using SAS version 8.12 (SAS Institute Inc., Cary, NC, USA).

Results and Discussion

Substrate characteristics

The C/N ratios of the different substrates and substrate mixtures in AD greatly influence biogas production [23,24]. A higher carbon content provides more carbon for CH_4 production, whereas a lower nitrogen content limits microbial activity because microbes need a considerable amount of nitrogen to maintain growth [8]. The ideal C/N ratios range from 9 to 30 for anaerobic digesters [25]. The chemical characteristics of substrates used in this study are shown in Table 1. The C/N ratio of GM was 17.97, which is too low for biogas production. However, the C/N ratios of WS, CS and RS were significantly higher (91.17, 88.13 and 92.91, respectively) than that of GM ($P<0.01$). This result suggested that CRs increased methane production when co-digested with GM under the optimal C/N ratio.

Table 1. Chemical characterization of substrates used in the co-digestion experiments.

	GM	WS	CS	RS
pH	7.94±0.15	ND	ND	ND
TS (%)	33.65±3.23, b	81.08±7.62, a	81.74±7.43, a	77.92±6.97, a
VS (%)	82.21±8.93, a	90.29±9.25, a	91.42±9.33, a	94.23±9.42, a
TC (%)	18.22±1.14, c	35.83±3.17, a	28.82±2.03, b	31.96±2.92, ab
TN (%)	1.014±0.11, a	0.393±0.02, b	0.327±0.04, b	0.344±0.02, b
C/N	17.97±0.84, b	91.17±3.44, a	88.13±4.65, a	92.91±3.10, a
Lignin (%)	ND	24.34±1.89, a	15.38±1.21, b	9.49±0.33, c

TS, total solid; VS, volatile solids; TC, Total carbon; TN, total nitrogen.
The values are the mean ± standard deviation of the triplicate measurements.
ND = not detected.
The ANOVA test was conducted to determine the differences between each cultivar. Values with the same letters indicate no significant difference at $P<0.01$.

Biogas yields and production rates at different GM/CR ratios

The daily biogas production by the co-digestion of GM and CRs during 55 d of digestion was calculated under different mixing ratios (Fig. 1). Samples from the mixing ratios of GM/WS 30:70, GM/CS 30:70 and GM/RS 50:50 were measured, and their peak yield values were 570, 585 and 525 mL/d on the 17th, 19th and 11th d, respectively (Fig. 1). The digestion of single GM substrate (100:0) produced biogas earlier than other combinations but had two relatively small peaks (402 and 500 mL/d) (Fig. 1). By contrast, the digestion of any single CR substrate (0:100) had only one peak (GM/WS 547, GM/CS 540 and GM/RS 477 mL/d) that occurred earlier than the other combinations (3rd d to 6th d) and decreased rapidly after the 16th d (Fig. 1). These results indicate that the co-digestion of GM and CRs could significantly delay the attainment of the highest gas production.

The final cumulative biogas productions by the co-digestion of GM and CRs at different mixing ratios are shown in Fig. 2. The cumulative biogas productions for GM/WS 10:90, 30:70, 50:50, 70:30 and 90:10 were 11890, 12765, 11253, 12685 and 9650 mL, respectively (Fig. 2A). These results showed an increase of 51.0%, 62.1%, 42.9%, 61.1% and 22.6% compared with single WS (7874 mL), and an increase of 51.0%, 62.1%, 42.9%, and 22.6% compared with single GM (10375 mL). However, the biogas production of GM/WS 90:10 (9650 mL) was lower than that of single GM (Fig. 2A). The same trends were observed for the GM/CS and GM/RS treatments, which had considerably higher increases (Fig. 2B and 2C). These data showed that the co-digestion of GM and CRs greatly improved biodegradability and biogas production at most mixing ratios compared with single substrate digestion. Our results supported those of Wu et al. [26], who found that co-digesting swine manure with CS, oat straw and WS significantly increase biogas production and net CH_4 volume at all C/N ratios.

To compare the effect of single substrate digestion and co-digestion with GM and CRs, the total biogas yield of each combination is shown in Fig. 3. The total biogas productions of most co-digestion systems were higher than the single digestion of either GM or CRs except those of GM/WS 90:10 and GM/RS 10:90. GM/CS 70:30 exhibited the highest total biogas yield of 16.02 L in all treatments, which was 83.02% and 54.44% higher than that of CS and GM alone, respectively. Among all the GM/RS treatments, the total biogas production of GM/RS 50:50 (15.70 L) was 111.28% and 51.31% higher than that of CS alone and GM alone, respectively. The co-digestion of GM/WS 30:70 was 62.12% and 23.04% higher than that of WS and GM alone, respectively. Compared with the TC contents of CS (28.82%) and RS (31.96%), the higher TC content of WS (35.83%) suppressed

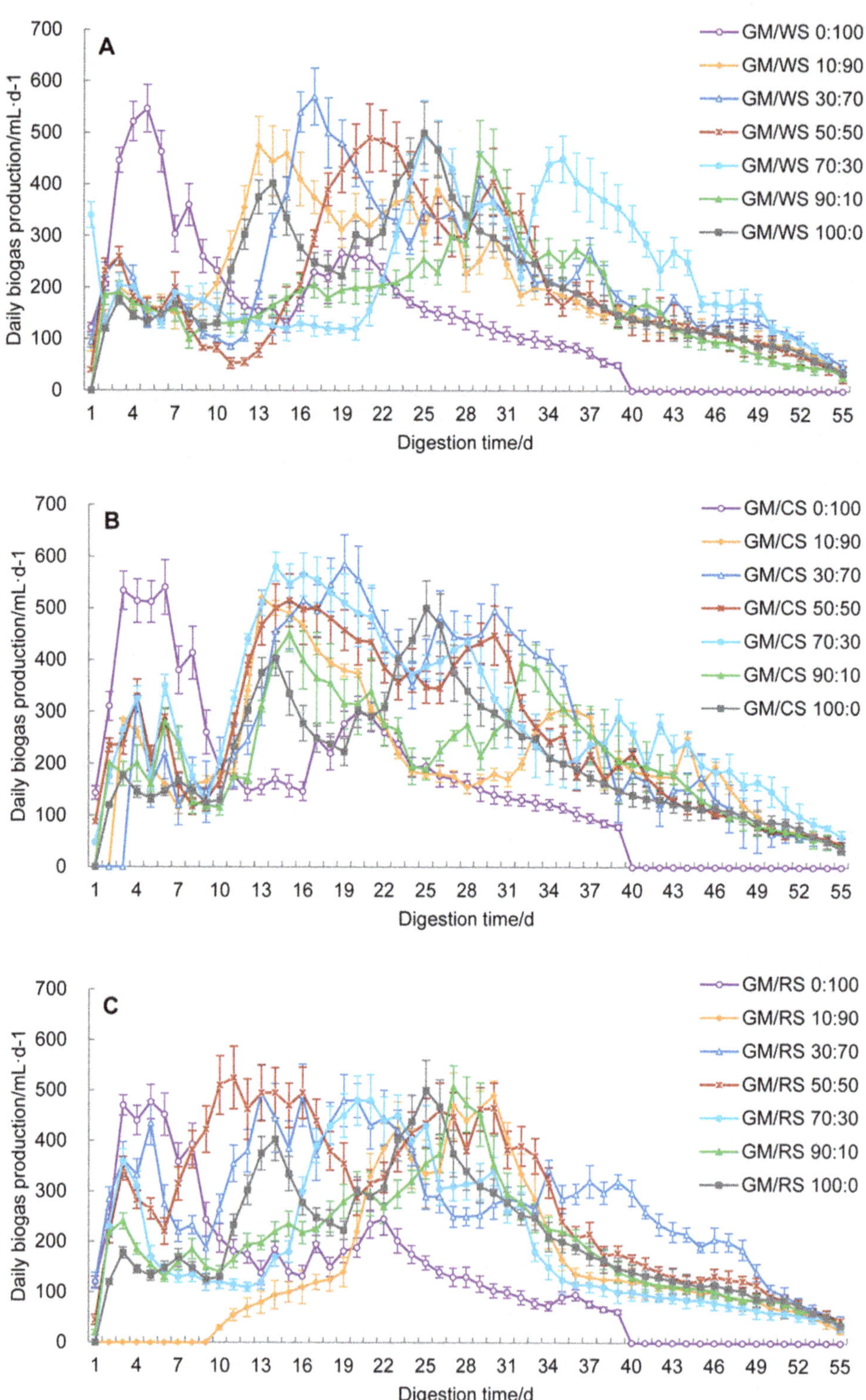

Figure 1. Daily biogas production from the co-digestion of GM and WS (A), CS (B) and RS (C) with different mixing ratios. Mean values originated from three independent replications. Vertical bars represent standard deviations.

microbial growth and methanogenesis because of the ammonium nitrogen deficiency and low pH [22,27,28].

These results indicated that co-digestion with suitable GM and CRs mixtures is an effective way to prolong the period of the highest gas production and improve biogas yield. The ANOVA indicated that the total biogas production of co-digestions were significantly higher ($P<0.01$) than the single digestion of GM or CRs (Fig. 3).

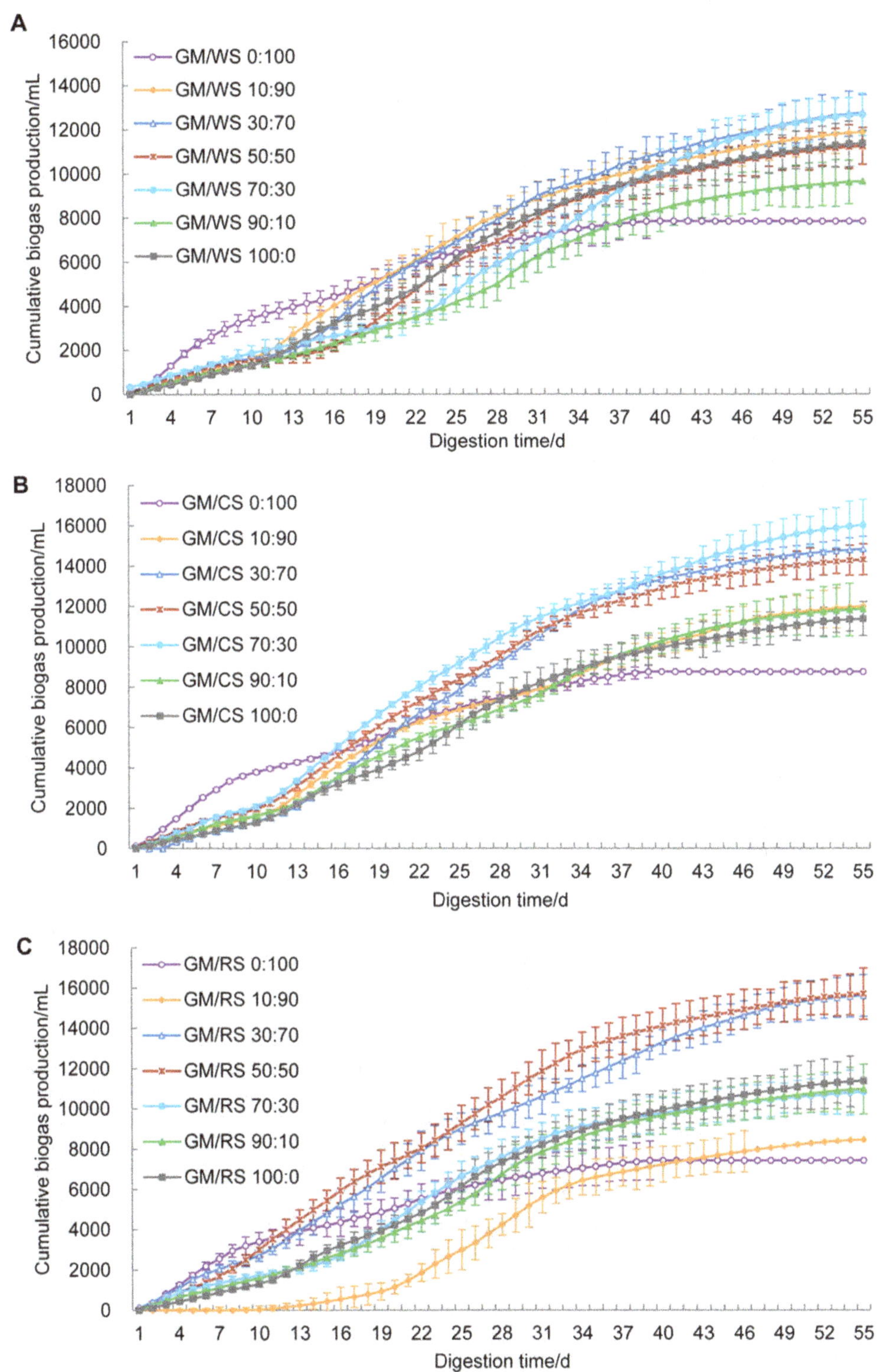

Figure 2. Cumulative biogas productions from co-digestion of GM and WS (A), CS (B) and RS (C) with different mixing ratios. Mean values originated from three independent replications. Vertical bars represent standard deviations.

Effect of C/N ratio on biogas production

The C/N ratio represents the relationship between the amount of carbon and nitrogen present in organic materials and is an important indicator for controlling biological treatment systems [23]. On one hand, a high C/N ratio indicates rapid nitrogen consumption by methanogens and leads to lower gas production. On the other hand, a low C/N ratio results in ammonia accumulation and an increase in pH values, which is toxic to methanogenic bacteria [29]. The mean value of C/N ratios for each co-digestion combinations and single digestion ranged from 92.79 to 17.97 (Table 2). The C/N ratios of co-digestions were significantly lower than those of CR materials ($P<0.01$, Table 1),

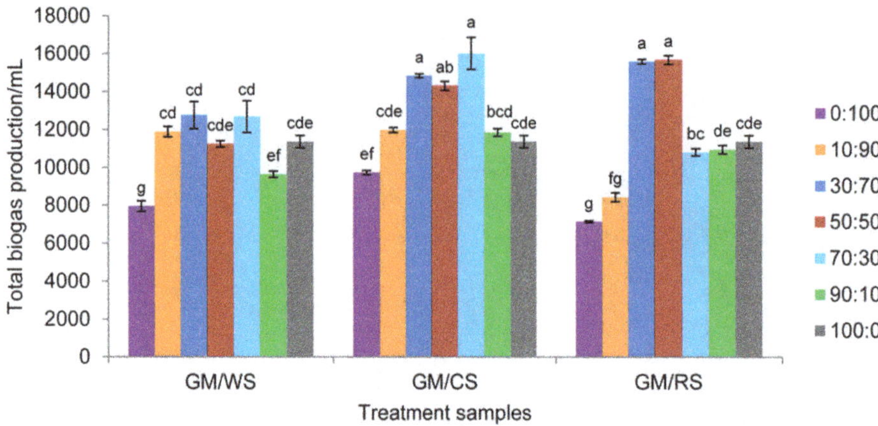

Figure 3. Total biogas productions from anaerobic co-digestion of GM with WS, CS and RS with different mixing ratios. Mean values originated from three independent replications. Vertical bars represent standard deviations. The ANOVA test was conducted to determine the differences between each cultivar. Values with the same letters indicate no significant difference at $P<0.01$.

thus indicating that co-digestion effectively reduced the C/N ratios of AD. Experimental data showed that the biogas yields of most co-digestions were higher than the corresponding single digestions. According to the cumulative biogas production (Fig. 3), the highest biogas yields (12765, 15698 and 16023 mL) at GM/WS 30:70 (C/N 35.64), GM/CS 70:30 (C/N 21.26) and GM/RS 50:50 (C/N 26.28) were 1.62, 2.11 and 1.83 times higher than that of CRs only, respectively. However, the total biogas yields of three GM/CR 10:90 treatments did not increase, and were even lower than that of single substrate. The reason for this result was that the C/N ratios of each GM/CR 10:90 treatment were less than 20 (Table 2). The results suggested that the ideal C/N ratio range is between 20 to 35 in the co-digestion of GM with CRs, which was consistent with the report of Verma [29], which revealed that the optimum C/N ratios in anaerobic digesters were between 20 to 30.

CRs typically contain high lignocellulosic contents. Problems such as low gas yield during the AD of these materials were usually associated with a high C/N ratio or high lignin content [30]. Although the C/N ratio was reduced by most co-digestions, no apparent increasing trend was observed in the biogas production of GM/WS, which even decreased slightly (GM/WS 90:10) compared with GM only. This phenomenon possibly resulted from the significantly higher lignin content (24.34%) of WS substrate than those of CS and RS (15.38% and 9.47%, respectively) ($P<0.01$, Table 1). To overcome the low degradability of lignin, reducing the particle size of CR substrate can increase the degradation rate of lignocelluloses and further improve biogas production [31].

Effects of pH and VFA

VFA and pH are the two key factors in AD [4]. The pH value and total VFA reflected the changing processes in the reactors (Fig. 4). The curves for the individual pH and total VFA of all mixtures and single substrates had similar trends. The growth of methanogens can be significantly influenced by the pH level [32]. The initial pH values of digesters gradually decreased from 6.5 to 6.0 with increasing CR percentage, and GM/RS 10:90 had the lowest pH value (5.5). The pH values increased from 6.5 as the percentage of GM increased in the 6th d, and then remained at approximately 6.8 until the 30th d. This stability confirmed that the daily biogas production of each mixture reached the methanogenesis stage, and that the pH value remained at approximately 6.8. Thereafter, the pH values dropped slightly to 6.0, thus indicating that the digestion changed in the later stages. However, the pH values of GM/CRs 0:100 decreased rapidly after the 18th d, thus showing the buffering capacity of GM. These results indicated that the best pH values for the co-digestion of GM and CRs ranged from 6.5 to 7.5.

VFAs are intermediate organic acid products, and the total VFA concentration is considered an important indicator of metabolic status in addition to the pH value during AD [33,34]. However, the VFA curves showed evidently contrasting trends with that of the pH values. VFA was initially approximately 7380 mg/L to 11767 mg/L for all treatments and then decreased to 4519 mg/L to 5484 mg/L at the 24th d. VFA increased again and finally decreased to 9812 mg/L to 11791 mg/L at the end of digestion (Fig. 3 and 4).

Table 2. Mean values for C/N ratios in the co-digestion of GM with three CRs.

Treatment	Co-digestion mixing ratios						
	0:100	10:90	30:70	50:50	70:30	90:10	100:0
GM/WS	91.05±3.44, a	58.24±0.48, b	35.64±0.58, c	29.71±1.22, d	22.06±0.82, e	19.12±0.83, f	17.97±0.84, f
GM/CS	88.51±4.65, a	53.43±2.50, b	32.64±1.46, c	25.13±1.13, d	21.26±0.97, e	18.90±0.87, e	17.97±0.84, e
GM/RS	92.79±3.10, a	57.46±0.30, b	34.82±0.61, c	26.28±0.77, d	21.80±0.82, e	19.05±0.83, f	17.97±0.84, f

The values are the mean ± standard deviation of triplicate measurements.
The ANOVA test was conducted to determine the differences between each cultivar. Values with the same letters indicate no significant difference at $P<0.01$.

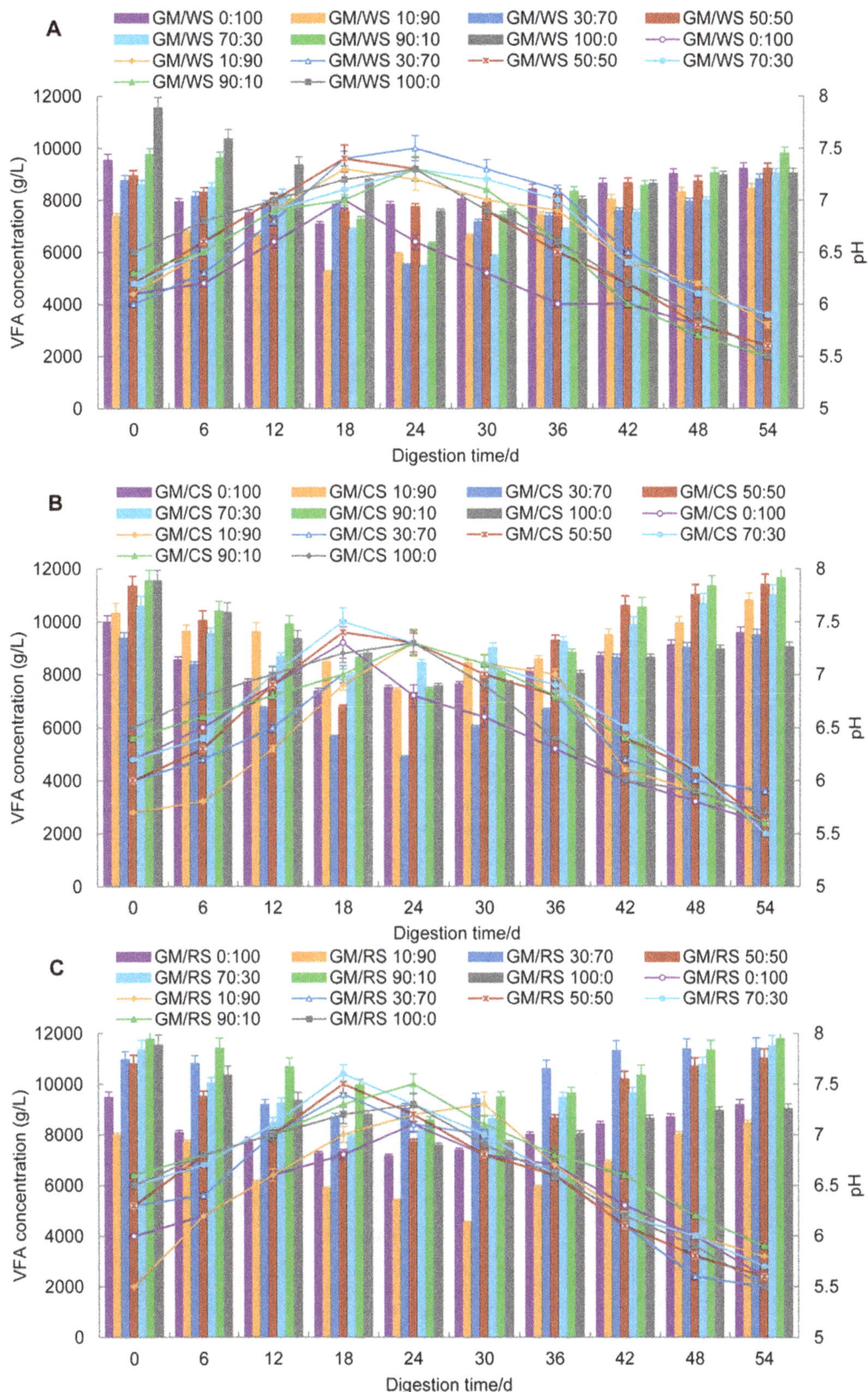

Figure 4. VFA concentrations and pH values from the co-digestion of GM and WS (A), CS (B) and RS (C) with different mixing ratios. Mean values originated from three independent replications. Columns represent VFA, lines represent pH values, and vertical bars represent standard deviations. VFA, volatile fatty acid.

The ammonia produced by the biological degradation of proteins and urea often results in VFA accumulation. The accumulation of VFA leads to the decrease of pH value, thus affecting the growth of methanogens during the AD process [6,24,30]. Our results showed that pH and VFA were co-related with biogas yield in AD. Thus, the pH values were proportional to biogas yield, whereas VFAs were inversely proportional. These results further indicated that pH decreased with increasing VFA accumulation. High concentrations of VFA are toxic to methanogens and inhibits hydrolysis rates in reactors [35]. The interaction between pH and VFA may lead to an "inhibited steady state" with a lower methane yield [30,36,37]. The extended gas production peaks in each mixing treatment might be explained by the co-digestion of GM and CRs, which relieves the inhibited steady state caused by pH and VFA effectively. The co-digestion of GM and CRs improves the buffering capacity to VFA accumulation and inhibits the acidogenesis process, which is consistent with the previous study [38].

Conclusion

The anaerobic co-digestion of GM with CRs is a promising way for improving biogas production. This co-digestion not only resolves the environmental problems caused by straws burning, but also overcomes C/N ratio imbalances in single digestion substrates and enhances the AD process.

Our results showed that the anaerobic co-digestions of GM with CS and RS were efficient and produced more cumulative biogas by reducing the C/N ratios of substrates. The best ratios were GM/CS 30:70, GM/CS 70:30, GM/RS 30:70 and GM/RS 50:50. However, the co-digestion of GM with WS did not improve the biogas yield significantly, which is consistent with the result in previous research [26]. The higher TC content of WS suppressed microbial growth and methanogenesis because of the deficiency of ammonium nitrogen and low pH. For the pH and VFA ranges in this study, pH decreased with increasing VFA accumulation, thus leading to the inhibition of biowaste hydrolysis rates.

Acknowledgments

We thank Dr. Xiaodong Wang, Dr Furong Liu and Dr. Yuheng Yang for critical reading of this manuscript and editorial guidance.

Author Contributions

Conceived and designed the experiments: TZ YZF GHY. Performed the experiments: TZ LLL ZLS. Analyzed the data: TZ GXR XHH. Wrote the paper: TZ GHY.

References

1. Jiang D, Zhuang D, Fu J, Huang Y, Wen K (2012) Bioenergy potential from crop residues in China: Availability and distribution. Renew Sustain Energy Rev 16: 1377–1382.
2. Field CB, Campbell JE, Lobell DB (2008) Biomass energy: the scale of the potential resource. Trend Ecol Evolut 23: 65–72.
3. Yang Z, Zhang H (2008) Strategies for development of clean energy in China. Petrol Sci 5: 183–188.
4. Madsen M, Holm-Nielsen JB, Esbensen KH (2011) Monitoring of anaerobic digestion processes: A review perspective. Renew Sustain Energy Rev 15: 3141–3155.
5. Song Z, Yang G, Guo Y, Zhang T (2012) Comparison of two chemical pretreatments of rice straw for biogas production by anaerobic digestion. BioResources 7: 3223–3236.
6. Weiland P (2010) Biogas production: current state and perspectives. Appl Microbiol Biotechnol 85: 849–860.
7. El-Mashad HM, Zhang R (2010) Biogas production from co-digestion of dairy manure and food waste. Bioresour Technol 101: 4021–4028.
8. Zhu D (2010) Co-digestion of Different Wastes for Enhanced Methane Production: The Ohio State University.
9. Zhang P, Yang Y, Tian Y, Yang X, Zhang Y, et al. (2009) Bioenergy industries development in China: Dilemma and solution. Renew Sustain Energy Rev 13: 2571–2579.
10. Wang FH, Ma WQ, Dou ZX, Ma L, Liu XL, et al. (2006) The estimation of the production amount of animal manure and its environmental effect in China. China Environ Sci 26: 614–617.
11. Jain M, Singh R, Tauro P (1981) Anaerobic digestion of cattle and sheep wastes. Agr Wastes 3: 65–73.
12. Kanwar S, Kalia A (1993) Anaerobic fermentation of sheep droppings for biogas production. World J Microbiol Biotechnol 9: 174–175.
13. Dai X, Duan N, Dong B, Dai L (2012) High-solids anaerobic co-digestion of sewage sludge and food waste in comparison with mono digestions: Stability and performance. Waste Management.
14. Creamer K, Chen Y, Williams C, Cheng J (2010) Stable thermophilic anaerobic digestion of dissolved air flotation (DAF) sludge by co-digestion with swine manure. Bioresour Technol 101: 3020–3024.
15. Luostarinen S, Luste S, Sillanpää M (2009) Increased biogas production at wastewater treatment plants through co-digestion of sewage sludge with grease trap sludge from a meat processing plant. Bioresour Technol 100: 79–85.
16. Bouallagui H, Lahdheb H, Ben Romdan E, Rachdi B, Hamdi M (2009) Improvement of fruit and vegetable waste anaerobic digestion performance and stability with co-substrates addition. J Environ Manage 90: 1844–1849.
17. Álvarez J, Otero L, Lema J (2010) A methodology for optimising feed composition for anaerobic co-digestion of agro-industrial wastes. Bioresour Technol 101: 1153–1158.
18. Macias-Corral M, Samani Z, Hanson A, Smith G, Funk P, et al. (2008) Anaerobic digestion of municipal solid waste and agricultural waste and the effect of co-digestion with dairy cow manure. Bioresour Technol 99: 8288–8293.
19. Xie S, Lawlor P, Frost J, Hu Z, Zhan X (2011) Effect of pig manure to grass silage ratio on methane production in batch anaerobic co-digestion of concentrated pig manure and grass silage. Bioresour Technol 102: 5728–5733.
20. Nguyen VCN, Fricke K (2012) Energy recovery from anaerobic co-digestion with pig manure and spent mushroom compost in the Mekong Delta. J Vietnamese Environ 3: 4–9.
21. APHA (1995) Standard methods for the examination of water and wastewater: Washington. DC, American Public Health Association.
22. Cuetos MJ, Fernández C, Gómez X, Morán A (2011) Anaerobic co-digestion of swine manure with energy crop residues. Biotechnol Bioprocess Eng 16: 1044–1052.
23. Wang X, Yang G, Feng Y, Ren G, Han X (2012) Optimizing feeding composition and carbon-nitrogen ratios for improved methane yield during anaerobic co-digestion of dairy, chicken manure and wheat straw. Bioresour Technol 120: 78–83.
24. Kayhanian M (1999) Ammonia inhibition in high-solids biogasification: an overview and practical solutions. Environ Technol 20: 355–365.
25. Siddiqui Z, Horan N, Anaman K (2011) Optimisation of C: N ratio for co-digested processed industrial food waste and sewage sludge using the BMP test. Int J Chem React Eng 9.
26. Wu X, Yao W, Zhu J, Miller C (2010) Biogas and CH_4 productivity by co-digesting swine manure with three crop residues as an external carbon source. Bioresour Technol 101: 4042–4047.
27. Panichnumsin P, Nopharatana A, Ahring B, Chaiprasert P (2010) Production of methane by co-digestion of cassava pulp with various concentrations of pig manure. Biomass Bioenerg 34: 1117–1124.
28. Carucci G, Carrasco F, Trifoni K, Majone M, Beccari M (2005) Anaerobic digestion of food Industry Wastes: Effect of codigestion on methane yield. J Environ Eng 131: 1037–1045.
29. Verma S (2002) Anaerobic digestion of biodegradable organics in municipal solid wastes: Columbia University.
30. Chen Y, Cheng JJ, Creamer KS (2008) Inhibition of anaerobic digestion process: A review. Bioresour Technol 99: 4044–4064.
31. Palmowskl L, Müller J (2000) Influence of the size reduction of organic waste on their anaerobic digestion. Water Sci Technol: 155–162.
32. Duarte A, Anderson G (1982) Inhibition modelling in anaerobic digestion. Water Sci Technol 14: 749–763.
33. Fernández A, Sanchez A, Font X (2005) Anaerobic co-digestion of a simulated organic fraction of municipal solid wastes and fats of animal and vegetable origin. Biochem Eng J 26: 22–28.
34. Habiba L, Hassib B, Moktar H (2009) Improvement of activated sludge stabilisation and filterability during anaerobic digestion by fruit and vegetable waste addition. Bioresour Technol 100: 1555–1560.
35. Veeken A, Hamelers B (1999) Effect of temperature on hydrolysis rates of selected biowaste components. Bioresour Technol 69: 249–254.
36. Angelidaki I, Ahring B (1992) Effects of free long-chain fatty acids on thermophilic anaerobic digestion. Appl Microbiol Biotechnol 37: 808–812.
37. Angelidaki I, Ahring B (1993) Thermophilic anaerobic digestion of livestock waste: the effect of ammonia. Appl Microbiol Biotechnol 38: 560–564.
38. Angelidaki I (1997) Anaerobic digestion in Denmark. Past, present and future. Servicio de Publicaciones. pp. 335–342.

17

Evaluation of the CENTURY Model Using Long-Term Fertilization Trials under Corn-Wheat Cropping Systems in the Typical Croplands of China

Rihuan Cong[1,2], Xiujun Wang[3,4], Minggang Xu[1]*, Stephen M. Ogle[5], William J. Parton[5]

1 Ministry of Agriculture Key Laboratory of Crop Nutrition and Fertilization, Institute of Agricultural Resources and Regional Planning, Chinese Academy of Agricultural Sciences, Beijing, China, **2** College of Resources and Environment, Huazhong Agricultural University, Wuhan, China, **3** State Key Laboratory of Desert and Oasis Ecology, Xinjiang Institute of Ecology and Geography, Chinese Academy of Sciences, Urumqi, China, **4** Earth System Science Interdisciplinary Center, University of Maryland, College Park, Maryland, United States of America, **5** Natural Resource Ecology Laboratory, Colorado State University, Fort Collins, Colorado, United States of America

Abstract

Soil organic matter models are widely used to study soil organic carbon (SOC) dynamics. Here, we used the CENTURY model to simulate SOC in wheat-corn cropping systems at three long-term fertilization trials. Our study indicates that CENTURY can simulate fertilization effects on SOC dynamics under different climate and soil conditions. The normalized root mean square error is less than 15% for all the treatments. Soil carbon presents various changes under different fertilization management. Treatment with straw return would enhance SOC to a relatively stable level whereas chemical fertilization affects SOC differently across the three sites. After running CENTURY over the period of 1990–2050, the SOC levels are predicted to increase from 31.8 to 52.1 Mg ha^{-1} across the three sites. We estimate that the carbon sequestration potential between 1990 and 2050 would be 9.4–35.7 Mg ha^{-1} under the current high manure application at the three sites. Analysis of SOC in each carbon pool indicates that long-term fertilization enhances the slow pool proportion but decreases the passive pool proportion. Model results suggest that change in the slow carbon pool is the major driver of the overall trends in SOC stocks under long-term fertilization.

Editor: Julio Vera, University of Erlangen-Nuremberg, Germany

Funding: Financial supports are from the National Science Foundation of China (41171239) and the National Basic Research Program (2011CB100501). The funders had no role in study design, data collection and analysis, decision to publish, or preparation of the manuscript.

Competing Interests: The authors have declared that no competing interests exist.

* E-mail: mgxu@caas.ac.cn

Introduction

Soil organic carbon (SOC) is one of the most important terrestrial pools for C storage. It is estimated that the total soil carbon pool is around 1400–1500 Pg C, which is approximately three times greater than the atmospheric pool (750 Pg C) [1,2]. The SOC pool represents a dynamic equilibrium resulting from changes in gains and losses. Even small changes in SOC at a site may potentially add up to significant changes in large-scale carbon cycling across a region [3]. Furthermore, SOC is relatively dynamic and can be greatly influenced by agricultural practices. Increases in SOC storage in cropland soils would benefit soil productivity and environmental health [4,5], and so alternative farming management practices have been evaluated to identify their potentials for increasing SOC in the agroecosystems [4–7].

Long-term experiments are crucial for determining fundamental crop, soil and ecological processes and their impacts on the environment [6–9]. Data from long-term experiments provide a unique resource to investigate long-term influences of climate, crop rotation and crop residue management on soil fertility [6–12]. However, SOC change is affected by complex interactions that vary across space and time depending on the environmental conditions and agricultural management practices. A weakness of long-term experiments is that they are typically restricted to small subset of the entire set of environmental conditions and management practices that exists [13].

Process-based models are an effective way to evaluate SOC changes across a broader set of environmental conditions and management practices [14]. In recent decades, the development and evaluation of soil organic matter models has improved the understanding of factors controlling SOC dynamics, and thus increased our ability to predict future SOC trends. A number of SOC models have been developed, but applying these models requires adequate evaluation with measured SOC trends from experimental for different environmental conditions and management practices [15]. For example, the CENTURY model [16] has been widely used to simulate SOC changes under different management conditions in long-term experiments (e.g., [17], [18] and [19]). With the development of CENTURY, the model has been successfully employed in long-term fertilizer, irrigation, pest management, and site-specific farming applications [20,21]. In China, CENTURY model has been used in grassland [22], forest [23], and regional farmland [24]. However, CENTURY modeling research was still limited in farmland especially under the double cropping rotations and in the acidic soil.

Here, we evaluate the CENTURY with data from three long-term experiments with wheat-corn cropping rotations and different fertilization practices. Specifically, our objectives were

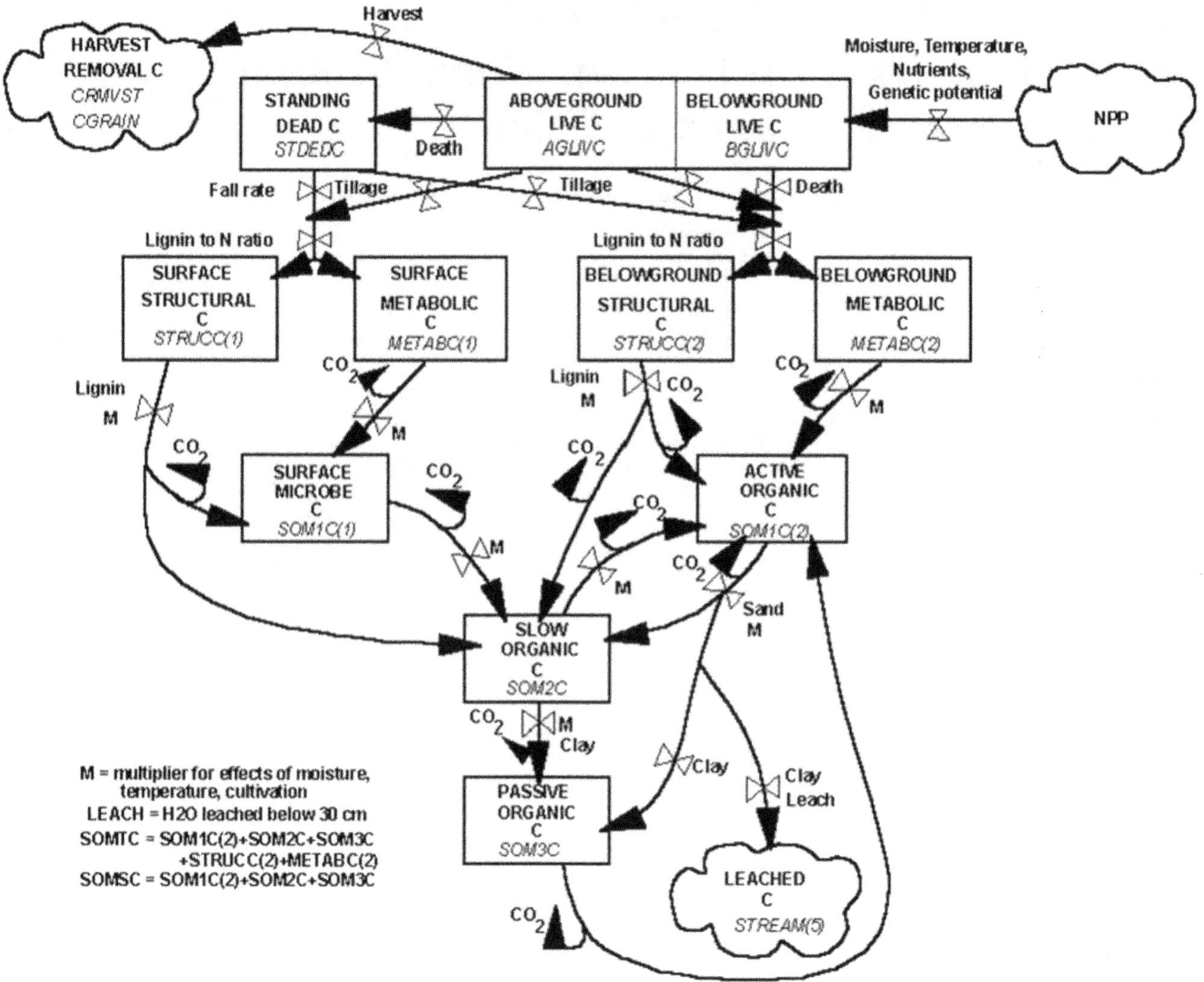

Figure 1. The pools and flows of carbon in the CENTURY model. The diagram showed the major factors which control the flows [29].

(i) to evaluate the performance of CENTURY with evaluation of modeled SOC stocks for different fertilizations and under acidic soil; (ii) to study the effect of fertilization practices on different SOC pools in the modeling framework; and (iii) to predict soil carbon potential under long-term fertilization.

Materials and Methods

Long-term Experiment

Three long-term experiments were utilized for this study, which were located at Changping (40°13′N, 116°15′E), Yangling (34°17′N, 108°00′E) and Qiyang (26°45′N, 111°52′E) in China. Climate conditions varied from semi-humid (Changping site) to

Table 1. Land management practices for the long-term experimental sites during the different blocks and periods used in the CENTURY model simulations.

No.	Periods	Management practices	Repeating sequence
1	up to 1900 (4000 yrs)	Native grassland, low-intensity grazing	1 year
2	1901–1960	Corn, low crop yield cultivar, ploughing, remove straw, applying manure	1 year
3	1961–1988	Wheat-corn rotation, low crop yield cultivar, ploughing, remove straw, applying chemical fertilizers and manure	2 years
4	1989–1989	Uniform tillage to set up long-term experiments	1 year
5	1990–2050	Long-term fertilization experiments	2 years

Table 2. Carbon input (kg $ha^{-1}yr^{-1}$) from manure and straw residue in each period used in the CENTURY model at Changping, Yangling, and Qiyang sites.

Sites	1901–1960	1961–1988	1989–1989	1990–2050 (long-term experiment)		
				NPKM	hNPKM	NPKS
Changping	500	500	–	3150	4725	1000
Yangling	500	500	–	3327	4991	1998
Qiyang	500	500	–	5838	8757	1052

humid warm-temperate (Yangling sites) to humid subtropical climate (Qiyang site). Annual mean temperature was 13.1°C at the Changping site, 14.9°C at the Yangling site, and 18.1°C at the Qiyang site. Annual precipitation was generally low at Changping (515 mm) and Yangling (525 mm) sites but 1445 mm at Qiyang. However, annual evaporation was much higher, varying from 993 mm to 1470 mm [25].

The experimental sites had double cropping systems, i.e., winter wheat and summer corn. Winter wheat was seeded in early November and harvested in early May at the Qiyang site. For the other two sites, winter wheat was seeded around October 20th, and harvested around June 1st. The wheat seeding rates ranged from 165 to 225 kg ha^{-1}. Summer corn was sown around June 10th, with the rate of 63 000–75 000 seeds per hectare, and harvested around October 1st at most sites. For the Qiyang site, summer corn was sown in holes between the wheat strips in early April, and harvested in the middle of July. Thus, there was a short period of inter-cropping for wheat and corn at the Qiyang site. However, in the CENTURY modeling, wheat was harvested in April and corn was sown in May at the Qiyang site since inter-cropping could not be fitted in the model. We believed that there would be little impact on the belowground biomass of wheat and corn, and thus to the soil organic carbon turnover.

Seven treatments were selected in this study: (i) control (no fertilizer); (ii) mineral nitrogen (N); (iii) mineral nitrogen and phosphorus combination (NP); (iv) mineral nitrogen, phosphorus and potassium combination (NPK); (v) NPK combinations with livestock or farmyard (i.e., livestock manure mixed with soil and/or crop residue) manure (NPKM); (vi) 1.5 times' application rate of NPKM (hNPKM); and (vii) mineral NPK combined with crop straw (NPKS). The total nitrogen applied (i.e., mineral plus organic) was equal (i.e., nitrogen balanced) in each of the fertilizer treatments (i.e., N, NP, NPK, NPKM and NPKS treatments). The NPKM and hNPKM treatments had 30% of total N applied from mineral fertilizer and the remaining 70% from organic manure. The treatment plots were initially randomized and isolated by 100-cm-cement baffle plates. There was no replicate for the treatments at these sites due to field availability. However, the plot size was relative large (196 m^2–400 m^2) at each study site. The durations of experiments were from 1990 to 2005 at the Changping site, from 1990 to 2008 at the Qiyang sites, and from 1990 to 2009 at the Yangling site.

Urea, calcium superphosphate, and potassium chloride were used as sources of mineral N, P, and K, respectively. All P and K fertilizers and nearly half of the mineral N fertilizer were applied as basal fertilizers prior to seeding. The remaining mineral N fertilizer was applied as top dressing during the growing season. Manure was generally applied once each year before wheat sowing. However, at the Qiyang site, 30% of manure was applied before wheat seeding and remaining 70% was applied before corn seeding. The sources of organic manure were farmyard manure mixed with crop residue at Changping site, cattle manure at Yangling site, and pig manure at Qiyang site [25]. For the NPKS treatment, half amount (2000 kg ha^{-1} yr^{-1}) of both wheat and

Table 3. Soil classification and initial physical and chemical properties (0–20 cm) in 1989.

Properties		Changping	Yangling	Qiyang
Soil classification (FAO)		Haplic Luvisol	Calcaric Regosol	Eutric Cambisol
Parent material		Diluvial Alluvium	Loess	Quaternary Red Clay
Clay mineral type		Hydromica, Montmorillonite	Hydromica, Montmorillonite	Kaolinite
Bulk density	(g cm^{-3})	1.58	1.30	1.19
Total porosity	(%)	40.4	49.6	51.7
Field capacity	(%)	24.8	21.2	23.7
Texture		silt loam	silt loam	clay
Clay (<2 μm)	(%)	10.2	16.8	40.9
Silt (2–50 μm)	(%)	72.0	51.6	27.7
Sand (50–2000 μm)	(%)	16.2	31.6	31.4
Soil pH		8.7	8.6	5.7
SOC	(g kg^{-1})	7.1	6.3	8.6
TN	(g kg^{-1})	0.80	0.83	1.07
C/N ratio		8.9	7.6	8.0

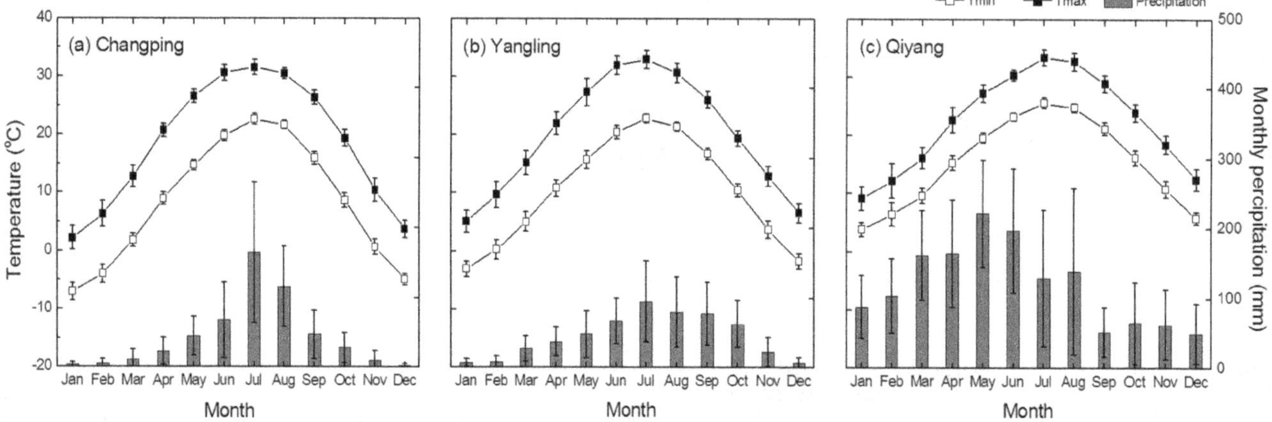

Figure 2. Average monthly precipitation, maximum (Tmax) and minimum (Tmin) temperatures from 1990 to 2010 at (A) Changping, (B) Yangling, and (C) Qiyang. Error bars are standard deviations for the mean values from 1990–2010.

corn straw was incorporated at the Qiyang site, whereas 2250 kg ha^{-1} yr^{-1} and 4500 kg ha^{-1} yr^{-1} of corn straw were incorporated at the Changping and Yangling sites, respectively [26]. Carbon and nutrient contents of manure are measured in 2009 and 2010. Nutrient contents of straw are measured annually over the period of experiment.

There was no irrigation at the Qiyang site since sufficient precipitation occurs during the growing season (i.e., from March to August). However, at the other two sites, irrigations were applied 2–3 times during wheat growing season: 5 mm before seeding, 4 mm for the over wintering stage if needed, and 5 mm at jointing stage. Irrigation was applied once during corn growing season, with 5 mm before seeding.

Plots were ploughed to a depth of 20 cm twice a year, usually in early June at most sites after wheat harvest and early October after corn harvest. In contrast, at the Qiyang site, the whole plot area was ploughed (20 cm depth) in late October before wheat seeding. Then, area between the wheat belts was ploughed (20 cm depth) in early April before the corn seeding.

Aboveground biomass was removed with negligible stubble left in the field. Thus, organic carbon input was mainly from manure application (i.e., NPKM and hNPKM treatments), residue

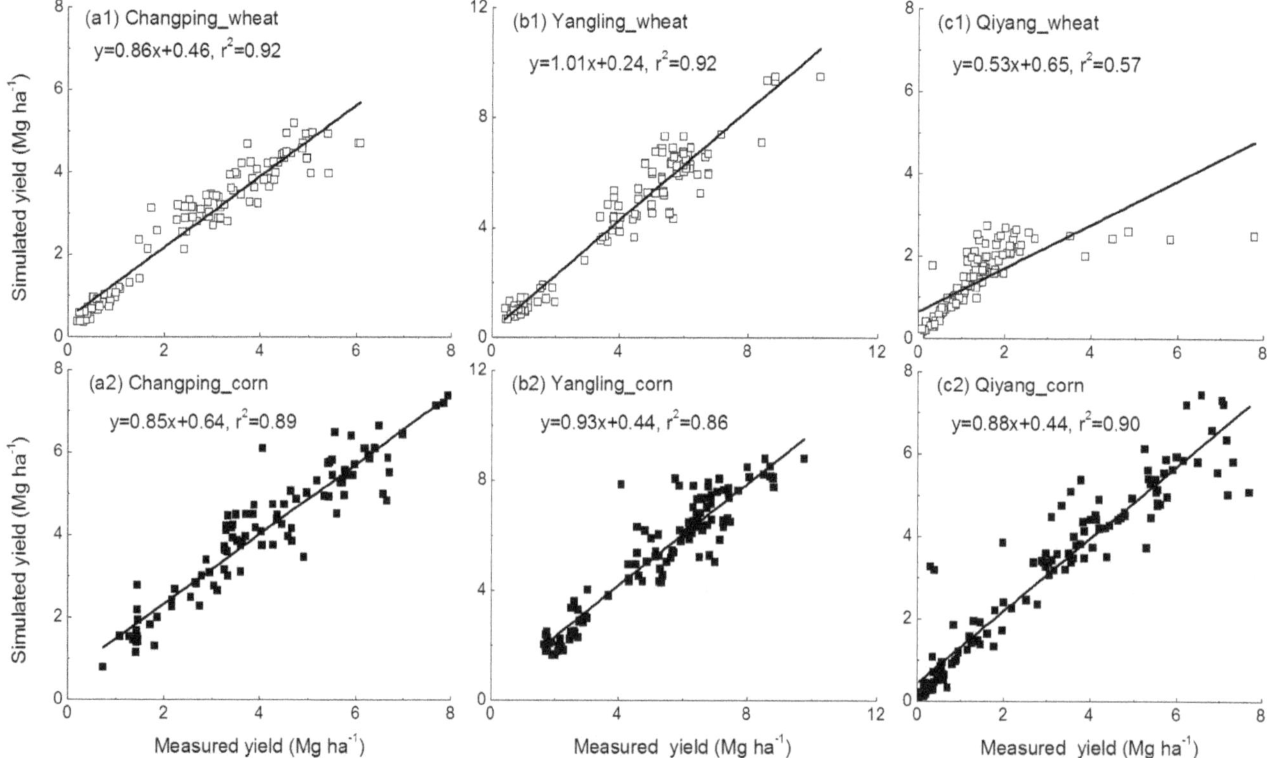

Figure 3. Correlationship between simulated and observed crop (i.e., □wheat and ■corn) grain yield data under the control, N, NP, NPK, NPKM, hNPKM and NPKS treatments at (A) Changping, (B) Yangling, and (C) Qiyang sites.

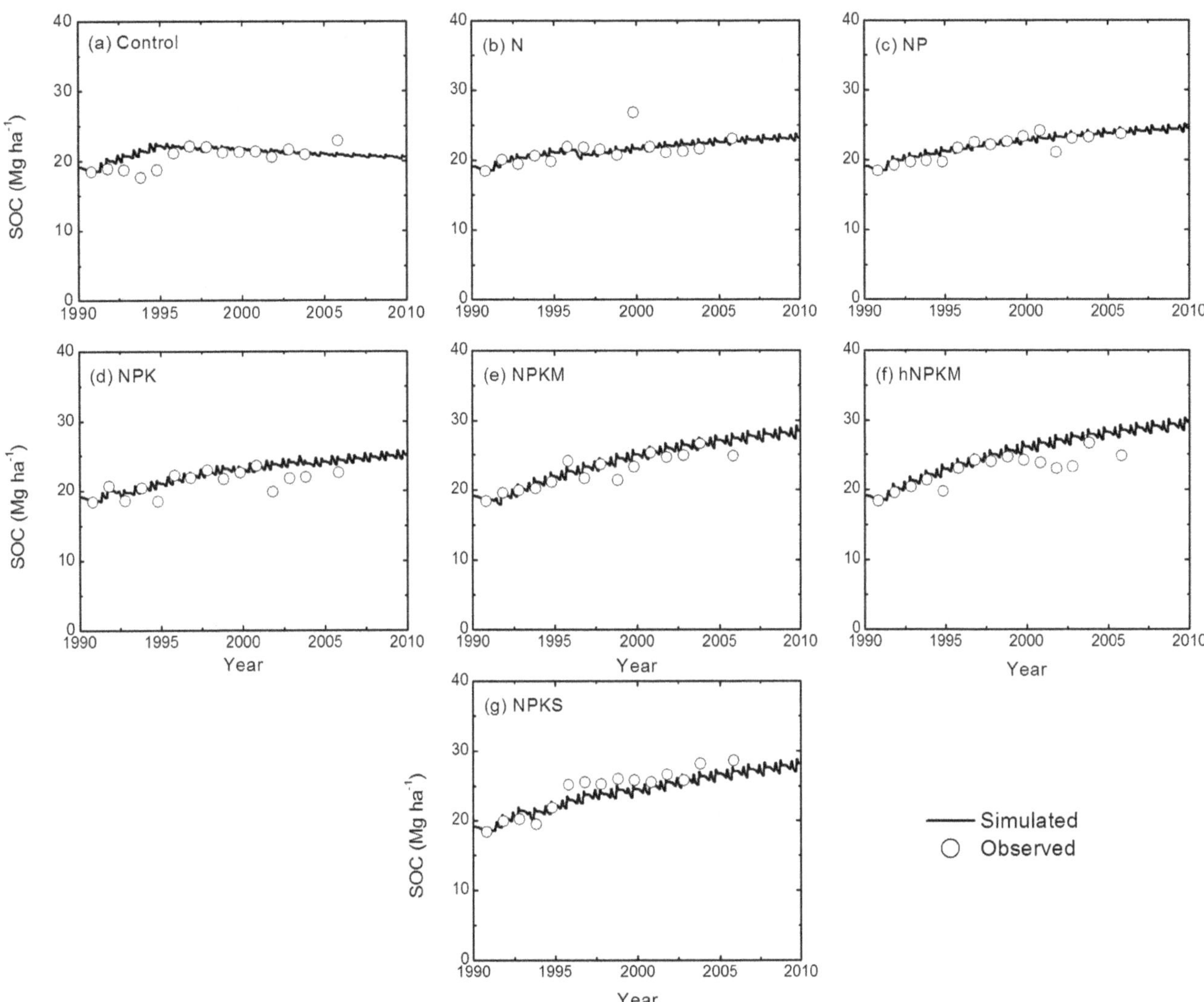

Figure 4. Simulated and measured soil organic carbon (SOC) stocks (0–20 cm) under the control, N, NP, NPK, NPKM, hNPKM and NPKS treatments at Changping.

addition (i.e., NPKS treatment) and from root system during crop growing season. All harvested biomass was removed from the plots for determining crop grain and residue yields. The grain and residue samples were air-dried, threshed, oven-dried at 70°C to a uniform moisture level, and then weighed separately.

During the experimental periods, surface soil (0–20 cm) was collected in each plot annually approximately 15 days after corn harvest. Five cores of soil from each plot were randomly taken, and soils were mixed thoroughly, and air dried for seven days after removing discrete plant residues and visible soil organism. Air-dried soil was sieved through 2 mm screen to determine pH (1:1w/v water). Representative subsamples were crushed to 0.25 mm for measurements of SOC [27]. Soil bulk density was measured in situ once every five or ten years. Surface soil SOC stock expressed as Mg ha^{-1} was calculated by multiplying the SOC content (g kg^{-1}) by averaged soil bulk density (g cm^{-3}) and depth (20 cm).

Model Description and Application

The simulation of SOC dynamics was performed with the CENTURY model (version 4.5) which was described by Parton and Rasmussen [28]. As shown in Fig. 1 [29], the arrows showing the CO_2 evolved in the transformations was indicative of the microbial growth efficiencies. The first-order decay rates for each of the pools corresponded to turnover times of roughly 3 and 0.5 years for the structural and metabolic components; 1.5 years for the active fraction, 25 years for the slow pool, and 1000 years for the passive pool [28]. The actual turnover times of each carbon pool was a function of the maximum turnover time of specific carbon pool and DEFAC. The value of DEFAC was calculated by multiplying the soil moisture factor (function of precipitation and stored soil water) and the soil temperature factor (function of the average monthly soil surface temperature). The turnover rate of active carbon pool was also a function of soil texture (higher for sandy soils), while the stabilization of active carbon into slow carbon was a function of the silt plus clay content [28].

Model simulations were set-up based on the historical crop rotations and farm practices' investigation from the local farmers at these sites. Five distinct periods were modeled (Table 1), including (1) an initialization period, i.e., 4000 years of native vegetation to reach an equilibrium; (2) the first cultivation period, i.e., plowing of the native grassland in 1901, and planting of corn

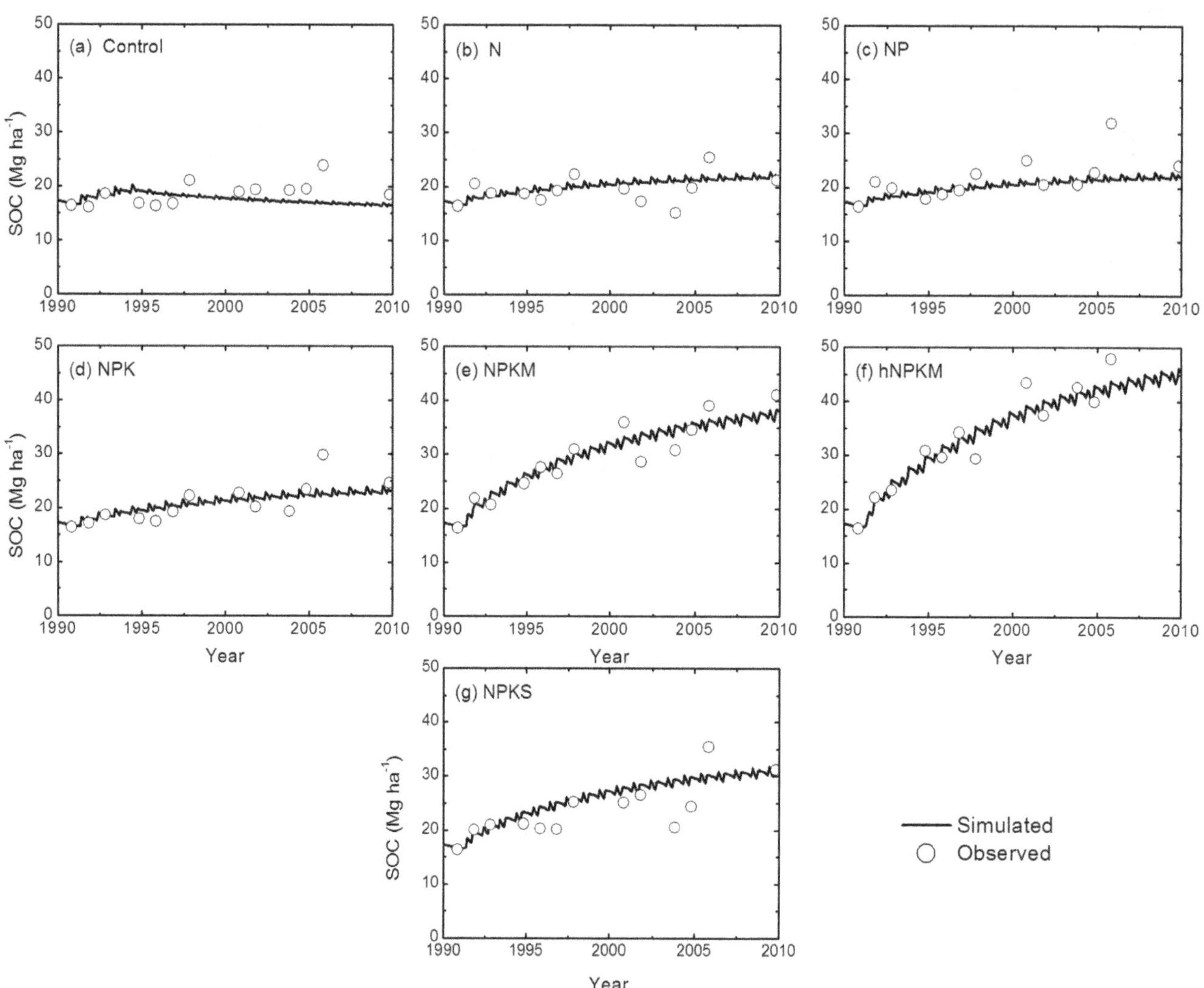

Figure 5. Simulated and measured soil organic carbon (SOC) stocks (0–20 cm) under the control, N, NP, NPK, NPKM, hNPKM and NPKS treatments at Yangling.

with removal of straw until 1960; (3) a second cultivation period, i.e., corn-wheat cropping with removal of straw from 1961 to 1988; (4) pre-experimental treatment period, i.e., uniform tillage with no crop in 1989 prior to establishing the long-term experiment; and (5) experiment period, i.e., the long-term corn-wheat fertility experiment from 1990 to 2050. Management during the experiment (1990–2010) was repeated for remainder of the experimental period until the end time of simulation to evaluate future trends. As shown in Table 2, carbon inputs from manure application were around 500 kg C ha^{-1} before the long-term experiment. Then carbon input varied based on measured data at different sites.

Weather data from 1990–2010 were obtained from China meteorological sharing service system (http://cdc.cma.gov.cn/). For the future simulations (i.e., 2011–2050), we used monthly mean climate data from the 1990–2010 period. Soil properties parameters were soil texture (sand, silt and clay content), soil pH, and bulk density (Table 3). Soils from the Changping and Yangling sites had high soil pH (~8.5) and silt loam texture with similar clay minerals (e.g., Hydromica and Montmorillonite). On the contrary, soil at the Qiyang site, developed from Quaternary red clay, had a lower soil pH (5.7) and heavy texture with the main clay mineral of Kaolinite.

As shown in Fig. 2, Changping and Yangling sites had similar climate condition whereas temperature and precipitation were much higher at the Qiyang site. The humid and warm climate at Qiyang site would accelerate decomposition of soil organic matter [28]. After setting up all the information (e.g., climate condition, soil texture and etc.) and modeling crop growth successfully, we modeled soil organic carbon pool by calibrating the dec4 value (i.e., maximum decomposition rate of soil organic matter with slow turnover) in the fix.100 file. By raising the parameter 10% higher for every test compared with the measured SOC, we finally got the sound fixed parameter as 0.0045 at the Qiyang site but 0.0023 (i.e., default value in the CENTURY model) at the other two sites [28].

For the plant production submodel, there were pools for live shoots and roots, and standing dead plant material. The potential production was a function of a genetic maximum (PRDX(1)) defined for each crop and 0–1 scalars depending on soil temperature, moisture status, shading by dead vegetation, and seedling growth. The maximum potential production of a crop, unlimited by temperature, moisture or nutrient stresses, was

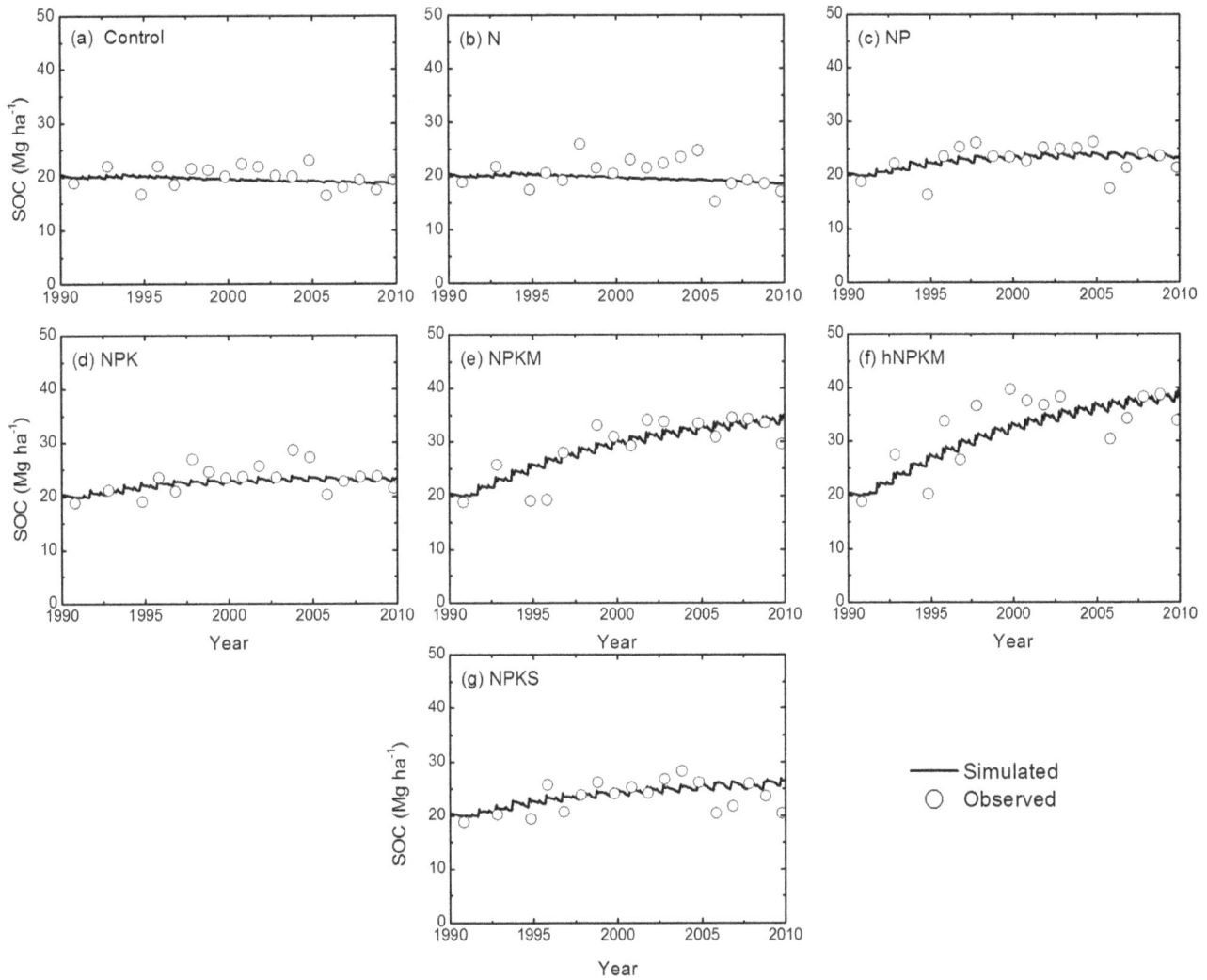

Figure 6. Simulated and measured soil organic carbon (SOC) stocks (0–20 cm) under the control, N, NP, NPK, NPKM, hNPKM and NPKS treatments at Qiyang.

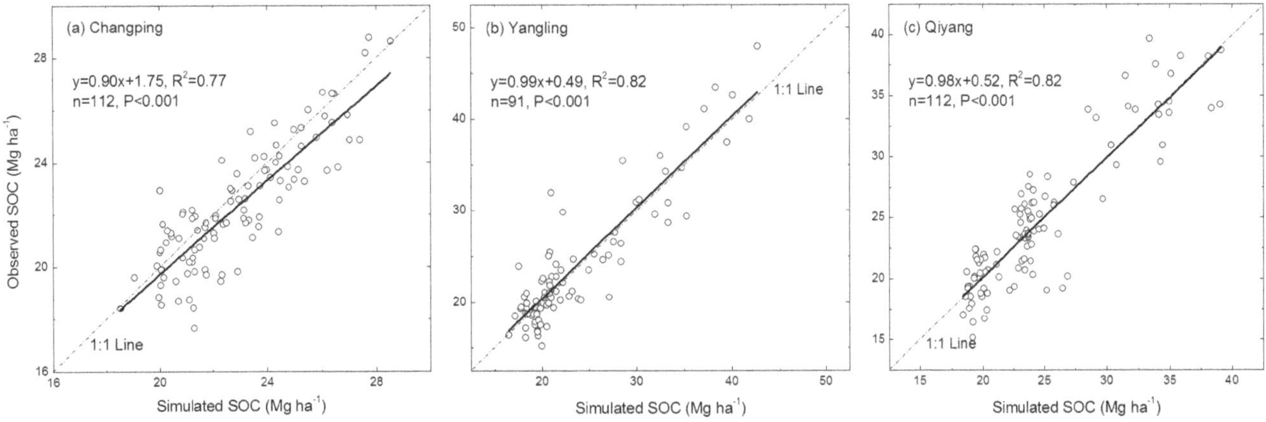

Figure 7. Correlation between simulated SOC and measured SOC under the control, N, NP, NPK, NPKM, hNPKM and NPKS treatments at (A) Changping, (B) Yangling, and (C) Qiyang sites.

Table 4. Statistics comparing simulated and measured SOC stocks at the three long-term experimental sites.

Site	Parameter[a]	Treatment						
		Control	N	NP	NPK	NPKM	hNPKM	NPKS
Changping	*n-RMSE*	6%	5%	5%	6%	7%	4%	6%
(n = 16×7)	*d*	0.57	0.81	0.9	0.81	0.91	0.97	0.91
	R^2	0.01	0.76	0.83	0.46	0.80	0.90	0.76
Yangling	*n-RMSE*	9%	12%	9%	12%	9%	9%	14%
(n = 13×7)	*d*	0.5	1.00	0.83	0.77	0.96	0.97	0.85
	R^2	0.39	0.19	0.43	0.59	0.89	0.88	0.61
Qiyang	*n-RMSE*	8%	7%	6%	9%	10%	11%	7%
(n = 16×7)	*d*	0.37	0.51	0.75	0.61	0.89	0.88	0.82
	R^2	0.07	0.16	0.47	0.34	0.69	0.64	0.61

[a]Note: *n-RMSE*, the normalized-root mean square err; *d*, the index of agreement; R^2, the sample correlation coefficient.

primarily determined by the level of photosynthetically active radiation, the maximum net assimilation rate of photosynthesis, the efficiency of conversion of carbohydrate into plant constituents, and the maintenance respiration rate [30]. Thus, the parameter for maximum potential production (PRDX(1)) had both genetic and environmental components. However, in CENTURY, the seasonal distribution of production was primarily controlled by the temperature response function rather than the seasonal variation in photosynthetically active radiation, so the maximum potential production parameter would reflect aboveground crop production in optimal summer conditions. This parameter would frequently be used to calibrate the predicted crop production for different environments, species, and varieties [31].

The CENTURY model had no function for pH effects on aboveground growth. Based on the 20 years of field measurement at the Qiyang site, soil pH had no impact on crop growth as long as the value remains above pH 5.5, and would completely stop growth in wheat and corn if the value declined to less than pH 4.0 [32]. We added the following soil pH factor (f_{pH}) in the crop growth sub-model to account for this effect:

$$\begin{cases} f_{pH}=0 & \text{if pH} \leq 4.0 \\ f_{pH}=\dfrac{(pH-\delta)}{(\delta-\beta)}+1 & \text{if } 4.0 \leq \text{pH} \leq 5.5 \\ f_{pH}=1 & \text{if pH} \geq 5.5 \end{cases} \qquad (1)$$

where δ and β were the maximum (i.e., $\delta = 5.5$) and minimum (i.e., $\beta = 4.0$) pH value that effected plant growth, respectively.

Model Evaluation

We compared the simulated SOC (0–20 cm) with field measured values. Both visual examination of graphic output and several statistical tests were used to evaluate the CENTURY model performance. Four statistical parameters were selected [33–36] for the evaluation, including (*i*) the sample coefficients of determination (R^2), (*ii*) the normalized-root mean square error (*n-RMSE*, equation 2), and (*iii*) the index of agreement (*d*, equation 3), as a descriptive measure of the average relative error [37].

$$n-RMSE=\frac{1}{\overline{M}}\sqrt{\sum_{i=1}^{n}(S_i-M_i)^2/n} \qquad (2)$$

where M_i were the measured values, S_i were the simulated values, $\bar{M}$ was the mean of the measured data and n was the number of the paired values. The index of agreement is computed with the following equation:

$$d=1-\frac{\sum_{i=1}^{n}(M_i-S_i)^2}{\sum_{i=1}^{n}(|S_i'|+|M_i'|)^2} \qquad (3)$$

where $S_i'=S_i-\overline{M}$ and $M_i'=M_i-\overline{M}$.

The classical R^2 statistic ($0<R^2<1$) provided the percentage of data variance that was accounted for by the model. The *RMSE* evaluated the difference between observed and modeled values in the original units of the data, while the normalized-*RMSE* (*n-RMSE*, equation 2) removed the influence of the units, and placed the results on a percentage scale for comparison of model performance among variables with different units. For an ideal fit,

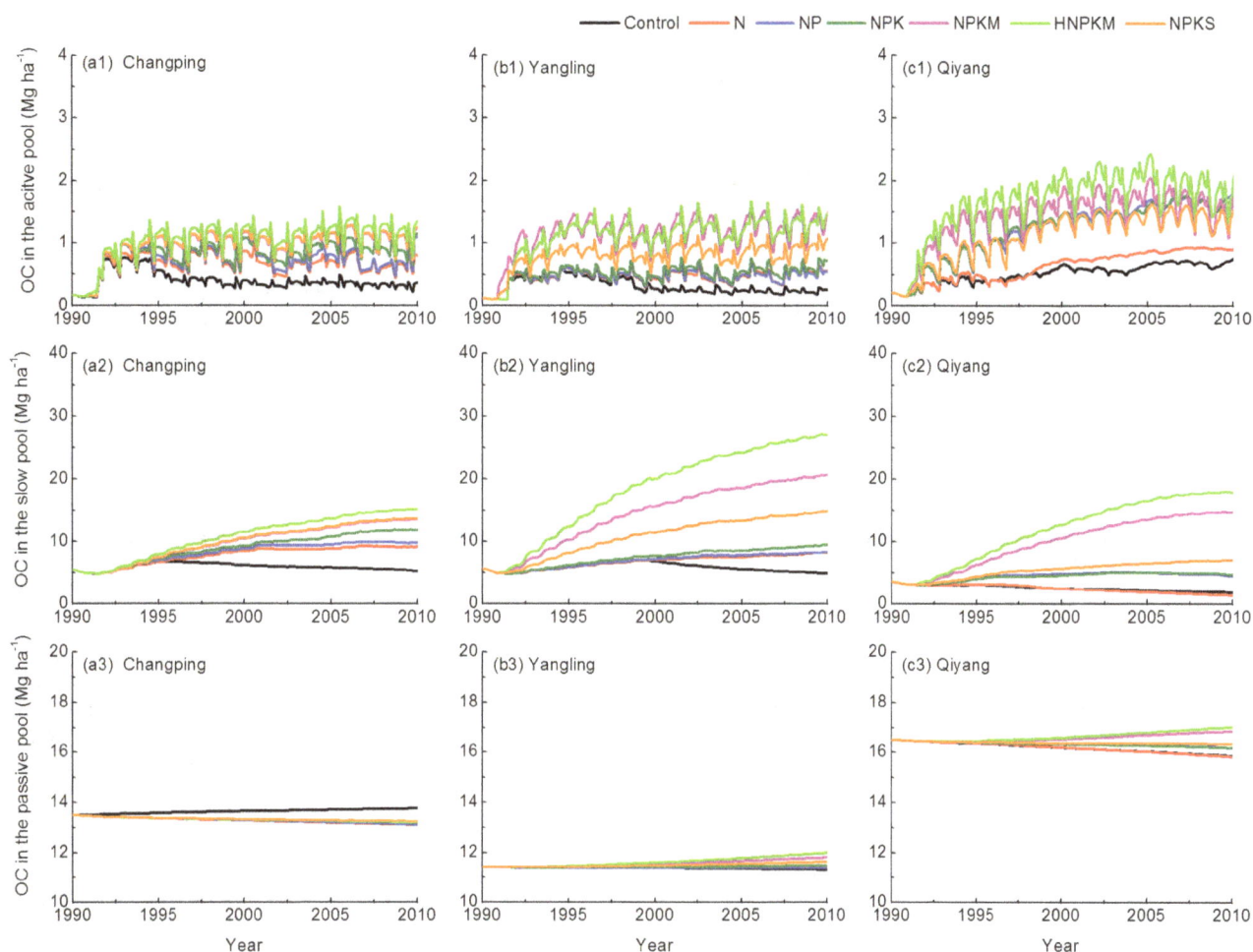

Figure 8. Soil organic carbon in the (1) active, (2) slow, and (3) passive pool under the control, N, NP, NPK, NPKM, hNPKM and NPKS treatments at (A) Changping, (B) Yangling, and (C) Qiyang sites.

R^2 would equal 100%, *RMSE* would equal zero and *d* would equal 1. In our study, *RMSE*<15%, *d*>0.5 and significant R^2 were used to bound the best case parameter sets for both calibration and validation simulations [38].

Results

Model Evaluation

We evaluated the model results for both grain yields and the SOC stock predictions. We were focusing on the SOC stock predictions of the model. In summary for the grain yield results (Fig. 3), we found that the model had correlation coefficients (R^2)

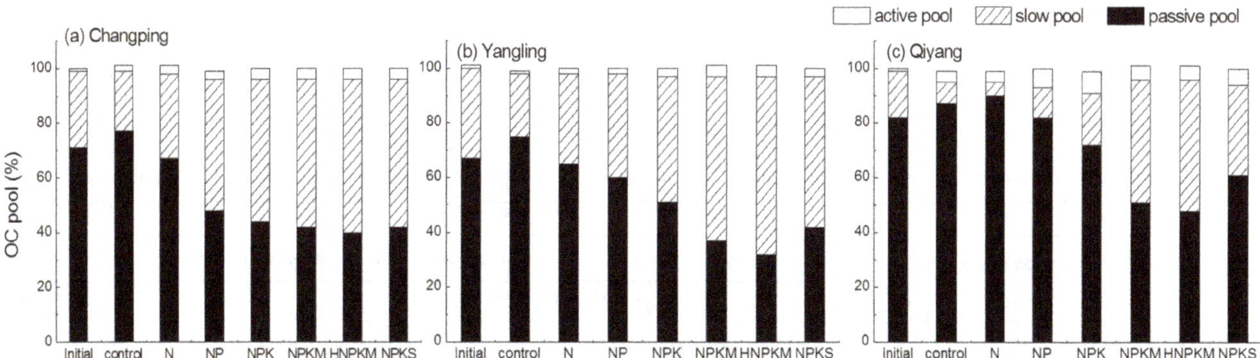

Figure 9. Proportion of changed SOC between the 1990 (initial) and 2010 in each soil organic matter pool under the control, N, NP, NPK, NPKM, hNPKM and NPKS treatments at (A) Changping, (B) Yangling, and (C) Qiyang sites.

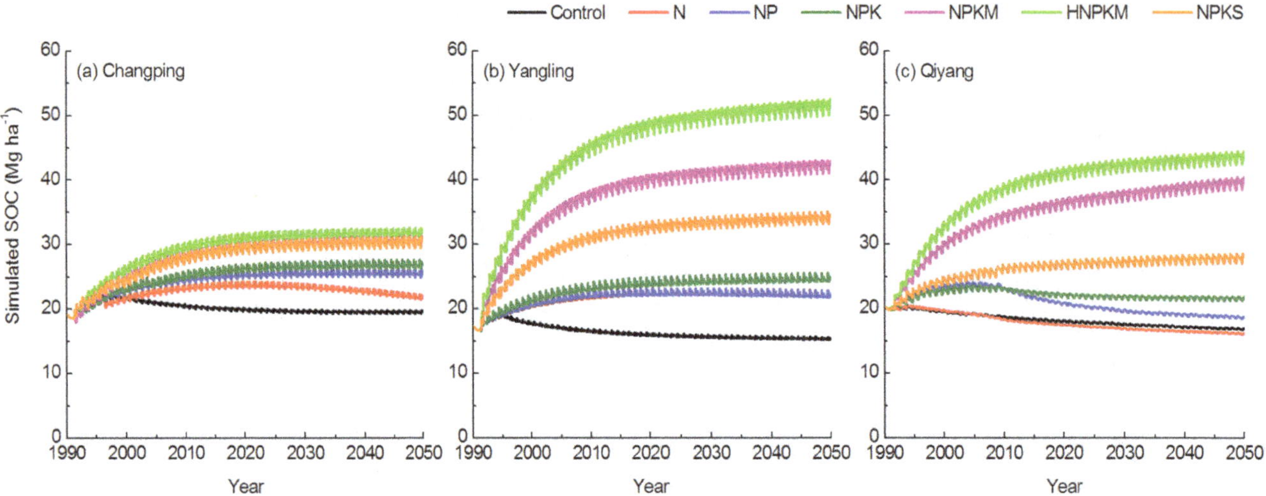

Figure 10. Soil organic carbon dynamics under the control, N, NP, NPK, NPKM, hNPKM and NPKS treatments at (A) Changping, (B) Yangling, and (C) Qiyang sites.

ranging from 0.57 to 0.92 ($p<0.05$), indicating that CENTURY model simulates the grain yields with reasonable accuracy.

The modeled SOC follows the trend of the measurements reasonably well in all treatments across the sites (Figs. 4, 5, and 6). Model results and measured SOC demonstrated that carbon levels decrease in the control treatment at all sites but increase under the fertilization treatments, except N only. For the N treatment, SOC is maintained at a stable level at most sites but did decline at the Qiyang site. The reason was related to a low soil pH that caused a decline in crop production and thus decline in residue input to the soil [25].

Linear regressions of observed vs. simulated SOC stocks (Fig. 7) are all highly significant ($P<0.001$) under various fertilization treatments. Approximately 77%–84% of the variability in the observed SOC stocks can be explained by the simulation. Therefore, in our experiment, the CENTURY model performs well and is suitable for predicting SOC dynamics for different types of N fertilizer (i.e., manure *versus* mineral) and N rates. The coefficients of determination (R^2) are higher at the Yangling and Qiyang sites than the Changping site.

RMSE is useful for evaluation of precision in model results [33]. Several studies have emphasized the need to estimate the precision in model predictions to evaluate model uncertainty [14,39,40]. In this study, we used several parameters for quantitative statistical analysis of measured and simulated SOC (Table 4). The *n-RMSE* ranges from 4% to 7% at the Changping site, 9%–14% at the Yangling site, and 6%–11% at the Qiyang site, respectively. Similarly, Fallon and Smith [41] calculated *n-RMSE* ranging from 1.8% to 16.4% when they modeled SOC in several long-term experiments in Europe. Also Álvaro et al. [20] had a low *n-RMSE* (below 6%) when they used the CENTURY model to evaluate tillage effects on SOC. Overall, the *n-RMSE* value is less than 15% for all the treatments in our study, demonstrating that the model is relatively precise in predicting the trends for different fertilization practices that are common in Chinese agricultural systems.

The *d* values and coefficients of determination (R^2) are highest in the NPKM, hNPKM and NPKS treatments at most sites. However, the R^2 value was much lower in the control (~0.07) or N (~0.19) treatments. The reason was that inter-annual fluctuations in crop yield were larger in the control and N treatments due to the long-term nutrient deficiency [25]. Therefore, carbon input from biomass was fluctuated and thus the simulation was not satisfied as other treatments.

Fertilization Effects on Organic Carbon Pools

Long-term fertilization has a variety of effects on SOC pools (Fig. 8 and 9). Generally, the active pool is the smallest carbon pool [16], and our results are consistent with past studies with an active pool containing only 1–3 Mg ha^{-1} across the sites (first row in Fig. 8). By 2010, the largest active pool is at the Qiyang site (Fig. 8_c1), followed by Changping (Fig. 7_a1) and Yangling (Fig. 8_b1) sites.

Long-term manure fertilization significantly increased the proportion of slow soil organic matter at all sites (Fig. 8). By the

Table 5. Estimated carbon sequestration potential (0–20 cm) under long-term hNPKM and NPKS fertilization treatments.

Site	SOC in 1990 (Mg ha^{-1})	SOC in 2050 (Mg ha^{-1})		Carbon sequestration potential (Mg ha^{-1})	
		NPKS	hNPKM	NPKS	hNPKM
Changping	22.4	30.5	31.8	8.1	9.4
Yangling	16.4	34.3	52.1	17.9	35.7
Qiyang	20.5	28.2	44.2	7.7	23.7

end of 2010, the slow pool accounted for ~42% of the total SOC at the Changping site, ~65% at the Yangling site, and ~48% at the Qiyang site (Fig. 9). In addition, trends in the slow pool are consistent with the overall change in total SOC stock at each of the sites (Figs. 4, 5, and 6). These results suggest that the carbon dynamics associated with the slow pool are key drivers of the trend in total SOC.

The percentage of SOC in the passive soil organic matter is 75%–87% under the control treatment but only 32%–48% under the hNPKM treatment (Fig. 9). Our modeling results suggest that a higher proportion of the organic carbon is contained in the passive pool with the slowest turnover rate if no fertilizer is added. However, long-term fertilization generally has a smaller effect on the passive pool compared to the slow pool according to the experimental data (third row in Fig. 8). The reason would be corresponded to turnover times of 25 years for the slow pool but 1000 years for the passive pool in the CENTURY [28].

Soil Carbon Dynamics and Carbon Sequestration Potential

Long-term fertilization has different effects on soil organic carbon across the three sites (Fig. 10). For example, the SOC stock (0–20 cm) increases and then decreases to a relative stable level in the control and N treatments at Changping site (Fig. 10a). Plots with balanced fertilization or manure application and straw return (i.e., NPK, NPKM, hNPKM and NPKS) have an increase in SOC and then remain relatively stable. By 2050, the SOC stocks reach 30.5–31.8 Mg ha^{-1} with organic fertilization and 26.8 Mg ha^{-1} with chemical fertilization (NPK).

For the Yangling site (Fig. 10b), SOC contents increase over time in the NPK, NPKM, hNPKM, and NPKS treatments. By the end of simulated period, SOC reaches 42.6 Mg ha^{-1} and 52.1 Mg ha^{-1} under the NPKM and hNPKM treatments, respectively. With straw return, the SOC level reaches 34.3 Mg ha^{-1}. Balanced chemical fertilization (NPK) also yields relatively high SOC contents at 24.6 Mg ha^{-1} by 2050. There is little difference in SOC stocks between N and NP treatments. SOC increases to 20.1 Mg ha^{-1} in 1994, then declines to 15.3 Mg ha^{-1} in 2050 for the control treatment.

Similar to the Yangling site, SOC levels increase through the entire period of simulation under the NPKM, hNPKM, and NPKS treatments for the Qiyang site (Fig. 10c). By 2050, SOC stocks are 40.2 Mg ha^{-1}, 44.2 Mg ha^{-1}, and 28.2 Mg ha^{-1} for the NPKM, hNPKM, and NPKS treatments, respectively. For all chemical fertilization treatments, we assume no decrease in pH after year 2010. Therefore, soil pH has not negative impact on carbon input due to a reduction in crop production associated with pH, and applying NPK fertilizers maintain SOC levels at 21.7 Mg ha^{-1}.

The SOC trends are relatively stable by the end of 2050 in the different treatments (Fig. 10). In order to estimating the carbon sequestration potential (i.e., the SOC increment by the end of simulation) with manure addition and regular straw incorporation, we further evaluated the SOC change under the hNPKM and NPKS treatments in the period of 1990–2050 (Table 5). The initial SOC level is highest at the Changping site (22.4 Mg ha^{-1}) and lowest at the Yangling site (16.4 Mg ha^{-1}). In contrasts, the highest simulated SOC level in 2050 happens at the Yangling site and lowest at the Changping site. Compared with the NPKS treatment, carbon sequestration potential is greater with hNPKM management, ranging from 9.4 Mg ha^{-1} to 35.7 Mg ha^{-1}. For the NPKS treatment, SOC increases to 17.9 Mg ha^{-1} at the Yangling site but only 7.1 Mg ha^{-1}–8.1 Mg ha^{-1} at the other two sites.

Discussion

Studies has proved that optimizing fertilization managements can raise crop yield as well as biomass, which would enhance biomass input into the soil from crop straw and roots [28,42,43]. Halvorson et al [44] reported that increasing nitrogen fertilizer rate would increase SOC levels in the 0–7.5 cm soil depth after 11 years cropping. In our study, chemical fertilization (expect N treatment) would also increase crop yield and thus increase SOC in the first 20 years of crop rotation at two of three sites. However, most chemical fertilization treatment would only maintain the SOC level in the long-term run (Fig. 10).

Straw incorporation would also increase carbon input to the soil. Li et al [45] found that cropland soil lost 1.6% of SOC due the less (25%) aboveground residue's returning. Using the DNDC model, Tang et al. [46] predicted that the soil carbon loss in China's cropland would be partly reversed if straw return and no tillage practices were expanded widely. In our study, averaged carbon sequestration rate would reach to 118 kg ha^{-1} yr^{-1} to 298 kg ha^{-1} yr^{-1} with straw returning during the period of 1990–2050 under current returning rate of straw residue (i.e., 2000–4500 kg ha^{-1} yr^{-1}). The results are lower than the estimated rate (108–728 kg ha^{-1} yr^{-1}) in the similar region by Lu et al [47], in which the estimated straw returning rates were also higher (4530–8110 kg ha^{-1} yr^{-1}).

Moreover, there are differences in the influence of manure on SOC trends among the sites. At the Changping site, SOC stocks reach a relatively stable level (30.0 Mg ha^{-1}–31.0 Mg ha^{-1}) by 2013 (Fig. 10a). In contrast, plots from the other two sites continue to increase in SOC over the entire simulated period. The reason for the site difference in the simulated trends is related to the manure quality between the sites, and the associated decomposition rates of the organic matter [28,48]. For the Changping site, farmyard manure is livestock manure mixed with soil and/or crop residue. The lignin content in farmyard manure is lower and decompose at a faster rate [49]. When we enhanced the lignin content of farmyard manure in CENTURY, the simulation of SOC rose to the much higher level (data not shown).

Many modeling research also focus on the regional estimation of SOC sequestration. Follett [50] estimated soil carbon sequestration potential would be 11–56 Tg C yr^{-1} for residue management in the USA. In China, Yan et al [51] estimated 32.5 Tg C yr^{-1} to be sequestered in croplands if 50% of the crop residue was returned to soils, and no tillage was adopted on 50% of the arable lands. However, these results are criticized for the models that are calibrated against few or no field experimental data. Our study estimated carbon sequestration potential using long-term experiment data, providing credible data to evaluate carbon sequestration response to different fertilization managements.

Conclusion

We used the CENTURY model (version 4.5) to simulate soil organic carbon (SOC) dynamics for long-term fertilization experiments with wheat-corn cropping systems in China. Our study indicated that CENTURY can simulate fertilization effects on SOC trends for different climatic conditions and soil properties that are common for wheat-corn systems in China. For most sites, the model was more precise in predicting the SOC trends for treatments with balanced nutrient additions (NPK, NPKM, hNPKM, and NPKS) than the control and unbalanced fertilization treatments (N and NP). After simulating SOC dynamics from 1990–2050, the SOC levels increased from 31.8–52.1 Mg ha^{-1} across the four sites. We estimate that carbon sequestration potential ranges from 9.4–35.7 Mg ha^{-1} under the high manure application practice (i.e., hNPKM treatment). Analysis of the proportion of organic carbon in each of the soil organic matter pools indicates that long-term fertilization enhances the slow pool and leads to a decline in the proportion of carbon in the passive pool. Model results suggest that changes in slow organic matter drive the overall trends in SOC stocks associated with long-term fertilization.

Acknowledgments

The authors would like to thank Dr. Sigen Chen and Ms. Cynthia Keough for the technical support on the CENTURY model. We acknowledge all the colleagues for their unremitting efforts on the long-term experiments from these sites.

Author Contributions

Conceived and designed the experiments: RHC MGX XJW. Performed the experiments: RHC SMO WJP. Analyzed the data: RHC MGX XJW. Contributed reagents/materials/analysis tools: SMO WJP. Wrote the paper: RHC.

References

1. Schlesinger WH (1997) Biogeochemistry: an analysis of global change: New York Academic Press.
2. Schuman GE, Janzen HH, Herrick JE (2002) Soil carbon dynamics and potential carbon sequestration by rangelands. Environmental pollution 116: 391–396.
3. Manzoni S, Porporato A (2009) Soil carbon and nitrogen mineralization: Theory and models across scales. Soil Biology & Biochemistry 41: 1355–1379.
4. Lal R, Follett F, Stewart BA, Kimble JM (2007) Soil carbon sequestration to mitigate climate change and advance food security. Soil Science 172: 943–956.
5. Lal R (2004) Soil Carbon Sequestration Impacts on Global Climate Change and Food Security. Science 304: 1623–1627.
6. Körschens M (2006) The importance of long-term field experimentsfor soil science and environmental research - a review. Plant, soil and environment 52: 1–8.
7. Zhang W, Xu M, Wang X, Huang Q, Nie J, et al. (2012) Effects of organic amendments on soil carbon sequestration in paddy fields of subtropical China. Journal of Soils and Sediments: 1–14.
8. Powlson D, Zhao BQ, Li XY, Li XP, Shi XJ, et al. (2010) Long-Term Fertilizer Experiment Network in China: Crop Yields and Soil Nutrient Trends. Agronomy Journal 102: 216–230.
9. Edmeades DC (2003) The long-term effects of manures and fertilisers on soil productivity and quality: a review. Nutrient Cycling in Agroecosystems 66: 165–180.
10. Thomsen IK, Christensen BT (2004) Yields of wheat and soil carbon and nitrogen contents following long-term incorporation of barley straw and ryegrass catch crops. Soil Use and Management 20: 432–438.
11. Bruun S, Christensen BT, Hansen EM, Magid J, Jensen LS (2003) Calibration and validation of the soil organic matter dynamics of the Daisy model with data from the Askov long-term experiments. Soil Biology & Biochemistry 35: 67–76.
12. Shirato Y, Yokozawa M (2005) Applying the Rothamsted Carbon Model for long-term experiments on Japanese paddy soils and modifying it by simple mining of the decomposition rate. Soil Science and Plant Nutrition 51: 405–415.
13. Yang Y, Luo Y, Finzi AC (2011) Carbon and nitrogen dynamics during forest stand development: a global synthesis. New Phytologist 190: 977–989.
14. Smith P, Smith JU, Powlson DS, McGill WB, Arah JRM, et al. (1997) A comparison of the performance of nine soil organic matter models using datasets from seven long-term experiments. Geoderma 81: 153–225.
15. Chilcott CR, Dalal RC, Parton WJ, Carter JO, King AJ (2007) Long-term trends in fertility of soils under continuous cultivation and cereal cropping in southern Queensland. IX*. Simulation of soil carbon and nitrogen pools using CENTURY model. Australian Journal of Soil Research 45: 206–217.
16. Denef K, Six J, Merckx R, Paustian K (2004) Carbon sequestration in microaggregates of no-tillage soils with different clay mineralogy. Soil Science Society of America Journal 68: 1935–1944.
17. Kelly RH, Parton WJ, Crocker GJ, Grace PR, Klir J, et al. (1997) Simulating trends in soil organic carbon in long-term experiments using the century model. Geoderma 81: 75–90.
18. Falloon P, Smith P (2002) Simulating SOC changes in long-term experiments with RothC and CENTURY: model evaluation for a regional scale application. Soil Use and Management 18: 101–111.
19. Bhattacharyya T, Pal DK, Williams S, Telpande BA, Deshmukh AS, et al. (2010) Evaluating the Century C model using two long-term fertilizer trials representing humid and semi-arid sites from India. Agriculture, Ecosystems & Environment 139: 264–272.
20. Álvaro-Fuentes J, López MV, Arrúe JL, Moret D, Paustian K (2009) Tillage and cropping effects on soil organic carbon in Mediterranean semiarid agroecosystems: Testing the Century model. Agriculture, Ecosystems & Environment 134: 211–217.
21. Cerri CEP, Easter M, Paustian K, Killian K, Coleman K, et al. (2007) Simulating SOC changes in 11 land use change chronosequences from the Brazilian Amazon with RothC and Century models. Agriculture, Ecosystems & Environment 122: 46–57.
22. Feng XM, Zhao YS (2011) Grazing intensity monitoring in Northern China steppe: Integrating CENTURY model and MODIS data. Ecological Indicators 11: 175–182.
23. Fang D, Zhou G, Jiang Y, Jia B, Xu Z, et al. (2012) Impact of fire on carbon dynamics of Larix gmelinii forest in Daxingan Mountains of Northeast China:A simulation with CENTURY model. The Journal of Applied Ecology 23: 2411–2421.
24. Wang SH, Shi XZ, Zhao YC, Weindorf DC, Yu DS, et al. (2011) Regional Simulation of Soil Organic Carbon Dynamics for Dry Farmland in East China by Coupling a 1:500 000 Soil Database with the Century Model. Pedosphere 21: 277–287.
25. Cong RH, Wang XJ, Xu MG, Zhang WJ, Xie LJ, et al. (2012) Dynamics of soil carbon to nitrogen ratio changes under long-term fertilizer addition in wheat-corn double cropping systems of China. European Journal of Soil Science 63: 341–350.
26. Cong R, Xu M, Wang X, Zhang W, Yang X, et al. (2012) An analysis of soil carbon dynamics in long-term soil fertility trials in China. Nutrient Cycling in Agroecosystems 93: 201–213.
27. Walkley A, Black IA (1934) An examination of Degtjareff method for determining soil organic matter and a proposed modification of the chromic acid titration method. Soil Science 37: 29–38.
28. Parton WJ, Rasmussen PE (1994) Long-Term Effects of Crop Management in Wheat-Fallow: II. CENTURY Model Simulations. Soil Science Society of America Journal 58: 530–536.
29. Alister K Metherell, Laura A Harding, C. Vernon Cole, Parton WJ (1993) CENTURY Soil Organic Matter Model Environment.
30. van Heemst HDJ (1986) Physiological principles. In: Keulen; HV, Wolf J, editors. Modelling of agricultural production: weather, soils and crops. Pudoc Wageningen. pp. 13–26.
31. Metherell AK, Harding LA, Cole CV, Parton WJ (1993) Plant Production Submodels. CENTURY Soil Organic Matter Model Environment Technical Documentation Agroecosystem Version 40.
32. Zhang H, Wang B, Xu M (2008) Effects of Inorganic Fertilizer Inputs on Grain Yields and Soil Properties in a Long-Term Wheat-corn Cropping System in South China. Communications in Soil Science and Plant Analysis 39: 1583–1599.
33. Smith JU, Smith P, Addiscott TM (1996) Quantitative methods to evaluate and compare soil organic matter (SOM) models. In: Powlson DS, Smith P, Smith JU, editors. Evaluation of Soil Organic Matter Models Using Existing, Long-term Datasets. Heidelberg: Springer-Verlag. pp. 181–200.
34. Cheng WX, Zhang QL, Coleman DC, Carroll CR, Hoffman CA (1996) Is available carbon limiting microbial respiration in the rhizosphere? Soil Biology & Biochemistry 28: 1283–1288.
35. Kong AYY, Six J, Bryant DC, Denison RF, van Kessel C (2005) The Relationship between Carbon Input, Aggregation, and Soil Organic Carbon Stabilization in Sustainable Cropping Systems. Soil Sci Soc Am J 69: 1078–1085.
36. Domke GM, Woodall CW, Walters BF, Smith JE (2013) From Models to Measurements: Comparing Downed Dead Wood Carbon Stock Estimates in the U.S. Forest Inventory. PLoS One 8: e59949.
37. Willmott CJ (1982) Some Comments on the Evaluation of Model Performance. Bulletin of the American Meteorological Society 63: 1309–1313.
38. Tonitto C, David MB, Drinkwater LE, Li CS (2007) Application of the DNDC model to tile-drained Illinois agroecosystems: model calibration, validation, and uncertainty analysis. Nutrient Cycling in Agroecosystems 78: 51–63.
39. Blagodatskaya E, Kuzyakov Y (2008) Mechanisms of real and apparent priming effects and their dependence on soil microbial biomass and community structure: critical review. Biology and Fertility of Soils 45: 115–131.
40. Ogle SM, Breidt FJ, Easter M, Williams S, Paustian K (2007) An empirically based approach for estimating uncertainty associated with modelling carbon sequestration in soils. Ecological Modelling 205: 453–463.
41. Falloon P, Smith P (2003) Accounting for changes in soil carbon under the Kyoto Protocol: need for improved long-term data sets to reduce uncertainty in model projections. Soil Use and Management 19: 265–269.
42. Halvorson AD, Wienhold BJ, Black AL (2002) Tillage, nitrogen, and cropping system effects on soil carbon sequestration. Soil Science Society of America Journal 66: 906–912.

43. Dumanski J, Desjardins RL, Tarnocai C, Monreal C, Gregorich EG, et al. (1998) Possibilities for future carbon sequestration in Canadian agriculture in relation to land use changes. Climatic Change 40: 81–103.
44. Halvorson AD, Reule CA, Follett RF (1999) Nitrogen fertilization effects on soil carbon and nitrogen in a dryland cropping system. Soil Science Society of America Journal 63: 912–917.
45. Li C, Zhuang Y, Frolking S, Galloway J, Harriss R, et al. (2003) Modeling Soil Organic Carbon Change in Croplands of China. Ecological Applications 13: 327–336.
46. Tang HJ, Qiu JJ, Van Ranst E, Li CS (2006) Estimations of soil organic carbon storage in cropland of China based on DNDC model. Geoderma 134: 200–206.
47. Lu F, Wang XK, Han B, Ouyang ZY, Duan XN, et al. (2009) Soil carbon sequestrations by nitrogen fertilizer application, straw return and no-tillage in China's cropland. Global Change Biology 15: 281–305.
48. Paustian K, Parton W, Persson J (1992) Modeling Soil Organic Matter in Organic-Amended and Nitrogen-Fertilized Long-Term Plots. Soil Science Society of America Journal 56: 476–488.
49. Zhang WJ, Wang XJ, Xu MG, Huang SM, Liu H, et al. (2010) Soil organic carbon dynamics under long-term fertilizations in arable land of northern China. Biogeosciences 7: 409–425.
50. R.F F (2001) Soil management concepts and carbon sequestration in cropland soils. Soil and Tillage Research 61: 77–92.
51. Yan HM, Cao MK, Liu JY, Tao B (2007) Potential and sustainability for carbon sequestration with improved soil management in agricultural soils of China. Agriculture Ecosystems & Environment 121: 325–335.

18

Carbon Sequestration Efficiency of Organic Amendments in a Long-Term Experiment on a Vertisol in Huang-Huai-Hai Plain, China

Keke Hua, Daozhong Wang*, Xisheng Guo*, Zibin Guo

Soil and Fertilizer Research Institute, Anhui Academy of Agricultural Sciences, Hefei, China

Abstract

Soil organic carbon (SOC) sequestration is important for improving soil fertility of cropland and for the mitigation of greenhouse gas emissions to the atmosphere. The efficiency of SOC sequestration depends on the quantity and quality of the organic matter, soil type, and climate. Little is known about the SOC sequestration efficiency of organic amendments in Vertisols. Thus, we conducted the research based on 29 years (1982–2011) of long-term fertilization experiment with a no fertilizer control and five fertilization regimes: CK (control, no fertilizer), NPK (mineral NPK fertilizers alone), NPK+1/2W (mineral NPK fertilizers combined with half the amount of wheat straw), NPK+W (mineral NPK fertilizers combined with full the amount of wheat straw), NPK+PM (mineral NPK fertilizers combined with pig manure) and NPK+CM (mineral NPK fertilizers combined cattle manure). Total mean annual C inputs were 0.45, 1.55, 2.66, 3.71, 4.68 and 6.56 ton/ha/yr for CK, NPK, NPKW1/2, NPKW, NPKPM and NPKCM, respectively. Mean SOC sequestration rate was 0.20 ton/ha/yr in the NPK treatment, and 0.39, 0.50, 0.51 and 0.97 ton/ha/yr in the NPKW1/2, NPKW, NPKPM, and NPKCM treatments, respectively. A linear relationship was observed between annual C input and SOC sequestration rate (SOCsequestration rate = 0.16 Cinput −0.10, R = 0.95, P<0.01), suggesting a C sequestration efficiency of 16%. The Vertisol required an annual C input of 0.63 ton/ha/yr to maintain the initial SOC level. Moreover, the C sequestration efficiencies of wheat straw, pig manure and cattle manure were 17%, 11% and 17%, respectively. The results indicate that the Vertisol has a large potential to sequester SOC with a high efficiency, and applying cattle manure or wheat straw is a recommendable SOC sequestration practice in Vertisols.

Editor: Xiujun Wang, University of Maryland, United States of America

Funding: This work was supported by the Natural Science Foundation of China (Grant No. 41401331), Special Fund for Agro-scientific Research in the Public Interest of China (Grant No. 201203030), the President Distinguished Youth Innovation Fund of Anhui Academic of Agricultural Science (Grants No. 14B1008 and 14B1007), and by the Anhui Province Science and Technology Program (Grant No. 1206c0805033). The funders had no role in study design, data collection and analysis, decision to publish, or preparation of the manuscript.

Competing Interests: The authors have declared that no competing interests exist.

* Email: wdzhong-3@163.com (DZW); 1078681598@qq.com (XSG)

Introduction

Soil organic carbon (SOC) sequestration contributes to the mitigation of greenhouse gas emissions and to the improvement of soil fertility [1]. Net SOC sequestration is the balance of organic C inputs into the soil (via crop residues, organic amendments in compost, animal manure, etc.) and organic C decomposition by soil microbes. SOC sequestration efficiency is commonly expressed by the relationship between annual C input and SOC accumulation rate, which is an indicator of soil C sequestration ability [2]. Therefore, information about the C sequestration efficiency is useful for seeking high efficiency management strategies of enhancing the SOC stock and soil fertility.

C sequestration efficiency is regulated by climate, the quantity, quality of added organic materials, and soil inherent properties [3–5]. Within a climate zone, soil inherent properties (i.e. initial SOC content, soil texture, clay type and aggregates) and cultivation practices have effects on SOC sequestration efficiency. A negative linear relationship has been reported between C sequestration efficiency and initial SOC content [6,7], mainly because SOC tends to increase faster if initial SOC content is far from its saturation level. Feller and Beare [8] reported a positive relationship between SOC sequestration and soil clays. Because clay type (i.e. 2:1 smectite clay minerals versus 1:1 allophanic clay minerals) has influence on the stabilization of SOC [9]. Elliott and Coleman [10] indicated that soil aggregates may physically protect SOC against decomposition by soil microbes. Recently, several studies have examined the SOC sequestration efficiency of different soil types based on long-term fertilization experiment [11–15]. However, no information is available about SOC sequestration efficiency and C sequestration efficiencies of organic amendments in Vertisols [16], which occupy approximately 4.0 million km^2 in Huang-Huai-Hai Plain of China. Vertisols are characterized by a low SOC content (less than 0.6% in the topsoil) a high clay content (more than 35%) and shrinking and swelling properties which contribute to self-mulching [17].

The purpose of this study was to evaluate the SOC sequestration efficiency under long-term fertilization practices in Vertisols, with a winter wheat–soybean double cropping system. Specific objectives were (1) to assess C input and SOC dynamics under various long-term fertilization practices (2) to estimate the SOC sequestration efficiency of different organic amendments.

Materials and Methods

Ethics Statement

The administration of the department of agricultural of Anhui Province gave permission for this research at the study site. We confirm that the field studies did not involve endangered or protected species.

Site description

The experiment is located at the Madian Agro-Ecological Station in Huang-Huai-Hai Plain, Eastern China (N33°13′, E116°37′). The areas has a sub-humid climate, with annual mean, maximum and minimum air temperature of 16.5°C, 36.5°C and −7.4°C, respectively (Fig. 1a). Annual precipitation ranged from about 457 to 1478 mm during the last 24 years (Fig. 1b), about 70% of which occurs from May to September. There are strong inter-annual and seasonal variations in precipitation. For example, a 90-day period of summer drought with a total precipitation of only 138 mm occurred in 1988. Precipitation in the soybean period ranged from 244 mm to 1049 mm with an average of 616 mm, which was much higher than that in the wheat period (259 mm).

The predominant Vertisols have developed in fluvial and lacustrine deposits. They are classified as Calcic Kastanozems, according to the soil classification system of the Food and Agriculture Organization (FAO). Soil pH ranges from 6.0 to 8.6 and SOC content from 5.8 to 7.5 g/kg in the area. The initial (1982) topsoil (0–20 cm) had a total SOC of 5.8±0.08 g/kg, total N 0.96±0.04 g/kg, total P 0.28±0.02 g/kg, bulk density 1.45 g/cm^3 and pH (1:2.5 w/v) 7.4, respectively. The topsoil under NPK treatment in 2013 had coarse sand (2 to 0.2 mm) 8.5±2.0 g/kg, fine sand (0.2 to 0.02 mm) 322.9±21.7 g/kg, silt (0.02 to 0.002 mm) 262.3±24.0 g/kg and clay (<0.002 mm) 406.3±24.0 g/kg, respectively.

Experimental design

The long-term field experiment was initiated in 1982. The experiment had six treatments (Table 1): no fertilizer (CK), mineral nitrogen-phosphorus-potassium fertilizers alone (NPK), mineral NPK fertilizers combined with 2.5 ton/ha/yr (dry base) of wheat straw (NPKW1/2), mineral NPK fertilizers combined with 5.0 ton/ha/yr of wheat straw (NPKW), mineral NPK fertilizers combined with 7.8 ton/ha/yr of pig manure (NPKPM), and mineral NPK fertilizers combined with 12.5 ton/ha/yr of cattle manure (NPKCM). The treatments were laid out in a randomized block design with four replications. The net plot size was 70 m^2 (14.9 m ×4.7 m). Each plot was isolated by 50 cm deep plates. Mineral N, P and K fertilization was urea, calcium superphosphate and potassium chloride, respectively. The amounts applied were similar to the amounts applied by local farmers, i.e., 180 kg N, 90 kg P_2O_5, and 135 kg K_2O/ha/yr. All fertilizers were applied as base fertilizer at the start of the wheat growing season. No fertilizers were applied to soybean growing season because of the high background value of soil N, P and K, which could supply sufficient available nutrients for the growth of soybean. Moreover, extra nitrogen was also obtained by nitrogen fixation by soybean itself. The mean chemical characteristics of pig manure were 360 g C/kg (dry base), 17 g N/kg and 8.9 g P/kg, and those of cattle manure were 370 g C/kg, 7.9 g N/kg and 4.2 g P/kg.

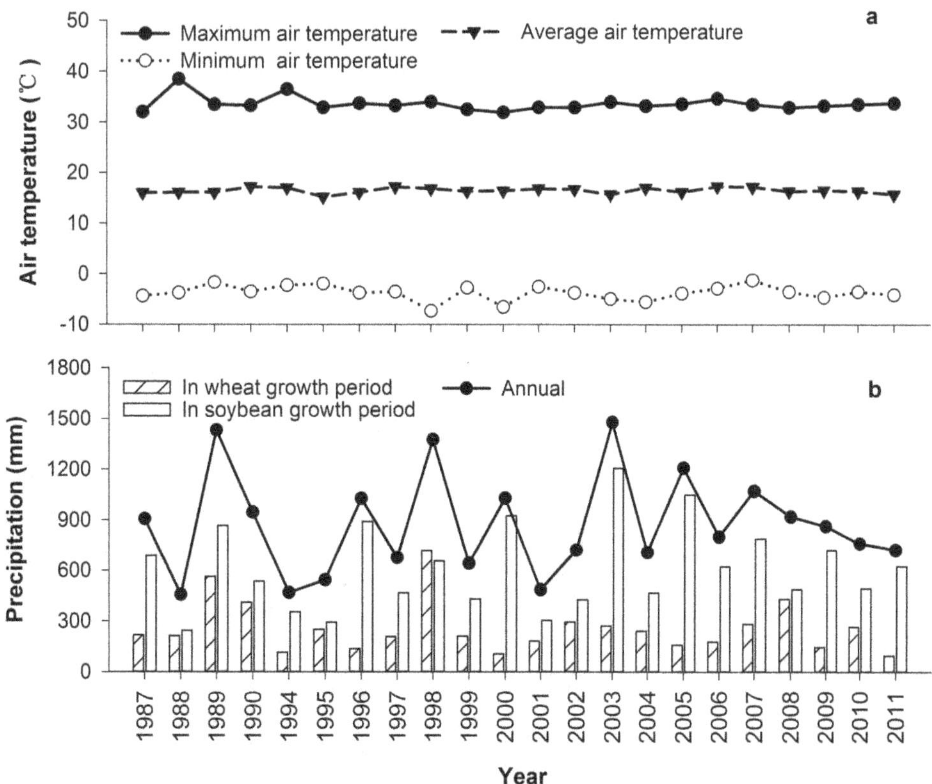

Figure 1. Air temperature and precipitation during the period 1987 to 2011 (no data for 1991 and 1993). (a) Maximum, minimum and average air temperatures. (b) Accumulated precipitation during the wheat growing period (from November to May), the soybean growing period (from June to October), and for the annual total.

Table 1. Mean application rates of mineral and organic fertilizers (kg/ha/yr) per treatment.

Treatment	N	P_2O_5	K_2O	Wheat straw	Pig manure	Cattle manure
CK	0	0	0	0	0	0
NPK	180	90	135	0	0	0
NPKW1/2	180	90	135	2500	0	0
NPKW	180	90	135	5000	0	0
NPKPM	180	90	135	0	7800	0
NPKCM	180	90	135	0	0	12500

Treatments were no fertilization (CK), mineral NPK fertilizers alone (NPK), mineral NPK fertilizers combined with 2.5 ton/ha/yr of wheat straw (NPKW1/2), mineral NPK fertilizers combined with 5.0 ton/ha/yr of wheat straw (NPKW), mineral NPK fertilizers combined with pig manure (NPKPM), and mineral NPK fertilizers combined with cattle manure (NPKCM).

A double cropping system of winter wheat-soybean is the common crop rotation in the region. At the Madian Agro-Ecological Station, plots were plowed to 20 cm depth after crop harvest. Winter wheat (*Triticum aestivum L.*), variety Yedan 13, was grown in rows from late October to May, and soybean (*Glycine max*), variety Zhonghuang 13, from June to September. Herbicides and pesticides were applied during the growth periods when it was needed. Weed residues in all the fields were manually removed after herbicides application. Hence, biomass of weeds left in the field was negligible. Wheat and soybean were harvested manually and all above-ground biomass, except for stubble, were removed from the experimental fields. Wheat grain yields were measured in 1983, 1985, and 1989–2011. Soybean grain yields were measured in 1983, 1985, 1989–1991 and 1998–2011. Grains were air dried, threshed and then weighted separately.

Soil sampling and analyses

Soil samples were collected from the top 20 cm each year after the soybean harvest (in October). Soil samples were randomly taken from three locations in each plot using a soil core sampler (inner diameter 7 cm). The samples were air dried and passed through an 8 mm sieve. Visible pieces of crop residues and roots were removed. Representative sub-samples were then passed through a 2 mm sieve and ground further to pass a 0.25 mm sieve for determination of SOC, total N, total P and other properties. The sieved soil was stored in glass jars until analysis. Soil bulk density was measured using the core method [18]. Soil pH was measured by the potentiometric method in a soil-water extract (1:2.5, w/v water) [18]. SOC content was determined following the vitriol acid-potassium dichromate oxidation method [19]. Total N was determined by using the method described by [19], and total P by Black [20]. Soil texture was measured following the gravitometer methods described by Lu [18]. Four replicates were carried out for each analysis.

Estimation of C inputs, and SOC sequestration rate and efficiency

C inputs into the topsoil (0–20 cm) included materials from crops residues (i.e., mainly roots and stubble), organic amendments (i.e. wheat straw, pig manure, and cattle manure). Annual amount of C inputs by roots from wheat and soybean were estimated at 30% [21] and 19% [22] of above-ground biomass, respectively. Annual C inputs by stubble from wheat and soybean were estimated at 15% of straw biomass. We used a straw to grain ratio of 1.2:1 and 1.6:1 to calculate wheat and soybean straw biomass, respectively [23]. The C contents of wheat and soybean were estimated at 400 g/kg and 453 g/kg (dry base), respectively [23]. The annual C input via crop residues was estimated from

$$C_{input} = ((Y_g + Y_s) \times R \times D_r + R_s \times Y_s) \times (1 - W) \times C/1000 \quad (1)$$

where Y_g, and Y_s were grain yield and straw yield (kg/ha); R was ratio of root biomass to above ground biomass (AGB); D_γ was ratio of root biomass in topsoil (0–20 cm) to root biomass in the soil profile, which were set at 0.753 for wheat [24] and at 0.984 for soybean [25]; R_s and W represented ratio of stubble to straw biomass and water content of air-dried grain (14% [23]); C was the organic C content of the crop (g/kg).

SOC stock ($SSOC$, ton/ha) in topsoil was calculated using the equation:

$$SSOC = SOC \times BD \times d/10 \quad (2)$$

where SOC was soil organic C content (g/kg), BD was the bulk density (g/cm^3), and d was the thickness of the soil layer (20 cm).

SOC sequestration rate (SSR, ton/ha/yr) was estimated for the topsoil by the following equation [4]:

$$SSR = (SSOC_t - SSOC_0)/t \quad (3)$$

where $SSOC_t$ and $SSOC_0$ were the stock of SOC at time t and in the initial year (1982); t was the duration of experiment. The SOC sequestration efficiency was derived from the slope of the linear regression between annual SOC sequestered and C input [14].

ΔSOC sequestration rates (ΔSSR, ton/ha/yr) via wheat straw, pig manure and cattle manure were estimated for topsoil by the following equation.

$$\Delta SSR = (SSR_{NPK+OM} - SSR_{NPK}) \quad (4)$$

where SSR_{NPK+OM} and SSR_{NPK} were the SOC sequestration rate in treatments with organic amendments combined with mineral NPK fertilizers (NPKW1/2, NPKW, NPKPM and NPKCM) and the treatment with mineral NPK fertilizers alone (NPK), respectively.

C sequestration efficiency of organic amendment (SEO) was calculated by the following equation.

$$SEO(\%) = \Delta SSR/CI \times 100\% \quad (5)$$

where CI was annual C input via organic amendment (ton/ha/yr).

Data analysis

All the statistical analyses were performed using SPSS 13.0 software package (SPSS, Inc., USA). Significant differences were analyzed using *LSD* test at significance level $P = 0.05$ or $P = 0.01$. Graphs were prepared with Sigma plot 10.0 software (Systat Software, Inc., Chicago, IL, USA).

Results

Grain yield and C inputs into the soil

Wheat and soybean yields are shown in Fig. 2. Average annual grain yields followed the order NPKCM> NPKPM> NPKW> NPKW1/2> NPK> CK. Grain yields tended to decrease under the CK treatment, but to increase under all fertilization treatments over time.

Similarly, the CK treatment had a low and decreasing crop C input, while the fertilization treatments had a high and increasing trend of C input via roots and stubble over time (Table 2). Average annual C input via roots and stubble was 2.4 times higher in the NPK treatment than in the CK treatment. Soil amendments with straw and manure slightly increased the C inputs via roots and stubble in the NPKW1/2, NPKW, NPKPM and NPKCW treatments, respectively, comparing to the NPK treatment. Similarly, CK had the lowest annual total C input t (0.45 ton/ha). Compared with CK, the average annual total C inputs in the NPK, NPKW1/2, NPKW, NPKPM and NPKCW treatments were increased by a factor of 2.4, 4.9, 7.2, 9.4, and 13.6, respectively.

SOC dynamics

Changes in SOC contents are shown in Fig. 3. As expected, CK had a low and slightly decreasing SOC content, whereas other treatments had an increasing SOC content. Changer over time in SOC content could be described by linear regression or linear plateau models. In CK, the changes in SOC content were best described by simple linear regression ($y = -0.03\,x+6.70$, $R = 0.59$, $P<0.01$), suggesting that SOC decreased by 3% per year. Changes in SOC contents of the NPK, NPKW1/2 and NPKW were also best described by simple linear regression, but here significant increases in SOC content were observed. Changes in SOC contents of the NPKPM and NPKCM were best described by linear plateau model. For NPKPM and NPKCW, SOC contents tended to increase rapidly in the first twenty years, but leveled off thereafter. However, a longer duration of soil C sequestration was observed in NPKPM (26 years) than in NPKCM (20 years).

C sequestration rate

Soil bulk density (BD) decreased in all fertilization treatments but not in the CK treatment (Table 3). The decrease was largest in the treatments receiving wheat straw and animal manures, which suggests that long-term application of organic amendments significantly improved soil physical conditions.

The mean SOC sequestration rate over the 29 years experimental period ranged from -0.08 ± 0.03 ton/ha/yr in CK to 0.97 ± 0.04 ton/ha/yr in NPKCM. Mineral fertilizer application (NPK) reversed the SOC decline in the CK treatment into a net SOC sequestration of 0.20 ± 0.02 ton/ha/yr. Mean SOC sequestration rates followed the order: NPKCM> NPKPM> NPKW> NPKW1/2> NPK. These findings are consistent with a large body of evidence indicating that long-term fertilizer, crop residues and manure applications increase the SOC content of arable soils [4,14]. Average ΔSOC sequestration rates (ΔSSR) of wheat straw at low input rate, wheat straw at high input rate, pig manure and cattle manure were 0.19 ± 0.03, 0.30 ± 0.04, 0.31 ± 0.03 and 0.77 ± 0.02 ton/ha, respectively. The result indicated that SOC sequestration rate changes were related to total C inputs.

Relationship between C input and SOC sequestration rate

There was a significantly positive linear correlation between C input and SOC sequestration rate (Fig. 4). The slope of the equation (0.16) indicates that on average 16% of the total C input into the soil was sequestered as SOC; the mean soil C sequestration efficiency was 16%. The results also indicate that 0.63 ton/ha/yr was needed to maintain SOC level constant at the initial level.

The mean relationships between C inputs and ΔSOC sequestration rates through organic amendments are also shown in Fig. 4. Interestingly, C sequestration efficiencies for wheat straw (c

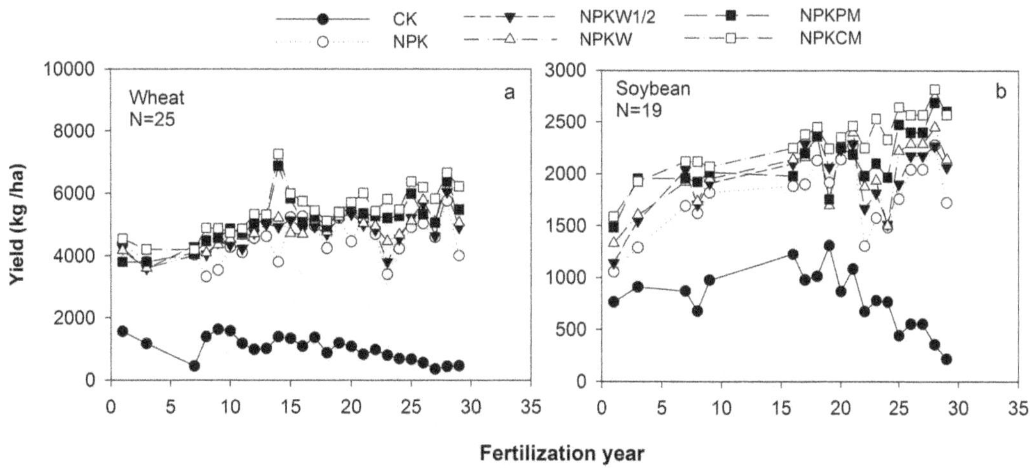

Figure 2. Average annual grain yields in six treatments during the period 1983 to 2011. (a) Wheat yields (b) Soybean yields. Treatments were no fertilization (CK), mineral NPK fertilizers alone (NPK), mineral NPK fertilizers combined with 2.5 ton/ha/yr of wheat straw (NPKW1/2), mineral NPK fertilizers combined with 5.0 ton/ha/yr of wheat straw (NPKW), mineral NPK fertilizers combined with 7.8 ton/ha/yr of pig manure (NPKPM), and mineral NPK fertilizers combined with 12.5 ton/ha/yr of cattle manure (NPKCM). Results are means (n = 4).

Table 2. Average annual crop C input (roots and stubble) and organic amendment C inputs for each treatment during the period 1987 to 2011.

Treatment	Annual crop C input (ton/ha/yr)				Organic amendment C input	Total
	Wheat	CV%	Soybean	CV%	(ton/ha/yr)	(ton/ha/yr)
CK	0.23c	38.3	0.22c	36.2	0	0.45
NPK	1.04b	14.3	0.51b	19.0	0	1.55
NPKW1/2	1.10b	11.8	0.56b	16.7	1.00	2.66
NPKW	1.13b	12.5	0.57b	16.1	2.00	3.71
NPKPM	1.20a	13.7	0.61a	14.0	2.87	4.68
NPKCM	1.28a	13.5	0.66a	12.3	4.62	6.56

Treatments were no fertilization (CK), mineral NPK fertilizers alone (NPK), mineral NPK fertilizers combined with 2.5 ton/ha/yr of wheat straw (NPKW1/2), mineral NPK fertilizers combined with 5.0 ton/ha/yr of wheat straw (NPKW), mineral NPK fertilizers combined with 7.8 ton/ha/yr of pig manure (NPKPM), and mineral NPK fertilizers combined with 12.5 ton/ha/yr of cattle manure (NPKCM). Data are present as means (n = 4). Those with the same letter are not significantly different ($P<0.05$ or 0.01).

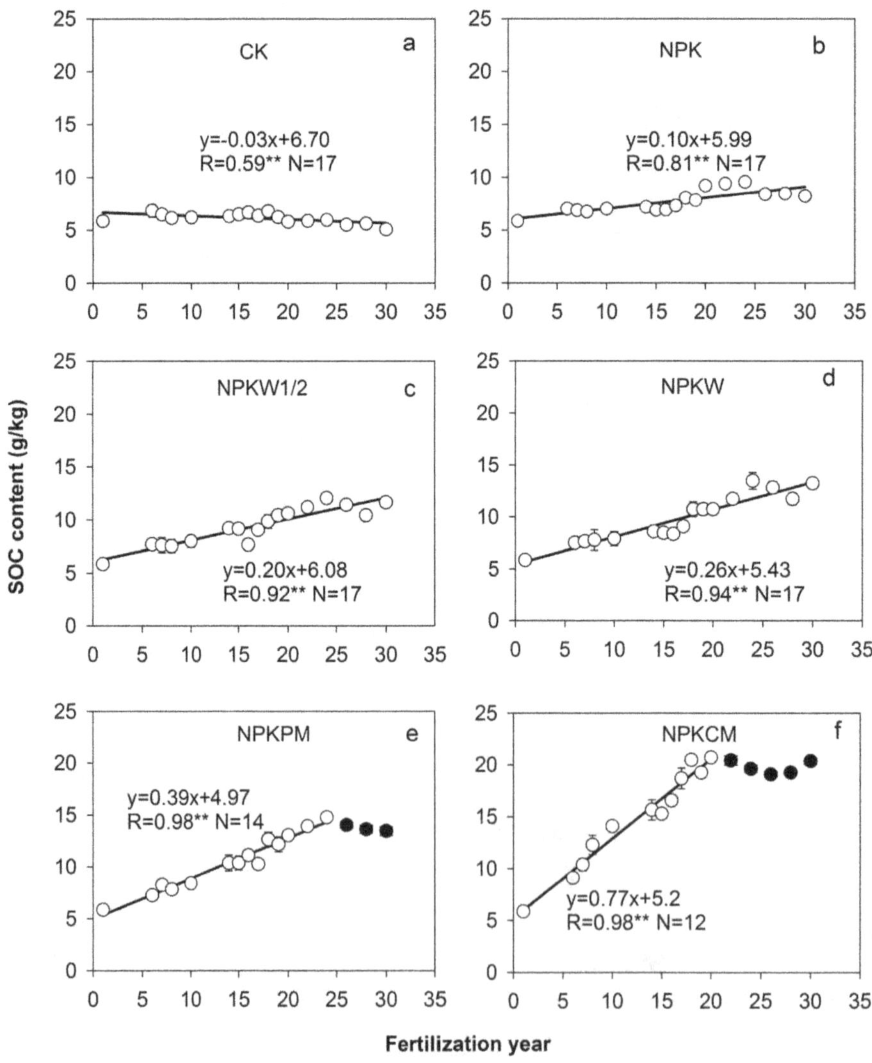

Figure 3. Changes in soil organic carbon (SOC) content over time in the six fertilization treatments during the period 1983 to 2011. CK: no fertilizer (a); NPK: mineral NPK fertilizers alone (b); NPKW1/2: mineral NPK fertilizers combined with 2.5 ton/ha/yr of wheat straw (c); NPKW: mineral NPK fertilizers combined with 5.0 ton/ha/yr of wheat straw (d); NPKPM: mineral NPK fertilizers combined with pig manure (e); NPKCM: mineral NPK fertilizers combined with cattle manure (f); ** indicates significant correlation at $P<0.01$, respectively. Error bars indicated for some years are standard errors (n = 4).

Table 3. Soil bulk density (BD), SOC contents, the estimated mean SOC sequestration rate (SSR) and mean ΔSOC sequestration rate (ΔSSR) during the 29-years period, for all treatments.

Treatment	1982			2011			SSR (ton/ha/yr)	ΔSSR (ton/ha/yr)
	BD (g/cm^3)	SOC (g/kg)	SSOC (ton/ha)	BD (g/cm^3)	SOC (g/kg)	SSOC (t/ha)		
CK	1.45	5.86	16.99	1.44±0.03a	5.08±0.17e	14.63±0.79e	−0.08±0.03e	-
NPK	1.45	5.86	16.99	1.38±0.01a	8.24±0.07d	22.74±0.47d	0.20±0.02d	-
NPKW1/2	1.45	5.86	16.99	1.21±0.02b	11.66±0.46c	28.22±1.58c	0.39±0.05c	0.19±0.03
NPKW	1.45	5.86	16.99	1.19±0.02b	13.23±0.55b	31.49±1.84b	0.50±0.06b	0.30±0.04
NPKPM	1.45	5.86	16.99	1.18±0.02b	13.46±0.42b	31.77±1.53b	0.51±0.05b	0.31±0.03
NPKCM	1.45	5.86	16.99	1.11±0.01c	20.36±0.32a	45.20±1.12a	0.97±0.04a	0.77±0.02

CK: no fertilizer; NPK: mineral fertilization; NPKW1/2: mineral fertilizer combined with 2.5 ton/ha/yr of wheat straw; NPKW: mineral fertilizer combined with 5.0 ton/ha/yr of wheat straw; NPKPM: mineral fertilizer combined with pig manure; NPKCM: mineral fertilizer combined with cattle manure. Data are presented as means (n = 4), and those designated with the same letter are not significantly different ($P<0.01$).

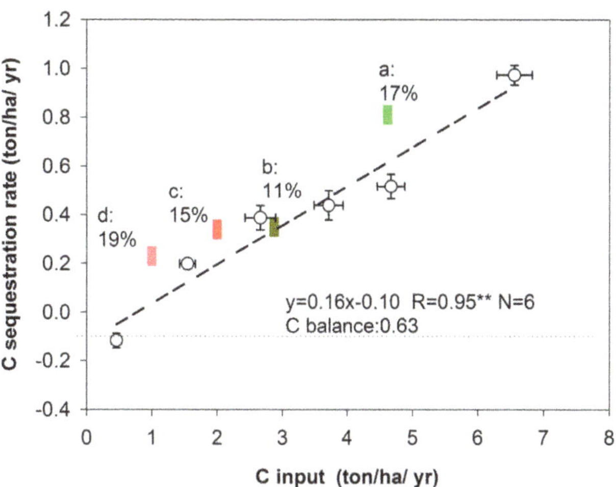

Figure 4. Relationship between C input and SOC sequestration rate. Solid line indicated linear regression curve between C inputs and SOC sequestration rates for the six treatments. a, b, c, and d represent C sequestration efficiencies for wheat straw at low input rate, wheat straw at high input rate, pig manure and cattle manure (calculated by equation (5)), respectively. ** indicated significant at $P<0.01$; error bars indicate standard errors (n = 4).

and d), pig manure (b) and cattle manure (a) differed. The C sequestration efficiencies for wheat straw was 17% (mean value), pig manure 11% and for cattle manure 17%. Hence, application cattle manure or wheat straw has a higher efficiency to enhance SOC.

Discussion

Soil C sequestration duration

Our results clearly show that trends in SOC depend on long-term fertilization regime. The changes of SOC content over time for the three treatments of NPK, NPK1/2W and NPKW are consistent with the observations reported by Zhou et al (2013) [26]. The time of soil C steady state occurring represents soil C sequestration duration [27]. West and Six [28] reported that the duration of C sequestration varies between ecosystems, climate regimes, and fertilization management (e.g., soil organic amendments inputs). West and Post [29] estimated that a change from conventional tillage to no-till results in an average 14% increase in soil C, during a mean experimental period of 15 years. Rui and Zhang [30] found that there was a negative correlation between soil C sequestration rates and duration of soil C sequestration. The C sequestration duration in the NPKW, NPKPM and NPKCM treatments reflect differences in the quantity and quality of C input. Our results indicate that the recommended mean C sequestration period of 20 years for National Greenhouse Gas according to the IPCC Guidelines [31] may not be suitable for all the organic amendments application.

Relation between C input and soil C sequestration

The relationship between C input into the soil and SOC sequestration may present in the SOC sequestration efficiency. Several studies showed that there was no general relationship between C input and SOC sequestration. Some soils might be C saturated [32–34]. Other studies demonstrated that SOC sequestration was linear or logarithmically related to the C input into soil [2,4,15]. In the present study, a clear linear relationship between annual C input and SOC sequestration rate was found, which was similar to findings in upland [4,35], but different from those findings in paddy soils [14,15,36]. The strong linear relationship indicated that the Vertisol had a large capacity to store SOC. Moreover, we did not find indications that the Vertisol had a clear, uniform saturation level for SOC sequestration. The results presented in Fig. 3b, c and d indicate that these maxima (equilibrium levels) have not been reached in 29 years. However, we observed that SOC sequestration rate leveled off after about 20 years in the NPKPM and NPKCM treatments at SOC levels of about 15 and 20 g/kg. The results suggest that SOC saturation level depends on C inputs of organic amendments.

Our results revealed that 0.63 ton/ha/yr (Fig. 4) was needed to maintain the initial SOC level, and that the unfertilized control treatment (CK) did not provide sufficient C input to maintain the initial SOC content. This finding is in agreement with the result from Yan et al. [15], who showed that the C input from unfertilized crops cannot sustain the native SOC content of upland soils. Interestingly, NPK fertilization did provide sufficient C input to contribute to a net SOC sequestration, indicating that the C input provided through the incorporation of only roots and stubble was sufficient to maintain and even enhance the C stock. This contrasts with result reported by Yang [37]. They reported that NPK fertilization was insufficient for maintaining the SOC level under conditions of traditional management with the removal of all aboveground crop biomass. These differences may relate to differences in crop yield levels and in SOC sequestration efficiency. Our study indicates C inputs from wheat and soybean stubbles and roots under conventional NPK fertilization seem sufficient to replenish the C loss through SOC decomposition in Vertisols.

C sequestration efficiency

The initial SOC content and soil texture (i.e. clay content) are important factors regulating SOC sequestration efficiency [9]. The Vertisol in our study had a low initial C content and a high clay content, which suggest that the SOC sequestration might be high. Indeed, we observed a relatively high efficiency irrespective of the type of C input (Fig. 4). Zhang [14] reported a C sequestration efficiency of 6.8% for alluvial soils in Zhengzhou, China, in the same climatic zone as our study. Stewart [38] argued that a greater SOC saturation deficit results in greater sequestration efficiency of added C. We estimated the soil C saturation deficit of the Vertisol at 80%, using the relationship between soil mineral (silt + clay) and C contents (g/kg soil) as proposed by Stewart [33]:

$$\textit{Csaturation deficit} = (1\text{-}\textit{initial SOC content/protective capacity}) \times 100\%$$

$$\textit{Protective capacity} = 0.21 \times (\textit{silt} + \textit{clay content}) + 14.76$$

Evidently, the high clay+silt content and the low initial SOC content contributed to the relatively high C saturation deficit. The protection of SOC by clay particles was well established [39,40]. Stewart [38] reported a larger enrichment of microbial derived carbohydrates in the clay fraction compared with that of the sand fraction. Hassink [41] reported that C associated with organo-mineral complexes in soil were chemically protected, and the protection increased with an increase in clay content. Six [9] found that soils with a high clay content had a low SOC decomposition rate, likely because SOC in clay soils was chemically stabilized and absorbed onto negatively charged clay minerals [42,43]. Therefore, the clay content has a positive effect on SOC sequestration efficiency [44].

Many factors have been suggested to affect the humification coefficient, including the type of input material, soil type, climate factors and soil nutrients status [45,46]. Maillard and Angers [47] reported that the quality of the organic amendment was the dominant driver determining C sequestration efficiencies of organic amendments, based on a meta-analysis of worldwide published reports. Likely, the quality of the organic amendment in our experiment was also the major reason explaining the differences in C sequestration efficiencies for wheat straw, pig manure and cattle manure. Especially lignin is considered to be one of the most chemical recalcitrant components [48]. In soils, lignins are synthesized from L-phenylalanine and cinnamic acids via various metabolic ways to form lignin precursors such as sinapyl and coniferylalcohols [49]. The lignin structure consists of aromatic rings with side chains and –OH and $-OCH_3$ groups linked by strong covalent bonds. Therefore, lignins are considered as stabilized component of SOC, influencing its pool-size and its turnover [50]. In our study, we observed that C sequestration efficiency of cattle manure was higher than that of pig manure, likely because the cattle manure was composted. Ghosh et al. [51] reported that the lignin content of cattle manure may reach 23.7%, which was significantly higher than that of pig manure. Therefore, application of composted cattle manure seems the preferred strategy for enhancing SOC sequestration in the Vertisol due to its high C sequestration efficiency.

There are a number of uncertainties in our study, which are quite common to many long-term field studies. We estimated the C inputs via roots and stubble from fixed percentages of aboveground biomass. Likely, the C input via roots and stubble depends on the level of fertilization, and varies from year to year due to variations in rainfall. Also, the C content of the pig and cattle manure was not measured each year, while there was quite a big annual variation in composition. Further, there are missing data, especially during the first half of the experiment, due to organizational matters. We obtained C sequestration efficiencies of wheat straw, pig manure and cattle manure based on the two main hypotheses. We assumed that the difference in C input via stubble and roots was negligible small between the treatment NPK and the treatments NPKW1/2, NPKW, NPKPM and NPKCM, and no significant priming effect caused by organic amendments application was existed. These uncertainties in primary data may also lead to uncertainties in the estimated mean SOC sequestration efficiency and C sequestration efficiencies of organic amendments, but do not undermine our main conclusions. Our future studies focus on the accurate quantification of C inputs via roots and stubble, effects of organic amendments application on the C inputs, and the accurate quantification of priming effect by organic amendments application using isotope technique in this ongoing field experiment. Also, SOC fractionation studies are undertaken to establish relationships between C input and SOC sequestration in different SOC pools (i.e. light and heavy C fractions).

Conclusions

The changes in SOC content over time were described by linear equations. However, the SOC sequestration rates clearly decreased in the NPKPM and NPKCM treatments after about 20 years. A significant linear relationship was observed between annual C input and SOC sequestration rate. The Vertisol in our study had a high SOC sequestration potential. Also, the overall mean SOC sequestration efficiency of the wide range of C inputs via roots, stubble, straw and manure was equally high, with an overall mean efficiency of 16% over the 29-years period. The C sequestration efficiencies of wheat straw and cattle manure are higher than pig manure, indicating that cattle manure or wheat straw application is a recommendable SOC sequestration practice in Vertisols.

Acknowledgments

We thank the staff of Madian Agro-Ecological Station and the Ministry of Agriculture, China. We sincerely thank Professor Oene Oenema from Environmental Sciences, Wageningen University, for the useful suggestions and comments. We are also greatly grateful for anonymous reviewers for their very constructive and valuable comments for our manuscript.

Author Contributions

Conceived and designed the experiments: DZW XSG. Performed the experiments: KKH DZW ZBG. Analyzed the data: KKH. Contributed reagents/materials/analysis tools: KKH DZW XSG ZBG. Wrote the paper: KKH.

References

1. Lal R (2004) Soil carbon sequestration impacts on global climate change and food security. Science 304: 1623–1627.
2. McLauchlan KK (2006) Effects of soil texture on soil carbon and nitrogen dynamics after cessation of agriculture, Geoderma 136: 289–299.
3. Tong C, Xiao H, Tang G, Wang H, Huang T, et al. (2009) Long-term fertilizer effects on organic carbon and total nitrogen and coupling relationships of C and N in paddy soils in subtropical China. Soil and Tillage Research 106: 8–14.
4. Zhang WJ, Xu MG, Wang XJ, Huang QH, Nie J, et al. (2012) Effects of organic amendments on soil carbon sequestration in paddy fields of subtropical China. Journal of Soils and Sediments 12: 457–470.
5. Freibauer A, Rounsevell MD, Smith P, Verhagen J (2004) Carbon sequestration in the agricultural soils of Europe. Geoderma 122: 1–23.
6. Li ZP, Lin XX, Che YP (2002) Analysis for the balance of organic carbon pools and their tendency in typical arable soils of Eastern China. Acta Pedologica Sinica 39: 351–360. (In Chinese).
7. Zhou P, Pag GX, Li LQ, Zhang XH (2009) SOC Enhancement in major types of paddy soils in a long-term agro-ecosystem experiment in South China. V. Relationship between carbon Input and soil carbon sequestration. Scientia Agricultura Sinica 42: 4260–4268. (In Chinese).
8. Feller C, Beare MH (1997) Physical control of soil organic matter dynamics in the tropics. Geoderma 79: 69–116.
9. Six J, Conant RT, Paul EA, Paustian K (2002) Stabilization mechanisms of soil organic matter: Implications for C-saturation of soils. Plant Soil 241: 155–176.
10. Elliott ET, Coleman DC (1988) Let the soil work for us. Ecological Bulletins 39: 23–32.
11. Gong W, Yan XY, Wang JY, Hu TX, Gong Y (2009) Long-term manuring and fertilization effects on soil organic carbon pools under a wheat-maize cropping system in North China Plain. Plant Soil 314: 67–76.
12. Zhang JB, Zhu TB, Cai ZC, Qin SW, Muller C (2012) Effects of long-term repeated mineral and organic fertilizer applications on soil nitrogen transformations. European Journal of Soil Science 63: 75–85.
13. Chander K, Goyal S, Mundra MC, Kapoor KK (1997) Organic matter, microbial biomass and enzyme activity of soils under different crop rotations in the tropics. Biology and Fertility of Soils 24: 306–310.
14. Zhang WJ, Wang XJ, Xu MG, Huang SM, Liu H, et al. (2010) Soil organic carbon dynamics under long-term fertilizations in arable land of northern China. Biogeosciences 7: 409–425.
15. Yan X, Zhou H, Zhu QH, Wang XF, Zhang YZ, et al. (2013) Carbon sequestration efficiency in paddy soil and upland soil under long-term fertilization in southern China. Soil and Tillage Research 130: 42–51.
16. Li DC, Zhang GL, Gong ZT (2011) On taxonomy of Shajiang Black Soils in China. Soils 43: 623–629. (In Chinese).
17. Guo ZB, Hua KK, Wang J, Guo XS, He CL, et al. (2014) Effects of different regimes of fertilization on soil organic matter under conventional tillage. Spanish Journal of Agricultural Research 2014 12: 801–808.
18. Lu RK (2000) Analytical methods of soil agricultural chemistry. China Agricultural Science and Technology Press, Beijing. (In Chinese).
19. Walkley A, Black IA (1934) An examination of the degtjareff method for determining soil organic matter and a proposed modification of the chromic acid titration method. Soil science 37: 29–38.
20. Black CA (1962) Methods of soil analysis. Part 2. Chemical and microbiological properties. Madison, Wisc: ASA.
21. Kuzyahov Y, Domenski G (2000) Carbon input by plants into the soil. Journal of Plant Nutrition and Soil science 163: 421–431.
22. IPCC (2006) IPCC guidelines for national greenhouse gas inventories. Prepared by the national greenhouse gas inventories programme. In: Eggleston HS, Buendia L, Miwa K, Ngara T and Tanabe K (Eds), Institute for Global Environmental Strategies. IPCC, Hayama, Japan.
23. NCATS (1994) Chinese organic fertilizer handbook. National Center for Agricultural Technology Service, Chinese Agricultural Publisher. (In Chinese).
24. Miao GY, Zhang YT, Jun Y, Hou YS, Pan XL (1989) A study on the development of root system in winter under inirrigated conditions in semi-arid Loss Plateau. Acta Agronomica Sina 15: 104–115. (In Chinese).
25. Lin WG, Wu JJ, Dong DJ, Zhong P, Wang JS, et al. (2012) Impact of different residue retention system on soybean root distribution in soil profile. Soybean Science 31: 584–588. (In Chinese).
26. Zhou ZC, Gan ZT, Shangguan ZP, Zhang FP (2013) Effects of long-term repeated mineral and organic fertilizer applications on soil organic carbon and total nitrogen in a semi-arid cropland. European Journal of Agronomy 45: 20–26.
27. Six J, Conant RT, Paul EA, Paustian K (2002) Stabilization mechanisms of soil organic matter: implications for C-saturation of soils. Plant Soil 241: 155–176.
28. West TO, Six J (2007) Considering the influence of sequestration duration and carbon saturation on estimates of soil carbon capacity. Climatic Change 80: 25–41.
29. West TO, Post WM (2002) Soil organic carbon sequestration rates by tillage and crop rotation: a global analysis. Soil Science Society of America Journal 66: 1930–1946.
30. Rui WY, Zhang WJ (2010) Effect size and duration of recommended management practices on carbon sequestration in paddy field in Yangtze Delta Plain of China: A meta-analysis. Agriculture, Ecosystems and Environment 135: 199–205.
31. Houghton JT, MeiraFilho LG, Lim B, Tréanton K, Mamaty I, et al. (eds) (1997) Revised 1996 IPCC Guidelines for National Greenhouse Gas Inventories, volumes 1–3. Hadley Centre Meteorological Office, United Kingdom.
32. Gulde S, Chung H, Amelung W, Chang C, Six J (2008) Soil carbon saturation controls labile and stable carbon pool dynamics. Soil Science Society of America Journal 72: 605–612.
33. Stewart CE, Paustian K, Conant RT, Plante AF, Six J (2009) Soil carbon saturation: Implications for measurable carbon pool dynamics in long-term incubations. Soil Biology and Biochemistry 41: 357–366.
34. Chung H, Kathie JN, Plantec A, Six J (2009) Evidence for carbon saturation in a highly structured and organic-matter-rich. Soil Science Society of America Journal 74: 130–138.
35. Kong AY, Six J, Bryant DC, Denison RF, Kessel CA (2005) The relationship between carbon input, aggregation, and soil organic carbon stabilization in sustainable cropping systems. Soil Science Society of America Journal 69: 1078–1085.
36. Cai ZC, Qin SW (2006) Dynamics of crop yields and soil organic carbon in a long-term fertilization experiment in the Huang-Huai-Hai Plain of China. Geoderma 136: 708–715.
37. Yang XM, Zhang XP, Fang HJ, Zhu P, Ren J, et al. (2003) Long-term effects of fertilization on soil organic carbon changes in continuous corn of northeast China: RothC model simulations. Environ Management 32: 459–465.
38. Stewart CE, Paustian K, Conant RT, Plante AF, Six J (2008b) Soil C saturation: evaluation and corroboration by long-term incubations. Soil Biology and Biochemistry 40: 1741–1750.
39. Feller C, Beare MH (1997) Physical control of soil organic matter dynamics in the tropics. Geoderma 79: 69–116.
40. Follett RF, Paul EA, Pruessner EG (2007) Soil carbon dynamics during a long-term incubation study involving ^{13}C and ^{14}C measurements. Soil Science 172: 189–208.
41. Hassink J (1997) The capacity of soils to preserve organic C and N by their association with clay and silt particles. Plant Soil 191: 77–87.
42. Guggenberger G, Christensen BT, Zech W (1994) Land-use effects on the composition of organic matter in particle-size separates of soil: 1. Lignin and carbohydrate signature. European Journal of Soil Science 45: 449–458.
43. Franzluebbers AJ, Haney RL, Hons FM, Zuberer DA (1996) Active fractions of organic matter in soils with different texture. Soil Biology and Biochemistry 28: 1367–1372.
44. Shi XZ, Wang HJ, Yu DS, Weindorf DC, Cheng XF, et al. (2009) Potential for soil carbon sequestration of eroded areas in subtropical China. Soil and Tillage Research 105: 322–327.
45. Lin XX, Wen QX (1987) Utilization of organic wastes in the Taihu region of Jiangsu Province of China. Resources and Conservation 13: 109–116.
46. Galantini JA, Rosell RA, Andriulo A, Miglierina A, Iglesias J (1992) Humification and nitrogen mineralization of crop residues in semi-arid Argentina. Science of the Total Environment 118: 263–270.
47. Maillard E, Angers DA (2014) Animal manure application and soil organic carbon stocks: a meta-analysis. Global Change Biology 20: 666–679.
48. Torres IF, Bastida F, Hernández T, Bombach P, Richnow HH, et al. (2014) The role of lignin and cellulose in the carbon-cycling of degraded soils under semiarid climate and their relation to microbial biomass. Soil Biology and Biochemistry 75: 152–160.
49. Higuchi T (1971) Formation and biological degradation of lignins. Advances in Enzymology and Related Areas of Molecular Biology 34: 207–283.

50. Thévenot M, Dignac MF, Rumpel C (2010) Fate of lignins in soils: a review. Soil Biology and Biochemistry 42: 1200–1211.
51. Ghosh PK, Ramesh P, Bandyopadhyay KK, Tripathi AK, Hati KM, et al. (2004) Comparative effectiveness of cattle manure, poultry manure, phosphor compost and fertilizer-NPK on three cropping systems in Vertisols of semi-arid tropics. I. Crop yields and system performance, Bioresource Technology 95: 77–83.

19

Mechanisms Controlling Arsenic Uptake in Rice Grown in Mining Impacted Regions in South China

Junhui Li[1]⁹, Fei Dong[1,2]⁹, Ying Lu[1]*, Qiuyan Yan[3], Hojae Shim[4]*

1 College of Natural Resources and Environment, South China Agricultural University, Guangzhou, China, **2** Agricultural Bureau of Xiangfen County, Shanxi Province, Xiangfen, China, **3** Institute of Wheat Research, Shanxi Academy of Agricultural Sciences, Linfen, China, **4** Department of Civil and Environmental Engineering, Faculty of Science and Technology, University of Macau, Macau SAR, China

Abstract

Foods produced on soils impacted by Pb-Zn mining activities are a potential health risk due to plant uptake of the arsenic (As) associated with such mining. A field survey was undertaken in two Pb-Zn mining-impacted paddy fields in Guangdong Province, China to assess As accumulation and translocation, as well as other factors influencing As in twelve commonly grown rice cultivars. The results showed that grain As concentrations in all the surveyed rice failed national food standards, irrespective of As speciation. Among the 12 rice cultivars, "SY-89" and "DY-162" had the least As in rice grain. No significant difference for As concentration in grain was observed between the rice grown in the two areas that differed significantly for soil As levels, suggesting that the amount of As contamination in the soil is not necessarily the overriding factor controlling the As content in the rice grain. The iron and manganese plaque on the root surface curtailed As accumulation in rice roots. Based on our results, the accumulation of As within rice plants was strongly associated with such soil properties such as silicon, phosphorus, organic matter, pH, and clay content. Understanding the factors and mechanisms controlling As uptake is important to develop mitigation measures that can reduce the amount of As accumulated in rice grains produced on contaminated soils.

Editor: Ivan Baxter, United States Department of Agriculture, Agricultural Research Service, United States of America

Funding: The authors have no support or funding to report.

Competing Interests: The authors have declared that no competing interests exist.

* Email: luying@scau.edu.cn (YL); hjshim@umac.mo (HS)

⁹ These authors contributed equally to this work.

Introduction

Arsenic (As) is a carcinogenic metalloid ubiquitous in the environment, and is obtained from natural and anthropogenic sources [1,2]. Anthropogenic activities such as metal mining and smelting, the use of As-containing pesticides, herbicides, wood preservatives, feed additives, and irrigation with As-rich groundwater, have resulted in elevated As levels in soil [1,3]. The transfer of As in soil-plant systems represents one of the principal pathways for human exposure to As [4]. A recent cohort study [5] showed that daily consumption of 500 g cooked rice containing As content above 200 μg/kg can give rise to genotoxic effects in humans.

Rice is the staff of life for 3 billion people, predominantly in Asia [6], contributing over 70% of the energy and 50% of the protein provided by their daily food intake [7]. China is the world's top rice producer, producing 36.9% of the world's rice yield on 22.8% of world rice cropping area [8], and a top rice consumer with more than 60% of the Chinese population relying on rice as a dietary staple [2]. Unfortunately, among grain crops, rice is particularly efficient in As accumulation as it is generally cultivated in flooded paddy fields where As is more soluble and available to plant uptake [9,10].

Some studies have revealed that As concentrations in rice grains were associated with the As concentrations in irrigated groundwater and/or soil [11–13], although rice grain can accumulate relatively large amounts of As even from soils not contaminated by As [14]. Others have shown that As in rice does not directly depend on total As concentration in soil and groundwater but may be due to various other factors controlling As solubility, bioavailability and uptake in the soil-rhizosphere-plant system [15–17]. Redox chemistry is one of the predominant factors controlling As speciation and solubility in soil [15]. Arsenic in paddy soil is taken up by plant roots via macro-nutrient transporters; arsenate via the phosphate transporters, and arsenite via silicon transporters [3,18]. Iron (Fe), through forming iron plaque on rice root surfaces, has strong influence on As-uptake by rice roots [19]. The soil physiochemical properties, e.g., redox condition, pH, organic matter, soil texture, Fe and Mn oxides, and sulfur, also affect the solubility and bioavailability of As [15,20]. In addition, the As concentration in various rice tissues varies between rice genotypes [21]. Understanding the genetics associated with grain As concentration is crucial for developing mitigation measures to counter the problem of food-chain contamination by As.

Arsenic is a natural component of Pb, Zn, Cu, and Au ores. Therefore, As is commonly found in soils in mine impacted regions at elevated levels, posing a risk to human and ecosystem health [1]. Paddy rice is one of the most important grain crops in South China [22]. The present study is centered on the Lechang and Renhua Pb-Zn mining regions located in the north of the Guangdong Province, South China. In this study, As levels for soil, root surface, root, straw, and grain were obtained so that grain As levels could be reviewed in regard to both soil and straw As levels. The objectives of this study were: to characterize concentrations of As in rice grains grown in the mining regions with elevated and non-elevated soil As levels, to explore the transfer of As from (rhizosphere) soil through the plant to grain, to identify rice cultivars with low As accumulation in grains, and to understand how the iron and manganese plaque on root surfaces, other macro- and micro-nutrients within the paddy soil and other edaphic properties influence As uptake, assimilation and redistribution, in order to develop potential strategies for reducing As accumulation in rice grains.

Materials and Methods

Ethics statement

No specific permits were required for the described field studies. No specific permissions were required for these locations. We confirm that the location is not privately-owned or protected in any way. We confirm that the field studies did not involve endangered or protected species.

Study area

The present study was conducted in two mining regions in the northern part of Guangdong Province, China (Fig. 1). This research area has a humid subtropical climate with a long-term average annual temperature of 19.6°C and an average annual precipitation of 1,522 mm [23]. The Fankou Pb-Zn mine is an extremely large mine located in Renhua County (Fig. 1). It is geologically situated in the northern part of the central Guangdong Hercynian trough of the South China parageosyncline. The mine was put into production in September 1968, and currently produces 4,500 t of ore per day. Major ore minerals in the Fankou ore mine are pyrite, sphalerite, and galena [24,25]. The mine is classified as a submarine hydrothermal spring effusion type lead/zinc mine, which is relevant to reformed sedimentary rock. [25,26]. Lechang Pb-Zn mine is located in Lechang County (Fig. 1). The major ore minerals are sphalerite, galena, pyrite, and chalcopyrite. As a conventional underground operation, this ore mine was opened in 1959 and is still in operation with a cover area of 1.5 km^2 and produces of 250,000 tons of waste rocks and 30,000 tons of tailings per year occupying respective 8,300 and 60,000 m^2 [23]. The ore of both mines is finely disseminated and complicated and the flotation technology to treat the core is so complex that it is difficult to remove contaminants and make use of the wastewater [27]. The surrounding paddy fields were seriously affected by the continuing year-round irrigation usage of untreated mining wastewater lifted from mines and filtrated from tailings [23]. The rice cropping system in the study area is double-season rice.

Soil and plant samples collection

A total of 28 soil and 28 rice plant (including root, straw, and grain) samples were collected at maturity from 28 paddy fields within or adjacent to the two Pb-Zn mining areas, i.e., eight from Lechang and twenty from Renhua (Fig. 1). The fields were chosen primarily to reflect different rice cultivars being commonly grown by local farmers in these regions. The fields were not irrigated and drained 5 days prior to harvest. Soil samples, 0–15 cm depth, were collected from the base of the rice stem using a soil auger at harvest. At the same time, individual plants of twelve commonly grown cultivars of rice, i.e., Shanyou (SY)-82, SY-86, SY-89, SY-122, SY-162, SY-428, Tianyou (TY)-10, Meixiangzhan (MXZ)-2, Mabei-Youzhan (MBYZ), Diyou (DY)-162, Jinyou (JY)-118, and Fengyou (FY)-998, were collected from the sites where soil samples were taken (Table S1). Composite soil and plant samples were derived by mixing sub-samples from 5 random sites within 25 m^2 per paddy field [22]. The six hybrid SY cultivars were genetically related in that they were all bred using Zhenshan 97A as the female parent. An entire single plant was dug up from 5 sites per each of the 28 fields. However, with the majority of the cultivars being grown only in Renhua or Lechang (three were grown in both), genotypic effects and regional effects were not distinct, but confounded. We therefore analyzed relationships between genotypic and environmental data across the two regions.

Sample preparation

After harvesting, collected plants were washed thoroughly in tap water, followed by deionized water, before extracting the iron plaque from fresh root surfaces using dithionite-citrate- bicarbonate (DCB) as described by Liu et al. [28]. After DCB extraction, plants were separated into their respective tissue components (ear, straw, and root) with stainless steel scissors, weighed to determine fresh weight, oven-dried at 80°C for 72 h, then weighed again (dry weight). Dry spikelets were dehusked by hand and divided into grain and husk. The oven-dried root, straw, and grain samples were powdered using a model MM200 ball mill (Retsch, Germany). Soil samples were air-dried, crushed to pass through a 2 mm nylon sieve (10-mesh), and homogenized. The sub-samples were then ground with an agate grinder to pass through a 150 μm nylon sieve (100-mesh).

Sample analyses

For soil digestion, 0.2 g soil samples were weighed into quartz glass tubes and 5 mL of 12.0 mol/L hydrochloric acid plus 5 mL of 15.2 mol/L nitric acid added, then left to stand overnight at room temperature before being digested on the block digester at 100°C for 1 h, then at 120°C for 1 h, and finally at 140°C for 4 h [22]. For digestion of plant tissues (root, straw, and grain), 0.2 g of grain samples, and 0.1 g of rice root and straw samples, were weighed into 50-mL polypropylene digest tubes and 2 mL of nitric acid added and left to stand overnight. Then 2 mL of hydrogen peroxide was added, and the samples digested using a microwave oven at 50% power (approximately 600 watts). The temperature was raised to 55°C held for 10 min, then to 75°C held for 10 min, and finally to 95°C for 30 min, and then allowed to cool to room temperature [22]. Total As concentrations in solution were determined by hydride generation atomic fluorescence spectrometry (AFS-8130, Beijing). Quality assurance and quality control of metal analyses were carried out by using duplicates (10% of the samples), reagent blanks and standard reference materials (National Environmental Monitoring Centre of China). The recoveries of As in standard reference materials were within ±10% of recommended values, and the relative standard deviation of duplicate measurements was less than 10%.

Selected soil properties, including soil pH, organic matter (OM), available phosphorus (AP), available silicon (ASi), cation exchange capacity (CEC), electrical conductivity (EC), and iron and manganese oxides, were analyzed according to the standard methods recommended by the Soil Science Society of China [29]. Free Fe and Mn, i.e., the bioavailable pool, were extracted by sodium sulfite-sodium citrate-sodium bicarbonate (DCB, mixture of 0.03 $mol \cdot L^{-1}$ $Na_3C_6H_5O_7 \cdot 2H_2O$, 0.125 $mol \cdot L^{-1}$ $NaHCO_3$ and 0.5 g $Na_2S_2O_4$) solution and determined by flame atomic absorption spectrophotometer (FAAS) (Hitachi Z-5300).

Iron plaque on fresh root surface was digested by DCB. For root digestion, 3 g mixed root sample per paddy was weighed into quartz glass tubes (100 mL) and steeped in 30 mL of DCB solution for 1 hr at 25°C before being transferred into quartz glass tubes

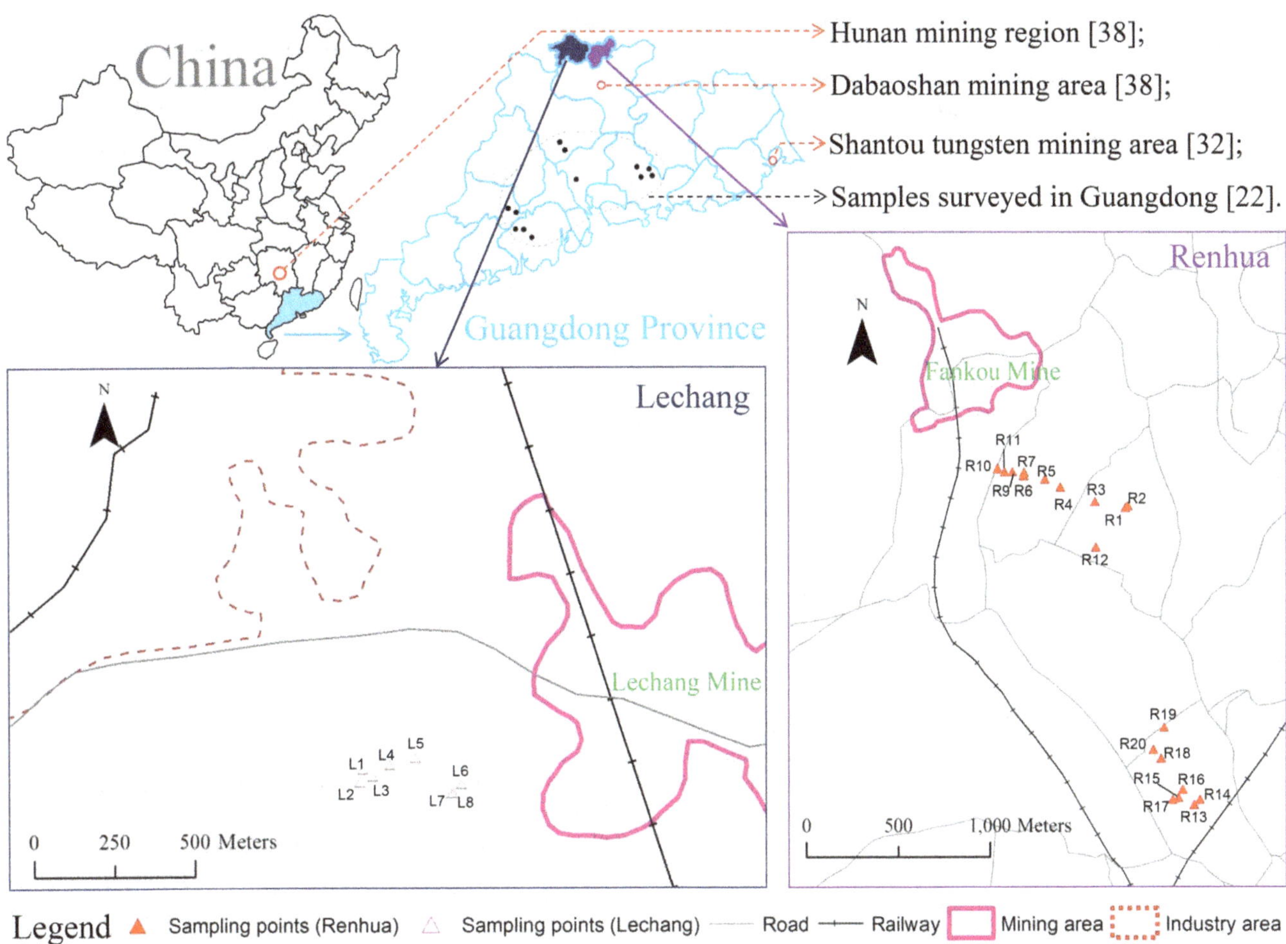

Figure 1. Location map of the study area and distribution of sampling sites.

(100 mL). The root surface As, Fe, and Mn digested by DCB were measured by AFS and FAAS respectively.

Calculation of As transfer factors

Transfer factors (TFs) were determined using the expression C_p/C_s, where C_p is the concentration of As in straw or grain and C_s is the concentration of As in corresponding soil or straw.

Straw/soil transfer factors (S^{traw}/S_{oil} TFs) = Straw As concentrations/Soil As concentrations.

Grain/soil transfer factors (G^{rain}/S_{oil} TFs) = Grain As concentrations/Soil As concentrations.

Grain/straw transfer factors (G^{rain}/S_{traw} TFs) = Grain As concentrations/Straw As concentrations.

Statistical analyses

Descriptive analyses were conducted with SPSS V13.0 for Windows. Principal component analysis (PCA), based on the correlation matrix, was carried out with XLStat-Pro 7.5.2 software, used as a Microsoft Excel plug-in. A probability level of $p<0.05$ was considered as significant difference.

Results and Discussion

Arsenic concentrations in soils

Soil characteristics are presented in Table 1. Arsenic concentrations in the surface paddy soil ranged from 3.7 to 120 mg·kg^{-1} with an average of 30.4 mg·kg^{-1} in Pb-Zn across the two mining areas in our survey (Table 1). Although uncommon, cultivated paddy soils in Hunan Province, China have been found to contain up to 1,226.5 mg·kg^{-1} [30]. The As concentration of soils in this survey would be in the lower part of the range, compared with other mining impacted arable land, e.g., the Hengyang Pb-Zn mine area in Hunan Province (with an average of 253 mg·kg^{-1}) [31], the Chenzhou Pb-Zn mine area in Hunan Province (405.7 mg·kg^{-1}) [30], the Shantou abandoned tungsten mine region in Guangdong Province, China (129 mg·kg^{-1}) [32], and the Rodalquilar Au-(Cu-Pb-Zn) mining district in Almería province, Southeastern Spain (180 mg·kg^{-1}) [33].

Mean As concentrations in soils collected from Renhua were 10.2 mg·kg^{-1} (Table 1), similar to previously reported background concentration of 10.4 mg·kg^{-1} in Guangdong surface soils [34] and the median surface soil As concentrations in the paddy fields in Guangdong [22]. In contrast, soil As concentrations collected from Lechang were elevated, averaging 80.8 mg·kg^{-1}, 8-fold higher compared to that of Renhua (Table 1). All eight soil samples collected from Lechang had As concentrations higher than the maximum allowable concentration (35 mg·kg^{-1} for soil with pH ≤5.5, 30 for soil with pH 5.5–6.5, 25 for soil with pH 6.5–7.5, and 20 for soil with pH ≥7.5) for agricultural soil in accordance with the Chinese Environmental Quality Standard for Soils [35]. Soil As concentrations exceeding 40 mg·kg^{-1} may be harmful to exposed organisms [36], and soil As exceeding

Table 1. Descriptive statistics of soil properties of Renhua and Lechang.

Characteristics	Probability	Renhua (n = 20)		Lechang (n = 8)		Total (n = 28)
		Mean±SD	Range	Mean±SD	Range	Mean±SD
pH	0.0096	4.8±0.4	*4.4*–6.0	6.7±0.7	5.9–**7.6**	5.4±1.0
As-soil ($mg{\cdot}kg^{-1}$)	<0.0001	10.2±4.2	*3.7*–20.0	80.8±25.9	53.7–**120.3**	30.4±35.2
Fe_2O_3 ($mg{\cdot}kg^{-1}$)	0.93	9.3±4.3	*3.6*–21.3	26.0±4.7	21.3–**35.6**	14.1±8.8
Mn_2O_3 ($mg{\cdot}kg^{-1}$)	<0.0001	17.8±14.0	*7.9*–62.1	214.8±100.4	94.8–**361.3**	74.1±104.7
AP ($mg{\cdot}kg^{-1}$)	0.012	26.5±14.9	*9.2*–**57.5**	17.1±5.5	10.4–25.0	23.8±13.5
ASi ($mg{\cdot}kg^{-1}$)	0.013	53.3±13.8	*29.8*–88.1	100.3±26.7	73.1–**150.0**	66.7±28.1
CEC ($cmol{\cdot}kg^{-1}$)	0.91	6.6±1.1	*4.9*–8.4	16.4±1.2	14.8–**18.3**	9.4±4.6
OM ($mg{\cdot}kg^{-1}$)	0.61	30.7±5.6	*18.3*–43.7	47.0±6.6	37.4–**57.3**	35.4±9.5
EC ($\mu s{\cdot}cm^{-1}$)	0.0054	106.9±47.1	*49.6*–210	460.1±110.2	307–**631**	207.8±176.4
Sand (%)	0.39	31.3±4.6	*17.4*–36.4	26.4±5.9	18.8–**37.4**	29.9±5.4
Silt (%)	0.040	50.3±4.7	*42.7*–**63.6**	49.5±1.9	46.4–52.0	50.1±4.1
Clay (%)	0.010	18.3±2.3	*14.7*–23.9	24.1±4.7	16.2–**30.6**	20.0±4.1

Results are presented as arithemic mean ± SD; probability indicates the differences between Renhua and Lechang; n represents Number of samples; the *italic* number represents the minimum value of the characteristic in all the 28 surveyed samples; the **bold** number represents the maximum value of the characteristic in all the 28 surveyed samples.

100 $mg{\cdot}kg^{-1}$ poses a severe risk to the pregnant women and their offspring [32,37]. All of the eight samples from Lechang were over 40 $mg{\cdot}kg^{-1}$ As, two of which were over 100 $mg{\cdot}kg^{-1}$ As in this survey, indicating severe As contamination in soils around the Lechang Pb-Zn mine.

Arsenic accumulation and translocation in tissues of rice plant

The overall mean total concentration of grain As was 0.26 $mg{\cdot}kg^{-1}$ (Table 2), which is comparable to field-collected unpolished rice from non-mining-impacted fields in Guangdong (0.29 $mg{\cdot}kg^{-1}$, n = 12, rice cultivar Peizha-Taifeng) [22] and Hunan mining region (0.30 $mg{\cdot}kg^{-1}$, n = 22) [38] (Fig. 1). In contrast, the average grain As in this survey was much higher than the field-collected unpolished rice from Dabaoshan mining areas in Guangdong (0.19 $mg{\cdot}kg^{-1}$, n = 95) [38] yet around half the grain As concentration reported in field-collected brown rice from Shantou tungsten mining area in Guangdong (0.56 $mg{\cdot}kg^{-1}$, n = 33) [32] (Fig. 1). All of the presently surveyed rice grains possessed As concentrations >0.17 $mg{\cdot}kg^{-1}$, exceeding Chinese maximum contaminant levels (MCLs) of 0.15 $mg{\cdot}kg^{-1}$, irrespective of As speciation [39], indicating that rice from this region would be a significant source of dietary As for the population. Chinese standards for As in rice are probably the strictest in the world, which have been designed to protect a nation with high rice intakes [38]. Compared to the global 'normal' range of 0.08–0.20 $mg{\cdot}kg^{-1}$ for As concentration in rice grain [12], 25 out of these 28 samples exceeded the 'normal' range. The mean As concentrations for these rice grain samples were much higher compared to that in rice from U.S. and Europe (both 0.198 $mg{\cdot}kg^{-1}$) [12].

It has been demonstrated that different rice cultivars showed significant differences for concentrations of As in straw, husk and grain [12,19]. While we did not detect significant differences for rice grain As among the Lechang varieties, we did find difference between rice cultivars harvested from Renhua (Table 3). Although the genetic differences for As-root and As-straw were not significant, the cultivars SY-89 and DY-162 showed the lowest As concentrations for all tissues when grown in Renhua (Table 3), while the cultivar SY-122 showed the highest or second highest concentrations for all three tissues in both Renhua and Lechang. What makes this especially interesting is that, though the soils within each site were not significantly different for As concentration, the SY-122 with the highest tissue As was grown in a field with relatively lower soil As per site. Despite the fact that Lechang was higher than Renhua for As in soils, higher As concentrations of soil, DCB extracts, and straw, the As concentration in grains from the two sites was comparable (0.28 and 0.25 $mg{\cdot}kg^{-1}$) (Table 2). When data were analyzed among just the varieties grown in both Renhua and Lechang, As concentrations in grain were again comparable, i.e., SY-428 0.27 (Renhua)-0.25 (Lechang) $mg{\cdot}kg^{-1}$, MBYZ 0.24–0.29, and SY-122 0.39-0.32, in spite of their having more As in soil, DCB extracts, and straw in samples from Lechang compared to Renhua (Table 3). Although the mean soil As concentration in Renhua was lower compared to the national soil background, the rice grain As concentrations exceeded the Chinese MCLs, i.e., rice grain can accumulate relatively large amounts of As even from soils having very low level of As. Williams et al. [41] also reported that there were elevated grain As concentrations even with background soil levels. It is clear that the amount of As added by contamination to soil is not necessarily the overriding factor controlling the As concentration in rice grain. Other researchers [11,12] reported that the high As levels in rice were associated with As-contaminated irrigation water. As uptake by rice mainly depends on As availability rather than total As in soil [14].

Similar to the total As in soils, the DCB-extracted As concentrations from the root surfaces were highly variable from one paddy field to another, and there was an approximate 3-fold difference in mean DCB-extracted As concentration between Lechang and Renhua (Table 2). Interestingly, SY-122 as one of the three cultivars grown in both locations had the highest recorded mean root As concentrations in both locations (Table 3), yet in Renhua it had the lowest plaque concentration of As (Table 3). The root As concentrations for SY-122 grown in Lechang (52.0 $mg{\cdot}kg^{-1}$) and Renhua (48.6 $mg{\cdot}kg^{-1}$) were simi-

Table 2. Descriptive statistics of rice plant accumulation and transfer factors.

Characteristics	Probability	Renhua (n = 20)		Lechang (n = 8)		Total (n = 28)
		Mean±SD	Range	Mean±SD	Range	Mean±SD
TF (soil-straw)	0.0047	0.36±0.18	0.13–**0.90**	0.079±0.041	*0.018*–0.13	0.28±0.20
TF (straw-grain)	0.62	0.084±0.30	*0.032*–**0.17**	0.057±0.023	0.036–0.11	0.076±0.030
TF (soil-grain)	0.0030	0.028±0.013	0.012–**0.057**	0.0037±0.0010	*0.0020*–0.0047	0.021±0.016
As-root ($mg \cdot kg^{-1}$)	0.49	25.6±12.5	*11.1*–**59.4**	35.0±9.4	24.8–52.0	28.3±12.3
As-straw ($mg \cdot kg^{-1}$)	0.029	3.3±1.3	*1.6*–6.9	5.8±2.8	2.1–**10.4**	4.0±2.1
As-grain ($mg \cdot kg^{-1}$)	0.37	0.25±0.051	0.18–**0.39**	0.28±0.067	*0.17*–0.38	0.26±0.06
As-DCB ($mg \cdot kg^{-1}$)	0.013	25.4±14.4	*6.4*–59.4	70.7±28.9	27.7–**101.8**	38.3±28.2
Mn-DCB ($mg \cdot kg^{-1}$)	<0.0001	25.6±22.5	*7.2*–93.4	189.1±109.4	75.6–**341.0**	72.3±95.5
Fe-DCB ($g \cdot kg^{-1}$)	0.044	29.7±11.4	*4.3*–**48.6**	36.3±5.0	26.5–44.6	31.6±10.4

Results are presented as arithemic mean ± SD; probability indicates the differences between Renhua and Lechang; n represents Number of samples; the *italic* number represents the minimum value of the characteristic in all the 28 surveyed samples; the **bold** number represents the maximum value of the characteristic in all the 28 surveyed samples.

lar, which is in stark contrast to the 10-fold differences for the corresponding DCB-extracted As and soil As levels observed in this survey (Table 3). In this regard, although the As concentrations in soil and root surface for Lechang were significantly higher, there was not a significant difference for concentration of As inside the root, suggesting that As in rice does not directly depend on the total As concentration in the soil and root surface but may be due to other factors and uptake mechanisms.

Regardless of rice cultivars and locations, the As concentrations in soil and DCB extracts were much higher than As concentrations in rice plants, excepting rice root, whereas no obvious trend was observed between the As concentrations in soil, DCB extracts, and roots (Tables 1–3). Rice roots contained considerably higher concentrations of As compared to any other parts of the plant, regardless of soil As concentration and rice cultivars (Tables 1–3). The levels of root As were found to be on average 7.7 times higher than their corresponding straw, a trend which was maintained throughout several orders of magnitude in grain As. Previous researches [31,41,42] also observed that much more As accumulated in rice root than other parts. In the current study, the levels of As in straw were found to be, on average, 15.4 times higher than their corresponding grain samples.

Both straw As concentration and mean As straw/soil transfer factors (S^{traw}/S_{oil} TFs) based on total As concentrations were highly variable between and/or within Renhua and Lechang locations (Table 2). Mean S^{traw}/S_{oil} TFs for Renhua and Lechang were significantly different, with the values of 0.36 and 0.079, respectively (Table 2), and in both location, the highest and the lowest mean S^{traw}/S_{oil} TFs were seen for SY-122 and SY-428, respectively (Table 3).

The range of grain/soil transfer factors (G^{rain}/S_{oil} TFs) were 0.012–0.057 for Renhua, and 0.0020–0.0047 for Lechang. There was an over 7-fold difference in mean G^{rain}/S_{oil} TFs between Renhua and Lechang respectively, probably related to the significant difference of soil As between the two locations (Table 1). Similar to S^{traw}/S_{oil} TF, the G^{rain}/S_{oil} TF for SY-122 was significantly higher compared to SY-428, regardless of location (Table 3). Mean As grain/straw transfer factors (G^{rain}/S_{traw} TFs) ranged from 0.038 to 0.11, averaging 0.076, which was a little higher compared to the As G^{rain}/S_{traw} TFs in rice surveyed in Guangdong [22]. Differences in As G^{rain}/S_{traw} TFs were not apparent between locations (Table 1) and cultivars respectively (Tables 3).

Factors affecting As transfer

As discussed above, As uptake by rice plants appears more affected by As availability than total As in the soil. The bioavailability of As to plants is governed by key edaphic physiochemical properties (e.g., pH, Eh, organic matter, texture, Fe/Mn-oxides/hydroxides, and phosphorus, silicon, and sulfur concentrations); environmental conditions and modification of the soil in the rhizosphere; these factors interact to influence As speciation in the soil [16,41]. Rice is normally cultivated in flooded paddy soil, an environment that leads to a mobilization and, hence, a much enhanced bioavailability of As to rice plants. Rice is also a strong accumulator of the macro-nutrient silicon, an element that plays an important role in the defense against a range of biotic and abiotic stresses [10]. The principal component analysis (PCA) (Fig. 2) was performed with the concentrations of As, Fe and Mn in DCB extracts from root surfaces, the concentration of total As in soil, the concentrations of Fe and Mn oxides, AP and ASi, selected soil properties, and the concentrations of As in rice tissues in order to analyze the relationships among these indices and identify the factors affecting As transfer. The first 2 principal components accounted for 67.6% of the variability observed among all the cultivars and across all harvest sites. The results from PCA is in agreement with predictions that plant As is determined more strongly by external soil properties affecting As availability than by differences in internal plant processes.

The Fe and Mn oxide phases are common in various soils and are very efficient in sorbing As [15]. Manganese plaque and Fe and Mn oxides positively correlated with As in rice tissues respectively (Fig. 2), indicating Mn plaque and Fe and Mn oxides in soil may inhibit As transfer from soil to rice plant. There were 1.2 times, 12.2 times, and 27.4 times difference in iron plaque, manganese plaque, and Mn_2O_3 level respectively between Lechang and Renhua (Table 1), which might account for fact that the higher As in the Lechang soils did not result in significantly greater As in plant tissues compared with the rices grown in Renhua region.

The mechanism for arsenate uptake, the dominant inorganic As species under aerobic conditions, is through phosphate transport-

Table 3. Cultivar means for each parameter as observed in Renhua and Lechang.

Location	Cultivar	S^{traw}/S_{oil}	G^{rain}/S_{traw}	G^{rain}/S_{oil}	As-root	As-straw	As-grain	As-DCB	Mn-DCB	Fe-DCB	As-soil
		TF	TF	TF	mg·kg^{-1}	mg·kg^{-1}	mg·kg^{-1}	mg·kg^{-1}	mg·kg^{-1}	g·kg^{-1}	mg·kg^{-1}
Renhua	SY-428	0.19±0.060b	0.11±0.053a	0.022±0.017b	22.6±11.2a	2.9±1.0a	0.27±0.052bc	25.4±7.8b	10.5±3.2a	28.0±4.4ab	16.0±6.8a
	MXZ	0.34±0.035ab	0.083±0.019a	0.028±0.0058ab	26.0±7.5a	2.8±0.4a	0.230.036cd	19.1±8.9b	35.3±30.3a	32.6±8.8ab	8.2±0.6b
	SY-162	0.37±0.10ab	0.086±0.015a	0.031±0.0047ab	22.2±5.2a	3.1±0.7a	0.26±0.016bc	24.9±16.3b	39.5±46.7a	29.8±19.9ab	8.7±0.8b
	MBYZ	0.57±0.46a	0.080±0.047a	0.034±0.010ab	38.1±30.1a	3.6±2.2a	0.24±0.0029bcd	14.5±5.6b	24.9±15.7a	32.9±13.9ab	7.2±2.0b
	TY-10	0.37±0.11ab	0.094±0.047a	0.037±0.027ab	19.5±5.3a	2.7±1.5a	0.22±0.012bcd	31.2±5.7b	31.9±11.4a	30.8±0.4ab	8.3±6.5ab
	SY-86	0.35±0.21ab	0.058±0.037a	0.016±0.00052b	30.0±14.7a	4.8±3.0a	0.22±0.00093bcd	57.5±2.7a	17.0±4.7a	38.0±6.1a	13.7±0.4ab
	SY-122	0.70a	0.081a	0.057a	48.6a	4.8a	0.39a	6.4b	16.5a	4.3b	6.9ab
	SY-89	0.21ab	0.084a	0.018b	11.1a	2.2a	0.19d	26.5b	21.5a	41.9a	10.8ab
	DY-162	0.23ab	0.079a	0.018ab	14.6a	2.3a	0.18d	23.5b	16.1a	23.1ab	9.9ab
	JY-118	0.40ab	0.064a	0.025ab	23.7a	4.8a	0.31ab	17.4b	18.9a	17.3ab	12.1ab
Lechang	SY-428	0.039±0.018c	0.077±0.026a	0.003±0.00062b	29.8±6.8b	3.6±2.2a	0.25±0.090a	68.2±33.1a	228.0±116.0a	37.4±1.3a	96.7±36.0a
	SY-82	0.12±0.010a	0.039±0.0043a	0.005±0.00011a	28.7±2.3b	8.6±2.6a	0.33±0.064a	64.8±52.4a	213.2±177.3a	35.6±12.8a	72.5±15.5a
	MBYZ	0.052bc	0.061a	0.003ab	43.0ab	4.7a	0.29a	67.0a	111.1a	35.7a	89.0a
	SY-122	0.12ab	0.038a	0.005a	52.0a	8.3a	0.32a	65.6a	75.6a	35.2a	68.3a
	FY-998	0.10ab	0.044a	0.005a	38.6ab	5.6a	0.25a	98.4a	215.4a	36.4a	53.7a

Results are presented as arithemic mean ± SD; means within a row for a certain genotype grown in Lechang or Renhua followed by different letters are significantly different at the 0.05 level; the comparisons are based on estimated marginal means.

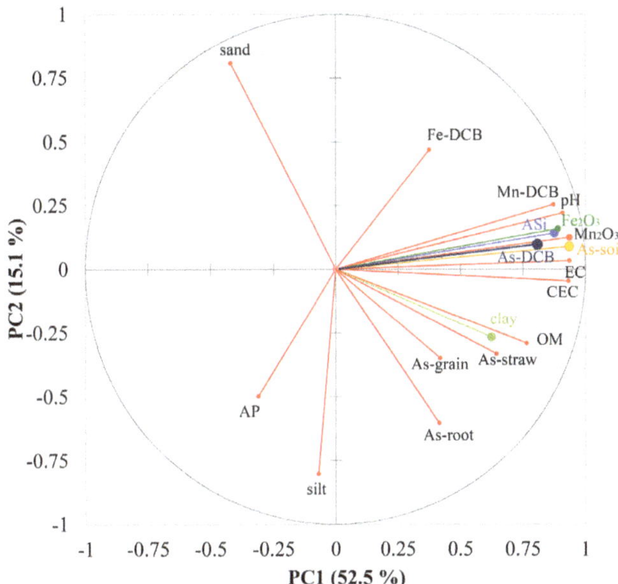

Figure 2. Plot of the first two principal components from Principal Component Analysis (PCA). The PCA was performed with As, Fe and Mn in DCB extracts, total As in soil, Fe_2O_3 and Mn_2O_3, available phosphorus (AP) and Si (ASi), selected soil properties, As in rice tissues.

ers, as arsenate is an analogue of phosphate [43]. Addition of phosphate to the soil may decrease arsenate uptake and consequently may reduce As toxicity, depending on soil conditions etc. [22]. In the current study, however, no remarkable relationships were observed between the AP in soil and rice As (Fig. 2), which is incompatible with the arsenate uptake mechanism discussed above, probably due to arsenite being the predominant form of As in flooded paddy soils [18], which doesn't compete with phosphate for transport as arsenate. Arsenite shares the same transport system responsible for silicon uptake, both influx and efflux transporters mediate transport of arsenite [28], and arsenite associated with iron plaque may be much more easily desorbed than arsenate [15,16]. Therefore, the application of silica fertilizer to soil can decrease the transfer of As from the soil and irrigation water to rice. The significantly higher available silicon in Lechang ($p = 0.013$), as shown in Table 1, may also be responsible for the fact that the higher soil As in Lechang did not result in significantly greater As in rice grain compared with the rices grown in Renhua region.

The PCA effects of root As, straw As, and grain As were close to each other, indicating the significant positive relationships between these rice tissues, while the As in rice tissues showed negative relationship with DCB-extracted As and soil As, respectively (Fig. 2). This further verifies that a rise in the soil As may not increase the accumulation of As in rice tissues.

Soil texture is another important factor affecting As bioavailability [15]. In general, soils with a clayey texture have less availability of As compared with sandy soils [17]. As observed in Table 1, the significant higher soil clay content in Lechang ($p = 0.010$) may decrease the availability of its soil As, so that even though these soils contained more As, they consequently inhibited its uptake by rice plants.

The solubility and bioavailability of As can be affected by soil pH because it controls the As speciation and leachability [15]. The soil pH differed significantly between Lechang and Rehua ($p = 0.0096$). The soils in Lechang ranged from neutral (pH 5.9) to slightly alkaline (pH 7.6) and in Renhua from strongly acidic (pH 4.4) to neutral (pH 6.0), respectively (Table 1). The soils collected from the Lechang mine region were expected to be more acidic and similar to those in Renhua. The industries in the northeast part of the paddy field (Fig. 1), including chemical plant, cement plant, textile mill, metal processing factory, plastic products factory, and bulb factory, might be affecting the soil pH. Arsenite solubility increases as the pH decreases within the range commonly found in soil (pH 3–9), while the pattern is reversed in the case of arsenate. Arsenite predominates in flooded paddy soils. In this regard, for the current study, a decline in soil pH can increase the mobilization of As in soils, which explain why we observed nearly equal concentrations of As in rice grain regardless of the soil As levels. The significant higher soil pH in Lechang may decrease the availability of its soil As.

By understanding the factors controlling bioavailability of As to rice plants and mechanisms of As uptake in plants, one could develop proper strategies for limiting As accumulation in rice grains. Possibilities include altering farm practices, e.g., growing rice aerobically in raised beds instead of in the traditional flooded paddy fields, which offers an opportunity to reduce the mobilization of soil arsenite and curtail As transfer from soil to grain. This approach would require, however, a fundamental change in farming practices in Asia [6,14], and aerobically produced rice is generally lower yielding [16,44] and more susceptible to rice blast disease and heat stress [45]. In addition, silica and phosphate fertilizations can be applied in soil to decrease As accumulation in rice, dependent on soil conditions [16]. Another tack would be the selection of rice cultivars with low accumulation of As in grains [40]. To be a success on the farm, any new cultivars will have to have decent yields. A hypothetical cancer risk pales in comparison with an empty stomach [6].

Conclusions

The results indicated both environmental and genetic effects caused diversity for grain As concentration among different rice cultivars grown in two locations in China with mining-contaminated soils. All the grain samples in this study exceeded national food standards for grain As. The fact that grain As levels were not significantly different in the rices from the two areas differing significant for soil As levels, suggests that As uptake by rice is determined more by As availability rather than by total As in the soil. The As behaviour in the soil-rice system was found associated with various factors, i.e., iron and manganese plaque, iron and manganese oxides in the soil, soil available silicon and phosphorus, soil pH, soil organic matter, and soil texture. Understanding the mechanisms controlling As uptake would improve our understanding of how soil As sometimes but not always increases rice grain As, and to develop genetic and physico-chemical strategies for reducing As accumulation in rice grains.

Author Contributions

Conceived and designed the experiments: FD YL JL QY. Performed the experiments: FD QY YL. Analyzed the data: JL YL HS FD QY. Contributed reagents/materials/analysis tools: YL HS. Wrote the paper: JL YL HS FD.

References

1. Smith E, Naidu R, Alston A (1998) Arsenic in the soil environment: a review. Adv Agron 64: 149–195.
2. Duan G, Liu W, Chen X, Hu Y, Zhu Y (2013) Association of arsenic with nutrient elements in rice plants. Metallomics 5: 784–792.
3. Zhao FJ, Ma JF, Meharg AA, McGrath SP (2009) Arsenic uptake and metabolism in plants. New phytol 181: 777–794.
4. Dave R, Singh P, Tripathi P, Shri M, Dixit G, et al. (2013) Arsenite tolerance is related to proportional thiolic metabolite synthesis in rice (*Oryza sativa* L.). Arch Environ Contam Toxicol 64: 235–242.
5. Banerjee M, Banerjee N, Bhattacharjee P, Mondal D, Lythgoe PR, et al. (2013) High arsenic in rice is associated with elevated genotoxic effects in humans. Sci Rep 3.
6. Stone R (2008) Arsenic and paddy rice: a neglected cancer risk? Science 321: 184–185.
7. Kennedy D (2002) The importance of rice. Science 296: 13–13.
8. Fan JB, Zhang YL, Turner D, Duan YH, Wang DS, et al. (2010) Root physiological and morphological characteristics of two rice cultivars with different nitrogen-use efficiency. Pedosphere 20: 446–455.
9. Xu XY, McGrath SP, Meharg AA, Zhao FJ (2008) Growing rice aerobically markedly decreases arsenic accumulation. Environ Sci Technol 42: 5574–5579.
10. Zhao FJ, Zhu YG, Meharg AA (2013) Methylated arsenic species in rice: geographical variation, origin, and uptake mechanisms. Environ Sci Technol 47: 3957–3966.
11. Williams PN, Islam MR, Adomako EE, Raab A, Hossain SA, et al. (2006) Increase in rice grain arsenic for regions of Bangladesh irrigating paddies with elevated arsenic in groundwaters. Environ Sci Technol 40: 4903–4908.
12. Zavala YJ, Duxbury JM (2008) Arsenic in rice: I. Estimating normal levels of total arsenic in rice grain. Environ Sci Technol 42: 3856–3860.
13. Meharg AA, Rahman MM (2003) Arsenic contamination of Bangladesh paddy field soils: implications for rice contribution to arsenic consumption. Environ Sci Technol 37: 229–234.
14. Ahmed ZU, Panaullah GM, Gauch H, McCouch SR, Tyagi W, et al. (2011) Genotype and environment effects on rice (*Oryza sativa* L.) grain arsenic concentration in Bangladesh. Plant Soil 338: 367–382.
15. Sahoo PK, Kim K (2013) A review of the arsenic concentration in paddy rice from the perspective of geoscience. Geosci J 17: 107–122.
16. Zhao FJ, McGrath SP, Meharg AA (2010) Arsenic as a food chain contaminant: mechanisms of plant uptake and metabolism and mitigation strategies. Annu Rev Plant Biol 61: 535–559.
17. Heikens A, Panaullah GM, Meharg AA (2007) Arsenic behaviour from groundwater and soil to crops: impacts on agriculture and food safety. Rev Environ Contam Toxicol 189: 43–87.
18. Ma JF, Yamaji N, Mitani N, Xu X-Y, Su Y-H, et al. (2008) Transporters of arsenite in rice and their role in arsenic accumulation in rice grain. P Natl Acad Sci 105: 9931–9935.
19. Liu WJ, Zhu YG, Hu Y, Williams PN, Gault AG, et al. (2006) Arsenic sequestration in iron plaque, its accumulation and speciation in mature rice plants (*Oryza sativa* L.). Environ Sci Technol 40: 5730–5736.
20. Xu J, Tian YZ, Zhang Y, Guo CS, Shi GL, et al. (2013) Source apportionment of perfluorinated compounds (PFCs) in sediments: Using three multivariate factor analysis receptor models. J Hazard Mat 260: 483–488.
21. Moreno-Jiménez E, Esteban E, Peñalosa J (2012) The fate of arsenic in soil-plant systems. In: Whitacre DM, editor: Springer New York. 1–37.
22. Lu Y, Dong F, Deacon C, Chen HJ, Raab A, et al. (2010) Arsenic accumulation and phosphorus status in two rice (*Oryza sativa* L.) cultivars surveyed from fields in South China. Environ Pollut 158: 1536–1541.
23. Yang QW, Lan CY, Shu WS (2008) Copper and Zinc in a paddy field and their potential ecological impacts affected by wastewater from a lead/zinc mine, P. R. China. Environm Monit Assess 147: 65–73.
24. Lu HZ (1983) Fluid inclusion study of Fankou Pb-Zn deposit, Fankou, Guangdong, China. Geochemistry 2: 45–57.
25. Wei X, Cao J, Holub RF, Hopke PK, Zhao S (2013) TEM study of geogas-transported nanoparticles from the Fankou lead–zinc deposit, Guangdong Province, South China. J Geochem Explor 128: 124–135.
26. Deng J, Yang L, Chen X, Wang Q, Liu Y (2005) Fluid system and ore-forming dynamics of the Yuebei Basin, China. In: Mao J, Bierlein F, editors. Mineral Deposit Research: Meeting the Global Challenge: Springer Berlin Heidelberg. 107–109.
27. Hu Y, Sun W, Liu R, Dai J (2012) Water recycling technology in Fankou lead-zinc mine of China. In: Drelich J, editor. Water in Mineral Processing. Englewood, Colorado, USA: Society for Mining Metallurgy & Exploration 371–388.
28. Liu WJ, Zhu YG, Smith FA, Smith SE (2004) Do iron plaque and genotypes affect arsenate uptake and translocation by rice seedlings (*Oryza sativa* L.) grown in solution culture? J Exp Bot 55: 1707–1713.
29. Lu RK (2000) Analytical methods for soils and agricultural chemistry. Beijing, China: China agricultural science and technology Press (In Chinese).
30. Liu HY, Probst A, Liao BH (2005) Metal contamination of soils and crops affected by the Chenzhou lead/zinc mine spill (Hunan, China). Sci Total Environ 339: 153–166.
31. Williams PN, Lei M, Sun G, Huang Q, Lu Y, et al. (2009) Occurrence and partitioning of cadmium, arsenic and lead in mine impacted paddy rice: Hunan, China. Environ Sci Technol 43: 637–642.
32. Liu CP, Luo CL, Gao Y, Li FB, Lin LW, et al. (2010) Arsenic contamination and potential health risk implications at an abandoned tungsten mine, southern China. Environ Pollut 158: 820–826.
33. Oyarzun R, Cubas P, Higueras P, Lillo J, Llanos W (2009) Environmental assessment of the arsenic-rich, Rodalquilar gold-(copper-lead-zinc) mining district, SE Spain: data from soils and vegetation. Environm Geol 58: 761–777.
34. Zhang HH, Yuan HX, Hu YG, Wu ZF, Zhu LA, et al. (2006) Spatial distribution and vertical variation of arsenic in Guangdong soil profiles, China. Environ Pollut 144: 492–499.
35. Chinese Environmental Protection Agency (2008) Environmental quality standards for soils. GB 15618–2008.
36. Dudka S, Miller WP (1999) Permissible concentrations of arsenic and lead in soils based on risk assessment. Water Air Soil Pollut 113: 127–132.
37. DeSesso JM, Jacobson CF, Scialli AR, Farr CH, Holson JF (1998) An assessment of the developmental toxicity of inorganic arsenic. Reprod Toxicol 12: 385–433.
38. Zhu YG, Sun GX, Lei M, Teng M, Liu YX, et al. (2008) High percentage inorganic arsenic content of mining impacted and nonimpacted Chinese rice. Environ Sci Technol 42: 5008–5013.
39. Chinese Food Standards Agency (2005) Maximum levels of contaminants in food. GB 2762–2005.
40. Norton GJ, Pinson SR, Alexander J, McKay S, Hansen H, et al. (2012) Variation in grain arsenic assessed in a diverse panel of rice (*Oryza sativa*) grown in multiple sites. New phytol 193: 650–664.
41. Williams PN, Villada A, Deacon C, Raab A, Figuerola J, et al. (2007) Greatly enhanced arsenic shoot assimilation in rice leads to elevated grain levels compared to wheat and barley. Environ Sci Technol 41: 6854–6859.
42. Liao XY, Chen TB, Xie H, Liu YR (2005) Soil As contamination and its risk assessment in areas near the industrial districts of Chenzhou City, Southern China. Environ Int 31: 791–798.
43. Norton GJ, Adomako EE, Deacon CM, Carey A-M, Price AH, et al. (2013) Effect of organic matter amendment, arsenic amendment and water management regime on rice grain arsenic species. Environ Pollut 177: 38–47.
44. Peng S, Bouman B, Visperas RM, Castañeda A, Nie L, et al. (2006) Comparison between aerobic and flooded rice in the tropics: Agronomic performance in an eight-season experiment. Field Crops Res 96: 252–259.
45. Farooq M, Siddique KHM, Rehman H, Aziz T, Lee D-J, et al. (2011) Rice direct seeding: Experiences, challenges and opportunities. Soil Till Res 111: 87–98.

Effects of Winter Cover Crops Straws Incorporation on CH_4 and N_2O Emission from Double-Cropping Paddy Fields in Southern China

Hai-Ming Tang*, Xiao-Ping Xiao*, Wen-Guang Tang, Ke Wang, Ji-Min Sun, Wei-Yan Li, Guang-Li Yang

Hunan Soil and Fertilizer Institute, Changsha, PR China

Abstract

Residue management in cropping systems is believed to improve soil quality. However, the effects of residue management on methane (CH_4) and nitrous oxide (N_2O) emissions from paddy field in Southern China have not been well researched. The emissions of CH_4 and N_2O were investigated in double cropping rice (*Oryza sativa* L.) systems with straw returning of different winter cover crops by using the static chamber-gas chromatography technique. A randomized block experiment with three replications was established in 2004 in Hunan Province, China, including rice-rice-ryegrass (*Lolium multiflorum* L.) (Ry-R-R), rice–rice–Chinese milk vetch (*Astragalus sinicus* L.) (Mv-R-R) and rice–rice with winter fallow (Fa-R-R). The results showed that straw returning of winter crops significantly increased the CH_4 emission during both rice growing seasons when compared with Fa-R-R. Ry-R-R plots had the largest CH_4 emissions during the early rice growing season with 14.235 and 15.906 g m^{-2} in 2012 and 2013, respectively, when Ry-R-R plots had the largest CH_4 emission during the later rice growing season with 35.673 and 38.606 g m^{-2} in 2012 and 2013, respectively. The Ry-R-R and Mv-R-R also had larger N_2O emissions than Fa-R-R in both rice seasons. When compared to Fa-R-R, total N_2O emissions in the early rice growing season were increased by 0.05 g m^{-2} in Ry-R-R and 0.063 g m^{-2} in Mv-R-R in 2012, and by 0.058 g m^{-2} in Ry-R-R and 0.068 g m^{-2} in Mv-R-R in 2013, respectively. Similar result were obtained in the late rice growing season, and the total N_2O emissions were increased by 0.104 g m^{-2} in Ry-R-R and 0.073 g m^{-2} in Mv-R-R in 2012, and by 0.108 g m^{-2} in Ry-R-R and 0.076 g m^{-2} in Mv-R-R in 2013, respectively. The global warming potentials (GWPs) from paddy fields were ranked as Ry-R-R>Mv-R-R> Fa-R-R. As a result, straw returning of winter cover crops has significant effects on increase of CH_4 and N_2O emission from paddy field in double cropping rice system.

Editor: Dafeng Hui, Tennessee State University, United States of America

Funding: This study was supported by the National Natural Science Foundation of China (No. 31201178), and the Public Research Funds Projects of Agriculture, Ministry of Agriculture of the P.R. China (No. 201103001). The funders had no role in study design, data collection and analysis, decision to publish, or preparation of the manuscript.

Competing Interests: The authors have declared that no competing interests exist.

* Email: hntfsxxping@163.com (XPX); tanghaiming66@163.com (HMT)

Introduction

With the current rise in global temperatures, numerous studies have focused on greenhouse gases (GHG) emissions [1–3]. Agriculture production is an important source of GHG emission [4]. In addition to carbon dioxide (CO_2), methane (CH_4) and nitrous oxide (N_2O) play important roles in global warming. The global warming potentials (GWPs) of CH_4 and N_2O are 25 and 298 times that of CO_2 in a time horizon of 100 years, respectively [5]. The concentrations of CH_4 and N_2O in the atmosphere are estimated to be increasing at the rates of 1% and 0.2–0.3% per year [6]. In addition to industrial emissions, farmland is another important source of atmospheric GHG [7–10]. Numerous results indicate that rice (*Oryza sativa* L.) paddy field is a significant source of CH_4 and N_2O emissions [10,11]. The anaerobic conditions in wetland rice field are favorable for fostering CH_4 emission [12]. Thus, the characteristics of CH_4 and N_2O emissions from paddy field and the reduction of emission have received attentions from scientists.

A considerable number of studies have shown that some farm operations can influence CH_4 and N_2O emission. For example, cropping system, crop type, water and nitrogen (N) management, organic matter application and tillage can regulate CH_4 and N_2O emission [13–15]. Tillage and crop straws retention have a great influence on CH_4 and N_2O emission through the changes of soil properties (e.g., soil porosity, soil temperature and soil moisture, etc.) [16–17]. In paddy soils, CH_4 is produced by archaea bacteria during the anaerobic degradation of organic matter and oxidized by methanotrophic bacteria [18]. Incorporation of organic material into soil can enhance the number and activity of archaea bacteria [19] and provide large amounts of active organic substrate for CH_4 production [20]. Soil amendment with organic material, such as crop straw [21] and green manure incorporation [22], has been well estimated to promote CH_4 emission in paddy fields. Biogenic N_2O production originates from nitrification and denitrification [23], which are processes involving microorganisms in the soil. N_2O flux in paddy fields was small in flooding condition, but peaked after drainage [24]. Some studies have indicated that the cropping system of winter fallow with cover crops has advantages of promoting soil quality, enhancing nutrient utilization, increasing crop yield, reducing soil erosion and chemical runoff, and inhibiting weed growth in paddy field [25–26].

Winter cover crops, which are grown during an otherwise fallow period, are a possible means of improving nutrient dynamics in the surface layer of intensively managed cropping systems. Chinese milk vetch (*Astragalus sinicus* L.) and ryegrass (*Lolium multiflorum* L.) are the main winter cover crops in Southern China. Growing these cover crops with straw mulching in the winter season after late rice harvest and incorporating them into soil as green manure before early rice transplanting next year is a traditional practice as well as rice straw incorporation. Hermawan and Bomke [26] suggested that growing winter cover crops such as annual ryegrass may protect aggregate breakdown during winter and result in a better soil structure after spring tillage, as opposed to leaving soil bare. Other potential benefits of winter cover crops are the prevention of nitrate leaching [27]; weed infestation [28]; and improvement of soil water retention, soil organic matter content and microbial activity [29]. Returning of crop straws have been suggested to improve overall soil conditions, reduce the requirement for N fertilizers and support sustainable rice productivity.

In recent years, many researches have studied the effects of winter cover crops on soil physical properties and crop productivity, methane emission, N availability and nitrogen surplus [30–32]. However, relatively few studies related to CH_4 and N_2O emissions and yields under different double cropping rice systems with different winter cover crops have been conducted in double–cropping paddy field in Southern China. Monitoring CH_4 and N_2O emissions of different winter cover crops–double cropping rice cultivation modes is important to maintain soil productivity, increase carbon (C) storage, and regulate the greenhouse effects. Therefore, the objectives of this research were: (1) to quantify CH_4 and N_2O emissions from paddy field and grain yield under different winter cover crops and double cropping rice systems, (2) to evaluate the GWPs of different winter cover crops–double cropping rice treatments in southern China.

Materials and Methods

Experimental site

The experiment was initiated in winter 2004 at the experimental station of the Institute of Soil and Fertilizer Research, Hunan Academy of Agricultural Sciences, China (28°11′58″ N, 113°04′47″ E). The typical cropping system in this area is double cropping rice. The soil type is a Fe–accumuli–Stagnic Anthrosol derived from Quaternary red clay (clay loam). The characteristics of the surface soil (0–20 cm) in 2004 are as follows: pH 5.40, soil organic carbon (SOC) 13.30 g kg^{-1}, total N 1.46 g kg^{-1}, available N 154.5 mg kg^{-1}, total phosphorous (P) 0.81 g kg^{-1}, available P 39.2 mg kg^{-1}, total potassium (K) 13.0 g kg^{-1}, and available K 57.0 mg kg^{-1}. All these data were tested before the experiment in 2004. This region has the subtropical monsoonal humid climate with a long hot period and short cold period. The average annual precipitation is approximately 1500 mm and the annual mean temperature is 17.1°C, the annual frost-free period is approximately from 270 days to 310 days. The daily precipitation and mean temperature data during the early and late rice growing season during 2012–2013 are presented in Fig. 1. The cropping system was that the early rice rotated with the late rice, and then planted winter cover crops till the next year's early rice transplanting.

Experimental design and field management

A randomized block experiment with three replications was established in 2004, and this study was conducted from 2012 to 2013. The experiment included three cropping systems: rice–rice–ryegrass (Ry-R-R), rice–rice–Chinese milk vetch (Mv-R-R), and rice–rice with winter fallow (Fa-R-R). The plot area was 1.1 m^2 (1 m × 1.1 m). After winter cover crops harvested, a moldboard plow was used to incorporate part of the crop straw into soil: both the ryegrass and Chinese milk vetch straw returned was 22500.0 kg ha^{-1}. All the plots were plowed once to a depth of 20 cm by using a moldboard plow 15 d before rice seedling transplanting. The early rice variety (*Oryza sativa* L.) Lingliangyou 211 and late rice variety (*Oryza sativa* L.) Fengyuanyou 299 were used as the materials in 2012 and 2013. One-month-old seedlings were transplanted with a density of 150,000 plants ha^{-1} (one seed per 16 cm × 16 cm) and 2–3 plants per hill. Gramoxone (paraquat) was applied to control weeds at 2 d before rice transplantation. The basal fertilizer of the early and late rice was applied at the rate of 150.0 kg N ha^{-1} and 180.0 kg N ha^{-1} as urea (60% for basal; 40% for top–dressed at the tillering stage), 75.0 kg P_2O_5 ha^{-1} as diammonium phosphate and 120.0 kg K_2O ha^{-1} as potassium sulfate. The different treatments during early and late rice season and field management were presented in Table 1.

Collection and measurement of CH_4 and N_2O

CH_4 and N_2O emitted from paddy field were collected using the static chamber–GC technique at 9:00–11:00 in the morning during the early and late rice growing season. The chamber (50 cm × 50 cm × 120 cm) was made of 5 mm PVC board with a PVC base. The base had a groove in the collar, in which the chamber could be settled. The chamber base was inserted into soil about 5 cm in depth with rice plant growing inside the base. The groove was 1 cm below flooded water, and the chamber was settled into the groove of the collar with water to prevent leakage and gas exchange. The chamber contained a small fan for stirring air, a thermometer sensor, and a trinal–venthole. From the second day after transplanting of early or late rice, gases were sampled weekly. Before sampling, the fan in the chamber started working to allow an even mix of air before extracting the air with a 50 ml injector at 0, 10, 20, and 30 min after closing the box. The air samples were transferred into 0.5 L sealed sample bags by rotating trinal venthole.

The quantities of CH_4 and N_2O emission were measured with a gas chromatograph (Agilent 7890A) equipped with flame ionization detector (FID) and electron capture detector (ECD). Methane was separated using 2 m stainless-steel column with an inner diameter of 2 mm 13XMS column (60/80 mesh), with FID at 200°C. Nitrous oxide was separated using a 1 m stainless-steel column with an inner diameter 2 mm Porapak Q (80/100 mesh) and ECD at 330°C.

Data analysis

Fluxes of CH_4 and N_2O were calculated with the following equation [33]:

$$F = ph \times \frac{273}{273+t} \times \frac{dc}{dt}$$

Where, F is the CH_4 flux (mg m^{-2} h^{-1}) or N_2O flux (μg m^{-2} h^{-1}); T is the air temperature (°C) inside the chamber; ρ is the CH_4 or N_2O density at standard state (0.714 kg m^{-3} for CH_4 and

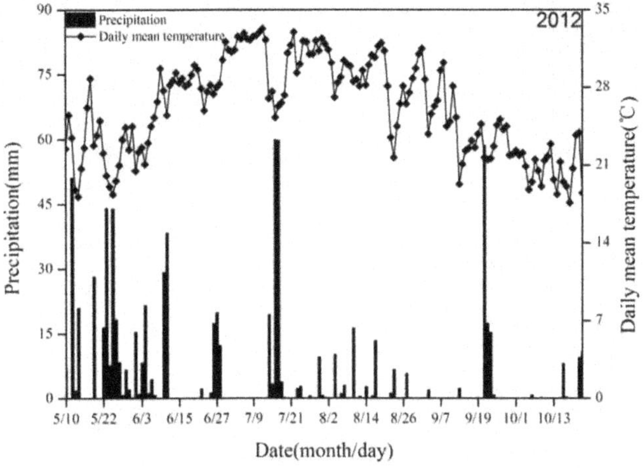

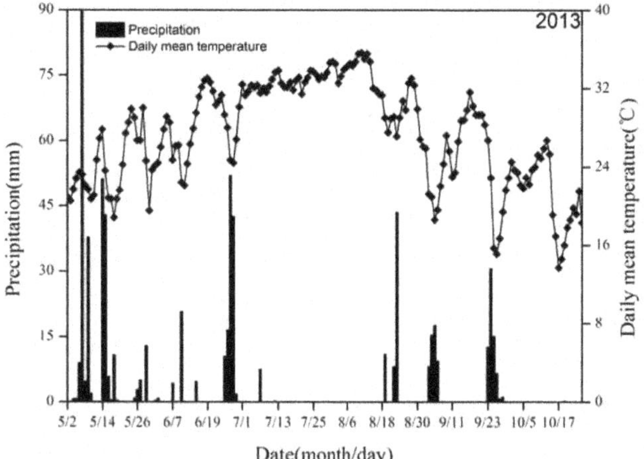

Figure 1. Daily precipitation and mean temperature at the study site in 2012 and 2013.

1.964 kg m^{-3} for N_2O); h is the headspace height of the chamber (m); and dc/dt is the slope of the curve of gas concentration variation with time.

The total emissions of CH_4 and N_2O were sequentially computed from the emissions between every 2 adjacent intervals of the measurements, based on a non–linear, least–squares method of analysis [34,35].

GWPs is defined as the cumulative radiative forcing both direct and indirect effects integrated over a period of time from the emission of a unit mass of gas relative to some reference gas. Carbon dioxide was chosen as this reference gas. The GWPs conversion parameters of CH_4 and N_2O (over 100 years) were adopted with 25 and 298 kg ha^{-1} CO_2-equivalent [5].

Statistical analysis

Data presented herein are means of 3 replicates in each treatment. All data were expressed as mean ± standard error. The data were analyzed as a randomized complete block, using the PROC ANOVA procedure of SAS [36]. Mean values were compared using the least significant difference (LSD) test, and a probability value of 0.05 was considered to indicate statistical significance.

Results

Characteristics of CH_4 emission flux from early and late rice fields

In the early rice season, the curve of CH_4 flux was low when early rice was newly transplanted, but increased quickly until the first peak about 2 weeks after transplanting, and then dramatically declined to a low level with relative stability with the second small peak appeared at 36 and 35 d after transplanting in 2012 and 2013, respectively (Fig. 2). The gradual increase of CH_4 emission after transplanting resulted from the decomposition of organic matter and the growth of rice. The second peak was mainly because of the continuous decomposition of organic matter under

Table 1. Management practices of different cropping systems.

Crop	Date (month/day)		Field management
	2012	2013	
Early rice	4/12	4/5	Sowing and seedling raising
	5/9	5/1	Paddy tillage
	5/10	5/2	Transplanting (16 cm×16 cm)
	5/18	5/10	Urea were applied at 130.0 kg ha^{-1} for top-dressed at tillering
	6/7–6/15	5/27–6/5	Drained out water and dried the soil at maximum tillering stage
	6/16–7/13	6/6–7/13	Wetting-drying alternation irrigation
	7/18	7/18	Grains were harvested
Late rice	6/25	6/27	Sowing and seedling raising
	7/21	7/19	Paddy tillage (The rate of early rice straw returning was 4 500.0 kg ha^{-1})
	7/22	7/20	Transplanting (16 cm×16 cm)
	7/30	7/28	Urea were applied at 156.5 kg ha^{-1} for top-dressed at tillering
	8/20–8/27	8/16–8/26	Drained out water and dried the soil at maximal tillering stage
	8/28–10/17	8/27–10/19	Wetting-drying alternation irrigation
	10/22	10/25	Grains were harvested

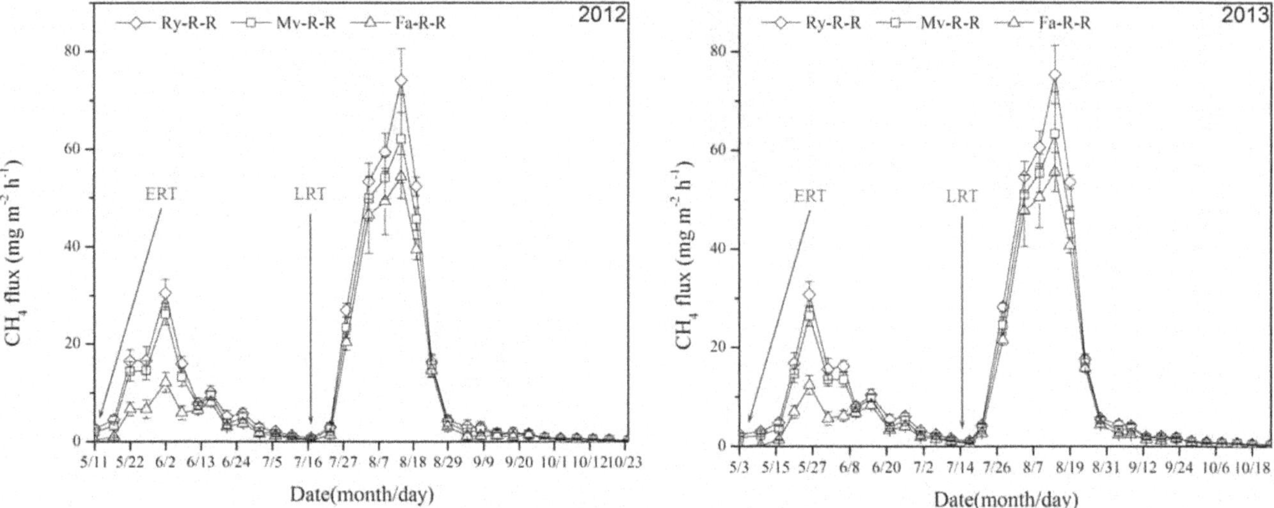

Figure 2. Effects of winter cover crops on CH_4 flux in early and late rice fields in 2012 and 2013. Ry-R-R: rice–rice–ryegrass cropping system; Mv-R-R: rice–rice–Chinese milk vetch cropping system; Fa-R-R: rice–rice cropping system with winter fallow. ERT: early rice transplanting; LRT: late rice transplanting. CH_4 emission rate is the mean of values measured within each treatment (n = 3).

high temperature. In the early rice season, the CH_4 flux values were significantly different among treatments with the order of Ry-R-R>Mv-R-R>Fa-R-R ($P<0.05$) (Fig. 2).

Methane emission in the late rice growing season mainly focused at tillering stage, and the peak value of CH_4 flux was observed at 23 and 24 d after transplanting in all treatments in 2012 and 2013, respectively. Then, the emission rate dramatically decreased to a low and stable level, especially from field drainage to harvest. The order of treatments in CH_4 emission was Ry-R-R>Mv-R-R>Fa-R-R (Fig. 2).

Characteristics of N_2O emission flux from early and late rice fields

The peak flux N_2O was emitted when the field was drained. Meanwhile, part of N_2O was emitted during wetting–drying alternation irrigation period. The first peak value of N_2O flux appeared at 7 and 15 d after transplanting in all treatments in 2012 and 2013, respectively, and then decreased. The order among treatments was Mv-R-R>Ry-R-R>Fa-R-R during the period from transplanting to field drainage, and Ry-R-R>Mv-R-R>Fa-R-R during wetting–drying alternation period. The N_2O flux in early rice paddy reached the highest peak at 32 and 35 d after transplanting in 2012 and 2013, respectively (Fig. 3).

In the late rice growing season, N_2O emission increased from field drainage to full heading stage, and mainly focused at booting stage. The order of N_2O emission fluxes among different treatments was Ry-R-R>Mv-R-R>Fa-R-R in the late rice growing season. In 2012, the average N_2O fluxes in the late rice growing season were 78.718 μg m^{-2} h^{-1} in Ry-R-R, 64.928 μg m^{-2} h^{-1} in Mv-R-R, and 32.275 μg m^{-2} h^{-1} in Fa-R-R. In 2013, the average N_2O fluxes in the late rice growing season were 81.453 μg m^{-2} h^{-1} in Ry-R-R, 67.662 μg m^{-2} h^{-1} in Mv-R-R, and 34.623 μg m^{-2} h^{-1} in Fa-R-R (Fig. 3).

Total CH_4 and N_2O emission from paddy fields in the growing durations of early and late rice

In the early rice growing season, the total CH_4 emissions of Ry-R-R and Mv-R-R were significantly higher than Fa-R-R ($P<0.05$), and the order of treatments was Ry-R-R>Mv-R-R>Fa-R-R (Table 2). The straws of winter cover crops incorporated into soil provided favorable soil condition and sufficient substance to be decomposed in the early rice season; therefore, the CH_4 emission quantities in straw returning treatments were higher than Fa-R-R ($P<0.05$). In 2012, the total CH_4 emissions from paddy fields during late rice entire growing season were 35.673 g m^{-2} in Ry-R-R, 31.542 g m^{-2} in Mv-R-R, 27.874 g m^{-2} in Fa-R-R. In 2013, the total CH_4 emissions from paddy fields during late rice whole growing season were 38.606 g m^{-2} in Ry-R-R, 34.358 g m^{-2} in Mv-R-R, 30.550 g m^{-2} in Fa-R-R. The order of treatments in total CH_4 emission was Ry-R-R>Mv-R-R>Fa-R-R (Table 2).

Compared to Fa-R-R, the other treatments increased total N_2O emissions in the early rice growing season, and the N_2O emissions increased by 0.05 g m^{-2} (131.58%) in Ry-R-R and 0.063 g m^{-2} (165.79%) in Mv-R-R in 2012, and by 0.058 g m^{-2} (138.1%) in Ry-R-R and 0.068 g m^{-2} (161.90%) in Mv-R-R in 2013, respectively. Similar results were observed in the late rice growing season in 2012, the total N_2O emissions increased by 0.104 g m^{-2} (144.44%) in Ry-R-R and 0.073 g m^{-2} (101.39%) in Mv-R-R. And the total N_2O emissions increased by 0.108 g m^{-2} (135.00%) in Ry-R-R and 0.076 g m^{-2} (95.00%) in Mv-R-R in 2013 (Table 2).

The emissions of CH_4 and N_2O were closely related to farming system, soil type, climate, and field management practices. Ry-R-R and Mv-R-R had larger total CH_4 emissions than Fa-R-R in the double rice growing season ($P<0.05$). Ry-R-R had the largest total N_2O emissions in the double rice growing season with the quantities of 0.264 g m^{-2} in 2012, and 0.288 g m^{-2} in 2013, respectively (Table 3).

Global warming potentials of CH_4 and N_2O

GWPs is an indicator to reflect the relative radioactive effect of a greenhouse gas, and the GWPs of CO_2 is defined as 1. In this study, the GWPs of CH_4 and N_2O from double cropping paddy fields varied with different winter cover crops, and the trend showed as Ry-R-R>Mv-R-R>Fa-R-R. In 2012, Ry-R-R had the largest GWPs (13281.79 kg CO_2–eq ha^{-1}) of total CH_4 and N_2O from double cropping paddy fields, followed by Mv-R-R (11657.44 kg CO_2–eq ha^{-1}), and Fa-R-R had the lowest GWPs

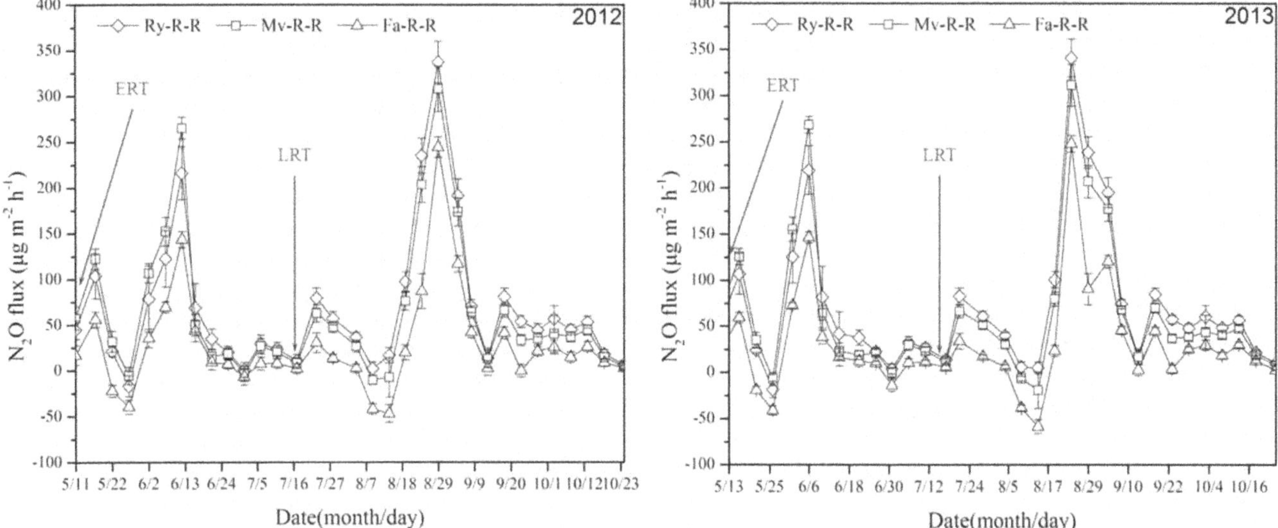

Figure 3. Effects of winter cover crops on N_2O flux in early and late rice fields in 2012 and 2013. Ry-R-R: rice–rice–ryegrass cropping system; Mv-R-R: rice–rice–Chinese milk vetch cropping system; Fa-R-R: rice–rice cropping system with winter fallow. ERT: early rice transplanting; LRT: late rice transplanting. N_2O emission rate is the mean of values measured within each treatment (n = 3).

of total CH_4 and N_2O (8993.12 kg CO_2–eq ha^{-1}). In 2013, Ry-R-R had the largest GWPs (14506.80 kg CO_2–eq ha^{-1}) of total CH_4 and N_2O from double cropping paddy fields, followed by Mv-R-R (12780.73 kg CO_2–eq ha^{-1}), and Fa-R-R had the lowest GWPs of total CH_4 and N_2O (9899.22 kg CO_2–eq ha^{-1}). According to GWPs, CH_4 from double cropping paddy fields had greater contribution to global warming than N_2O (Table 3).

Double rice grain yield of Mv-R-R was the highest, the lowest was Fa-R-R (Table 3). We also estimated per yield GWPs which was calculated as GWPs divided by rice grain yield. As is shown in Table 3, per yield GWPs of Ry-R-R was significantly higher than Mv-R-R and Fa-R-R ($P<0.05$), and the lowest was Fa-R-R.

Discussion

CH_4 emission

Methane emission is complex processes including production, oxidation, and emission. Chidthaisong et al. [37] reported that the highest CH_4 peaks were observed at flowering and heading stages, which could be related to the development of intense reducing conditions in the rice rhizosphere. In this study, we found that CH_4 emission was low in paddy fields after transplanting during early rice–growing season, and increased with the decomposition of organic matters and growth of rice. In addition, CH_4 emission was influenced by soil temperature and soil redox potential (Eh). Yu et al. [38] reported that CH_4 emission showed an exponential decrease by an Eh increase. In this study, the CH_4 flux and total CH_4 emission from paddy fields during the early and late rice growing season were much larger in Ry-R-R and Mv-R-R compared to Fa-R-R, which was similar to the result by Lee et al. [22]. The reasons for above result may be: first, microbial activities were improved after returning straws of winter cover crops into the soil due to the supplements of carbon source and energy for microbial activities to accelerate consumption of soil oxygen and decrease of soil Eh; second, methanogens became active due to the large quantities of C source, which provided reactive substrate for CH_4 emission from paddy fields. In the early rice growing season, the order of CH_4 flux and total CH_4 emission from paddy fields varied among treatments, which were highly related to the returning straw type, and straw decomposition rate. During the late rice growing season, the CH_4 emission increased gradually with the decomposition of organic matters and growth of rice after transplanting, and reach the peak value at tillering stage in all treatments. However, CH_4 emissions in both rice seasons were reduced in a large extent after field drying, because (1) soil aeration was improved during this period, and the activities of methanogens were restricted; and (2) the physiological activity of rice plant decreased, thereby limiting the ability for transportation and emission of CH_4 [39].

Although straw returning helps to maintain soil fertility and protect environment, but it enhances CH_4 emission simultaneously. Pandey et al. [12] showed that CH_4 emission was positively related to straw returning amount under permanent flooding condition, whereas N_2O emission had a reverse relationship with the amount of straw returning. In this study, we found that CH_4 flux in the late rice growing season was much higher than that in the early rice growing season, and peak appeared earlier. As straws of early rice (4500 kg ha^{-1}) returned to field before transplanting of late rice, the paddy soil of late rice was under anoxic condition after transplanting, which was favorable for CH_4 production and emission. Temperature was the major reason for the differences in the CH_4 emission pattern between the early and the late rice season. Soil temperature had a predictive functional relationship with CH_4 emission. Zhang et al. [40] reported that there was a strong positive correlation between CH_4 emission and soil temperature. In this experimental area, the late rice season was the hottest time in summer (Fig. 1). Therefore, high temperatures enhanced the decomposition of crop straws in the moist environment. In contrast to the warm temperatures of the late rice season, the air temperatures of the early rice season were lower, which resulted in slower crop straws decomposition and little CH_4–substrate. Hence, these differences in weather factors (e.g., temperature) resulted in the different characteristics of CH_4 between the early and the late rice seasons. However, there were significantly differences among treatments although they had similar trends. This indicated that CH_4 flux and emission from paddy fields were affected by different winter cover crops.

Table 2. Effects of winter cover crops on CH_4 and N_2O emission from rice fields during whole growing season of early and late rice (g m^{-2}).

Year	Treatment	CH_4			N_2O		
		Early rice	Late rice	Total	Early rice	Late rice	Total
2012	Ry-R-R	14.235±0.411a	35.673±1.030a	49.908±1.441a	0.088±0.003a	0.176±0.05a	0.264±0.008a
	Mv-R-R	12.092±0.349b	31.542±0.912b	43.634±1.260b	0.101±0.003b	0.145±0.04b	0.246±0.007a
	Fa-R-R	6.732±0.194c	27.874±0.805c	34.606±0.999c	0.038±0.001c	0.072±0.02c	0.110±0.003b
2013	Ry-R-R	15.906±0.459a	38.606±1.115a	54.512±1.574a	0.100±0.003b	0.188±0.006a	0.288±0.008a
	Mv-R-R	13.523±0.390b	34.358±0.992b	47.882±1.382b	0.110±0.003a	0.156±0.005b	0.266±0.008a
	Fa-R-R	7.535±0.218c	30.550±0.882c	38.085±1.099c	0.042±0.002c	0.080±0.002c	0.122±0.004b

Ry-R-R: rice–rice–ryegrass cropping system; Mv-R-R: rice–rice–Chinese milk vetch cropping system; Fa-R-R: rice–rice cropping system with winter fallow.
Values are presented as mean ± SE (n = 3). Means in each column with different letters are significantly different at the $P < 0.05$ level.

Table 3. Double rice grain yield, global warming potentials (GWPs) of CH_4 and N_2O and per yield GWPs from rice fields under different cropping patterns.

Year	Treatment	CH_4 emission	N_2O emission	GWPs of CH_4	GWPs of N_2O	GWPs of CH_4 and N_2O	Double rice grain yield	Per yield GWPs CO_2
		(g m^{-2})	(g m^{-2})	(kg CO_2-eq ha^{-1})	(kg CO_2-eq ha^{-1})	(kg CO_2-eq ha^{-1})	(kg ha^{-1})	(kg kg^{-1})
2012	Ry-R-R	49.908±1.441a	0.264±0.008a	12494.38±360.68a	787.41±22.73a	13281.79±383.41a	13800.23±398.38a	0.96±0.03a
	Mv-R-R	43.634±1.260b	0.246±0.007a	10923.85±315.34b	733.58±21.18a	11657.44±336.52b	15089.30±435.59a	0.77±0.02b
	Fa-R-R	34.606±0.999c	0.110±0.003b	8663.66±250.10c	329.46±9.51b	8993.12±259.61c	14359.00±414.51a	0.63±0.02c
2013	Ry-R-R	54.512±1.574a	0.288±0.008a	13646.99±393.95a	859.81±24.82a	14506.80±418.76a	14738.87±425.47a	0.98±0.03a
	Mv-R-R	47.882±1.382b	0.266±0.008a	11987.20±346.04b	793.53±22.91a	12780.73±368.95b	14896.57±430.03a	0.86±0.02b
	Fa-R-R	38.085±1.099c	0.122±0.004b	9534.57±275.24c	364.64±10.53b	9899.22±285.77c	13625.16±322.60a	0.73±0.02c

Ry-R-R: rice–rice–ryegrass cropping system; Mv-R-R: rice–rice–Chinese milk vetch cropping system; Fa-R-R: rice–rice cropping system with winter fallow.
Values are presented as mean ± SE (n = 3). Means in each column with different letters are significantly different at the $P<0.05$ level.

N_2O emission

The emissions of N_2O are closely related to soil moisture, oxygen, temperature, content of soil organic matter and pH [4,11,17]. Great positive interaction has been reported between N_2O emission and green manure or chemical nitrogen fertilizer in early rice growing season [41]. In this study, we found that N_2O emission in the early rice growing season focused in the period of field drainage, and the Ry-R-R and Mv-R-R with winter cover crops had more N_2O emissions than Fa-R-R in both rice growing seasons (Fig. 3). N_2O emission from paddy field is promoted with the amount of straw returning via increasing soil denitrification, which provides the soil microbial substrates and energy for soil nitrification and denitrification process [42]. Different ranking of treatments in N_2O flux and total N_2O emission might be related to the decomposition rates of winter crop species during the rice growing season. In the late rice growing season, the total N_2O emissions of treatments Ry-R-R and Mv-R-R were significantly higher than Fa-R-R ($P<0.05$). This possibly results from that soil nitrification and denitrification process has been facilitated after the early rice straw returning through carbon and energy resource regulation (Table 1); a small amount of winter crop straw remains in the soil until the growing season of late rice; and tillage practice before late rice transplanting helps the incorporation of straws into soil, which may improve the soil nitrification and denitrification process.

Global warming potentials of CH_4 and N_2O

Global warming potential can be used as an index to estimate the potential effects of different greenhouse gases on the global climate system. Bhatia et al. [5] estimated that GWPs of rice–wheat system increased by 28% on full substitution of organic N by chemical N. Zhu et al. [43] reported that the highest GWPs was found in Chinese milk vetch incorporation in double cropping rice system, which was 21–325% higher than the other three treatments. In this study, the GWPs of CH_4, N_2O or both had different orders. For a comprehensive consideration, GWPs of both CH_4 and N_2O is more important to assess the effect of a farming system on climate warming. Therefore, it is necessary to make a combined estimate of global warming effects of CH_4 and N_2O emitted from each treatment. Thus, we introduced the GWPs and per yield GWPs into this study for global warming calculations. Although the global warming effect of N_2O is 12 times as large as that of CH_4, CH_4 emissions were nearly 370 times that of N_2O, resulting in the majority of GWPs originating from CH_4 (Table 3). Therefore, it is certain that the GWPs and per yield GWPs values for Ry-R-R and Mv-R-R were larger than Fa-R-R ($P<0.05$), due to their greater CH_4 emissions. But the GWPs of CH_4 and N_2O and per yield GWPs of Mv-R-R was significantly lower than Ry-R-R ($P<0.05$). It should be mentioned that, the cultivation of ryegrass, Chinese milk vetch and its incorporation is a process involving C accumulation from the atmosphere to the soil, while the production of synthetic nitrogen fertilizer consumes fossil fuels that release C and contribute to greenhouse gas emissions. Therefore, we recommend Mv-R-R pattern in double cropping rice areas in the Middle and Lower reaches of Yangtze River in China, which correspond to Chinese milk vetch as winter cover crop + double rice.

Conclusions

The emissions of CH_4 and N_2O from double cropping paddy fields were significantly enhanced by returning different winter cover crops. The effects on CH_4 and N_2O fluxes and emissions were different among treatments, and the emission characteristics varied greatly between early and late rice growing season. The orders of treatments were Ry-R-R>Mv-R-R>Fa-R-R for total emissions of CH_4 and N_2O during double rice seasons, and Ry-R-R>Mv-R-R>Fa-R-R for GWPs of total CH_4 and N_2O from double cropping paddy fields. Compared with Ry-R-R, Mv-R-R and Fa-R-R reduced CH_4 emission during rice growing seasons. The GWPs (based on CH_4 emission) under Mv-R-R and Fa-R-R was significantly ($P<0.05$) lower than Ry-R-R. Although the cumulative N_2O emission under Ry-R-R and Mv-R-R were higher than that from Fa-R-R ($P<0.05$), GWPs of N_2O was relatively low compared to that of CH_4. The GWPs (based on CH_4 and N_2O) of Mv-R-R and Fa-R-R is lower than that of Ry-R-R ($P<0.05$). Meanwhile, the GWPs of CH_4 and N_2O and per yield GWPs of Mv-R-R was significantly lower than Ry-R-R ($P<0.05$). Thus, Mv-R-R is beneficial in GHG mitigation and it can be extended as an excellent cropping pattern in double rice cropped regions.

Author Contributions

Conceived and designed the experiments: XPX GLY. Performed the experiments: HMT. Analyzed the data: HMT WGT. Contributed reagents/materials/analysis tools: JMS KW WYL. Wrote the paper: HMT.

References

1. Levy PE, Mobbs DC, Jones SK, Milne R, Campbell C, et al. (2007) Simulation of fluxes of greenhouse gases from European grasslands using the DNDC model. Agric Ecosyst Environ 121: 186–192.
2. Saggar S, Hedley CB, Giltrap DL, Lambie SM (2007) Measured and modeled estimates of nitrous oxide emission and methane consumption from a sheep-grazed pasture. Agric Ecosyst Environ 122: 357–365.
3. Hernandez-Ramirez G, Brouder SM, Smith DR, Van Scoyoc GE (2009) Greenhouse gas fluxes in an eastern corn belt soil: Weather, nitrogen source, and rotation. J Environ Qual 38: 841–854.
4. Wassmann R, Neue HU, Ladha JK, Aulakh MS (2004) Mitigating greenhouse gas emissions from rice-wheat cropping systems in Asia. Environ Devel Sustain 6: 65–90.
5. Bhatia A, Pathak H, Jain N, Singh PK, Singh AK (2005) Global warming potential of manure amended soils under rice-wheat system in the Indo-Gangetic plains. Atmospheric Environ, 39(37): 6976–6984.
6. Verge XPC, Kimp CD, Desjardins RL (2007) Agricultural production, greenhouse gas emissions and mitigation potentia l. Agric Forest Meteorol 142: 255–269.
7. Lokupitiya E, Paustian K (2006) Agricultural soil greenhouse gas emissions: A review of national inventory methods. J Environ Qual 35: 1413–1427.
8. Verma A, Tyagi L, Yadav S, Singh SN (2006) Temporal changes in N_2O efflux from cropped and fallow agricultural fields. Agric Ecosyst Environ 116: 209–215.
9. Liu H, Zhao P, Lu P, Wang YS, Lin YB, et al. (2008) Greenhouse gas fluxes from soils of different land-use types in a hilly area of South China. Agric Ecosyst Environ 124: 125–135.
10. Tan Z, Liu S, Tieszen LL, Tachie-Obeng E (2009) Simulated dynamics of carbon stocks driven by changes in land use, management and climate in a tropical moist ecosystem of Ghana. Agric Ecosyst Environ 130: 171–176.
11. Kallenbach CM, Rolston DE, Horwath WR (2010) Cover cropping affects soil N_2O and CO_2 emissions differently depending on type of irrigation. Agric Ecosyst Environ 137: 251–260.
12. Pandey D, Agrawal M, Bohra JS (2012) Greenhouse gas emissions from rice crop with different tillage permutations in rice-wheat system. Agric Ecosyst Environ 159: 133–144.
13. Yagi K, Minami K (1990) Effect of organic matter application on methane emission from some Japanese paddy fields. Soil Sci Plant Nutr 36: 599–610.
14. Yagi K, Tsuruta H, Kanda KI, Minami K (1996) Effect of water management on methane emission from a Japanese rice paddy field: Automated methane monitoring. Global Biogeochem Cycles 10: 255–267.
15. Nishimura S, Sawamoto T, Akiyama H, Sudo S, Yagi K (2004) Methane and nitrous oxide emissions from a paddy field with Japanese conventional water

management and fertilizer application. Global Biogeochem Cycles 18, GB2017, doi:10.1029/2003GB002207
16. Al-Kaisi MM, Yin X (2005) Tillage and crop residue effects on soil carbon and carbon dioxide emission in corn-soybean rotations. J Environ Qual 34: 437–445.
17. Yao Z, Zheng X, Xie B, Mei B, Wang R, et al. (2009) Tillage and crop residue management significantly affects N-trace gas emissions during the non-rice season of a subtropical rice-wheat rotation. Soil Biol Biochem 41: 2131–2140.
18. Groot TT, VanBodegom PM, Harren FJM, Meijer HAJ (2003) Quantification of methane oxidation in the rice rhizosphere using ^{13}C-labelled methane. Biogeochemistry 64: 355–372.
19. Yue J, Shi Y, Liang W, Wu J, Wang CR, et al. (2005) Methane and nitrous oxide emissions from rice field and related microorganism in black soil, northeast China. Nutr Cy Agroecosyst 73: 293–301.
20. Sethunathan N, Kumaraswamy S, Rath AK, Ramakrishnan B, Satpathy SN, et al. (2000) Methane production, oxidation, and emission from Indian rice soils. Nutr Cy Agroecosyst 58: 377–388.
21. Ma J, Xu H, Yagi K, Cai ZC (2008) Methane emission from paddy soils as affected by wheat straw returning mode. Plant Soil 313: 167–174.
22. Lee CH, Park KD, Jung KY, Ali MA, Lee D, et al. (2010) Effect of Chinese milk vetch (*Astragalus sinicus* L.) as a green manure on rice productivity and methane emission in paddy soil. Agric Ecosyst Environ 138: 343–347.
23. Bouwman AF (1998) Nitrous oxides and tropical agriculture. Nature 392: 866–867.
24. Cai ZC, Lanughlin RJ, Stevens RJ (2001) Nitrous oxide and dinitrogen emissions from soil under different water regimes and straw amendment. Chemosphere 42: 113–121.
25. Rittera WF, Scarborough RW, Chirnside AEM (1998) Winter cover crops as a best management practice for reducing nitrogen leaching. J Contam Hydrol 34: 1–15.
26. Hermawan B, Bomke AA (1997) Effects of winter cover crops and successive spring tillage on soil aggregation. Soil Tillage Res 44: 109–120.
27. McCracken DV, Smith MS, Grove JH, MacKown CT, Blevins RL (1994) Nitrate leaching as influenced by cover cropping and nitrogen source. Soil Sci Soc Am J 58: 1476–1483.
28. Barnes JP, Putnam AR (1983) Rye residues contribute weed suppression in no-tillage cropping systems. J Chem Ecol 9: 1045–1057.
29. Powlson DS, Prookes PC, Christensen BT (1987) Measurement of soil microbial biomass provides an early indication of changes in total soil organic matter due to straw incorporation. Soil Biol Biochem 19(2): 159–164.
30. Mitchell JP, Shennan C, Singer MJ, Peters DW, Miller RO, et al. (2000) Impacts of gypsum and winter cover crops on soil physical properties and crop productivity when irrigated with saline water. Agr Water Manag 45: 55–71.
31. Chang HL, Ki DP, Ki YJ, Muhammad AA, Dokyoung L, et al. (2010) Effect of Chinese milk vetch (*Astragalus sinicus* L.) as a green manure on rice productivity and methane emission in paddy soil. Agr Ecosyst Environ 138: 343–347.
32. Salmeróna M, Isla R, Cavero J (2011) Effect of winter cover crop species and planting methods on maize yield and N availability under irrigated Mediterranean conditions. Field Crops Res 123: 89–99.
33. Zheng X, Wang M, Wang Y, Shen R, Li J, et al. (1998) Comparison of manual and automatic methods for measurement of methane emission from rice paddy fields. Adv Atmos Sci 15: 569–579.
34. Parashar DC, Gupta PK, Rai J, Sharma RC, Singh N (1993) Effect of soil temperature on methane emission from paddy field. Chemosphere 26: 247–250.
35. Singh JS, Singh S, Raghubanshi AS, Saranath S, Kashyap AK (1996) Methane flux from rice/wheat agroecosystem as affected by crop phenology, fertilization and water lever. Plant Soil 183: 323–327.
36. SAS Institute (2003) SAS Version 9.1.2 2002–2003. SAS Institute Inc., Cary, NC.
37. Chidthaisong A, Obata H, Watanabe I (1999) Methane formation and substrate utilization in anaerobic rice soils as affected by fertilization. Soil Biol Biochem 31: 135–143.
38. Yu K, Bohme F, Rinklebe J, Neue HU, DeLaune RD (2007) Major biogeochemical processes in soils-A microcosm incubation from reducing to oxidizing conditions. Soil Sci Soc Am J 71: 1406–1417.
39. Yang X, Shang Q, Wu P, Liu J, Shen Q, et al. (2010) Methane emissions from double rice agriculture under long-term fertilizing systems in Hunan, China. Agric Ecosyst Environ 137: 308–316.
40. Zhang HL, Bai XL, Xue JF, Chen ZD, Tang HM, et al. (2013) Emissions of CH_4 and N_2O under different tillage systems from double-cropped paddy fields in Southern China. PLoS ONE 8(6): e65277. doi:10.1371/journal.pone.0065277.
41. Petersen SO, Mutegi JK, Hansen EM, Munkholm LJ (2011) Tillage effects on N_2O emissions as influenced by a winter cover crop. Soil Biol Biochem 43: 1509–1517.
42. Huang Y, Zou JW, Zheng XH, Wang YS, Xu XK (2004) Nitrous oxide emissions as influenced by amendment of plant residues with different C: N ratios. Soil Biol Biochem 36: 973–981.
43. Zhu B, Yi LX, Hu YG, Zeng ZH, Tang HM, et al. (2012) Effects of Chinese Milk Vetch (*Astragalus sinicus* L.) residue incorporation on CH_4 and N_2O emission from a double-rice paddy soil. J Integrative Agric 11(9): 1537–1544.

Engineering *Aspergillus oryzae* A-4 through the Chromosomal Insertion of Foreign Cellulase Expression Cassette to Improve Conversion of Cellulosic Biomass into Lipids

Hui Lin[1]☯, Qun Wang[2]☯, Qi Shen[2], Junwei Ma[1], Jianrong Fu[1]*, Yuhua Zhao[2]*

1 Institute of Environment Resource and Soil Fertilizer, Zhejiang Academy of Agriculture Science, Hangzhou, China, **2** Institute of Microbiology, College of Life Sciences, Zhejiang University, Hangzhou, China

Abstract

A genetic modification scheme was designed for *Aspergillus oryzae* A-4, a natural cellulosic lipids producer, to enhance its lipid production from biomass by putting the spotlight on improving cellulase secretion. Four cellulase genes were separately expressed in A-4 under the control of *hlyA* promoter, with the help of the successful development of a chromosomal genetic manipulation system. Comparison of cellulase activities of PCR-positive transformants showed that these transformants integrated with *celA* gene and with *celC* gene had significantly ($p<0.05$) higher average FPAase activities than those strains integrated with *celB* gene and with *celD* gene. Through the assessment of cellulosic lipids accumulating abilities, *celA* transformant A2-2 and *celC* transformant D1-B1 were isolated as promising candidates, which could yield 101%–133% and 35.22%–59.57% higher amount of lipids than the reference strain A-4 (WT) under submerged (SmF) conditions and solid-state (SSF) conditions, respectively. Variability in metabolism associated to the introduction of cellulase gene in A2-2 and D1-B1 was subsequently investigated. It was noted that cellulase expression repressed biomass formation but enhanced lipid accumulation; whereas the inhibitory effect on cell growth would be shielded during cellulosic lipids production owing to the essential role of cellulase in substrate utilization. Different metabolic profiles also existed between A2-2 and D1-B1, which could be attributed to not only different transgene but also biological impacts of different integration. Overall, both simultaneous saccharification and lipid accumulation were enhanced in A2-2 and D1-B1, resulting in efficient conversion of cellulose into lipids. A regulation of cellulase secretion in natural cellulosic lipids producers could be a possible strategy to enhance its lipid production from lignocellulosic biomass.

Editor: Jae-Hyuk Yu, University of Wisconsin - Madison, United States of America

Funding: This study was supported by the National Natural Science Foundation of China (31070079; 41271335), the High Technology Research and Development Program of China (863 Program) (2012AA06A203), and the National Key Technology R & D Program (2012BAC17B04), the Science and Technology Project of Zhejiang Province (2011C13016; 2013C3303), and the Environmental Science Project of Zhejiang Province (2012B006). The funders had no role in study design, data collection and analysis, decision to publish, or preparation of the manuscript.

Competing Interests: The authors have declared that no competing interests exist.

* Email: yhzhao225@zju.edu.cn (YZ); fujr@mail.zaas.ac.cn (JF)

☯ These authors contributed equally to this work.

Introduction

Biodiesel is widely recognized as a type of green fuel, which has advantages of low sulfur content, being non-toxic and biodegradable, lack of aromatics, and excellent lubricity [1]. Today, the shortage of less-cost oils, which can be used as a raw material in biodiesel production, has become a major obstacle in the development of biodiesel market. It is essential and urgent to explore sustainable and low-priced oil feedstocks for the wide use of biodiesel.

Single cells oils (SCOs) are triacylglycerols from renewable biomass, which have been well documented to be a promising feedstock for biodiesel production. SCOs can be produced by some oleaginous microorganisms, such as yeast, fungi, bacteria and microalgae, through microbial fermentation. A lot of efforts have been done all over the world to reduce the production cost during SCOs production and to explore cheap and abundant substrates for oleaginous microorganism cultivation. The social and economic benefits of producing SCOs from lignocellulosic biomass instead of crops are widely appreciated. Testing strategies to establish more efficient, less-cost and sustainable technologies based on microbial fermentation for SCOs production from lignocellulosic biomass is under current investigation by various start-up biotechnology companies and research centers [2,3,4,5,6]. Among these researches and technologies, direct microbial conversion of lignocellulosic biomass into lipids, as an example of consolidated bioprocessing (CBP), is considered to be economically attractive for "third generation" biofuel production due to its simple feedstock processing and low energy inputs [7]. To do so, hydrolysis of cellulose and hemicelluloses in biomass and

Table 1. Genes and plasmids used in this study.

Gene/plasmid	Function/comments	Source/reference
Gene		
celA (AO090026000102)[a]	Cellulase gene	[14,24]
celB (AO090010000314)[a]	Cellulase gene	[14,24]
celD (AO090012000941)[a]	Cellulase gene	This work
celC (AO090001000348)[a]	Cellulase gene	This work
eGFP	Enhanced Green Fluorescent Protein gene	[29]
hlyA promoter	Promoter region of *hlyA* (a hemolysin-like protein gene, AO090010000018[a])	[14]
amyA terminator	Terminator region of *amyA* (a taka-amylase gene, AO090023000944[a])	[29,30]
Plasmid		
pPTRI	A chromosomal integrating and *E. coli-Aspergillus* shuttle vector	[17]
pPTRI-Tamy	Vector pPTRI containing *amyA* terminator	This work

[a]DOGAN accession number (DOGAN, http://www.bio.nite.go.jp/dogan/).

production of valuable product, which were currently accomplished in different reactors or different organisms, are combined in a single process step in CBP. During the lipid production process in CBP, the substantial capital and material expense occurred in the enzymes preparation could be avoided.

Integration of cellulose hydrolysis and lipid biosynthesis by mixing cellulolytic enzymes with auxiliary nutrients in a single bioreactor is an example of CBP for lipid production from lignocellulosic materials [4,8]. However, the high cost of the cellulolytic enzymes is the primary hindrance in this cellulosic lipids production. Isolation or engineering of microorganisms, which could de-polymerize biomass polysaccharides to fermentable sugars efficiently and convert this mixed-sugar hydrolysate into cellular lipids effectively, is recognized to be another example of CBP to produce lipids directly from plant biomass. In recent years, microbial species capable of converting cellulose into microbial lipid directly, such as *Aspergillus oryzae* A-4 [9], *Microsphaeropsis* sp. [6], *Colletotrichum* sp. and *Alternaria* sp. [10] have been successfully isolated from natural environments. Several recombinant oleaginous CBP producers are also being reported after metabolic engineering to incorporate the feature of naturally biomass polysaccharides utilization into oleaginous cell factories [3,11,12]. In the report of Hetzler et al. [12], introduction of a cellobiose utilization pathway enabled *Rhodococcus opacus* PD630 to accumulate fatty acids up to 39.5±5.7% (wt/wt) of cell dry mass from cellobiose substrates. One-step production of lipids from birch cellulose was demonstrated to be feasible by co-fermentation of the recombinant cellulose degrading strain of *R. opacus* PD630 and the recombinant cellubiose-utilizing strain of *R. opacus* PD630 [3]. Nevertheless, the reports regarding construction of oleaginous CBP producers are still limited. It is indispensible to carry out more studies for the purpose of exploring new strains, improving currently available strains and optimizing the fermentation conditions.

In our previous work [9], a cellulolytic and oleaginous strain of *Aspergillus oryzae* A-4 was isolated from soil environments. The biochemical approaches by controlling the nutritional and cultivation conditions to channel metabolic flux into lipid biosynthesis has been performed to optimize the lipid production of *A. oryzae* A-4 under solid-state conditions with wheat straw and bran mixture as substrates. Although these attempts are promising, the current lipid conversion efficiency of *A. oryzae* A-4 is still far from the requirements from commercially viable production. Considering various aspects of a SCOs CBP scheme, here we sought to modify the cellulase secretion of strain A-4 for the exploration of recombinant CBP producers with enhanced lipid production from lignocelluloses. It is known that the biochemical event associated with the lipid production in direct microbial conversion system with cellulosic wastes as substrates is composed of three steps: polymers decomposition, monomer uptake, and lipid biosynthesis. In such cases, the hydrolytic activity of the microbes, which is related to the polymers decomposition, could associate closely with the storage lipid accumulation [9,13]. Thus, it could be speculated that insufficient available sugars for *A. oryzae* A-4 to take up is a major obstacle for its high lipid yielding. Heterogeneous cellulase expression in this study was controlled under the promoter from a hemolysin-like protein gene (*hlyA* gene promoter). It has been demonstrated that high-level protein expression under the control of *hlyA* promoter can not be repressed by glucose not only in solid-state culture of *A. oryzae* but also in liquid culture [14]. *A. oryzae* A-4 is a good platform that can be engineered for an efficient CBP scheme after only several modifications rather than de novo introduction of complete metabolism pathway. Actually, it is difficult to engineer a microorganism with all of the desired features necessary for an efficient CBP scheme so far [15].

Materials and Methods

Strains

The oleaginous fungus *A. oryzae* A-4, which was previously isolated by Lab of Microbiology, Zhejiang University [9], was used as a recipient for transformation and chromosomal DNA preparation in this work. *A. oryzae* A-4 were maintained on Czapek-Dox (CD) plate (0.6% $NaNO_3$, 0.052% KCl, 0.152% KH_2PO_4, 0.052% $MgSO_4 \cdot 7H_2O$, 1% glucose and pH 6.5) at 4°C before use. The *Escherichia coli* DH5α used for plasmid recovery and cloning experiments was grown in Luria-Bertani broth (LB).

Plasmids construction

The genes and plasmids used in this study are summarized in Table 1. Primers were listed in Table S1. The construction procedure of the recombinant expression vector for *A. oryzae* A-4 was described below. The DNA fragment of *amyA* terminator was first generated by polymerase chain reaction (PCR) from the genomic DNA of *A. oryzae* ATCC 42149. The vector pPTRI-Tamy was then constructed by the ligation of the *amyA* terminator fragment with the parent vector pPTR I (*Takara* Bio Inc.). Construction of plasmids expressing target genes under the control of the *hlyA* promoter was performed by using overlap PCR, which comprised of two PCR steps. In the first PCR step, the target gene fragment and its corresponding *hlyA* promoter fragment was generated by PCR, respectively. For each molecule, the primer at the end to be joined is constructed such that it has a 5′ overhang complementary to the end of the other molecule. In the second step, the target gene fragment and its corresponding *hlyA* promoter fragment are mixed, and a PCR was carried out with only the primers for the far ends. The overlapping complementary sequences introduced will serve as primers and the two sequences will be fused. To test if the expression vector could be used for foreign protein expression, eGFP expression vector was first constructed for transformation. To obtain the eGFP expression vector, the primer set of FP2-GFP-overlap/RP2-GFP-Rec and FP1-hlyA-Rec/RP1-hGFP-overlap was used for the isolation of the *eGFP* fragment (AGX13949) from vector PET-eGFP (previously constructed in our lab) and its corresponding *hlyA* promoter fragment from *A. oryzae* ATCC 42149 genomic DNA, separately. Similarly, the primer sets of FP2-CelA-overlap/RP2-CelA-Rec + FP1-hlyA-Rec/RP1-hCelA-overlap; FP1-hlyA-Rec/RP1-hCelB-overlap + FP2-CelB-overlap/RP2-CelB-Rec; FP1-hlyA-Rec/RP1-hCelD-overlap + FP2-CelD-overlap/RP2-CelD-Rec and FP1-hlyA-Rec/RP1-hCelC-overlap + FP2-CelC-overlap/RP2-CelC-Rec were designed for the construction of CelA, CelB, CelD and CelC expression vectors, respectively. All DNA fragments for the construction of CelA, CelB, CelD and CelC expression vectors were cloned from *A. oryzae* ATCC 42149 genomic DNA. The ligation of the vector-specific fragment with the vector pPTRI-Tamy was achieved by using CloneEZ® PCR Cloning Kit (GenScript). In the final construct, the separate target gene (i.e. *celA*, *celB*, *celD*, *celC* or *eGFP*) was cloned downstream of the *hlyA* promoter and upstream of the *amyA* terminator in vector pPTR I. All PCRs were carried out with PrimeSTAR HS DNA Polymerase (Takara Bio Inc.). Nucleotide sequences of constructed plasmids were sequenced in Shanghai Sangong Co., Ltd.

Protoplasts preparation, transformation and PCR checking

The pyrithiamine-resistant transformation system was applied for the strain A-4, and each expression vector was chromosomally integrated into A-4 according to Protoplast-PEG method as previously described but with slight modification [16,17]. Spores of *A. oryzae* A-4 were inoculated on PDA plates at 30°C for 3–4 days. 1 ml of spores suspension with a spore concentration of 1×10^7 ml^{-1} was then collected from the plates and inoculated into 100 ml of CD liquid medium. Incubation was performed in an orbital shaker at 180 rpm and 30°C for 20–24 hours. The germinated spores were washed with sterilized distilled water and then resuspended in 5 ml of protoplast forming solution (0.8 M NaCl, 10 mM Na phosphate buffer, pH 6.0) containing Yatalase (Takara Bio Inc.) at a concentration of 20 mg ml^{-1}. The suspension was incubated at 30°C for 2–2.5 hours to allow the release of the protoplasts. The protoplasts were subsequently collected and washed twice with 0.8 M NaCl. The collected protoplasts were resuspended in S1 (0.8 M NaCl, 10 mM $CaCl_2$, 10 mM Tris-HCl, pH 8.0) plus 1/5 of the final volume of S2 (40% (w/v) PEG 4000, 50 mM $CaCl_2$, 50 mM Tris-HCl, pH 8.0), at a concentration between 1×10^7 and 1×10^9 protoplasts ml^{-1}. 0.2 ml of the protoplast suspension were mixed with 10–15 µl (1–10 µg) of plasmid and then maintained on ice for 30 min. After that, 1 ml of S2 was added and the mixture was incubated at room temperature for 15 min. The mixture was diluted with 8.5 ml of S1 and the protoplast was collected by centrifugation before resuspended in 0.2 ml of S1. The protoplast suspension was poured as an overlay on CD selection soft plates containing 0.1 mg l^{-1} pyrithiamine hydrobromide (PT-h, Sigma). The plates were incubated at 30°C for 5–7 days in order to allow the regeneration of the protoplasts. Spores from single colonies were collected as described above and stored at −80°C for further analysis. PCR checking was carried out for initial screening of positive transformants. In most cases, the fusion fragments consisting of *hlyA* promoter and target genes could be obtained from the chromosomal DNA of the positive transformant by using primers in overlap PCR for the far ends (Table S1), while no products could be observed from that of host strain using the same primers.

Microscopic observations

Conidia of recombinant strains harboring the eGFP expression cassette were inoculated into CD selection plates with 0.1 mg l^{-1} PT-h. Cells were grown at 30°C for 3 days, after that the fungal hyphae was placed onto a slide for observation under fluorescence to measure the fluorescence expression of *eGFP* transformants. Lipid accumulation of the wild-type A-4 can be initiated on a nitrogen-limited solid medium as described in our previous work [9]. After 4–6 days' cultivation, the fungal mycelia was taken and mixed with Nile red solution (final concentration in acetone: 0.25 µg ml^{-1}). Fluorescence observation was performed immediately on the mycelia stained with Nile red. All the observations were performed by using an Eclipse 80i microscope (Nikon) equipped with Plan APO VC 100X/1.40 oil objective.

Southern blot analysis

The purified genomic DNA was digested with *Hind* III and separated by agarose gel electrophoresis. The genomic DNA fragments were transferred onto a positively charged nylon membrane, Hybond-N+ (Amersham Pharmacia Biotech). A 0.7-kb PCR amplified fragment containing the open reading frame (ORF) of Ampicillin resistance gene (*Amp*^{r}) was used as a probe. Probe labeling and blot detection were performed using the DIG High Prime DNA Labeling and Detection Starter Kit II (Roche, Mannheim, Germany) according to the manufacturer's instructions.

Fermentation experiments

Submerged fermentation (SmF) for SCOs production was carried out in basal medium [18] containing (g L^{-1}): KH_2PO_4, 2.0; $(NH_4)_2SO_4$, 2.1; $MgSO_4\cdot7H_2O$, 0.3; $CaCl_2\cdot7H_2O$, 0.3; $MnSO_4\cdot H_2O$, 0.00156; $ZnSO_4\cdot7H_2O$, 0.0014; $CoCl_2\cdot6H_2O$, 0.00266; yeast extract, 0.5; pH 5.5; supplemented with 2% of carbon source. Carbon sources used for SmF experiments involved glucose, maltose, avicel and wheat straw. All the SmF experiments were carried out with four replications and at 30°C and 180 rpm for 4 days. The SSF medium consisted of 3.6 g wheat straw, 0.4 g wheat bran, and 4 ml basal medium in Petri dishes (U = 9 cm). After sterilization, the media was cooled down and inoculated with 0.5 ml inoculums per gram of dry mass and static cultivated at

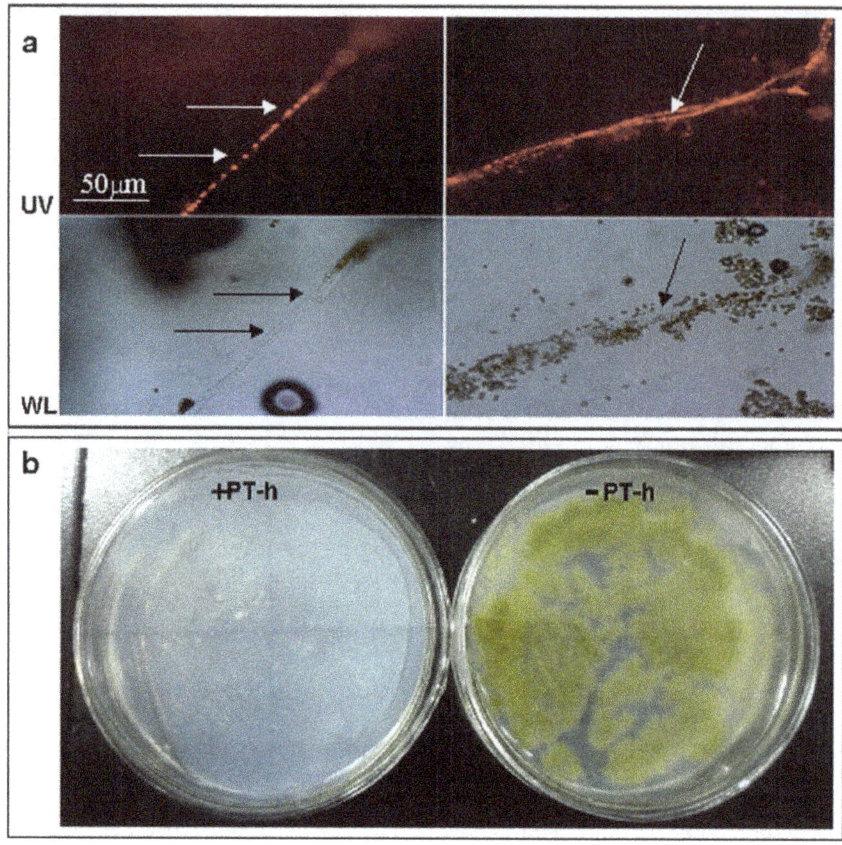

Figure 1. Lipid droplets accumulated in the mycelium of the reference strain A-4 (A) and the cell growth of strain A-4 grown on CD-plate without (-PT-h) and with 0.1 mg l^{-1} PT-h (+PT-h) (B). (A) Strain A-4 was cultivated on wheat straw and bran mixture for 6 days. The Nile red stained fermented products were then microscopic observed under ultraviolet (UV) and white light (WL); (B) Inhibitory effects of PT-h on the cell growth of A-4. Incubation was performed at 30°C for 5 days.

30°C, 50%–80% humidity. All the SSF experiments were carried out with three replications. Spore suspension (1×10^6 spore ml^{-1}) was used as the inoculums in this work. Wheat straw and bran were dried at 80°C for 4 h and milled to 20–40 meshes before use.

Ethics statements. Wheat straw, collected from Bozhou city, Anhui province (GPS coordinates: 33.923136, 115.83361) was the agricultural waste, and no specific permissions were required for the sampling activity. The field studies did not involve endangered or protected species.

Analytical methods

The supernatants of the fermented SmF media were collected for enzyme assay after centrifugation, while the solid residues were gathered and dried for a lipid assay [19] and a biomass assay based on glucosamine estimation of the fungal cell wall [20]. The total lipid in SSF was extracted and determined as described in our previous report [9]. The percentage of oil in relation to the dry matter (w/w) was expressed as lipid content. The fungal biomass in SSF was determined by the same method [20] as described in SmF, and the biomass from SSF experiments was depicted as mg glucosamine per gram dry substrate (mg gds^{-1}). For enzyme assay, the crude enzymes should be first extracted from the solid-state fermented products. Extraction was performed at 37°C for 2 hours by mixing 1 gram of fermented residue with 20 ml of citric acid buffer (pH 4.8). The cellulase activity in the crude enzyme extract was measured according to the method of Ghose [21] by determination of filter paper cellulase (FPA) activity and carboxymethylcellulase (CMCase) activity. The release of reducing sugars was assayed using 3, 5-dinitrosalicylic acid (DNS) method. One unit of FPA activity (IU) and CMCase activity (IU) was defined as the amount of enzymes required to release 1 μmol of substrate per minute under assay conditions. The protein concentration in the crude enzyme extract was quantified by the Bradford assay [22]. Loss in dry matter (LDM) in SSF was obtained using the following equation:

$$\%\text{LDM} = \frac{\text{Initial dry weight of substrates} - \text{dry weight of fermented products}}{\text{Initial dry weight of substrates}} \times 100\%$$

Results and Discussion

Development of a system for chromosomal manipulation in *A. oryzae* A-4

In order to construct a transformation system for genetic manipulation, it is necessary to screen a drug exhibiting effective inhibitory effects on host strains and to obtain its corresponding resistance gene. Compared with bacteria, filamentous fungi have a stronger tolerance to many types of drugs [23], which would result in some difficulties for their genetic manipulation. Pyrithiamine is

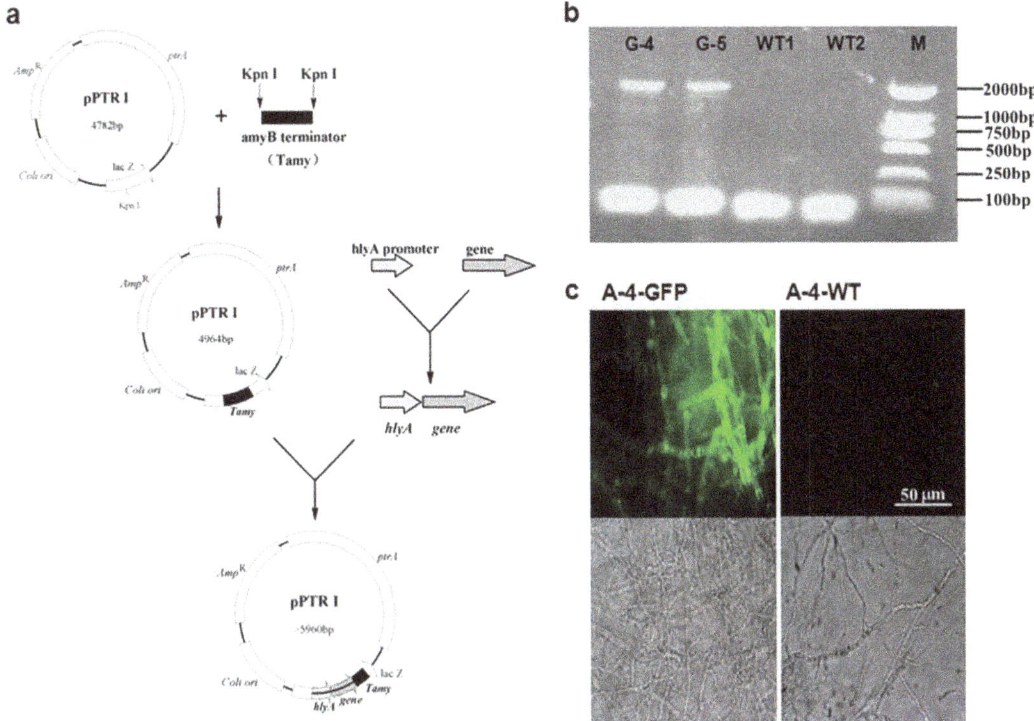

Figure 2. Recombinant strains engineered to express eGFP protein. (A) Construction of the recombinant expression vector for *A. oryzae* A-4; (B) Isolation of positive transformants by PCR checking; (C) Microscopic observations of mycelium samples exposed to ultraviolet (UV) and white light (WL), which collected from the transformant G-4 (A-4-GFP) and the wild-type A-4 (A-4-WT), respectively.

a potent antagonist of thiamine. In 2002, Japanese researchers found that cell growth of a majority of detected filamentous fungi, including *Aspergillus oryzae* RIB138, *Aspergillus niger* IAM2561, *Aspergillus nidulans* FGSC89 and *Trichoderma reesei* IFO31326, could be suppressed when more than 0.1 mg l^{-1} of pyrithiamine was added into the culture medium [17]. Kubodera et al. [17] then constructed the *E. coli-Aspergillus* shuttle vector pPTR I and pPTR II, and the target filamentous fungi were found to be insensitive to pyrithiamine after the transformation of pPTR I or pPTR II. In this work, pyrithiamine hydrobromide (PT-h) was used instead of pyrithiamine as the selection pressure to construct a transformation system for *A. oryzae* A-4, which can accumulate a high quantity of cellular lipids using cellulosic materials as the major substrate (Fig. 1a). As shown in Fig. 1b, the cell growth of wild-type A-4 presented well on the CD solid medium without PT-h but was completely inhibited with 0.1 mg l^{-1} PT-h. The result indicated that PT-h exhibited a similar result to pyrithiamine. 0.1 mg l^{-1} PT-h can be used as selection pressure for *A. oryzae* A-4.

The eGFP expression vector of pPTRI-Tamy-hlyA-eGFP, which comprised of the *hlyA* promoter, the *amyB* terminator, the pyrithiamine resistant gene (*ptrA*) and the target *eGFP* gene, was constructed according to the procedure described in Fig. 2a. Transformants harboring eGFP expression cassette, which exhibited well growth performances on the CD solid media with PT-h, were subsequently isolated and used for PCR checking. No PCR products of the expect size, namely the fusion fragment consisting of *hlyA* promoter and *eGFP* gene (*PhlyA-eGFP* fusion gene), can be amplified from the genomic DNA of strains without the integration of the vector pPTRI-Tamy-hlyA-egfp (Fig. 2b). The transformants with PCR products amplified from their genomic DNA were considered as the positive transformants (Fig. 2b). Further investigation of the PCR-positive transformants using fluorescence microscopy (Fig. 2c) showed that an obvious eGFP-fluorescence could be observed in the mycelia of transformant G-4 but hardly found in the mycelia of the reference strain A-4. The result demonstrated that the above-described manipulation procedure can be used for the chromosomal genetic manipulation in *A. oryzae* A-4. Indeed, a proper genetic manipulation system is necessary not only for heterogeneous protein expression in *A. oryzae* A-4 but also for the regulatory mechanisms elucidation during cellulosic lipids accumulation.

Construction of recombinant strains with increased cellulase activities

According to the procedure described in Fig. 2a, four recombinant vectors separately harboring *celA* gene, *celB* gene, *celD* gene and *celC* gene were constructed and introduced into the host strain *A. oryzae* A-4. For each recombinant vector, more than ten transformants were selected after several generations. The genomic DNAs from all transformants and the wild-type strain A-4 were extracted and used as templates for PCR checking to demonstrate whether the cellulase expression cassette was successfully integrated into the host genome. The size of PCR products (*PhlyA-celA/celB/celD/celC* fusion gene) obtained from the genomic DNA of transformants harboring *celA* gene, *celB* gene, *celD* gene and *celC* gene were approximately 1.8 kb, 2.3 kb, 2.6 kb and 2.4 kb, respectively; no PCR products could be obtained from wild-type A-4 (Fig. S1). Overall, five positive *celA* transformants, five positive *celB* transformants, six positive *celD* transformants and six positive *celC* transformants were isolated for further investigations owing to the PCR results.

SmF of wheat straw by these isolated transformants was subsequently performed. Cellulase activity determination was

Table 2. Cellulase activities, lipid production and cell growth of A2-E, A2-2, B11-C2, D1-B1 and D1-2(3).

Strains		Cellulase activity			Lipid production			
					SmF*		SSF**	
		FPAase (IU ml^{-1})	CMCase (IU ml^{-1})	Protein (μg ml^{-1})	Lipid yield (g l^{-1})	Biomass (mg l^{-1})	Lipid yield (g l^{-1})	LDM (%)***
Control[+]		5.01 (0.27)	19.71 (0.90)	89.64 (6.36)	-	-	-	-
WT[++]		5.20 (0.43)	32.94 (0.65)	102.17 (4.43)	0.63^{b} (0.04)	15.01 (0.27)	36.14^{D} (3.98)	26.05 (2.33)
celA	A2-E	7.32 (0.57)	134.69 (3.06)	141.45 (3.56)	0.69^{b} (0.07)	23.33 (3.33)	35.14^{D} (0.83)	32.05 (0.63)
	A2-2	8.02 (0.27)	312.56 (13.23)	181.93 (6.39)	1.47^{a} (0.05)	23.24 (3.02)	53.20^{B} (0.29)	31.20 (0.71)
celB	B11-C2	6.38 (0.17)	237.79 (16.18)	141.69 (7.73)	0.74^{b} (0.04)	17.50 (1.70)	45.62^{C} (5.95)	33.50 (2.68)
celC	D1-B1	8.29 (0.34)	107.33 (13.44)	146.27 (7.94)	1.27^{a} (0.24)	23.16 (3.60)	63.30^{A} (4.63)	25.85 (1.34)
	D1-2(3)	7.55 (0.51)	137.87 (14.69)	121.21 (0.17)	0.74^{b} (0.15)	13.73 (0.82)	41.26^{C} (1.31)	27.00 (2.12)

Incubation for cellulase activity measurement was conducted under submerged (SmF) conditions using wheat straw as substrate for 4 days. Lipid production and cell growth of transformants were determined under both submerged and solid-state conditions.

[+]The control transformant introduced with the negative vector pPTRI without cellulase expression cassette.

[++]The wild-type *A. oryzae* A-4.

*Submerged fermentation from wheat straw after 4 days.

**Solid state fermentation from wheat straw and bran mixture after 4 days.

***Loss in dry matter (LDM).

One-way ANOVA is used to test for differences: a, b means $p<0.05$; A, B, C, D means $p<0.05$; Values in brackets are standard errors.

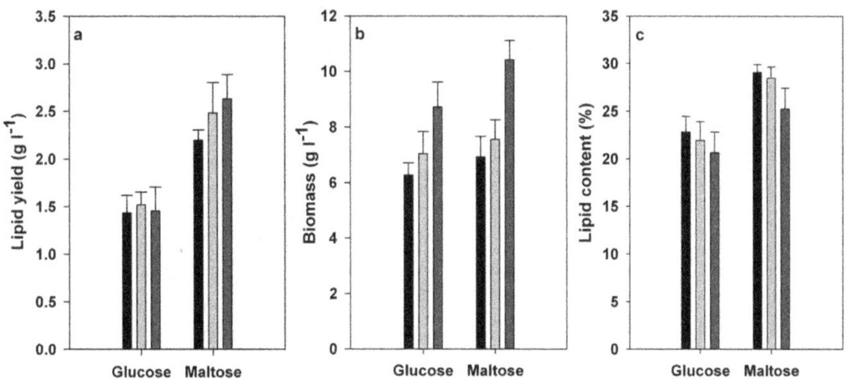

Figure 3. The lipid production (A, C) and cell growth (B) of A2-2 (black column), D1-B1 (gray column) and wild-type A-4 (dark gray column) in SmF experiments using glucose and maltose based media, respectively.

carried out to investigate whether the introduction of exogenous cellulase expression cassette will alter or enhance the cellulase production in PCR-positive transformants (Table S2). Among 22 positive transformants, the *celA* transformants A2-E and A2-2, the *celB* transformant B11-C2 and the *celC* transformants D1-B1 and D1-2(3) exhibited apparently and significantly ($p<0.05$) higher

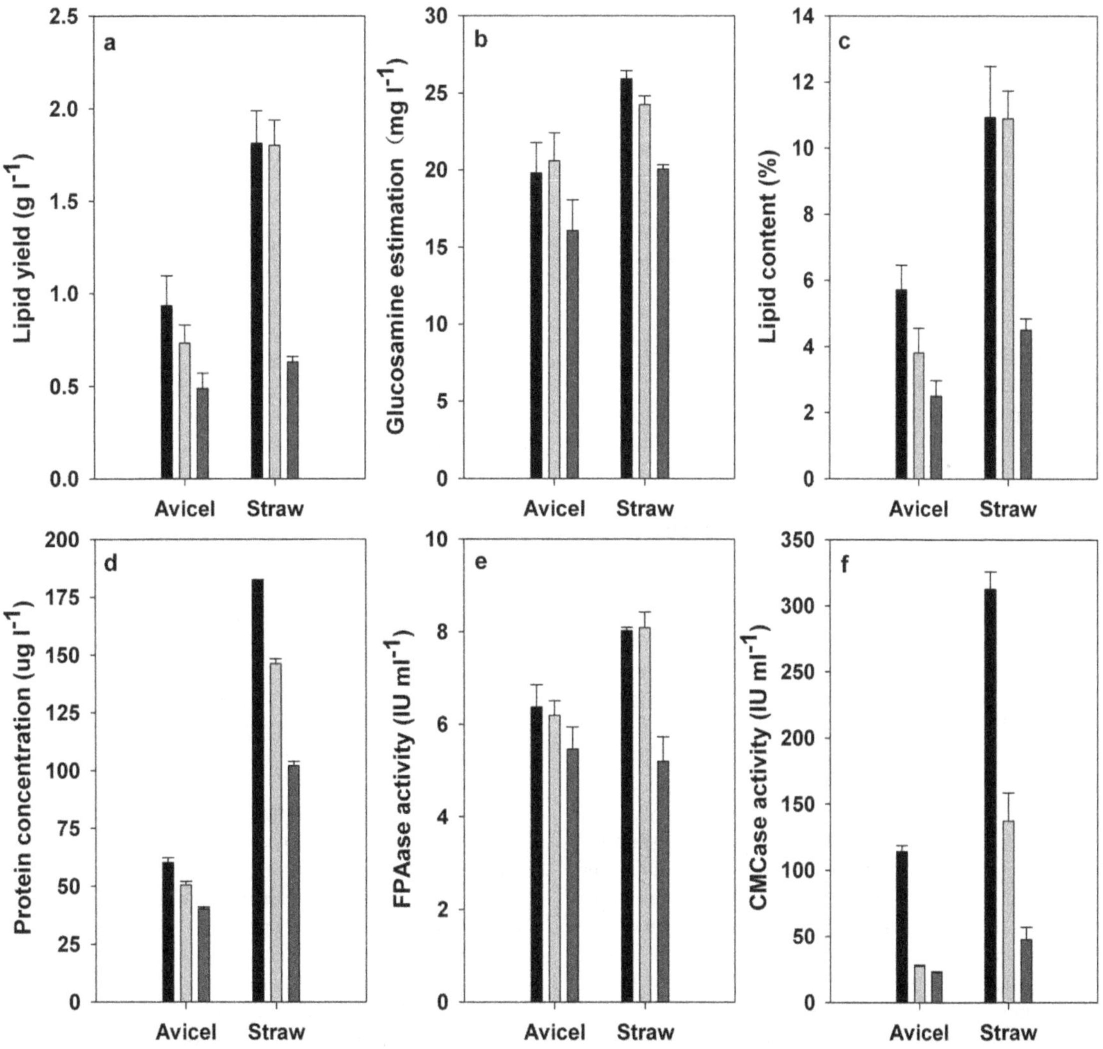

Figure 4. The lipid production (A, C), cell growth (B), cellulase secretion (E, F) and extracellular protein secretion (D) of A2-2 (black column), D1-B1 (gray column) and wild-type A-4 (dark gray column) in SmF experiments using avicel and straw based media, respectively.

cellulase activities than the wild-type A-4 (Table 2). The FPAase activity of the above-mentioned five transformants increased by 22.69%–59.42% compared with that of wild-type strain. CMCase activity of the above-mentioned recombinant transformants was 2.26–8.49 times higher than that of wild-type A-4. Consistently, the fermented media of A2-E, A2-2, B11-C2, D1-B1 and D1-2(3) exhibited higher extracellular protein concentrations than that of wild-type A-4 and the control transformant harboring blank vector pPTR I.

Differences in cellulase secretion among 22 PCR-positive transformants harboring different cellulase genes were also concluded based on the results of Table S2. Both *celA* gene and *celB* gene encode endoglucanases, a kind of cellulase tending to hydrolyze amorphous cellulose such as carboxymethyl cellulose effectively [24]. The over-expression of endoglucanase offered the *celA* transformants and the *celB* transformants significantly ($p<0.05$) higher CMCase activities than other detected strains. It could be noted that insignificant ($p>0.05$) difference in CMCase activity was found between the *celA* transformants and the *celB* transformants. However, only the introduction of *celC* expression cassette into A-4 strain gave a promising positive effect on cellulase production though both *celD* gene and *celC* gene encode cellubiohydralase. Post Hoc comparisons using the Fisher LSD test further demonstrated that transformants integrated with *celA* gene and with *celC* gene exhibited significantly ($p<0.05$) higher average FPAase activity than other strains including *celB* transformants, *celD* transformants, the reference strain A-4 and the negative control transformant (transformed with blank vector pPTR I). Since the total cellulase activity is represented by FPAase activity in most cases, it could be speculated that it is more possible to get recombinant strains with improved cellulase secretion by introduction of *celA* gene and *celC* gene rather than *celB* gene and *celD* gene into *A. oryzae* A-4.

Cellulosic lipids production of transformants with increased cellulase activity

The cellulosic lipids accumulating abilities of transformants A2-E, A2-2, B11-C2, D1-B1 and D1-2 (3) were assessed by using SmF media with wheat straw as the main substrate and SSF media with wheat straw and bran mixtures as the main substrate. Fermentations were performed for 4 days and the lipid production results were shown in Table 2. Under the SmF condition, the lipid yields of A2-2 and D1-B1 were 1.47 g l^{-1} and 1.27 g l^{-1}, both of which were significantly ($p<0.05$) higher than those of other transformants and the wild-type A-4. Likewise, almost all transformants exhibited a significantly ($p<0.05$) higher lipid yield than the wild-type A-4 under the SSF condition, except A2-E. Among these transformants, D1-B1 exhibited a most robust lipid production and released a lipid yield of 63.30 mg gds^{-1} after 4 days' fermentation under the SSF condition. In our previous report, wild-type *A. oryzae* A-4 gave a maximum lipid yield of 62.87 mg gds^{-1} after 6 days' fermentation under an optimized SSF condition. The fermentation period of D1-B1 was shortened compared with that of the wild-type A-4. Besides, the lipid production of D1-B1 here was not under the optimized SSF condition, so it is potentially possible to be further improved after biochemical optimization. In conclusion, A2-2 and D1-B1 produced 1.33 times and 1.02 times higher amounts of lipids than wild-type A-4 after SmF of cellulose, while the lipid yields of A2-2 and D1-B1 increased by 35.22% and 59.57% compared to wild-type strains after SSF of wheat straw and bran mixtures. A2-2 and D1-B1 have an enhanced cellulosic lipid production and can be selected for subsequent characterization.

Variability in metabolism associated to the introduction of cellulase gene

SmF using monosaccharides based carbon sources. For traditional lipid production from either glucose or maltose, the cellulase over-expression exhibited little positive effects on the increase in lipid yields; the lipid production of A2-2 and D1-B1 on maltose media were ever poorer than that of wild-type A-4 (Fig. 3a). As shown in Fig. 3b, the overall biomass yields for both A2-2 and D1-B1 were lower than that for wild-type A-4 when cultured on glucose media and maltose media, which suggested that the introduction and expression of cellulase gene did inhibit the cell growth of the transformants cultured on *monosaccharide* based media. The metabolic burdens imposed by the cellulase over-expression on host strains could be expected to be a main cause for the reduced cell growth as cellulase expression is unnecessary for the cell growth of strain A-4 using glucose or maltose as carbon source. The simlar metabolic burdens imposed on host strains by the over-expression of foreign proteins have been previously reported by other researchers [25,26].

The lipid content (%) presented the ratio of lipid yield in Fig. 3a and the corresponding biomass in Fig. 3b. Both A2-2 and D1-B1 had higher lipid contents than wild-type A-4 (Fig. 3c), which indicated that their lipid yields per unit of biomass have been improved. Therefore, it can be concluded that A2-2 and D1-B1 had a more robust lipid accumulation than wild-type A-4 when using glucose media and maltose media. The increased lipid accumulation in A2-2 and D1-B1 suggested that the insertion and over-expression of foreign cellulase gene affected the flux of carbon sources, and more carbon sources would be diverted into lipid biosynthesis.

SmF using cellulose based carbon sources. The fermentation using cellulase based substrates did exhibit a different profile from that using monosaccharides based carbon sources. As shown in Fig. 4, both A2-2 and D1-B1 yielded a significantly higher amount of lipids than wild-type A-4. The introduction and expression of cellulase gene in strain A-4 enhanced not only lipid accumulation (Fig. 4c) but also the cell growth (Fig. 4b). The extracellular protein concentration, FPAase activity and CMCase activity from fermentation trails of A2-2 and D1-B1 by using both avicel and wheat straw were found to be significantly ($p<0.05$) and apparently higher than those from the trial of wild-type A-4 (Fig. 4d, 4e and 4f). The A2-2 with endoglucanase overexpressed exhibited higher CMCase activity than D1-B1 with cellubiohydralase overexpressed (Fig. 4f). However, insignificant ($p<0.05$) differences in the FPAase activities were found between D1-B1 and A2-2 (Fig. 4c). It is well-known that sufficient available monosaccharide is indispensible for robust cell growth. Cellulase is necessary for cellulose decomposition in CBP from lignocellulose. Thus, it could be suggested that the enhanced cell growth and lipid production (Fig. 4a) in A2-2 and D1-B1 were attributed to their improved cellulase secretion owing to the over-expression of foreign cellulase. Moreover, we noted that positive effects of the cellulase gene integration on fermentation were more apparent in wheat straw trails than that in avicel trails. It appeared that the improving effects of cellulase over-expression on the lipid production were more effective when using wheat straw as substrates, although more researches should be performed for this conclusion. Overall, our work indicates the promising utilization of A2-2 and D1-B1 for lipid production from wheat straw.

SSF using wheat straw and bran mixtures. As shown in Fig. 5, the ferment strains exhibited cellulase secretion and lipid production both in the order of D1-B1>A2-2>wild-type A-4. SSF profiles using wheat straw and bran mixtures as substrates were

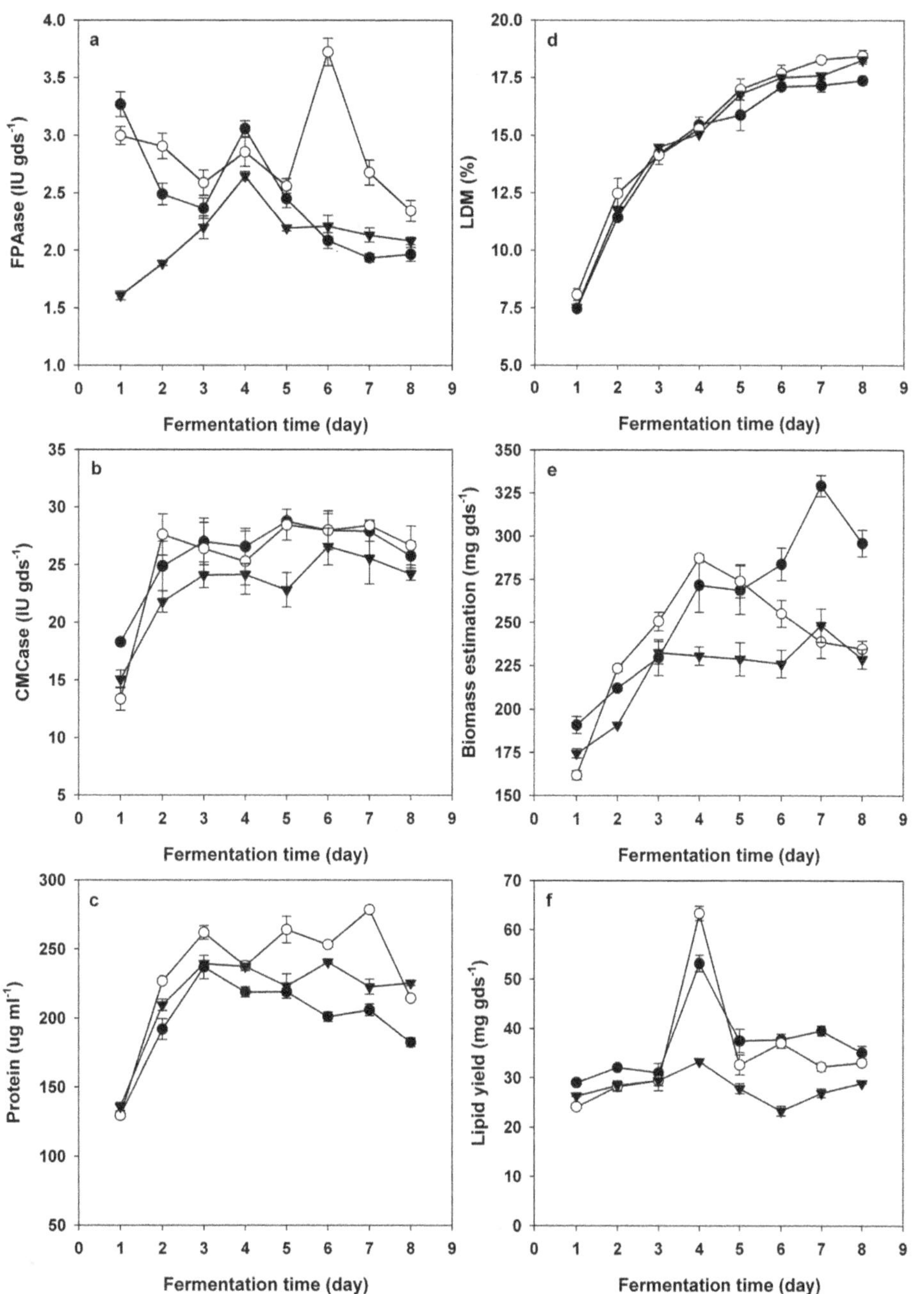

Figure 5. Solid-state fermentation of wheat straw and bran mixture by D1-B1 (blank circle), A2-2 (closed circle) and wild-type A-4 (closed triangle). (A) Time courses of the filter paper activity (FPAase activity); (B) Time courses of the CMCase activity; (C) Time courses of the extracellular protein concentration; (D) Time courses of the loss in dry matter (LDM); (E) Time courses of the cell growth; (F) Time courses of the lipid yield.

characterized in detail for A2-2, D1-B1 and wild-type A-4 as follows.

In terms of FPAase activity (Fig. 5a), introduction of cellulase gene into cells changed the cellulase secretion in microbes. In the early stage of fermentation (within 4 days), both A2-2 and D1-B1 showed an extremely high cellulase activity at the first 24 hours, which then decreased gradually. Different from that observed in A2-2 and D1-B1, the cellulase activity of the wild-type A-4 increased slowly and gradually throughout the early fermentation stage. It seemed that the introduced cellulase genes in both A2-2 and D1-B1 gave a robust expression in the early fermentation stage; this may be the reason why A2-2 harboring endoglucanase gene showed a higher CMCase activity than either D1-B1 harboring cellubiohydralase gene or wild-type A-4 at the first 24 hours (Fig. 5b). The strong cellulase over-expression that tends to occur in the early fermentation stage such as within 24 hours could be attributed to the characteristics of the *hlyA* promoter in most cases. The similar phenomenon had been reported by Bando et al. [14], who firstly showed that the endoglucanase gene expressed under the control of the *hlyA* promoter would give a continued increase of endoglucanase activity within the 48 hours but a gradual decrease after 48 hours. Besides, the introduction of *celC*

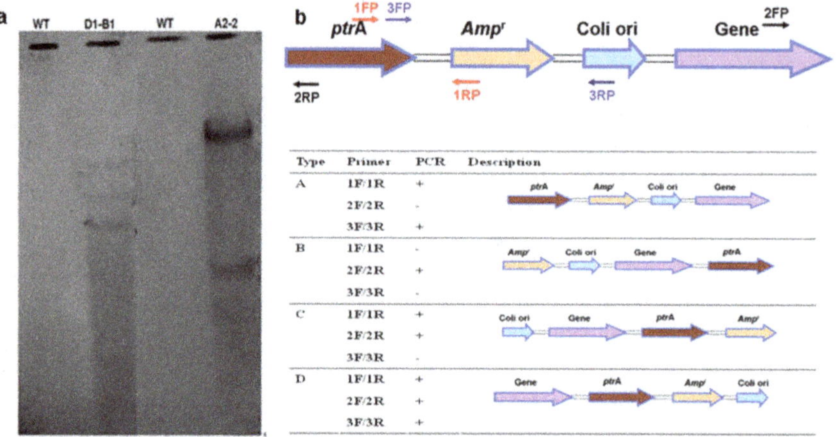

Figure 6. Preliminarily study of cellulase gene integration profiles in A2-2 and D1-B1, respectively. (A) Southern blot analysis. A part of *Amp*^{r} gene (0.7-kb) was amplified by PCR and used as the probe. (B) Primers designed for the insertion analysis of the target gene in A2-2 (*celA*) and D1-B1 (*celC*), and the theoretical results of PCR using the designed primers.

gene attributed to a particular phenomenon in FPAase activity of D1-B1 that FPAase activity of D1-B1 showed a sharp increase during the later fermentation stage from the 5th day to the 7th day. Thus, the detail metabolic profiles between A2-2 harboring *CelA* gene and D1-B1 harboring *CelC* gene has become different though the cellulase overexpression led to increased cellulase secretion and lipid production in both A2-2 and D1-B1.

The extracellular protein concentrations were in the order of D1-B1>WT>A2-2 (Fig. 5c). According to those found in fermentations by using monosaccharide based carbon source, it is supposed that the lower extracellular protein concentration in A2-2 could attribute to its reduced endogenous protein secretion due to the metabolic burden imposed by the over-expression of foreign cellulase genes.

There is no apparent difference in LDM among the three detected strains (Fig. 5d). However, A2-2 and D1-B1 have higher growth rates than the reference strain A-4. The biomass of wild-type strains increased slowly with the proceeding of the time and decreased after the 7th day, while the biomass of D1-B1 increased rapidly and reached the maximum on the 4th day. Different from that of both D1-B1 and the reference strain, the cell growth of A2-2 increased rapidly and stably with the increase of fermentation time. Although the growth rate of A2-2 was lower than D1-B1, the maximum biomass of A2-2 was found to be higher than that of D1-B1 (Fig. 5e). A2-2, D1-B1 and the reference strain showed similar changing trends of lipid yields, and the lipid yields of all strains increased during the early fermentation and reached the maximum on the 4th day. Furthermore, we noted that that the fermentation period from 72 h to 96 h might be important for the lipid accumulation (Fig. 5f). The higher increasing rates of lipid yield from 72 h to 96 h related closely with the final lipid yields in A2-2 and D1-B1.

Generally, the transformation and over-expression of foreign cellulase gene attributed to the enhancement of lipid accumulation in A2-2 and D1-B1. Although over-expression of foreign cellulase gene in strain A-4 imposed metabolic burden on cells, it offers A2-2 and D1-B1 higher biomass and lipid yields during cellulosic lipids production. Polymers decomposition mainly depends on the hydrolytic activity of the microbes [13], so it is understandable that more sufficient available carbon sources provided for microbes will lead to higher theoretical yields of lipids. As the conventional substrate for SCOs production was glucose, the current reports regarding genetic manipulations for enhanced SCOs production mainly centered on the modification of lipid biosynthesis pathway such as regulating the TAG biosynthesis pathway enzymes and blocking competing pathways (e.g., β-oxidation) [27]. According to the previous reports and the current investigation, we proposed that combined manipulation of the hydrolytic ability and lipid biosynthesis pathway in microbes would be more effective to promote cellulosic lipid production in CBP system.

Preliminarily study of the cellulase gene integration profiles in transformants

Although both A2-2 and D1-B1 showed enhanced lipid production, the detail metabolic profiles of A2-2 harboring *celA* gene and D1-B1 harboring *celC* gene were found to be different. Integration profiles of cellulase gene in the two transformants were investigated preliminarily in this work. The pPTRI based vector is integrated with the host genomic DNA by non-homologous recombination. For non-homologous integration of transforming vectors in the fungus, the complete vector backbone would often be integrated into the genomic DNA. Thus, Ampicillin resistance gene (*Amp*^{r}) was used instead of cellulase gene as the probe for southern blot analysis. For southern blot analysis, genomic DNA extracted from A2-2 and D1-B1 was digested with *Hind*III and subsequent hybridization with *Amp*^{r} probe. The hybridization results would provide an estimate of transgene copy number and the number of insertion loci. According to the results shown in Fig. 6a, it can be estimated that A2-2 contains two copies of the transgene while D1-B1 has three copies (Fig. 6a). Subsequently, three primer sets were designed to investigate the integration type of "cellulase expression cassette" in transformants (Fig. 6b). According to PCR results using genomic DNA of A2-2, D1-B1 and wild-type strain A-4 as templates (Fig. S2), the insertion type of "cellulase expression cassette" in A2-2 was suggested to be type A. However, one strain can receive several plasmids and every plasmid could integrate with the host genome in different types. Thus, the insertion type in D1-B1 is difficult to determine using PCR, and the integration type of "cellulase expression cassette" in D1-B1 could be type D or the combination of several types. Overall, identification of the insertion type of exogenous expression cassette by using one method such as PCR could be insufficient for some transformants (e.g. D1-B1). Although more researches and technologies are needed to answer this question, the obtained information of the insertion type of "cellulase expression cassette" in both A2-2 and D1-B1 is important for the elucidation of the insertion loci of the transgene, which could be guidance for primer design in Genome Walking. Besides, it has been reported that different insertion of genes on the chromosome has significant effects on expression levels of foreign enzymes, which is mediated through the interferences from local sequences, gene orientation, and insertion position relative to the chromosomal origin of replication [28]. Thus, it could be proposed that the different metabolic profiles between A2-2 and D1-B1 during cellulosic lipid production could attributed to not only their different transgene but also biological impact of different integration profiles in transformants.

Conclusion

This work indicated that the enhancement of lipid production can be promisingly feasible by regulating cellulase secretion of natural CBP producers, although lipid production was not affected and controlled by cellulase activity alone. For strain A-4, it is more possible to get strains with increased cellulase activity by the introduction of *celA* and *celC* gene than *celB* and *celD* gene. *celA* transformant A2-2 and *celC* transformant D1-B1 were demonstrated to be good candidates for efficient conversion of biomass into lipids, exhibiting significantly higher lipid yields than the reference strain using cellulosic substrates. The foreign cellulase introduction contributed to not only enhanced simultaneous saccharification but also improved lipid accumulation in A2-2 and D1-B1.

Supporting Information

Figure S1 PCR results using the genomic DNA of transformants and wild-type A-4.

Figure S2 PCR detection of the genomic DNA of the A2-2, D1-B1 and wild-type A-4 by using the primers as described in Table S1.

Table S1 Primers used in this study.

Table S2 Cellulase activities of 22 PCR-positive transformants and the reference strains.

Author Contributions

Conceived and designed the experiments: HL QW. Performed the experiments: HL QW QS. Analyzed the data: HL QW JM. Contributed reagents/materials/analysis tools: HL QW QS JM. Contributed to the writing of the manuscript: HL QW QS JF YZ. Obtained permission for use of fluorescence microscope: JF YZ JM.

References

1. Atabani AE, Silitonga AS, Badruddin IA, Mahlia TMI, Masjuki HH, et al. (2012) A comprehensive review on biodiesel as an alternative energy resource and its characteristics. Renewable & Sustainable Energy Reviews 16: 2070–2093.
2. Zhan J, Lin H, Shen Q, Zhou Q, Zhao Y (2013) Potential utilization of waste sweetpotato vines hydrolysate as a new source for single cell oils production by *Trichosporon fermentans*. Bioresour Technol 135: 622–629.
3. Hetzler S, Broker D, Steinbuchel A (2013) Saccharification of cellulose by recombinant *Rhodococcus opacus* PD630 strains. Appl Environ Microbiol 79: 5159–5166.
4. Gong ZW, Shen HW, Wang Q, Yang XB, Xie HB, et al. (2013) Efficient conversion of biomass into lipids by using the simultaneous saccharification and enhanced lipid production process. Biotechnology for Biofuels 6.
5. Steen EJ, Kang YS, Bokinsky G, Hu ZH, Schirmer A, et al. (2010) Microbial production of fatty-acid-derived fuels and chemicals from plant biomass. Nature 463: 559–U182.
6. Peng XW, Chen HZ (2008) Single cell oil production in solid-state fermentation by *Microsphaeropsis* sp from steam-exploded wheat straw mixed with wheat bran. Bioresource Technology 99: 3885–3889.
7. Carere CR, Sparling R, Cicek N, Levin DB (2008) Third generation biofuels via direct cellulose fermentation. International Journal of Molecular Sciences 9: 1342–1360.
8. Liu W, Wang Y, Yu Z, Bao J (2012) Simultaneous saccharification and microbial lipid fermentation of corn stover by oleaginous yeast *Trichosporon cutaneum*. Bioresour Technol 118: 13–18.
9. Lin H, Cheng W, Ding HT, Chen XJ, Zhou QF, et al. (2010) Direct microbial conversion of wheat straw into lipid by a cellulolytic fungus of *Aspergillus oryzae* A-4 in solid-state fermentation. Bioresource Technology 101: 7556–7562.
10. Dey P, Banerjee J, Maiti MK (2011) Comparative lipid profiling of two endophytic fungal isolates-*Colletotrichum* sp. and *Alternaria* sp. having potential utilities as biodiesel feedstock. Bioresour Technol 102: 5815–5823.
11. Zhao CH, Cui W, Liu XY, Chi ZM, Madzak C (2010) Expression of inulinase gene in the oleaginous yeast *Yarrowia lipolytica* and single cell oil production from inulin-containing materials. Metab Eng 12: 510–517.
12. Hetzler S, Steinbuchel A (2013) Establishment of cellobiose utilization for lipid production in *Rhodococcus opacus* PD630. Appl Environ Microbiol 79: 3122–3125.
13. Geisseler D, Horwath WR (2009) Relationship between carbon and nitrogen availability and extracellular enzyme activities in soil. Pedobiologia 53: 87–98.
14. Bando H, Hisada H, Ishida H, Hata Y, Katakura Y, et al. (2011) Isolation of a novel promoter for efficient protein expression by *Aspergillus oryzae* in solid-state culture. Appl Microbiol Biotechnol 92: 561–569.
15. Deng Y, Fong SS (2011) Metabolic engineering of *Thermobifida fusca* for direct aerobic bioconversion of untreated lignocellulosic biomass to 1-propanol. Metab Eng 13: 570–577.
16. Gomi K, Iimura Y, Hara S (1987) Integrative Transformation of *Aspergillus oryzae* with a Plasmid Containing the *Aspergillus nidulans argB* Gene. Agric Biol Chem 51: 2549–2555.
17. Kubodera T, Yamashita N, Nishimura A (2002) Transformation of *Aspergillus* sp. and *Trichoderma reesei* using the pyrithiamine resistance gene (*ptrA*) of *Aspergillus oryzae*. Biosci Biotechnol Biochem 66: 404–406.
18. Singh A (1991) Lipid production by a cellulolytic strain of *Aspergillus-Niger*. Letters in Applied Microbiology 12: 200–202.
19. Bligh EG, Dyer WJ (1959) A rapid method of total lipid extraction and purification. Canadian journal of biochemistry and physiology 37: 911–917.
20. Roopesh K, Ramachandran S, Nampoothiri KM, Szakacs G, Pandey A (2006) Comparison of phytase production on wheat bran and oilcakes in solid-state fermentation by *Mucor racemosus*. Bioresource Technology 97: 506–511.
21. Ghose TK (1987) Measurement of cellulase activities. Pure and Applied Chemistry 59: 257–268.
22. Bradford MM (1976) A rapid and sensitive method for the quantitation of microgram quantities of protein utilizing the principle of protein-dye binding. Anal Biochem 72: 248–254.
23. Kubodera T, Yamashita N, Nishimura A (2000) Pyrithiamine resistance gene (*ptrA*) of *Aspergillus oryzae*: cloning, characterization and application as a dominant selectable marker for transformation. Biosci Biotechnol Biochem 64: 1416–1421.
24. Kitamoto N, Go M, Shibayama T, Kimura T, Kito Y, et al. (1996) Molecular cloning, purification and characterization of two endo-1,4-beta-glucanases from *Aspergillus oryzae* KBN616. Appl Microbiol Biotechnol 46: 538–544.
25. Kim YS, Seo JH, Cha HJ (2003) Enhancement of heterologous protein expression in *Escherichia coli* by co-expression of nonspecific DNA-binding stress protein, Dps. Enzyme and Microbial Technology 33: 460–465.
26. van Rensburg E, den Haan R, Smith J, van Zyl WH, Gorgens JF (2012) The metabolic burden of cellulase expression by recombinant *Saccharomyces cerevisiae* Y294 in aerobic batch culture. Appl Microbiol Biotechnol 96: 197–209.
27. Courchesne NMD, Parisien A, Wang B, Lan CQ (2009) Enhancement of lipid production using biochemical, genetic and transcription factor engineering approaches. Journal of Biotechnology 141: 31–41.
28. Block DH, Hussein R, Liang LW, Lim HN (2012) Regulatory consequences of gene translocation in bacteria. Nucleic Acids Res 40: 8979–8992.
29. Shoji JY, Maruyama J, Arioka M, Kitamoto K (2005) Development of *Aspergillus oryzae* thiA promoter as a tool for molecular biological studies. FEMS Microbiol Lett 244: 41–46.
30. Mabashi Y, Kikuma T, Maruyama J, Arioka M, Kitamoto K (2006) Development of a versatile expression plasmid construction system for *Aspergillus oryzae* and its application to visualization of mitochondria. Biosci Biotechnol Biochem 70: 1882–1889.

22

Effects of Temperature and Carbon-Nitrogen (C/N) Ratio on the Performance of Anaerobic Co-Digestion of Dairy Manure, Chicken Manure and Rice Straw: Focusing on Ammonia Inhibition

Xiaojiao Wang[1]*, Xingang Lu[2], Fang Li[1], Gaihe Yang[1]

1 College of Agronomy, Northwest A&F University, Yangling, Shaanxi, People's Republic of China, **2** School of Chemical Engineering, Northwest University, Xian, Shaanxi, People's Republic of China

Abstract

Anaerobic digestion is a promising alternative to disposal organic waste and co-digestion of mixed organic wastes has recently attracted more interest. This study investigated the effects of temperature and carbon-nitrogen (C/N) ratio on the performance of anaerobic co-digestion of dairy manure (DM), chicken manure (CM) and rice straw (RS). We found that increased temperature improved the methane potential, but the rate was reduced from mesophilic (30~40°C) to thermophilic conditions (50~60°C), due to the accumulation of ammonium nitrogen and free ammonia and the occurrence of ammonia inhibition. Significant ammonia inhibition was observed with a C/N ratio of 15 at 35°C and at a C/N ratio of 20 at 55°C. The increase of C/N ratios reduced the negative effects of ammonia and maximum methane potentials were achieved with C/N ratios of 25 and 30 at 35°C and 55°C, respectively. When temperature increased, an increase was required in the feed C/N ratio, in order to reduce the risk of ammonia inhibition. Our results revealed an interactive effect between temperature and C/N on digestion performance.

Editor: Wenjun Li, National Center for Biotechnology Information (NCBI), United States of America

Funding: This work was supported by science and technology support projects 'the biological technology integration and demonstration of high yield biogas digestion from the mix ingredients' (2011 BAD15B03) from Ministry of Science and Technology Department of the People's Republic of China and Research Fund for the Doctoral Program of Higher Education of Northwest A & F University, China(2013BSJJ057). The funders had no role in study design, data collection and analysis, decision to publish, or preparation of the manuscript.

Competing Interests: The authors have declared that no competing interests exist.

* E-mail: w-xj@nwsuaf.edu.cn

Introduction

Anaerobic digestion is an effective way of converting agricultural waste into biogas that can be used to generate energy, which is especially efficient in rural western China. In the past decade, this technology has received great attention in both scientific research and practice. However, the efficiency of anaerobic digestion may be limited by inadequate amount and diversity of waste from a single resource, which is insufficient for large-scale digesters, as well as the drawbacks of using single substrates, such as improper carbon-nitrogen (C/N) ratios, low pH of the substrate itself, poor buffering capacity, and high concentrations of ammonia [1,2,3]. Therefore, co-digestion of mixture substrates for biogas production has recently attracted more interest.

Co-digestion of various biosolid wastes, a process that utilizes the nutrients and bacterial diversity in those wastes to optimize the digestion process, is an attractive approach for improving the efficiency of biotransformation [4]. A primary advantage of co-digestion is that it could efficiently balance feedstock carbon and nitrogen and a balanced C/N ratio of feedstock is likely to improve methane production. An early study conducted by Wu *et al.* revealed that swine manure co-digested with corn stalks at a C/N ratio of 20 obtained increased cumulative biogas production up to 11-fold and increased cumulative net methane volume up to 16-fold, when compared to swine manure digested alone [5]. Recent study by Wang *et al.* also suggested that co-digestion of dairy manure, chicken manure and wheat straw, had better digestion performance with stable pH and low concentrations of total ammonium nitrogen (TAN) and free ammonia (FA) at adjusted C/N ratios of 25 and 30 [6]. Similar observations were also reported by Hills for dairy manure, demonstrating that the greatest methane production was achieved when the C/N ratio was adjusted to 25 using glucose [7]. By optimizing the substrate C/N ratio, co-digestion of wastes of different C/N characteristics can greatly enhance the efficiency of biogas digestion.

Although many studies indicated that the optimal C/N ratios in methane fermentation were 25~30 [8,9,10], the depletion of carbon and nitrogen could be affected by operating conditions, such as temperature, resulting in the occurrence of inhibitory effects. It has been reported that the high FA concentration could inhibit thermophilic more seriously than mesophilic digestion [11,12,13]. A decrease in operating temperature from 60°C to 37°C in anaerobic digesters with a high ammonia concentration provided relief from FA inhibition, leading to increase in biogas yield [14,15]. FA concentration under mesophilic digestion is already inhibitory in the range of 80~150 mg L^{-1} at a pH of 7.5 [16,17,18]. However, under thermophilic conditions, when the concentration of FA was increased to 620 mg L^{-1} in the ammonia

Table 1. Chemical characteristics of raw materials used in this study.

Substrate	[a] TS content/%	VS content/%	pH	Total carbon/g $kg^{-1}VS$	Total Kjeldahl nitrogen/g $kg^{-1}VS$	C/N
DM	15.8±0.34	81.5±1.41	7.26±0.03	65.8±1.19	2.96±0.05	22.2±0.22
CM	29.9±0.67	65.3±1.26	6.93±0.11	58.6±1.77	6.11±0.08	9.6±0.16
RS	89.2±1.59	92.3±1.34	-	328±5.67	6.34±0.11	51.7±1.62

[a]±shows the standard error.

toxicity test, a gradual decrease of 21% was observed in biogas [19]. Another study also indicated that thermophilic flora tolerated at least twice as much FA compared to mesophilic flora [20]. Because the concentrations of TAN and FA originally depend on the content of organic nitrogen in the reactor and on C/N ratios, the indicator of substrate carbon and nitrogen content may also interact with temperature and that interaction results in different concentrations of ammonia and FA, as well as inhibitory effects.

Base on previous studies mentioned above, there are interactive effects between temperature and ammonia in the digestion process and the digestion efficiency is dramatically affected by the temperature and C/N ratio. Thus, to investigate this interaction, we first examined the effect of a series of temperatures on the mixtures of certain ratios of C/N (25), and secondly, compared the digestion performance of mixtures with a series of C/N ratios by adjusting the proportions of each substrate, dairy manure (DM), chicken manure (CM) and rice straw (RS) under mesophilic and thermophilic conditions.

Materials and Methods

Substrate characteristics

DM and CM were collected from a livestock farm located in Yangling, China. RS was obtained from a local villager. Before being put into the reactor, the air-dried RS was cut into pieces (2~3 cm). The substrates were individually homogenized and subsequently stored at 4°C for further use. The chemical characterization of each substrate tested in this study is shown in Table 1. All samples were collected and tested in triplicate, and the averages of the three measurements are presented.

Ethics statement

The collections of DM and CM were permitted by livestock farms belonging to 'Besun' group in Yangling, China. The RS was provided voluntarily by a local villager in Qishan, Baoji, China. The inoculum was obtained from a household biogas digester in a biogas demonstration village named Cuixigou in Yangling and the collection was permitted by the hosts. The all experimental procedures conformed to the regulations established by the Ethics Committee of the Research Center of Recycle Agricultural Engineering and Technology of Shaanxi Province, China.

Experimental design and set-up

Experiment 1: Three mixture sets were investigated in this experiment: set A (DM+ RS), set B (CM+RS), and set C (DM+ CM+RS). For set A and set B, the C/N ratio was 25, achieved by adjusting the DM/RS or CM/RS ratio. For set C, based on a DM/CM ratio of 1:1, multi-component substrates were prepared by adding RS to the DM-CM mixtures in order to adjust the C/N ratio to 25. The proportions of all substrates in each mixture were in a volatile solid (VS) state. The operation temperatures were 20, 30, 40 (mesophilic), 50, and 60°C (thermophilic), respectively.

Experiment 2: For all mixture sets, RS was added into the DM-CM mixtures with a VS ratio of 1:1, in order to adjust the C/N ratio to selected levels. C/N ratios of 15, 20, 25, 30 and 35 were selected in tests at a temperature of 35°C, but ratios of 20, 25, 30, 35 and 40 were selected in tests at a temperature of 55°C.

The initial VS ratio of substrate to inoculum was kept at 1:2 for all experimental setups. Each reactor had a 1 L capacity and contained 600 mL of total liquid, including 200 mL of inoculum and mixed substrate of 15gVS/L. The inoculum used for digestion at 20, 30, 35 and 40°C was digested cattle manure, taken from a lab-scale reactor operated at 35°C with a hydraulic

Table 2. Effects of temperature on pH value in anaerobic co-digestion with a C/N ratio of 25.

Temperature (°C)	DM+RS		CM+RS		DM+CM+RS	
	[a] Average	Final	Average	Final	Average	Final
20	6.12±0.13	6.64±0.14	5.42±0.11	5.01±0.07	5.92±0.10	6.58±0.11
30	6.89±0.15	7.12±0.16	6.42±0.03	6.92±0.02	7.11±0.11	7.35±0.14
40	7.21±0.09	7.44±0.13	7.19±0.08	7.38±0.12	7.48±0.13	7.67±0.11
50	7.58±0.11	7.61±0.13	7.66±0.12	7.79±0.07	7.56±0.12	7.74±0.03
60	7.69±0.15	7.88±0.15	7.82±0.10	8.11±0.11	7.72±0.16	7.92±0.13
[b]$LSD_{0.05}$= 0.47 [c]$LSD_{0.05}$=0.61						

[a]±shows the standard error

[b]LSD value at the 5% level based on all average values from three mixture sets at all operation temperatures

[c]LSD value at the 5% level based on all final values from three mixture sets at all operation temperatures

Table 3. Effects of temperature on total ammonia content in anaerobic co-digestion with a C/N ratio of 25.

Temperature (°C)	DM+RS		CM+RS		DM+CM+RS	
	[a] Average	Final	Average	Final	Average	Final
20	182.3±3.9	229.3±4.8	495.2±7.2	532±9.2	477±8.3	521±1.88
30	260.2±5.8	518.5±8.5	552.5±2.0	674±14.3	531±10.2	778±12.8
40	421.2±7.5	772.7±13.4	768.4±12.6	995±22.2	737±10.7	921±17.7
50	541.5±9.4	968.6±18.7	938.3±14.8	1261±18.3	869±3.1	1116±16.3
60	593.6±11.4	1052.4±18.8	951.8±12.5	1201±4.8	906±14.9	1256±20.6
[b]$LSD_{0.05}$= 52.6 [c]$LSD_{0.05}$=81.9						

[a]±shows the standard error
[b]LSD value at the 5% level based on all average values from three mixture sets at all operation temperatures
[c]LSD value at the 5% level based on all final values from three mixture sets at all operation temperatures.

retention time (HRT) of 15 days. Additionally, digestion at 50, 55 and 60°C was inoculated with digested cattle manure from the lab-scale reactor operated at 55°C with a HRT of 15 days. A control with only inoculum was used to determine biogas production due to endogenous respiration. Each treatment was performed in triplicate. All reactors were tightly closed with rubber septa and screw caps. The headspace of each reactor was flushed with nitrogen gas for about 3 min to assure anaerobic conditions prior to starting the digestion tests. To provide mixing of the reactor contents, all reactors were shaken manually for about 1 min, once a day prior to measurement of biogas volume.

Analytical techniques

Total solids, VS, pH, total Kjeldahl nitrogen (TKN), and total ammonium nitrogen (TAN) analysis were performed according to APHA Standard Methods [21]. Total organic carbon was determined by the method described by Cuetos *et al.* [22]. For all treatments, FA concentration was calculated in accordance with Hansen *et al.* [23]. The volume of biogas was measured by displacement of water. Methane content in the produced biogas was analyzed with a fast methane analyzer (Model DLGA-1000, Infrared Analyzer, Dafang, Beijing, China). The C/N ratio was determined by dividing the total organic carbon content by the total nitrogen content, according to the following equation.

$$C/N=\frac{W1\times C1+W2\times C2+W3\times C3}{W1\times N1+W2\times N2+W3\times N3}$$

Where W1, W2 and W3 were the VS weight in a single substrate in the mixture, C1, C2 and C3 were the organic carbon content (g kg^{-1}VS) in each substrate and N1, N2 and N3 were the nitrogen content (g kg^{-1}VS) in each substrate.

Results

Effects of temperature on the performance of anaerobic co-digestion based on experiment 1

Increased temperature resulted in pH increases in all three mixtures (Table 2). The pH values in digesters at 20°C, with average values of 6.12, 5.42 and 5.92 in the mixtures of DM+RS, CM+RS and DM+CM+RS, respectively, were far lower than those under other temperatures. From 30 to 60°C, the average pH values were in the range of 6.42 ~7.82.

A linear correlation between TAN and temperature (20 – 60°C) was observed and the highest TAN value was 1,261 mg L^{-1} in the mixture of CM+RS at 50°C (Table 3). The relationship between FA (Y, mg L^{-1}) and temperature (T, °C) was evaluated by the following equations: $Y=0.0302e^{1.82T}$ in the mixture of DM+RS, $Y=0.0216e^{2.0T}$ in the mixture of CM+RS and $Y=0.101e^{1.65T}$ in the mixture of DM+CM+RS. On average, the mixture of DM+CM+RS had significantly higher TAN and FA concentrations than the mixture of DM+RS, but was lower than the mixture of CM+RS (Tables 3 and 4).

With the increase of temperature, methane potential continuously increased, but the increasing rate was lower under thermophilic than under mesophilic conditions (Fig. 1). The mixture of DM+CM+RS had a little higher methane potential than the mixtures of DM+RS and CM+RS.

Table 4. Effects of temperature on free ammonia content in in anaerobic co-digestion with a C/N ratio of 25.

Temperature (°C)	DM+RS		CM+RS		DM+CM+RS	
	[a]Average	Final	Average	Final	Average	Final
20	0.1±0.004	0.4±0.009	0.1±0.002	0.9±0.01	0.2±0.001	0.8±0.01
30	1.6±0.04	5.4±0.08	1.2±0.03	4.4±0.06	5.4±0.09	13.8±0.1
40	12.7±0.5	31.4±0.6	17.9±0.8	35.5±1.6	32.8±1.1	63.7±2.1
50	65.4±2.2	101.2±4.8	108.6±3.7	189.2±4.3	82.0±1.8	153.3±3.2
60	142.5±3.6	324.5±4.9	240.7±5.2	479.2±3.2	192.6±3.2	376.7±8.2
[b]$LSD_{0.05}$ = 23.2 [c]$LSD_{0.05}$ = 45.6						

[a]± shows the standard error
[b]LSD value at the 5% level based on all average values from three mixture sets at all operation temperatures
[c]LSD value at the 5% level based on all final values from three mixture sets at all operation temperatures

Effects of C/N ratio on the performance of anaerobic co-digestion based on experiment 2

The pH value and the concentrations of TAN and FA were significantly influenced by C/N ratios at 35°C. For digesters with C/N ratios of 15 and 20, the pH values were higher than 7.0 during the whole digestion process, and the final pH values reached to 8.09 and 7.68, respectively (Fig. 2A).The average pH value was as low as 6.67 when the C/N ratio increased to 35. C/N ratios of 25 and 30 resulted in average pH values of 7.12 and 7.02, respectively. In addition, the contents of TAN and FA decreased with increased C/N ratios (Fig. 2B and C). Low C/N ratios of 15 and 20 resulted in TAN and FA concentrations as high as 2610, 2258 mg L^{-1} and 314, 108 mg L^{-1}, respectively. Treatments with C/N ratios of 25, 30 and 35 resulted in low and stable TAN and FA during the anaerobic process. The average concentrations of TAN were 985, 739 and 568 mg L^{-1} when C/N ratios were of 25, 30, and 35, respectively and the average concentrations of FA were 9.1, 7.5 and 2.2 mg L^{-1} when C/N ratios were of 25, 30, and 35, respectively.

Under 55°C, pH values were between 7.0 and 7.92 in treatments with C/N ratios of 20 and 25. Stable pH values around 7.0 were observed when C/N ratios were of 30 and 35. When the C/N ratio was increased to 35, the pH value was lower, at around 6.2 (Fig. 3A). The concentrations of TAN in treatments with C/N ratios of 20 and 25 increased up to 1500 mg L^{-1} by day 10 and reached peaks as high as 2415 and 1932 mg L^{-1}, respectively (Fig. 3B). FA increased continuously in digestion with final concentrations of 461 and 235 mg L^{-1} when C/N ratios were of 20 and 25. For C/N ratios between 30 and 40, TAN and FA concentrations were in the range of 430~1426 mg L^{-1} and 2~131 mg L^{-1}, respectively (Fig. 3B and C).

Methane potential increased first and then decreased with increases of C/N ratios. The highest methane potential was observed with a C/N ratio of 25 at 35°C with 272 mL g^{-1}VS and with a C/N ratio of 30 at 50°C with 286 mL g^{-1}VS, respectively (Fig. 4). The quadratic models for methane potential in terms of the C/N ratio as a variable were significant and the equations at 35°C (1) and 55°C (2) were expressed as follows:

$$Y = -0.8475X^2 + 45.36X - 345.3,\ R^2 = 0.9652 \quad (1)$$

$$Y = -1.16X^2 + 71.16X - 781.4,\ R^2 = 0.8922 \quad (2)$$

Where Y was methane potential and X was the C/N ratio. The optimum conditions for maximum methane potential were calculated as a C/N of 26.76 at 35°C and a C/N ratio of 30.67 at 55°C, respectively. Accordingly, the highest methane potential was estimated as 265.7 and 309.9 mL g^{-1} VS.

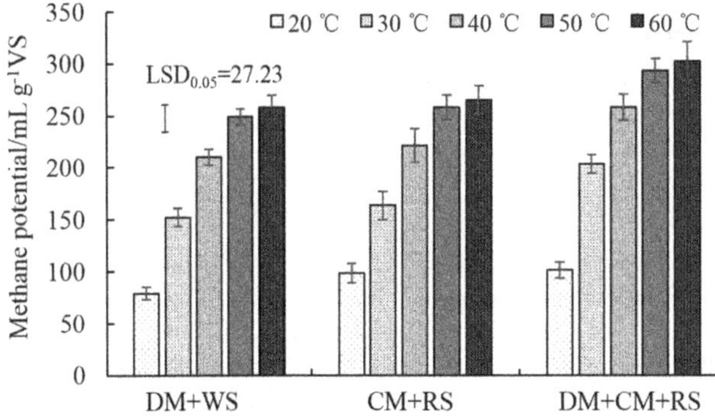

Figure 1. Effects of temperature on methane potential in mixtures with a C/N ratio of 25. Values are presented as the mean ±standard error of three replicates (n = 3). Vertical bars represent LSD at the 5% level.

Discussion

According to the study by Calli *et al.*, ammonia inhibition occurs in the range of 1500~3000 mg L^{-1} TAN when the pH value is over 7.4 [24]. Then, TAN concentrations of three mixtures were in a safe range below 1261 mg L^{-1} at temperatures between 20 and 60°C (Table 3). Compared with ammonium nitrogen, FA has been suggested as the active component causing ammonia inhibition, since it is freely membrane-permeable [25]. It has been reported that a range between 80 and 150 mg L^{-1} FA was inhibitory for methanogens [16,26]. In our study, FA concentrations were in this range at 50°C and far higher than 150 mg L^{-1} at 60°C (Table 4), indicating the occurrence of ammonia inhibition. Based on experiment 1, temperature obviously played a greater role in methane production in the range of 20 ~40°C than in the range of 40 ~60°C. Methane potentials in three mixtures were an average of 2.49 times higher at 40 than 20°C, but only 1.20 times higher at 60 than at 40°C. And no significant difference was found in methane potential between 50 and 60°C.

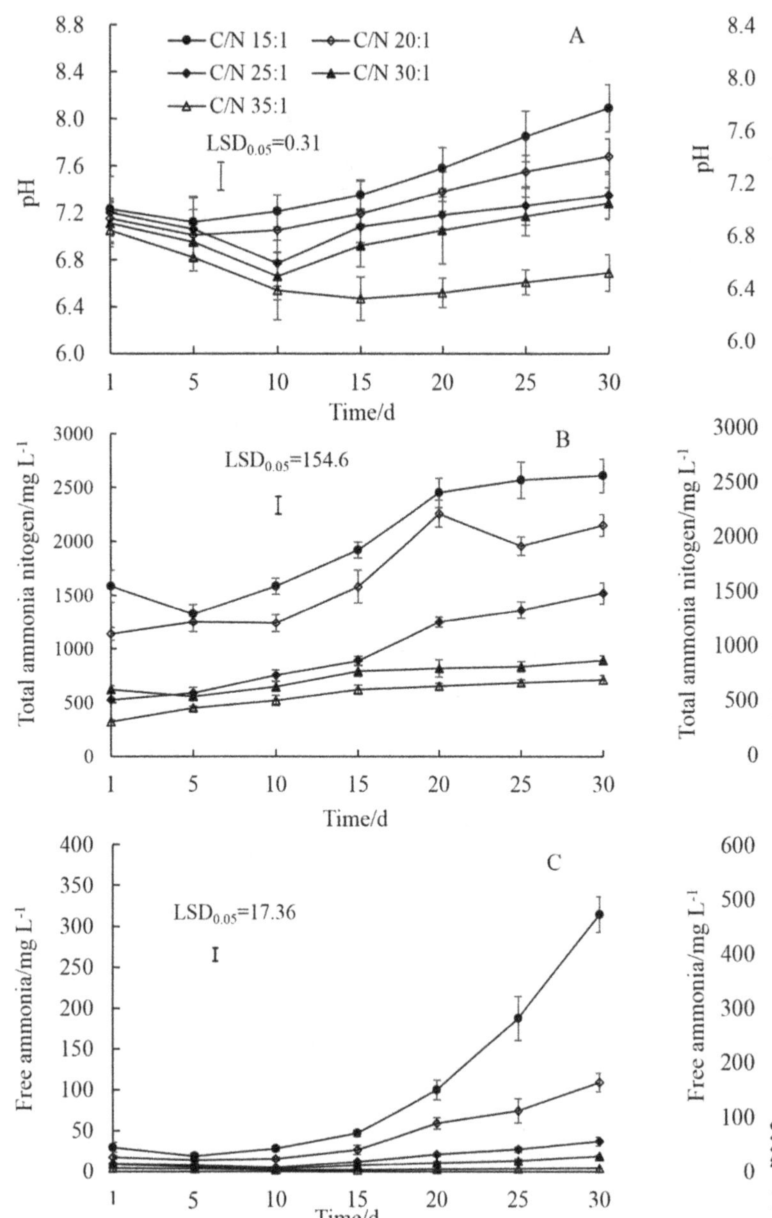

Figure 2. Changes of pH, total ammonium nitrogen, and free ammonia with different C/N ratios in the mixture of dairy manure (DM), chicken manure (CM), and rice straw (RS) in anaerobic co-digestion at 35°C. Values are presented as the mean ±standard error of three replicates (n = 3). Vertical bars represent LSD at the 5% level.

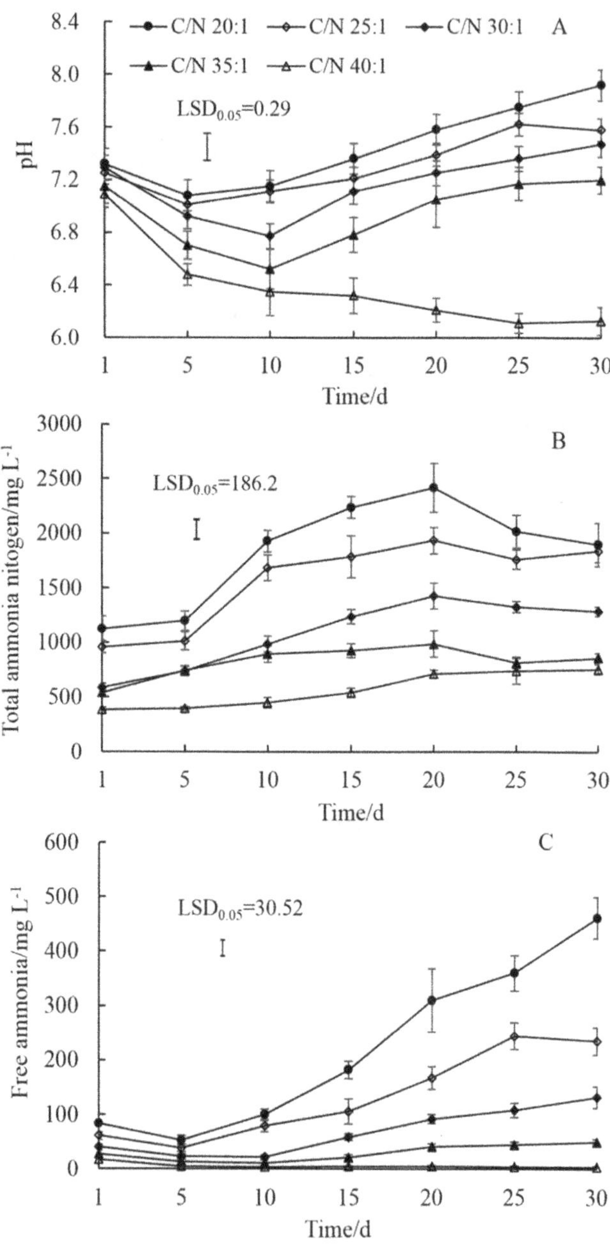

Figure 3. Changes of pH, total ammonium nitrogen, and free ammonia with different C/N ratios in the mixture of dairy manure (DM), chicken manure (CM), and rice straw (RS) IN anaerobic co-digestion at 55°C. Values are presented as the mean ±standard error of three replicates (n = 3). Values are presented as the mean ±standard error of three replicates (n = 3). Vertical bars represent LSD at the 5% level.

These results also suggest the existence of an inhibitory effect by ammonia under thermophilic conditions. However, in the anaerobic digestion of organic wastes, it has been reported that methane production was inhibited up to 50% by 220 mg L^{-1} FA at 37°C and by 690 mg L^{-1} FA at 55°C [20]. That is, thermophilic flora tolerated at least twice as much FA as compared to mesophilic flora. The higher methane potential under thermophilic conditions suggested that increased ammonia did not completely inhibit the digestion process and did not offset the advantage of increased temperature in thermodynamics and kinetics, which might result from proper C/N ratios of mixture substrates.

Due to the potential role of the C/N ratio in regulating the inhibitory effects of ammonia, digestions with different C/N ratios were tested in experiment 2 under mesophilic and thermophilic conditions to further obtain optimal C/N ratios with less ammonia inhibition. We found that the mixture of DM+CM+RS had better digestion performance in methane potential than the mixtures of DM+RS and CM+RS (Fig. 1), which might be due to the increased buffering capacity and the synergistic effect, which was inconsistent with the result reported by Wang *et al.* [6]. The mixture of DM+CM+RS was then selected for follow-up studies.

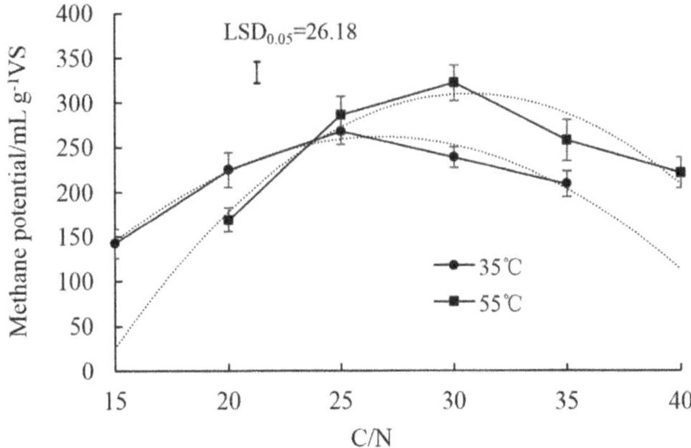

Figure 4. Changes of methane potential with different C/N ratios in the mixture of dairy manure (DM), chicken manure (CM), and rice straw (RS) in anaerobic co-digestion at 35°C and 55°C. The dotted lines were fitting curves for both temperatures. Values are presented as the mean ±standard error of three replicates (n = 3). Vertical bars represent LSD at the 5% level.

Substrates that have low C/N ratios contain relatively high concentrations of ammonia, exceeding concentrations necessary for microbial growth, and probably inhibiting anaerobic digestion [3,23]. TAN concentrations were as high as 2500 mg L^{-1} and FA increased up to final concentrations of 314 and 461 mg L^{-1}, when the C/N ratio was of 15 at 35°C and was 20 at 55°C, respectively (Figs. 2 and 3). Therefore, methane potential was reduced down to 142 $mLg^{-1}VS$ at 35°C and 169 $mLg^{-1}VS$ at 55°C, accounting for just 53.0% and 52.5%, compared with their maximum values (Fig. 4). Under both temperatures, with the increase of C/N ratios, TAN and FA concentrations decreased. For example, the average FA concentrations at 35°C were reduced 56.5, 83.7, 90.7 and 97.1% from a C/N ratio of 15 to 20, 25, 30 and 35, respectively. Previous reports suggested that using a feedstock C/N ratio from 27 to 32 promotes steady digester operation at optimum ammonia nitrogen levels and feedstock with a C/N ratio of 32 producing a lower concentration of ammonia nitrogen and FA [25,27]. Thus, the digestion system was sensitive to the feed C/N ratio and a higher C/N ratio reduced the protein solubilization rate and hence produced lower TAN and FA concentration within the system, which was found to be advantageous.

Ammonia inhibition under mesophilic and thermophilic conditions has been compared in previous studies. It has been observed that an increase in temperature resulted in a reduction of the biogas yield, due to the increased inhibition of FA under higher temperature [14,15,28]. In our study, ammonia inhibition occurred with a C/N ratio of 20 at 55°C, whereas a C/N ratio of 15 experienced inhibition at 35°C, suggesting that higher temperature improved the degradation efficiency of organic nitrogen to ammonia nitrogen. However, when C/N ratios were higher than 25, methane potential at 55°C was higher than at 35°C (Fig. 4), indicating higher C/N ratios reduced the risk of ammonia inhibition under thermophilic conditions. Moreover, the optimal C/N ratios were obtained at 26.76 and 30.67 under mesophilic and thermophilic conditions, respectively, by optimizing the quadratic models between methane potential and C/N ratio. These results showed that ammonia inhibition occurring under thermophilic conditions might be avoided by optimizing the C/N ratio in co-digestion of different substrates. However, a very high C/N ratio promotes the growth of methanogen populations that are able to meet their protein requirements and will, therefore, no longer react with the remaining carbon content of the substrate, resulting in a low production of gas.

Conclusions

This study demonstrated an interactive effect between C/N ratio and temperature on the performance of anaerobic co-digestion of dairy manure, chicken manure and rice straw. Our results suggest that increased temperature from mesophilic to thermophilic conditions resulted in ammonia inhibition, however, this kind of inhibition could be reduced or avoided by increasing the C/N ratio of mixed feedstock to an appropriate level. In anaerobic co-digestion of DM, CM and RS, the optimal C/N level was 26.76 at 35°C and 30.67 at 55°C. Adjusting the proportions of mixture substrates in anaerobic co-digestion to obtain suitable feed characteristics, such as the C/N ratio, pH and nutrients, is an effective way to achieve desired digestion performance.

Author Contributions

Conceived and designed the experiments: XJW XGL GHY. Performed the experiments: XJW FL. Analyzed the data: XJW XGL GHY. Wrote the paper: XJW.

References

1. Banks C, Humphreys P (1998) The anaerobic treatment of a ligno-cellulosic substrate offering little natural pH buffering capacity. Water Sci Technol 38:29–35.
2. Zhang T, Liu LL, Song ZL, Ren GX, Feng ZY, et al.(2013) Biogas Production by Co-Digestion of Goat Manure with Three Crop Residues. PLoS ONE 8(6): e66845.
3. Procházka J, Dolejš P, Máca J, Dohányos M (2012) Stability and inhibition of anaerobic processes caused by insufficiency or excess of ammonia nitrogen. Appl Microbiol Biotechnol 93:439–447.
4. Wang XJ, Yang GH, Feng YZ, Ren GX, Han XH (2012) Optimizing feeding composition and carbon-nitrogen ratios for improved methane yield during anaerobic co-digestion of dairy chicken manure and wheat straw. Bioresour Technol 120:78–83.
5. Wu X, Yao WY, Zhu J, Miller C (2010) Biogas and CH(4) productivity by co-digesting swine manure with three crop residues as an external carbon source. Bioresour Technol 101: 4042–4047.
6. Wang XJ, Yang GH, Li F, Feng YZ, Ren GX, et al.(2013) Evaluation of two statistical methods for optimizing the feeding composition in anaerobic co-digestion: Mixture design and central composite design. Bioresour Technol 131:172–178.
7. Hills DJ (1979) Effects of carbon: nitrogen ratio on anaerobic digestion of dairy manure. Agr Wastes 1:267–278.
8. Kayhanian M, Tchobanoglous G (1992) Computation of C/N ratios for various organic fractions. Biocycle 33:58–60.
9. Marchaim U, Krause C (1993) Propionic to acetic-acid ratios in overloaded anaerobic- digestion. Bioresour Technol 43:195–203.
10. Yen HW, Brune DE (2007) Anaerobic co-digestion of algal sludge and waste paper to produce methane. Bioresour Technol 98:130–134.
11. Zeeman G, Wiegant W, Koster-Treffers M, Lettinga G (1985) The influence of the total-ammonia concentration on the thermophilic digestion of cow manure. Agr Wastes 14:19–35.
12. Wiegant W, Zeeman G (1986) The mechanism of ammonia inhibition in the thermophilic digestion of livestock wastes. Agr Wastes 16:243–253.
13. Angelidaki I, Ahring B (1993) Thermophilic anaerobic digestion of livestock waste: the effect of ammonia. Appl Microbiol Biot 38:560–564.
14. Angelidaki I, Ahring B (1994) Anaerobic thermophilic digestion of manure at different ammonia loads: effect of temperature. Water Res 28:727–731.
15. Hansen KH, Angelidaki I, Ahring BK (1999) Improving thermophilic anaerobic digestion of swine manure. Water Res 33:1805–1810.
16. Braun R, Huber P, Meyrath J (1981) Ammonia toxicity in liquid piggery manure digestion. Biotechnol Lett 3:159–164.
17. Kroeker E, Schulte D, Sparling A, Lapp H (1979) Anaerobic treatment process stability. J Water Pollut Control Fed 718–727.

18. Siles JA, Martin MA, Chica AF, Martin A (2010) Anaerobic co-digestion of glycerol and wastewater derived from biodiesel manufacturing. Bioresour Technol 101:6315– 6321.
19. Gallert C, Winter J (1997) Mesophilic and thermophilic anaerobic digestion of source- sorted organic wastes: effect of ammonia on glucose degradation and methane production. Appl Microbiol Biot 48:405–410.
20. PHA (1995) Standard methods for the examination of water and wastewater. Washington. DC, American Public Health Association.
21. Cuetos MJ, Fernandez C, Gomez X, Moran A (2011) Anaerobic co-digestion of swine manure with energy crop residues. Biotechnol Bioprocess Eng 16:1044–1052.
22. Hansen KH, Angelidaki I, Ahring BK (1998) Anaerobic digestion of swine manure: Inhibition by ammonia. Water Res 32:5–12.
23. Calli B, Mertoglu B, Inanc B, Yenigun O (2005) Effects of high free ammonia concentrations on the performances of anaerobic bioreactors. Process Biochem 40:1285–1292.
24. Kayhanian M (1999) Ammonia inhibition in high-solids biogasification: An overview and practical solutions. Environ Technol 20:355–365.
25. Ahring BK, Angelidaki I, Johansen K (1992) Anaerobic treatment of manure together with industrial-waste. Water Sci Technol 25:311–318.
26. Koster I, Lettinga G (1988) Anaerobic digestion at extreme ammonia concentrations. Biol Waste 25:51–59.
27. Zeshan Karthikeyan OP, Visvanathan C (2012) Effect of C/N ratio and ammonia-N accumulation in a pilot-scale thermophilic dry anaerobic digester. Bioresour Technol 113:294–302.
28. Garcia ML, Angenent LT (2009) Interaction between temperature and ammonia in mesophilic digesters for animal waste treatment. Water Res 43:2373–2382.

23

Discovery of Microorganisms and Enzymes Involved in High-Solids Decomposition of Rice Straw Using Metagenomic Analyses

Amitha P. Reddy[1,2❡], Christopher W. Simmons[1,2,3❡], Patrik D'haeseleer[1,5], Jane Khudyakov[1,5], Helcio Burd[1,7], Masood Hadi[1,6¤], Blake A. Simmons[1,6], Steven W. Singer[1,4], Michael P. Thelen[1,5], Jean S. VanderGheynst[1,2]*

1 Joint BioEnergy Institute, Emeryville, California, United States of America, **2** Biological and Agricultural Engineering, University of California Davis, Davis, California, United States of America, **3** Food Science, University of California Davis, Davis, California, United States of America, **4** Earth Sciences Division, Lawrence Berkeley National Laboratory, Berkeley, California, United States of America, **5** Physical and Life Sciences Directorate, Lawrence Livermore National Laboratory, Livermore, California, United States of America, **6** Biological and Materials Science Center, Sandia National Laboratories, Livermore, California, United States of America, **7** Physical Biosciences Division, Lawrence Berkeley National Laboratory, Berkeley, California, United States of America

Abstract

High-solids incubations were performed to enrich for microbial communities and enzymes that decompose rice straw under mesophilic (35°C) and thermophilic (55°C) conditions. Thermophilic enrichments yielded a community that was 7.5 times more metabolically active on rice straw than mesophilic enrichments. Extracted xylanase and endoglucanse activities were also 2.6 and 13.4 times greater, respectively, for thermophilic enrichments. Metagenome sequencing was performed on enriched communities to determine community composition and mine for genes encoding lignocellulolytic enzymes. *Proteobacteria* were found to dominate the mesophilic community while *Actinobacteria* were most abundant in the thermophilic community. Analysis of protein family representation in each metagenome indicated that cellobiohydrolases containing carbohydrate binding module 2 (CBM2) were significantly overrepresented in the thermophilic community. *Micromonospora*, a member of *Actinobacteria*, primarily housed these genes in the thermophilic community. In light of these findings, *Micromonospora* and other closely related *Actinobacteria* genera appear to be promising sources of thermophilic lignocellulolytic enzymes for rice straw deconstruction under high-solids conditions. Furthermore, these discoveries warrant future research to determine if exoglucanases with CBM2 represent thermostable enzymes tolerant to the process conditions expected to be encountered during industrial biofuel production.

Editor: Mark R. Liles, Auburn University, United States of America

Funding: This work was supported by the University of California Laboratory Fees Research Program #12-LR-237496 and performed as part of the DOE Joint BioEnergy Institute (http://www.jbei.org) supported by the U.S. Department of Energy, Office of Science, Office of Biological and Environmental Research, through contract DE-AC02-05CH11231 between Lawrence Berkeley National Laboratory and the U.S. Department of Energy. The funders had no role in study design, data collection and analysis, decision to publish, or preparation of the manuscript.

Competing Interests: The authors have declared that no competing interests exist.

* E-mail: jsvander@ucdavis.edu

¤ Current address: Synthetic Biology Program, Space BioSciences Division, NASA Ames Research Center, Moffett Field, California, United States of America

❡ These authors contributed equally to this work.

Introduction

Considerable efforts are underway to identify plant sources and conversion technologies to enable economical and sustainable production of fuels and chemicals from plant biomass and meet renewable fuel standards [1–4]. Agricultural residues are a promising resource because they do not compete with land used for food production [5–8]. Residues of particular interest are the hulls and straw associated with rice cultivation, harvest and processing. In 2010 worldwide rice production exceeded 690 million tons on 159 million ha of land [9] with estimated rice straw generation of 5.6–6.7 t/ha (890–1,065 million dry tons in 2010) [5,10,11]. While rice straw could be a significant resource for biofuel feedstock, challenges related to pretreatment and enzymatic hydrolysis have prevented its widespread conversion to biofuel. The development of cost-effective enzymes that efficiently hydrolyze plant cell wall polysaccharides under industrially relevant conditions would enable biofuel production from plant biomass feedstocks like rice straw [12,13].

Microbial communities that decompose plant cell wall polymers (lignocellulose) in extreme environments have been identified as a promising source of hydrolyzing enzymes [14,15]. Discovery of enzymes in these types of environments is particularly challenging due to a number of factors, including the tendency of carbohydrate-active enzymes to bind to substrates and interference by compounds present in lignocellulosic biomass when analyzing proteins and other metabolites. Approaches based on nucleic acid analyses offer alternatives that may overcome traditional methods of microorganism and enzyme discovery [16,17].

Table 1. Experimental design for enzyme extraction from rice straw and corresponding enzyme activities.

	Coded Design Setting			Xylanase	Endogucanase
Treatment	NaCl*	Tween 80**	Ethylene Glycol***	(IU gdw^{-1})	(IU gdw^{-1})
1	0	0	0	0.81	0.21
2	+1	−1	−1	0.76	0.21
3	0	0	0	0.70	0.19
4	+1	+1	+1	1.27	0.23
5	−1	−1	−1	0.60	0.22
6	−1	+1	+1	1.25	0.34
7	+1	−1	+1	1.08	0.23
8	−1	−1	+1	0.88	0.22
9	+1	+1	−1	0.83	0.16
10	0	0	0	0.77	0.18
11	−1	+1	−1	0.71	0.12
12	Sodium Acetate buffer control			0.85	0.19

*NaCl: −1 = 0.1 wt%, 0 = 0.8 wt%, +1 = 1.5 wt%.
**Tween 80: −1 = 0.01 wt%, 0 = 0.08 wt%, +1 = 0.15 wt%.
***Ethylene Glycol: −1 = 0 wt%, 0 = 25 wt%, +1 = 50 wt%.

The goal of this research was to use a combination of enrichment and metagenomic approaches to discover promising organisms and enzymes for the efficient hydrolysis of rice straw. Recognizing that bioconversion processes may occur over a range of temperatures and in high-solids environments, enrichments were completed as solid fermentations at 35°C and 55°C.

Materials and Methods

High Solids Incubations

Finished green waste compost was obtained from a commercial facility that composts agricultural residues including tree and vine prunings, with permission from Greg Kelly (Northern Recycling, Zamora, CA). Compost was solar-dried and stored at 4°C until applied as inocula. Fresh rice straw (*Oryza sativa L.*, California rice M206) was collected as described previously [18]. The dried straw was extracted with ethanol for 1.5 days and water for 2 days in a soxhlet extractor, dried in a vacuum oven for 4 days to 3.2% moisture on a dry basis (3.1% on a wet basis), and stored in zipper lock bags at 4°C until needed.

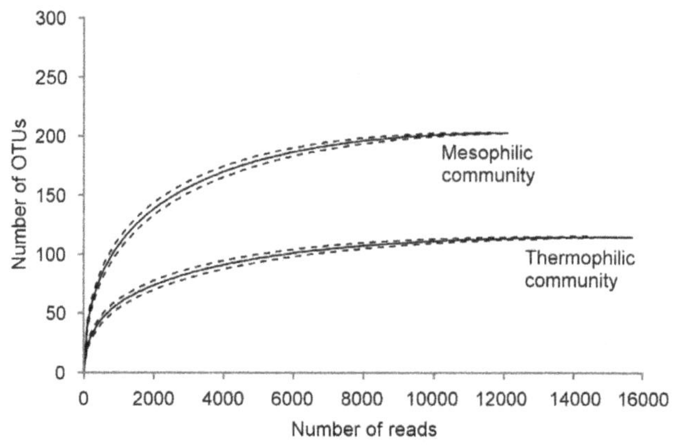

Figure 1. Rarefaction curves from pyrotag data for enriched mesophilic and thermophilic microbial communities. Dashed lines indicate ±1 standard error.

Table 2. Cumulative carbon dioxide evolution rate after 7 days of incubation (cCER) and extracted enzyme activity after each incubation period.

	cCER		Xylanase*		Endoglucanase*	
	(mg CO_2 (g dry feedstock)$^{-1}$)		(IU gdw^{-1})		(IU gdw^{-1})	
Sampling point	35°C	55°C	35°C	55°C	35°C	55°C
T1	35	109	**	**	**	**
T2	36	95	3.82 (0.06)	2.0 (0.07)	0.70 (0.03)	0.68 (0.03)
T3	48	246	2.53 (0.05)	11.5 (0.4)	0.40 (0.06)	1.9 (0.3)
T4***	39 (4)	292 (19)	2.07 (0.04)	7.4 (0.1)	0.094 (0.007)	1.35 (0.02)

*Values for xylanase and endoglucanase activities are given as means for triplicate assays with one standard deviation given in parentheses.
**data not available.
***n = 3 for T4 respiration measurements. Values in parentheses represent one standard deviation.

High-solids incubations were conducted as described previously with minor modifications [15]. Briefly, bioreactors with a 0.2 L working volume were loaded with 5–10 g dry weight of rice straw and inocula mixture. Prior to incubation, rice straw was wetted with minimal media [19] to a moisture content of 400 wt% dry basis (g water g dry solid^{-1}) and equilibrated at 4°C overnight. For the initial enrichment in each experiment, wetted rice straw was inoculated with 10 wt% (g dry compost (g dry solid)$^{-1}$) compost. Every 6 to 7 days, fresh feedstock was inoculated with 10 wt% (g dry enriched sample (g total dry weight)$^{-1}$) of the enriched community and transferred to a new bioreactor.

Incubator temperature was maintained at 35°C for mesophilic incubations. For the first enrichment of thermophilic incubations, the incubator temperature was maintained at 35°C for 1 day, ramped to 55°C over one day, and held at 55°C for the duration of the experiment. Water lost during incubation was replaced and each bioreactor was mixed every 3.5 days.

Microbial community respiration rate, represented as CO_2 evolution rate (CER), was measured for all incubated samples. Carbon dioxide concentration was measured on the influent and effluent air of the bioreactors using an infrared CO_2 sensor (Vaisala, Woburn, MA) and flow was measured with a thermal mass flow meter (Aalborg, Orangeburg, NY). Carbon dioxide and flow data were recorded every 20 min using a data acquisition system. Carbon dioxide evolution rate and cumulative respiration (cCER) were calculated as described previously [20].

Two sets of enrichments were completed. Enrichments for selection of an enzyme extraction buffer ran for 5 weeks. The second set of enrichments ran for four weeks, yielding a total of four sampling points (T1, T2, T3 and T4) for enzyme activity measurement. The T4 sampling point consisted of three replicate enrichments while T1, T2 and T3 where individual enrichments. Samples from the T4 sampling point were collected for DNA extraction.

Enzyme Extraction from Solid Samples

Buffer components for enzyme extraction were selected using a full factorial experiment (Table 1). Extractions were conducted with ethylene glycol (0–50 wt%), Tween 80 (0.01–0.15 wt %) and NaCl (0.1–1.5 wt %) and a sodium acetate buffer (50 mM, pH = 5.0) control.

To extract enzymes, three grams (wet weight) of freshly harvested colonized feedstock was shaken with 27 g of buffer for 60 minutes at 150 RPM and room temperature. Samples were centrifuged at 4°C and 10,000×g for 20 min and then vacuum filtered using 0.2 μm membranes. The extraction buffer was exchanged with sodium acetate buffer using VivaSpin columns with a PES membrane and a 5 kDa molecular weight cut off (VWR, West Chester, PA). Endoglucanase and xylanase activities in dialyzed extracts were measured as described previously [15]. JMP statistical software (v. 8.0.1, SAS Institute, Cary, NC) was used to perform stepwise regression and determine significant buffer components.

For samples T2–T4, enzymes were extracted using 50 wt% ethylene glycol, 0.15 wt% Tween 80 and 1.5 wt% NaCl and assayed according to methods described elsewhere [15]. All assays were completed in triplicate. Activities were reported as IU gdw^{-1} where one IU = μmol product min^{-1}.

DNA Extraction

Samples from the T4 time point were frozen in liquid nitrogen, homogenized with an oscillating ball mill (MM400, Retsch Inc., Newtown, PA), and stored with LifeGuard Soil Preservation Solution (Mo Bio Laboratories, Inc., Carlsbad, CA) in a ratio of 1:2.5 (sample:LifeGuard) at −80°C. Samples were thawed on ice and processed with the MoBio PowerSoil DNA Isolation kit (Mo Bio Laboratories, Inc., Carlsbad, CA).

16S rDNA Library Construction, Sequencing, and Binning

A fragment of the 16S small-subunit rRNA gene was PCR-amplified from DNA extracts using the primer sequences 926F and 1392R containing 454 adapters and barcodes using a previously described method [21]. AMPure Solid Phase Reversible Immobilization (SPRI) beads (Beckman Coulter) were used to purify amplicons. Emulsion PCR was performed using a GS FLX Titanium MV emPCR Kit (Roche). A Genome Sequencer FLX instrument and associated Titanium series kits (Roche) were used for sequencing of amplicons. Sequencing reads were analyzed using the methods of Kunin et al. [22]. In brief, PyroTagger software (Joint Genome Institute) was used to quality trim base calls, trim primer sequences from reads, remove duplicate reads, and bin reads by performing blastn alignments against the Greengenes database using default settings [23].

Metagenome Sequencing, Assembly, and Annotation

DNA fragments for 454 and Illumina sequencing were created using the Joint Genome Institute standard library generation protocols for Roche 454 GS FLX Titanium and Illumina HiSeq 2000 platforms. Metagenome sequencing was performed using a Roche GS FLX Titanium sequencing kit on a Roche/454 FLX-Ti system. Illumina sequencing was performed on a HiSeq 2000 system. Combined sequencing reads from 454 and Illumina runs were quality trimmed using a quality threshold of 10. Trimmed reads were assembled with SOAPdenovo [24] and Newbler [25] for contigs >1800 bp. A minimum overlap identity of 98% and a minimum overlap length of 80 bases were used. Contigs longer than 1800 bp and contigs resulting from Newbler assembly were assembled into a single assembly using Minimus [26] with a minimum overlap length of 80 bases, a minimum overlap identity of 98%, and a consensus error of 0.06 for joining. Burrows-Wheeler Aligner [27] was used to map reads back to contigs in order to confirm proper placement and calculate read depth for contigs. Annotation of contigs was performed using the Joint Genome Institute's Integrated Microbial Genomes with Microbiomes-Expert Review (IMG/M-ER) pipeline [28].

Table 3. Ecological measures for microbial communities from mesophilic and thermophilic enrichments on rice straw.

Method	Enrichment	Shannon index, H	Richness, S	Pielou index, J
Pyrotag sequencing	Mesophilic	3.62	204	0.68
	Thermophilic	2.49	115	0.52

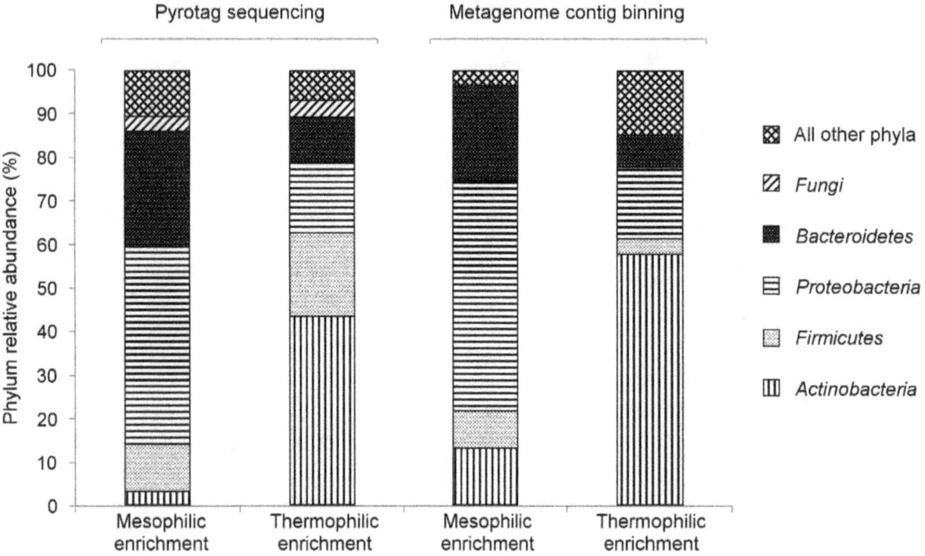

Figure 2. Pylum composition of microbial communities from mesophilic and thermophilic enrichments on rice straw.

Contig Binning

All contigs were scanned for genes within phylogenetic marker COGs using the IMG/M toolset [29]. The IMG/M pre-set list of marker COGs was used. Amino acid sequences of detected marker COG genes were imported into the Galaxy platform [30–32]. The blastp function of Galaxy was used to align marker COG genes against the NCBI protein database with an E-value cutoff of 1e-10. The best blast hit for each marker COG gene was used to bin its contig of origin at the genus level. For contigs with more than one marker COG gene, the taxonomy for over 50% of marker COG genes had to agree for the contig to be binned. Contigs with marker COG genes stemming from the same genus were collated into binning training sets. ClaMS software was used for supervised binning of metagenome contigs seeded with genus training sets [33]. Within ClaMS, De Bruijn chain signatures were used as the metric for binning with a kmer length of 2 and a signature cutoff value of 0.005.

Metagenome Analysis

R software running the VEGAN package [34] was used to determine the Shannon index, richness, and Pielou index of each community based on pyrotag data. Rarefaction curves were generated from pyrotag data using PAST software [35]. Similarity percentage (SIMPER) analysis was executed as described previously [36]. IMG/M was used for comparative genomics. The abundance profile search tool was used to find differences in protein family representation between the two metagenomes. Protein families from the Pfam database were used [37]. For the search, gene counts were normalized by the total number of genes in a given metagenome. Gene counts refer to the number of homologs in a metagenome for a given gene and do not factor in gene copy number. As a result, gene counts indicate how many different versions of a particular gene exist within a metagenome and are not skewed by how abundant the source microorganisms are in the community. Search criteria were set to find only protein families for which normalized gene counts were at least twice as abundant in the thermophilic enrichment community compared

Table 4. SIMPER analysis of genera accounting for >75% of dissimilarity between thermophilic and mesophilic microbial communities based on metagenome binning.

Taxonomy	Thermophilic enrichment relative abundance	Mesophilic enrichment relative abundance	% contribution
Micromonospora (Actinobacteria)	30.5	0.0	19.3
Mycobacterium (Actinobacteria)	20.4	0.0	12.9
Chryseobacterium (Bacteroidetes)	0.0	14.7	9.3
Pseudoxanthomonas (Proteobacteria)	11.6	24.3	8.1
Conexibacter (Actinobacteria)	0.0	9.9	6.3
Phenylobacterium (Proteobacteria)	0.0	8.6	5.5
Thermobifida (Actinobacteria)	6.8	0.0	4.3
Brevundimonas (Proteobacteria)	0.0	6.4	4.1
Candidatus Solibacter (Acidobacteria)	5.5	0.0	3.5
Brevibacillus (Firmicutes)	0.0	4.3	2.7

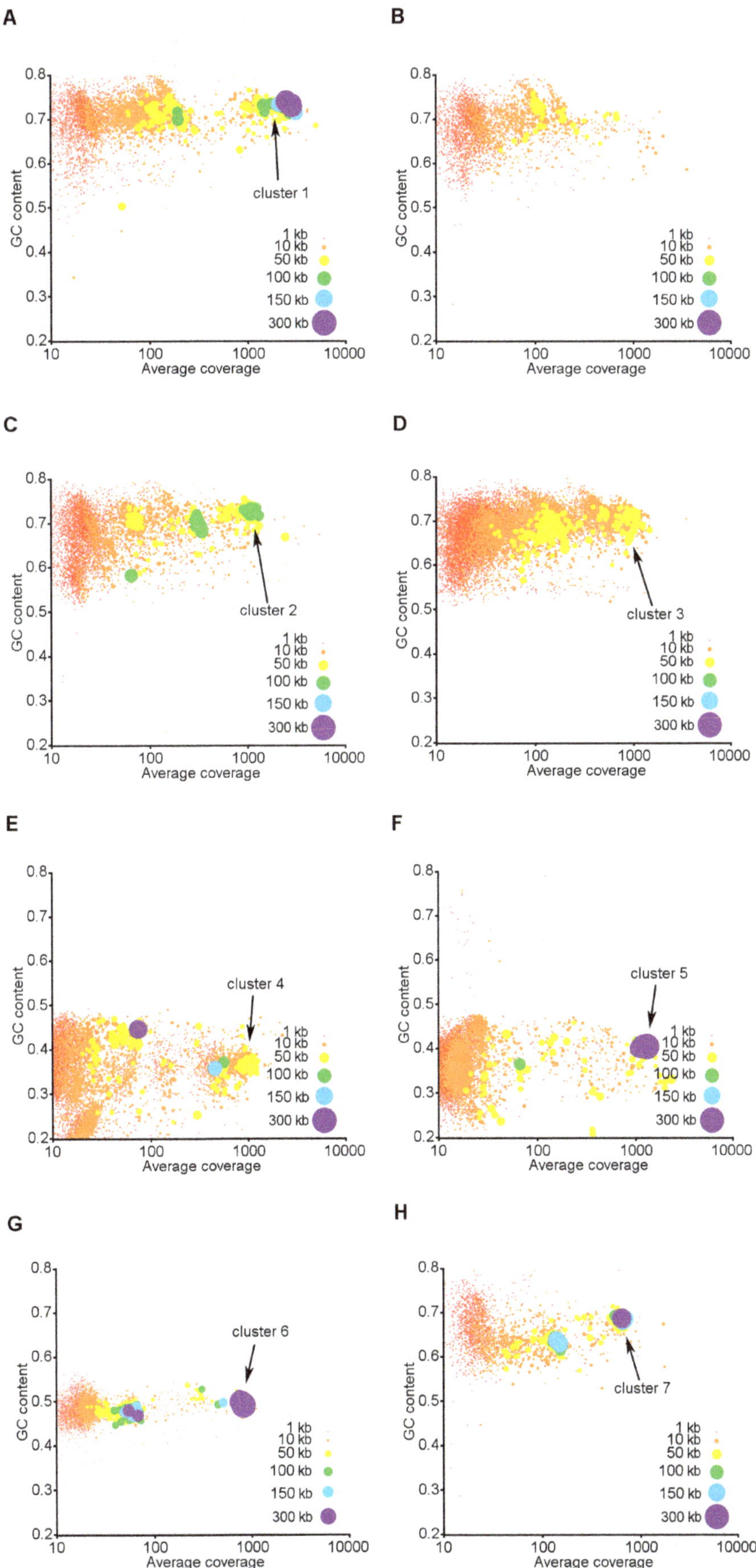
A
B
C
D
E
F
G
H
GC content
Average coverage
cluster 1
cluster 2
cluster 3
cluster 4
cluster 5
cluster 6
cluster 7
1 kb
10 kb
50 kb
100 kb
150 kb
300 kb

Figure 3. Scatterplots of contig properties for select genus bins in thermophilic and mesophilic communities. Plotted contigs correspond to (A) *Micromonospora* (*Actinobacteria*) in thermophilic community, (B) *Mycobacterium* (*Actinobacteria*) in thermophilic community, (C) *Pseudoxanthomonas* (*Proteobacteria*) in thermophilic community, (D) *Pseudoxanthomonas* (*Proteobacteria*) in mesophilic community, (E) *Chryseobacterium* (*Bacteroidetes*) in mesophilic community, (F) *Niabella* (*Bacteroidetes*) in thermophilic community, (G) *Niastella* (*Bacteroidetes*) in mesophilic community, and (H) *Chelativorans* (*Proteobacteria*) in thermophilic community. Genera presented in A–E account for >50% of total dissimilarity between thermophilic and mesophilic communities. Notable clusters with high abundance or large contigs are labeled for reference in subsequent analyses.

to the mesophilic enrichment community. Gene counts between communities were compared using the D-score statistic [29] with a minimum gene count threshold of 5. A false discovery rate of 0.05 was used for determining statistical significance.

For select protein families identified through the abundance search, genes were analyzed with respect to their clusters of orthologous genes (COG) group. Genes from deconstruction-relevant COGs were aligned using MUSCLE [38] and processed in Phylip [39] to perform bootstrapping with 1000 replicates, generate F84 distance matrices, and perform neighbor-joining. Phylip was used to find the consensus tree using a majority rule to retain branches present in ≥50% of bootstrap replicates. The cellobiohydrolase CelD gene from *Aspergillus fumigatus* was used as an outgroup.

Data Archiving

Metagenome raw reads, assembled scaffolds, and gene annotations can be accessed through IMG/M. The metagenomes are listed as Taxon Object ID 2199352012 (Mesophilic rice straw/compost enrichment metagenome: eDNA_1 (Mesophilic 454/Illumina Combined June 2011 assem)) and Taxon Object ID 2199352008 (Thermophilic rice straw/compost enrichment metagenome: eDNA_2 (Thermophilic 454/Illumina Combined June 2011 assem)).

Results

Identification of Extraction Buffer

The enzyme activities extracted from incubated rice straw are presented in Table 1. Xylanase activities from rice straw varied between 0.85 IU g dw^{-1} for sodium acetate extraction to 1.25–1.27 IU $(g\ dw)^{-1}$ for extractions containing 50 wt% ethylene glycol and 0.15 wt% Tween 80 in the presence of either 0.1 wt% NaCl or 1.5 wt% NaCl. Endoglucanase extraction also varied with the composition of the buffer, but differences were much smaller compared to xylanase. Like xylanase, the highest activity, 0.34 IU $(g\ dw)^{-1}$ was observed with extractions containing 50 wt% ethylene glycol and 0.15 wt% Tween 80.

Ethylene glycol had a significant positive effect on xylanase ($p<0.001$) and endoglucanase ($p\text{-value}<0.02$) extractions. For both xylanase and endoglucanase extraction, the interaction between Tween 80 and ethylene glycol was significant. When ethylene glycol was at 50 wt% in the buffer, increasing Tween 80 from 0.01 wt% to 0.15 wt% increased xylanase extraction ($p\text{-value}=0.036$) and endoglucanase extraction ($p\text{-value}=0.029$). Sodium chloride had a significant positive effect on xylanase activity extracted from rice straw ($p\text{-value}=0.021$), but had no effect on endoglucanase activity ($p\text{-value}>0.05$).

Temperature Effects on Microbial Activity and Extracted Endoglucanase and Xylanase Activities

Microbial respiration and extracted enzymatic activity were greater for thermophilic compared to mesophilic incubations

Table 5. Contig cluster properties for selected clusters (Figure 3) with high abundance or large contigs in thermophilic and mesophilic communities.

Cluster	Community*	Taxonomy	Number of marker COG genes (out of 70)	Mean count for detected marker COG genes	Total sequence length in cluster (Mb)	Average genome size in IMG database (Mb)
1	T	*Micromonospora*	70		6.6	6.9
		(*Actinobacteria*)		1.29		
2	T	*Pseudoxanthomonas*	23		1.6	3.4
		(*Proteobacteria*)		1.04		
3	M	*Pseudoxanthomonas*	48		3.3	3.4
		(*Proteobacteria*)		1.38		
4	M	*Chryseobacterium*	68		4.4	5.6
		(*Bacteroidetes*)		1.46		
5	T	*Niabella*	64		3	n/a
		(*Bacteroidetes*)		1.03		
6	M	*Niastella*	70		6.6	n/a
		(*Bacteroidetes*)		1.03		
7	T	*Chelativorans*	38		3.1	4.9
		(*Proteobacteria*)		1.08		

*T, thermophilic community; M, mesophilic community.

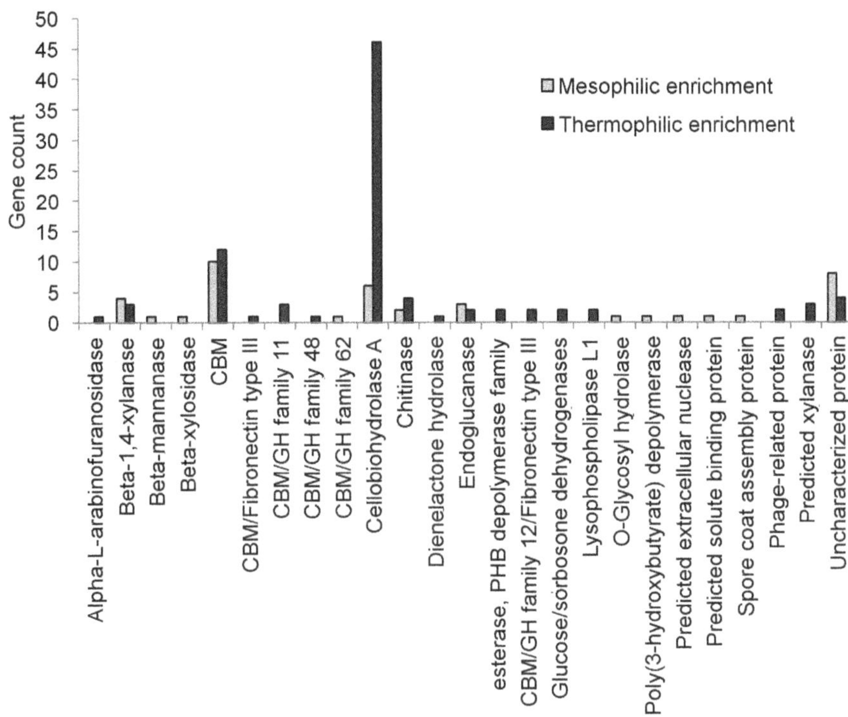

Figure 4. COG classifications of genes containing CBM2 motifs in microbial communities from thermophilic and mesophilic enrichments.

(Table 2). For the T4 sampling point, cumulative respiration was 7.5 times greater at 55°C compared to 35°C, while extracted xylanase and endoglucanase activities were 2.6 and 13.4 times greater, respectively. For 35°C incubations, there was little change in cumulative respiration and extracted enzyme activities with enrichment. In contrast, respiration increased by a factor of 3 between enrichments T2 and T4 at 55°C. Similar changes were observed in the activity of extracted enzymes. Xylanase and endoglucanase activity increased by factors of 2.7 and 1 between enrichments T2 and T4, respectively.

Metagenome Sequencing and Assembly

Illumina and 454 sequencing of mesophilic and thermophilic communities yielded total read counts of 447,683,681 and 448,669,837, respectively. Of these reads, 94.3% of reads from the mesophilic community passed quality filtering, while 91.1% of thermophilic community reads passed. Assembly of filtered reads resulted in 264,109 contigs for the mesophilic community and 512,311 contigs for the thermophilic community.

Microbial Community Composition

Rarefaction curves generated from pyrotag reads showed a clear asymptote for both communities, indicating sufficient sampling to capture most operational taxonomic units (OTUs) within communities (Figure 1). Pyrotag sequencing revealed that microbial communities from the thermophilic enrichment were less diverse than those from the mesophilic enrichment (Table 3). Decreased diversity in the thermophilic community, as indicated by a lower Shannon index relative to the mesophilic community, stemmed from decreased richness and evenness during thermophilic enrichment. Differences in microbial community structure between mesophilic and thermophilic enrichments were primarily a result of differences in abundance for *Actinobacteria*, *Firmicutes*, *Proteobacteria*, and *Bacteroidetes* bacteria (Figure 2). Both pyrotag sequencing and abundance data for metagenome contigs containing 16S rRNA genes indicated enrichment of *Actinobacteria* under thermophilic conditions relative to mesophilic conditions. Alternately, the data showed decreases in relative abundance for *Proteobacteria* and *Bacteroidetes* in the thermophilic culture compared to the mesophilic culture. SIMPER analysis of binned metagenome contigs revealed that genera within *Actinobacteria* were the largest contributors to dissimilarity between the thermophilic and mesophilic communities (Table 4). Increased abundance of *Micromonospora* and *Mycobacterium* in the thermophilic community accounted for approximately one third of the Bray-Curtis dissimilarity between the two communities. Decreased abundance of *Chryseobacterium* and *Pseudoxanthomonas* (members of *Bacteroidetes* and *Proteobacteria*, respectively) in the thermophilic community was also a major contributor to the overall dissimilarity between the thermophilic and mesophilic communities.

Plotting of contig properties for these genus bins allowed for an approximate count of the species or strains present within each genus (Figure 3). Within the scatterplots, contigs that form distinct clusters share similar GC content and coverage within the metagenome and can be assumed to originate from the same organism or organisms that are closely related and have similar abundance within the community. The presence of multiple distinct clusters within some genus bins suggests that multiple unique species or strains within that genus were present in the community. In particular, there was a high-abundance *Micromonospora* cluster accompanied by several lower abundance clusters in the thermophilic community (Figure 3A). The *Mycobacterium* bin in the thermophilic community lacked a high abundance cluster on the order observed for *Micromonospora* but did contain multiple lower abundance clusters (Figure 3B). Contigs from these bins had high GC content. *Pseudoxanthomonas* clusters were also predomi-

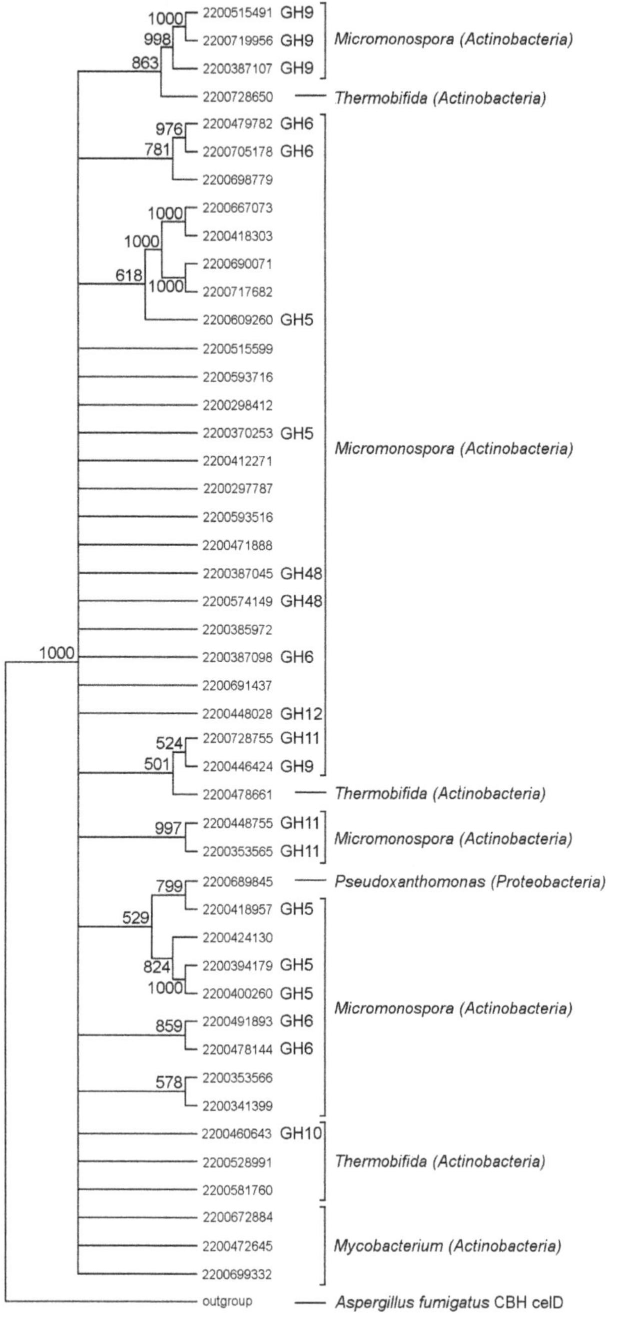

Figure 5. Consensus neighbor-joining tree of CBH-A genes with CBM2 in the thermophilic microbial community. Genes are represented by their IMG gene object ID numbers. For genes that had a glycoside hydrolase (GH) family ascribed to them during annotation, GH family number is indicated next to the gene object ID. Numbers at nodes denote the percentage of trees that support that node out of 1000 bootstrap replicates.

nantly high GC and prominent clusters were observed in both the thermophilic and mesophilic communities (Figure 3C–D). The highest abundance *Pseudoxanthomonas* cluster in each community shared similar GC contents, suggesting they may correspond to the same or similar organisms. *Chryseobacterium* clusters in the mesophilic community had GC contents generally below 50% and a single high abundance cluster was observed (Figure 3E). Although they were not major contributors to the overall dissimilarity between the two communities, notable clusters were observed in several other genus bins. For example, *Niabella* in the thermophilic community contained one highly abundant cluster with large contigs, suggesting well-assembled sequences (Figure 3F). Similarly, *Niastella* and *Chelativorans* in the mesophilic and thermophilic communities, respectively, contained high abundance, well-assembled clusters (Figure 3G–H).

Clusters were screened for ribosomal and non-ribosomal phylogenetic marker COG genes to gauge the completeness of their genomes. The Joint Genome Institute's list of seventy such marker COG genes was used. These genes are expected to be broadly conserved and contain species-specific sequences. As these conserved marker genes are typically spread out across microbial genomes, a cluster was considered to have captured the majority of an organism's genome if it contained a complete set of marker COG genes and had a total sequence length comparable to published genomes in the same genus. This analysis revealed varying levels of genome content in each cluster (Table 5). The high abundance *Micromonospora* and *Niastella* clusters in the thermophilic and mesophilic communities, respectively, both contained a complete set of marker COG genes. Additionally, the total sequence length within the *Micromonospora* cluster is similar to other genomes within this genus. These data suggest that metagenome sequenced potentially captured a near complete genome for this particular *Micromonospora* species. Although, no other *Niastella* genomes have been sequenced for comparison, the presence of a complete set of marker COG genes and total sequence length comparable to bacterial genomes suggest most of the genome for this organism may have been captured as well. Similarly, the high abundance mesophilic *Chyseobacterium* and thermophilic *Niabella* clusters have near complete sets of marker genes, suggesting the majority of their genome sequence are represented in their respective clusters. The high abundance *Psuedoxanthomonas* clusters in both the thermophilic and mesophilic communities, as well as the thermophilic *Chelativorans* cluster, were only partially assembled, with marker genes counts indicating that 33–69% of these genomes are represented in the cluster sequences. Furthermore, as most marker COG genes are expected to occur as a single copy per genome, the average gene count across all marker COGs with at least one hit was used as an indicator of how many species or strains were represented within each cluster. For all clusters, the average gene count within all detected marker COGs was less than 1.5, suggesting the presence of only a single species or strain within each cluster and minimal errors in assembly or binning.

Protein Families in Metagenomes Relevant to Rice Straw Deconstruction

Metagenomes were compared to find protein families overrepresented in the thermophilic community relative to the mesophilic community. For these particular communities, a critical p-value of 4.46e-3 denoted statistical significance. Among the four most overrepresented protein families in the thermophilic enrichment, carbohydrate-binding module family 2 (CBM2) was significantly enriched in the thermophilic community ($p < 1e\text{-}15$) with 91 hits (equaling a normalized frequency of 208.4) in the thermophilic community versus 39 (equaling a normalized frequency of 45.9) in the mesophilic community. The second and third most abundant CBMs in the thermophilic community were CBM48 and CBM4/9 with 86 and 34 hits, respectively. In contrast to CBM2, both CBM48 and CBM4/9 were significantly underrepresented in the thermophilic community compared to the mesophilic community ($p = 1.2e\text{-}3$ for CBM48 and $p = 1.87e\text{-}7$ for CBM4/9).

Table 6. Glycoside hydrolase genes relevant to lignocellulose deconstruction in high-abundance organisms within thermophilic and mesophilic community metagenomes.

Protein family	GH family	Dominant types	Number of hits within cluster sequence						
			Cluster 1	Cluster 2	Cluster 3	Cluster 4	Cluster 5	Cluster 6	Cluster 7
pfam00150	5	β-mannosidase endo-β-1,4-glucanase endo-β-1,4-mannosidase endo-β-1,4-xylanase β-1,4-cellobiosidase β-1,3-mannanase xyloglucan-specific endo-β-1,4-glucanase exo-β-1,4-glucanase	2	0	1	0	1	5	0
pfam00232	1	β-glucosidase	2	0	0	0	0	1	0
pfam00331	10	endo-1,4-β-xylanase endo-1,3-β-xylanase	5	0	1	0	2	0	0
pfam00457	11	xylanase	2	0	1	0	0	0	0
pfam00722	16	endo-1,3-β-glucanase endo-1,3(4)-β-glucanase xyloglucanase	2	0	1	2	0	6	0
pfam00759	9	endoglucanase cellobiohydrolase β-glucosidase	2	1	3	0	1	2	0
pfam00933	3	β-glucosidase 1,4-β-xylosidase exo-1,3-1,4-glucanase α-L-arabinofuranosidase	10	2	4	3	2	3	1
pfam01270	8	cellulose endo-1,4-β-xylanase reducing-end-xylose releasing exo-oligoxylanase	0	1	1	0	0	0	0
pfam01341	6	endoglucanase cellobiohydrolase	4	0	0	0	0	0	0
pfam01670	12	endoglucanase xyloglucan hydrolase β-1,3-1,4-glucanase	1	0	0	0	0	0	0
pfam01915	3C	β-glucosidase 1,4-β-xylosidase exo-1,3-1,4-glucanase α-L-arabinofuranosidase	5	1	5	2	1	3	0
pfam02011	48	reducing end-acting cellobiohydrolase endo-β-1,4-glucanase	1	0	0	0	0	0	0
pfam02156	26	β-mannanase β-1,3-xylanase	0	0	0	0	0	3	0
pfam03648	67N	α-glucuronidase xylan α-1,2-glucuronidase	1	0	1	0	1	3	0
pfam03664	62	α-L-arabinofuranosidase	1	0	0	0	0	0	0
pfam04616	43	β-xylosidase α-L-arabinofuranosidase arabinanase xylanase	3	0	4	0	11	11	0
pfam07477	67C	α-glucuronidase xylan α-1,2-glucuronidase	1	0	2	0	1	1	0
pfam07488	67M	α-glucuronidase xylan α-1,2-glucuronidase	1	0	2	0	1	1	0

COG classifications for all genes containing CBM2 revealed that overrepresentation of CBM2-containing genes in the thermophilic community stemmed primarily from overrepresentation of cellobiohydrolase A (CBH-A) genes (COG 5297) (Figure 4). Several CBH-A genes exhibited similarity with respect to their glycoside hydrolase (GH) family (Figure 5). Out of 46 genes, 45 were housed on contigs binned to *Actinobacteria*. Of the 45 CBH-A genes binned to *Actinobacteria*, 37 were binned to the genus

Micromonospora, 5 were binned to *Thermobifida*, and 3 were binned to *Mycobacterium*. Notably, 22 of the 46 CBH-A genes with CBM2 in the thermophilic community mapped back to the high abundance *Micromonospora* cluster (cluster 1 in Figure 3A). Several of the CBH-A gene sequences were fragmented. As a result, while there was enough sequence present to facilitate annotation as a CBH-A and genus binning, the fragmentation prevented meaningful neighbor-joining of these genes. Furthermore, such fragmentation may have also prevented assignment of these sequences to GH families. These fragmented sequences are largely reflected in the similarity tree presented in Figure 5 as genes that only branch with respect to the outgroup.

Abundant and well-assembled clusters were screened for all GH protein families (Table 6). GH hits were compared to the CAZy database [40] to isolate those relevant to lignocellulose deconstruction. Clusters contained a range of GH genes spanning cellulases and hemicellulases. The high abundance thermophilic *Micromonospora* cluster contained endoglucanases and several types of hemicellulases in addition to the overrepresented CBH genes noted previously. Hemicellulases corresponded to GHs active on xylan, mannan, and arabinan. High abundance *Pseudoxanthomonas* clusters in both the mesophilic and thermophilic communities also exhibited a variety of cellulases and hemicellulases. The thermophilic cluster registered fewer GHs, perhaps owing to a less complete assembly than that in the mesophilic community. The high abundance *Chryseobacterium* cluster in the thermophilic community contained mostly hemicellulases. *Bacteroidetes* clusters in the thermophilic and mesophilic communities contained cellulases and hemicellulases. Compared to other clusters, there were more hemicellulases in the GH 43 family present in the *Bacteroidetes* clusters.

Discussion

Microbial Activity and Extracted Endoglucanase and Xylanase Activities

Thermophilic incubations on rice straw yielded higher microbial activity and extracted enzyme activity levels than mesophilic incubations. The higher activity at 55°C is consistent with other observations of plant biomass decomposition. For instance, food waste and green waste composts decomposed two times faster at 45°C compared to 35°C [41]. The higher decomposition and enzyme activity levels observed at 55°C support the use of thermophilic environments for discovery of organisms and enzymes for biomass deconstruction.

Enzyme extraction from incubated rice straw increased with increasing concentrations of ethylene glycol in the extraction buffer. The results suggest secreted enzymes have strong hydrophobic interactions with rice straw polysaccharides. Such interactions have been identified for carbohydrate binding modules associated with cellulases [42,43]. Similar effects of ethylene glycol were observed for the extraction of xylanase and endoglucanase from corn stover and switchgrass incubated under high-solids thermophilic conditions [15]. These observations indicate thermophilic, high-solids biomass deconstruction systems favor organisms that secrete enzymes with strong hydrophobic plant cell wall interactions and that enzyme binding plays an important role in these systems.

Microbial Community Composition

The largest contributor to dissimilarity between the thermophilic and mesophilic communities was *Micromonospora*. Certain species within *Micromonospora* have been characterized as cellulose degraders and thermophiles [44–47]. Enrichment of *Micromonospora* under thermophilic conditions in this study is consistent with these prior observations. Moreover, several *Micromonospora* species have been shown to be capable of deconstructing rice straw in liquid culture and compost systems [48–50]. These data support the possibility that enrichment of *Micromonospora* species in this study under thermophilic conditions corresponds to these species taking a more active role in rice straw deconstruction compared to mesophilic conditions. Both mesophilic and thermophilic communities contained high-abundance genera within *Proteobacteria* and *Bacteroidetes* phyla. Prevalence of *Pseudoxanthomonas* in both enrichments is in agreement with previous studies that have found *Pseudoxanthomonas* species to be major components of bacterial consortia with high cellulolytic activity under mesophilic and thermophilic conditions [51–53]. In contrast, other *Proteobacteria* genera like *Chelativorans*, which was detected in high abundance in the thermophilic community, are not well studied with respect to cellulolytic activity and have not been reported as lignocellulose degraders.

Bacteroidetes genera found in the thermophilic and mesophilic communities have previously been isolated from arboreal and greenhouse soils. Notably, *Niabella* species isolated from soils grew only under mesophilic conditions [54–56]. Tolerance to temperatures as high as 55°C, as observed here, has not been reported previously for this genus. Moreover, no *Niabella* isolates to date have exhibited the ability to hydrolyze carboxymethylcellulose [54,55,57]. Alternately, several *Chryseobacterium* species have been detected in cellulose-degrading gut communities [58,59]. Likewise, species within the genus *Niastella* have been isolated from arboreal soils and several have exhibited the ability to hydrolyze carboxymethylcellulose [60]. *Niastella* isolates are typically mesophiles [56,60], which may explain the decrease in *Niastella* abundance seen under thermophilic conditions in this study. The data presented here demonstrate that temperature may significantly impact the community members and enzymes responsible for lignocellulose degradation, a vital consideration when using metagenomics to discover lignocellulolytic enzymes for biofuel production.

Community Metagenomes and Determinants of Lignocellulolytic Activity

Lignocellulolytic enzymes detected in high-abundance and well-assembled metagenome contig clusters compliment organism abundance data to help elucidate each species' potential role in rice straw deconstruction. Comparison of genes containing protein family domains relevant to lignocellulose deconstruction in each metagenome provides an avenue for identifying promising enzymes for biofuels applications. In this study, the thermophilic metagenome was screened for protein families that were represented in significantly greater quantities compared to the mesophilic metagenome. Such overrepresented protein families may indicate specific genes that confer a selective advantage to their host organism under thermophilic conditions. Moreover, overrepresented genes encoding lignocellulolytic enzymes present targets for further investigation, as they may represent thermotolerant enzymes that maintain activity under high-solids conditions similar to those necessary for biofuel production. In this research, one such deconstruction-relevant protein family, carbohydrate-binding module family 2, was significantly overrepresented in the thermophilic community. If cellulases containing CBM2 do confer an advantage to the *Actinobacteria* that produce them during high-solids culture on rice straw, it may be due to increased activity of these enzymes under thermophilic conditions. A variety of cellulose-binding CBMs exist in nature with varying affinities to different plant cell walls, potentially exploiting various structural changes that result from cellulose interactions with other cell wall components [61]. Cellulases with CBM2 may be better able to bind cellulose within the unique structure of rice straw cell walls. Additionally, binding of cellulases with CBM2 to cellulose may be more stable compared to other CBMs under thermophilic conditions or the structure of CBM2 itself may be more thermostable. Further characterization of CBM2 is needed to assess these possibilities. Since most CBM2-containing enzymes in the thermophilic community were cellobiohydrolases derived from *Actinobacteria* (high-abundance *Micromonospora* in particular), these CBHs also warrant further investigation as thermophilic enzymes for high-solids rice straw deconstruction.

Although they were not a major contributor to overall community dissimilarity nor did they contain significantly overrepresented protein families, *Niabella* bacteria present in high abundance in the thermophilic community did contain more family 43 GHs compared to other deconstruction-relevant GHs. Analysis of marker COG genes suggests that most of this *Niabella* species' genome is represented in the metagenome sequence and, as a result, it can be reasonably assumed that the observed asymmetry in GHs is truly representative of this organism's genome. As family 43 GHs are active on hemicellulose, *Niabella* may primarily utilize hemicellulose during rice straw decomposition. Previous characterization of *Niabella* species have only examined activity on cellulose and have neglected hemicellulose polysaccharides [54,55,57]. As none of these previously studied isolates were active on cellulose, the observed abundance of hemicellulase genes in this work provides motivation to investigate hemicellulolytic activity in this genus and to determine if *Niabella* family 43 GHs represent enzymes for hemicellulose deconstruction in high-solids environments.

This work demonstrates the usefulness of the metagenomic approach for identifying genes of interest in microbial communities enriched to select for organisms capable of deconstructing rice straw under industrially relevant conditions. This technique can presumably be applied to other microbial community systems to identify target genes with industrially-applicable capabilities. The metagenomic approach gauges the abundance of specific organisms and provides insight into their potential capabilities by revealing the genes they possess. However, it must be noted that metagenomics provides no indication of whether organisms actually express these genes. As a result, additional metatranscriptomic and metaproteomic analyses are required to ultimately confirm their activity within enrichment cultures.

Acknowledgments

We thank Chao Wei Yu for assistance with rice straw collection, Dean C. Dibble for soxhlet extraction of rice straw, Josh Claypool and Lauren Jabusch for assistance with bioreactors and Hannah Woo for shipping samples. We also thank Tijana Glavina del Rio, Susannah Tringe and Stephanie Malfatti of the DOE Joint Genome Institute for their assistance in obtaining sequencing data.

Author Contributions

Conceived and designed the experiments: APR JSV. Performed the experiments: APR. Analyzed the data: APR CWS PD JK JSV. Contributed reagents/materials/analysis tools: HB MH BAS SWS MPT JSV PD. Wrote the paper: CWS JSV APR.

References

1. Parker N, Tittmann P, Hart Q, Nelson R, Skog K, et al. (2010) Development of a biorefinery optimized biofuel supply curve for the Western United States. Biomass and Bioenergy 34: 1597–1607.
2. Somerville C, Youngs H, Taylor C, Davis SC, Long SP (2010) Feedstocks for Lignocellulosic Biofuels. Science 329: 790–792.
3. EPA (2012) Regulation of Fuels and Fuel Additives: 2012 Renewable Fuel Standards. Federal Register 77: 1320–1358.
4. Simmons B, Loque D, Blanch H (2008) Next-generation biomass feedstocks for biofuel production. Genome Biology 9: 1–6.
5. Kim S, Dale BE (2004) Global potential bioethanol production from wasted crops and crop residues. Biomass and Bioenergy 26: 361–375.
6. Matteson GC, Jenkins BM (2007) Food and processing residues in California: Resource assessment and potential for power generation. Bioresource Technology 98: 3098–3105.
7. Mendu V, Shearin T, Campbell JE, Stork J, Jae J, et al. (2012) Global bioenergy potential from high-lignin agricultural residue. Proceedings of the National Academy of Sciences 109: 4014–4019.
8. Tuck CO, Pérez E, Horváth IT, Sheldon RA, Poliakoff M (2012) Valorization of Biomass: Deriving More Value from Waste. Science 337: 695–699.
9. FAO (2012) Food and Agricultural Organization of the United Nations (FAOSTAT). Food and Agricultural Organization of the United Nations.
10. Kadam KL, Forrest LH, Jacobson WA (2000) Rice straw as a lignocellulosic resource: collection, processing, transportation, and environmental aspects. Biomass and Bioenergy 18: 369–389.
11. Summers MD, Hydeb PR, Jenkins BM (2001) Yields and property variations for rice straw in California. 5th International Biomass Conference of the Americas Orlando, Florida, USA.
12. Klein-Marcuschamer D, Oleskowicz-Popiel P, Simmons BA, Blanch HW (2012) The challenge of enzyme cost in the production of lignocellulosic biofuels. Biotechnology and Bioengineering 109: 1083–1087.
13. Rubin EM (2008) Genomics of cellulosic biofuels. Nature 454: 841–845.
14. Gladden JM, Allgaier M, Miller CS, Hazen TC, VanderGheynst JS, et al. (2011) Glycoside Hydrolase Activities of Thermophilic Bacterial Consortia Adapted to Switchgrass. Appl Environ Microbiol: AEM.00032-00011.
15. Reddy AP, Allgaier M, Singer SW, Hazen TC, Simmons BA, et al. (2011) Bioenergy feedstock-specific enrichment of microbial populations during high-solids thermophilic deconstruction. Biotechnology and bioengineering 8: 2088–2098.
16. Allgaier M, Reddy AP, Park JI, Ivanova N, D'haeseleer P, et al. (2010) Targeted Discovery of Glycoside Hydrolases from a Switchgrass-Adapted Compost Community. PLoS One 5: e8812.
17. DeAngelis KM, Gladden JM, Allgaier M, D'haeseleer P, Fortney JL, et al. (2010) Strategies for Enhancing the Effectiveness of Metagenomic-based Enzyme Discovery in Lignocellulolytic Microbial Communities. Bioenergy Research 3: 146–158.
18. Cheng YS, Zheng Y, Yu CW, Dooley TM, Jenkins BM, et al. (2010) Evaluation of High Solids Alkaline Pretreatment of Rice Straw. Applied Biochemistry and Biotechnology 162: 1768–1784.
19. DeAngelis KM, Gladden JM, Allgaier M, D'haeseleer P, Fortney JL, et al. (2010) Strategies for Enhancing the Effectiveness of Metagenomic-based Enzyme Discovery in Lignocellulolytic Microbial Communities. Bioenergy Research.
20. Reddy AP, Jenkins BM, VanderGheynst JS (2009) The critical moisture range for rapid microbial decomposition of rice straw during storage. Transactions of the Asabe 52: 673–677.
21. Engelbrekston A, Kunin V, Wrighton K, Zvenigorodsky N, Chen F, et al. (2010) Experimental factors affecting PCR-based estimates of microbial species richness and evenness. The ISME Journal 4.
22. Kunin V, Engelbrekston A, Ochman H, Hugenholtz P (2010) Wrinkles in the rare biosphere: pyrosequencing errors can lead to artificial inflation of diversity estimates. Environmental Microbiology 12: 118–123.
23. DeSantis T, Hugenholtz P, Larsen N, Rojas M, Brodie E, et al. (2006) Greengenes, a chimera-checked 16s rRNA gene database and workbench compatible with ARB. Applied Environmental Microbiology 72: 5069–5072.
24. Li R, Zhu H, Ruan J, Qian W, Fang X, et al. (2010) De novo assembly of human genomes with massively parallel short read sequencing. Genome Res 20: 265–272.
25. Chaisson M, Pevzner P (2007) Short read fragment assembly of bacterial genomes. Genome Res 18: 324–330.
26. Sommer D, Delcher A, Salzberg S, Pop M (2007) Minimus, a fast, lightweight genome assembler. BMC Bioinformatics 8: 64.
27. Li H, Durbin R (2010) Fast and accurate long-read alignment with Burrows-Wheeler transform. Bioinformatics 26: 589–595.
28. Markowitz VM, Chen IMA, Palaniappan K, Chu K, Szeto E, et al. (2010) The integrated microbial genomes system: an expanding comparative analysis resource. Nucleic Acids Research 38: D382–D390.
29. Markowitz V, Ivanova N, Palaniappan K, Szeto E, Korzeniewski F, et al. (2006) An experimental metagenome data management and analysis system. Bioinformatics 22: e359-e367.
30. Giardine B, Riemer C, Hardison R, Burhans R, Elnitski L, et al. (2005) Galaxy: a platform for interactive large-scale genome analysis. Genome Res 15: 1451–1455.

31. Blankenberg D, Von Kuster G, Coraor N, Ananda G, Lazarus R, et al. (2010) Galaxy: a web-based genome analysis tool for experimentalists. Curr Protoc Mol Biol 89: 19.10.11–19.10.21.
32. Goecks J, Nekrutenko A, Taylor J, Team TG (2010) Galaxy: a comprehensive approach for supporting accessible, reproducible, and transparent computational research in the life sciences. Genome Biology 11: R86.
33. Pati A, Heath LS, Kyrpides NC, Ivanova N (2011) ClaMS: A Classifier for Metagenomic Sequences.
34. Dixon P (2003) VEGAN, a package of R functions for community ecology. Journal of Vegetation Science 14: 927–930.
35. Hammer Ø, Haper DAT, Ryan PD (2001) PAST: Paleontological statistics software package for education and data analysis. Palaeontologia Electronica 4: 9.
36. Clark K (1993) Non-parametric multivariate analyses of changes in community structure. Australian Journal of Ecology 18: 117–143.
37. Finn R, Tate J, Mistry J, Coggill P, Sammut S, et al. (2008) The Pfam protein families database. Nucleic Acids Res 36: D281–288.
38. Edgar R (2004) MUSCLE: a multiple sequence alignment method with reduced time and space complexity. BMC Bioinformatics 5: 113.
39. Felsenstein J (1989) PHYLIP – Phylogeny Inference Package (Version 3.2). Cladistics 5: 164–166.
40. Cantarel BL, Coutinho PM, Rancurel C, Bernard T, Lombard V, et al. (2009) The Carbohydrate-Active EnZymes database (CAZy): an expert resource for Glycogenomics. Nucleic Acids Research 37: D233–D238.
41. Aslam DN, VanderGheynst JS, Rumsey TR (2008) Development of models for predicting carbon mineralization and associated phytotoxicity in compost-amended soil. Bioresource Technology 99: 8735–8741.
42. Beckham GT, Matthews JF, Bomble YJ, Bu L, Adney WS, et al. (2010) Identification of Amino Acids Responsible for Processivity in a Family 1 Carbohydrate-Binding Module from a Fungal Cellulase. The Journal of Physical Chemistry B 114: 1447–1453.
43. Georgelis N, Yennawar NH, Cosgrove DJ (2012) Structural basis for entropy-driven cellulose binding by a type-A cellulose-binding module (CBM) and bacterial expansin. Proceedings of the National Academy of Sciences 109: 14830–14835.
44. Erikson D (1952) Temperature/Growth Relationships of a Thermophilic Actinomycete, Micromonospora vulgaris. Journal of General Microbiology 6: 286–294.
45. Fergus CL (1969) The cellulolytic activity of thermophilic fungi and Actinomycetes. Mycologia 61: 120–129.
46. Gallagher J, Winters A, Barron N, McHale L, McHale AP (1996) Production of cellulase and β-glucosidase activity during growth of the actinomycete Micromonospora chalcae on cellulose-containing media. Biotechnology Letters 18: 537–540.
47. Menezes ABd, Lockhart RJ, Cox MJ, Allison HE, McCarthy AJ (2008) Cellulose Degradation by Micromonosporas Recovered from Freshwater Lakes and Classification of These Actinomycetes by DNA Gyrase B Gene Sequencing. Appl Environ Microbiol 74: 7080–7084.
48. Chowdhury NA, Moniruzzaman M, Nahar N, Choudhury N (1991) Production of cellulases and saccharification of lignocellulosics by A. Micromonospora sp. World Journal of Microbiology and Biotechnology 7: 603–606.
49. Abdulla HM, El-Shatoury SA (2007) Actinomycetes in rice straw decomposition. Waste Management 27: 850–853.
50. Kausar H, Sariah M, Mohd Saud H, Zahangir Alam M, Razi Ismail M (2011) Isolation and screening of potential actinobacteria for rapid composting of rice straw. Biodegradation 22: 367–375.
51. Haruta S, Cui Z, Huang Z, Li M, Ishii M, et al. (2002) Construction of a stable microbial community with high cellulose-degradation ability. Applied Microbiology and Biotechnology 59: 529–534.
52. Kato S, Haruta S, Cui ZJ, Ishii M, Igarashi Y (2005) Stable Coexistence of Five Bacterial Strains as a Cellulose-Degrading Community. Applied and Environmental Microbiology 71: 7099–7106.
53. Okeke B, Lu J (2011) Characterization of a defined cellulolytic and xylanolytic bacterial consortium for bioprocessing of cellulose and hemicelluloses. Applied Biochemistry and Biotechnology 163: 869–881.
54. Kim B-Y, Weon H-Y, Yoo S-H, Hong S-B, Kwon S-W, et al. (2007) Niabella aurantiaca gen. nov., sp. nov., isolated from a greenhouse soil in Korea. International Journal of Systematic and Evolutionary Microbiology 57: 538–541.
55. Weon H-Y, Yoo S-H, Kim B-Y, Son J-A, Kim Y-J, et al. (2009) Niabella ginsengisoli sp. nov., isolated from soil cultivated with Korean ginseng. International Journal of Systematic and Evolutionary Microbiology 59: 1282–1285.
56. Wang Y, Cai F, Tang Y, Dai J, Qi H, et al. (2011) Flavitalea populi gen. nov., sp. nov., isolated from soil of a Euphrates poplar (Populus euphratica) forest. International Journal of Systematic and Evolutionary Microbiology 61: 1554–1560.
57. Wang H, Zhang YZ, Man CX, Chen WF, Sui XH, et al. (2009) Niabella yanshanensis sp. nov., isolated from the soybean rhizosphere. International Journal of Systematic and Evolutionary Microbiology 59: 2854–2856.
58. Ramin M, Alimon AR, Abdullah N (2009) Identification of cellulolytic bacteria isolated from the termite *Coptotermes curvignathus* (Holmgren). Journal of Rapid Methods & Automation in Microbiology 17: 103–116.
59. Honein K, Kaneko G, Katsuyama I, Matsumoto M, Kawashima Y, et al. (2012) Studies on the Cellulose-Degrading System in a Shipworm and its Potential Applications. Energy Procedia 18: 1271–1274.
60. Weon H-Y, Kim B-Y, Yoo S-H, Lee S-Y, Kwon S-W, et al. (2006) Niastella koreensis gen. nov., sp. nov. and Niastella yeongjuensis sp. nov., novel members of the phylum Bacteroidetes, isolated from soil cultivated with Korean ginseng. International Journal of Systematic and Evolutionary Microbiology 56: 1777–1782.
61. Blake AW, McCartney L, Flint JE, Bolam DN, Boraston AB, et al. (2006) Understanding the Biological Rationale for the Diversity of Cellulose-directed Carbohydrate-binding Modules in Prokaryotic Enzymes. Journal of Biological Chemistry 281: 29321–29329.

Permissions

All chapters in this book were first published in PLOS ONE, by The Public Library of Science; hereby published with permission under the Creative Commons Attribution License or equivalent. Every chapter published in this book has been scrutinized by our experts. Their significance has been extensively debated. The topics covered herein carry significant findings which will fuel the growth of the discipline. They may even be implemented as practical applications or may be referred to as a beginning point for another development.

The contributors of this book come from diverse backgrounds, making this book a truly international effort. This book will bring forth new frontiers with its revolutionizing research information and detailed analysis of the nascent developments around the world.

We would like to thank all the contributing authors for lending their expertise to make the book truly unique. They have played a crucial role in the development of this book. Without their invaluable contributions this book wouldn't have been possible. They have made vital efforts to compile up to date information on the varied aspects of this subject to make this book a valuable addition to the collection of many professionals and students.

This book was conceptualized with the vision of imparting up-to-date information and advanced data in this field. To ensure the same, a matchless editorial board was set up. Every individual on the board went through rigorous rounds of assessment to prove their worth. After which they invested a large part of their time researching and compiling the most relevant data for our readers.

The editorial board has been involved in producing this book since its inception. They have spent rigorous hours researching and exploring the diverse topics which have resulted in the successful publishing of this book. They have passed on their knowledge of decades through this book. To expedite this challenging task, the publisher supported the team at every step. A small team of assistant editors was also appointed to further simplify the editing procedure and attain best results for the readers.

Apart from the editorial board, the designing team has also invested a significant amount of their time in understanding the subject and creating the most relevant covers. They scrutinized every image to scout for the most suitable representation of the subject and create an appropriate cover for the book.

The publishing team has been an ardent support to the editorial, designing and production team. Their endless efforts to recruit the best for this project, has resulted in the accomplishment of this book. They are a veteran in the field of academics and their pool of knowledge is as vast as their experience in printing. Their expertise and guidance has proved useful at every step. Their uncompromising quality standards have made this book an exceptional effort. Their encouragement from time to time has been an inspiration for everyone.

The publisher and the editorial board hope that this book will prove to be a valuable piece of knowledge for researchers, students, practitioners and scholars across the globe.

List of Contributors

Tao Ren
College of Resources and Environmental Science, China Agricultural University, Beijing, China
College of Resources and Environment, Huazhong Agricultural University, Wuhan, China

Jingguo Wang, Qing Chen and Fusuo Zhang
College of Resources and Environmental Science, China Agricultural University, Beijing, China

Shuchang Lu
Department of Agronomy, Tianjin Agricultural University, Tianjin, China

Jifu Li, Jianwei Lu, Xiaokun Li, Tao Ren, Rihuan Cong and Li Zhou
College of Resources and Environment, Huazhong Agricultural University, Wuhan, China
Key Laboratory of Arable Land Conservation (Middle and Lower Reaches of Yangtse River), Ministry of Agriculture, Wuhan, China

Stephen M. Ogle and William J. Parton
Natural Resource Ecology Laboratory, Colorado State University, Fort Collins, Colorado, United States of America

Christopher J. Kucharik
Department of Agronomy, University of Wisconsin-Madison, Madison, Wisconsin, United States of America
Nelson Institute Center for Sustainability and the Global Environment, University of Wisconsin-Madison, Madison, Wisconsin, United States of America
Department of Energy Great Lakes Bioenergy Research Center, University of Wisconsin-Madison, Madison, Wisconsin, United States of America

Andy VanLoocke
Department of Atmospheric Sciences, University of Illinois, Urbana, Illinois, United States of America

John D. Lenters
School of Natural Resources, University of Nebraska-Lincoln, Lincoln, Nebraska, United States of America

Melissa M. Motew
Nelson Institute Center for Sustainability and the Global Environment, University of Wisconsin-Madison, Madison, Wisconsin, United States of America

Yi Liu, Jing Wang, Yong Tao, Juan Xie, Junfeng Pan and Zhiguo Li
Laboratory of Aquatic Botany and Watershed Ecology, Wuhan Botanical Garden, Chinese Academy of Sciences China, Wuhan, China

Dongbi Liu
Institute of Plant Protection and Soil Fertilizer, Hubei Academy of Agricultural Sciences, Wuhan, China

Guoshi Zhang and Fang Chen
Laboratory of Aquatic Botany and Watershed Ecology, Wuhan Botanical Garden, Chinese Academy of Sciences China, Wuhan, China
China Program, International Plant Nutrition Institute (IPNI), Wuhan, China

Liqun Zhu, Naijuan Hu and Zhengwen Zhang
College of Agriculture, Nanjing Agricultural University, Nanjing, China

Minfang Yang and Xinhua Zhan
College of Resources and Environmental Science, Nanjing Agricultural University, Nanjing, China

Linnéa Asplund, Göran Bergkvist and Martin Weih
Department of Crop Production Ecology, Swedish University of Agricultural Sciences, Uppsala, Sweden

Matti W. Leino
Swedish Museum of Cultural History, Julita, Sweden
IFM -Biology, Linköping University, Linköping, Sweden

Anna Westerbergh
Department of Plant Biology and Forest Genetics, BioCenter, Swedish University of Agricultural Sciences, Uppsala, Sweden

Peng Zhang, Ting Wei, Zhikuan Jia, Qingfang Han and Xiaolong Ren
The Chinese Institute of Water-Saving Agriculture, Northwest A&F University, Yangling, Shaanxi, China
Key Laboratory of Crop Physi-Ecology and Tillage Science in Northwestern Loess Plateau, Ministry of Agriculture, Northwest A&F University, Yangling, Shaanxi, China

Yongping Li
Guyuan Institute of Agricultural Sciences, Guyuan, Ningxia, China

Laura E. Webb, Imke J. M. de Boer and Eddie A. M. Bokkers
Animal Production Systems Group, Wageningen University, Wageningen, Netherlands

Margit Bak Jensen
Department of Animal Sciences, Aarhus University, Tjele, Denmark

Bas Engel
Biometris, Wageningen University, Wageningen, Netherlands

Cornelis G. van Reenen
Livestock Research, Wageningen University and Research Centre, Lelystad, Netherlands

Walter J. J. Gerrits
Animal Nutrition Group, Wageningen University, Wageningen, Netherlands

Zilin Song, Xiaofeng Liu, Yuexiang Yuan and Yinzhang Liao
Chengdu Institute of Biology, Chinese Academy of Science, Chengdu, Sichuan, PR China

Zhiying Yan and GaiheYang
Research Center of Recycle Agricultural Engineering Technology of Shaanxi Province, Northwest A&F University, Yangling, Shaanxi, PR China

Hui Lin
Institute of Microbiology, College of Life Sciences, Zhejiang University, Hangzhou, China
Institute of Environment, Resource, Soil and Fertilizer, Zhejiang Academy of Agricultural Sciences, Hangzhou, China

Qi Shen, Qun Wang and Yu-Hua Zhao
Institute of Microbiology, College of Life Sciences, Zhejiang University, Hangzhou, China

Ju-Mei Zhan
Institute of Plant Science, College of Life Sciences, Zhejiang University, Hangzhou, China

Qing-zhong Zhang, Xing-ren Liu, Yi-ding Wang, Jian Huang and Ning Lu
Key Laboratory of Agricultural Environment, Ministry of Agriculture, Sino-Australian Joint Laboratory For Sustainable Agro-Ecosystems, Institute of Environment and Sustainable Development in Agriculture, Chinese Academy of Agricultural Sciences, Beijing, China

Feike A. Dijkstra
Centre for Carbon, Water and Food, Department of Environmental Sciences, The University of Sydney, Camden, New South Wales, Australia

Yanfang Xue, Shanchao Yue, Wei Zhang, Dunyi Liu, Zhenling Cui, Xinping Chen and Chunqin Zou
Center for Resources, Environment and Food Security, China Agricultural University, Beijing, China

Youliang Ye
College of Resources and Environmental Sciences, Henan Agricultural University, Zhengzhou, China

Bing Gao, Xiaotang Ju, Fang Su, Fengbin Gao, Qingsen Cao, Peter Christie, Xinping Chen and Fusuo Zhang
College of Resources and Environmental Sciences, China Agricultural University, Beijing, China

Oene Oenema
Wageningen University and Research Center, Alterra, Wageningen, The Netherlands

Xiaoming Li, Qirong Shen, Xinlan Mei, Wei Ran, Yangchun Xu and Guanghui Yu
Agricultural Ministry Key Lab of Plant Nutrition and Fertilization in Low-Middle Reaches of the Yangtze River, Nanjing, PR China
College of Resources and Environmental Sciences, Nanjing Agricultural University, Nanjing, PR China

Dongqing Zhang
Baoshan Environmental Protection Bureau, Shanghai, PR China

Tomáš Větrovský and Petr Baldrian
Laboratory of Environmental Microbiology, Institute of Microbiology of the ASCR, v.v.i., Praha, Czech Republic

Kari Timo Steffen
Department of Applied Chemistry and Microbiology, University of Helsinki, Helsinki, Finland

Tong Zhang and Zilin Song
College of Forestry and the Research Center of Recycle Agricultural Engineering and Technology of Shaanxi Province, Northwest A&F University, Yangling, Shaanxi, People's Republic of China

Guangxin Ren, Yongzhong Feng, Xinhui Han, Gaihe Yang and Linlin Liu
College of Agronomy and the Research Center of Recycle Agricultural Engineering and Technology of Shaanxi Province, Northwest A&F University, Yangling, Shaanxi, People's Republic of China

Rihuan Cong
Ministry of Agriculture Key Laboratory of Crop Nutrition and Fertilization, Institute of Agricultural Resources and Regional Planning, Chinese Academy of Agricultural Sciences, Beijing, China
College of Resources and Environment, Huazhong Agricultural University, Wuhan, China

Xiujun Wang
State Key Laboratory of Desert and Oasis Ecology, Xinjiang Institute of Ecology and Geography, Chinese Academy of Sciences, Urumqi, China
Earth System Science Interdisciplinary Center, University of Maryland, College Park, Maryland, United States of America

Keke Hua, Daozhong Wang, Xisheng Guo and Zibin Guo
Soil and Fertilizer Research Institute, Anhui Academy of Agricultural Sciences, Hefei, China

Junhui Li and Fei Dong
College of Natural Resources and Environment, South China Agricultural University, Guangzhou, China
Agricultural Bureau of Xiangfen County, Shanxi Province, Xiangfen, China

Ying Lu
College of Natural Resources and Environment, South China Agricultural University, Guangzhou, China

Qiuyan Yan
Institute of Wheat Research, Shanxi Academy of Agricultural Sciences, Linfen, China

Hojae Shim
Department of Civil and Environmental Engineering, Faculty of Science and Technology, University of Macau, Macau SAR, China

Minggang Xu
Ministry of Agriculture Key Laboratory of Crop Nutrition and Fertilization, Institute of Agricultural Resources and Regional Planning, Chinese Academy of Agricultural Sciences, Beijing, China

Hai-Ming Tang, Xiao-Ping Xiao, Wen-Guang Tang, Ke Wang, Ji-Min Sun, Wei-Yan Li and Guang-Li Yang
Hunan Soil and Fertilizer Institute, Changsha, PR China

Hui Lin, Junwei Ma and Jianrong Fu
Institute of Environment Resource and Soil Fertilizer, Zhejiang Academy of Agriculture Science, Hangzhou, China

Qun Wang, Qi Shen and Yuhua Zhao
Institute of Microbiology, College of Life Sciences, Zhejiang University, Hangzhou, China

Xiaojiao Wang, Fang Li and Gaihe Yang
College of Agronomy, Northwest A&F University, Yangling, Shaanxi, People's Republic of China

Xingang Lu
School of Chemical Engineering, Northwest University, Xian, Shaanxi, People's Republic of China

Amitha P. Reddy and Jean S. VanderGheynst
Joint BioEnergy Institute, Emeryville, California, United States of America
Biological and Agricultural Engineering, University of California Davis, Davis, California, United States of America

Christopher W. Simmons
Joint BioEnergy Institute, Emeryville, California, United States of America
Biological and Agricultural Engineering, University of California Davis, Davis, California, United States of America
Food Science, University of California Davis, Davis, California, United States of America

Patrik D'haeseleer, Jane Khudyakov and Michael P. Thelen
Joint BioEnergy Institute, Emeryville, California, United States of America
Physical and Life Sciences Directorate, Lawrence Livermore National Laboratory, Livermore, California, United States of America

Helcio Burd
Joint BioEnergy Institute, Emeryville, California, United States of America
Physical Biosciences Division, Lawrence Berkeley National Laboratory, Berkeley, California, United States of America

Masood Hadi and Blake A. Simmons
Joint BioEnergy Institute, Emeryville, California, United States of America
Biological and Materials Science Center, Sandia National Laboratories, Livermore, California, United States of America

Steven W. Singer
Joint BioEnergy Institute, Emeryville, California, United States of America
Earth Sciences Division, Lawrence Berkeley National Laboratory, Berkeley, California, United States of America

Index

www.ingramcontent.com/pod-product-compliance
Lightning Source LLC
Chambersburg PA
CBHW080525200326
41458CB00012B/4334
9781632399489